Mathematics and Geosciences: Global and Local Perspectives
Volume II

Edited by
María Charco
Jesús Ildefonso Díaz
José Fernández
Rafael Orive
María-Luisa Osete

Previously published in *Pure and Applied Geophysics*
(PAGEOPH), Volume 173, No. 2 and 3, 2016

 Birkhäuser

Editors

María Charco
Instituto de Geociencias (IGEO)
(CSIC, UCM)
Madrid, Spain

Rafael Orive
Instituto de Ciencias Matemáticas
CSIC-UAM-UCM-UC3M
Madrid, Spain

Jesús Ildefonso Díaz
Facultad de Ciencias Matemáticas
Instituto de Matemática
Interdisciplinar
Madrid, Spain

María-Luisa Osete
CSIC, UCM, Ciudad Universitaria
Instituto de Geociencias
Madrid, Spain

José Fernández
Instituto de Geociencias (CSIC-UCM)
Madrid, Spain

ISBN 978-3-319-32704-4   ISBN 978-3-319-32705-1 (eBook)
DOI 10.1007/978-3-319-32705-1

Library of Congress Control Number: 2015945248

Mathematics Subject Classification (2010): 86-XX, 60-XX, 62-XX, 65-XX, 76-XX

Cover illustration: © 2016 Cannavo and Palano

Cover design: deblik, Berlin

Printed on acid-free paper

This book is published under the trade name Birkhäuser

The registered company is Springer International Publishing AG, Switzerland

www.birkhkauser-science.com

Contents

Introduction to Mathematics and Geosciences: Global and Local Perspectives, Volume II

María Charco,[1] Rafael Orive,[2,3] Jesús Ildefonso Díaz,[4,5] María Luisa Osete,[1,6] and José Fernández[1]

Science makes better understanding of the world we live. Earth, our home, is a dynamic system continuously evolving: seas advance and recede, volcanoes erupt and become extinct, mountain chains build and erode away, these are some examples that provide changes on landscape formation. In the course of humanity, processes that occur both in the solid Earth and its fluid components have had a great influence on life on Earth, causing natural disasters, changes in climate, etc., that have conditioned the development and evolution of different species that inhabit it. Our way of life depends on its resources. How to ensure sustainable use of the planet is one of the current challenges facing humanity. These issues, among others, fed the idea that understanding the Earth System and its interaction with humanity is vital for our survival.

The value of the Geosciences to help us in this way is unquestioned. Nevertheless, these challenges are not just geoscience research problems, but also important research problems in Mathematics, Statistics and Computer Science. In this way, Mathematics provides the rigor, language and theoretical basics of every scientific research. For instance, we can use Plate Tectonics theory to show the step that Mathematics provides from intuition to certainty. Plate Tectonic theory had its beginnings in the first decade of the twentieth century when Alfred Wegener proposed his theory of "continental drift" (WEGENER 1912). At such time, Wegener's idea was very controversial in part because he could not provide a mechanism for continents moving, just that there were some observational evidences. McKenzie applied his physical–mathematical knowledge to study the viscosity of the lower mantle (MCKENZIE 1966) providing the physical–mathematical basis to the existing two layers in the mantle, each of them in motion, which contribute to continental drift. He refuted two other conceptual models of the Earth: the Earth was a homogeneous sphere and the theory that Earth had an inviscid core and a homogeneous mantle. McKenzie's work showed that Earth is far more dynamic than previously thought and added to the growing awareness that convection in mantle was driving continental drift.

Geoscientists use a wide range of modern tools for observing the Earth and for understanding its dynamic evolution. For example, the high spatio-temporal resolution of remote sensing data can provide an overwhelming volume of data in exquisite detail. Nevertheless, some processes can only be observed indirectly and so infrequently. Having in mind that the Earth is a dynamic system with different parts and the previously described, it is clear we need new conceptual approaches to characterize complex systems that vary strongly in space and time in ways not accounted for in current paradigms. In such a context, modeling is the core of Geosciences and Mathematics interaction. Models are approximations primarily based on physical arguments that

[1] Instituto de Geociencias (IGEO) (CSIC, UCM), Plaza de Ciencias 1 & 3, 28040 Madrid, Spain. E-mail: m.charco@igeo. ucm-csic.es; mlosete@fis.ucm.es; jft@mat.ucm.es

[2] Instituto de Ciencias Matemáticas (ICMAT), CSIC-UAM-UCM-UC3M, C. Nicolás Cabrera 13-15, 28049 Madrid, Spain. E-mail: rafael.orive@icmat.es

[3] Departamento de Matemáticas, Facultad de Ciencias, Universidad Autónoma de Madrid, 28049 Madrid, Spain.

[4] Instituto de Matemática Interdisciplinar, Facultad de Ciencias Matemáticas, Universidad Complutense de Madrid, Plaza de Ciencias 3, 28040 Madrid, Spain. E-mail: jidiaz@ucm.es

[5] Departamento de Matemática Aplicada, Facultad de Ciencias Matemáticas, Universidad Complutense de Madrid, Plaza de Ciencias 3, 28040 Madrid, Spain.

[6] Departamento de Física de la Tierra I, Facultad de Ciencias Físicas, Universidad Complutense de Madrid, Plaza de Ciencias, 1, 28040 Madrid, Spain.

© Springer International Publishing Switzerland 2016
M. Charco, J.I. Díaz, J. Fernández, R. Orive, M.-L. Osete (eds.),
Mathematics and Geosciences: Global and Local Perspectives,
DOI 10.1007/978-3-319-32705-1_1

require a rigorous mathematical approach: approximating solutions, checking the reliability of the model to describe physical phenomena, using models to predict quantities on which some conclusions can be drawn, etc. In addition, new mathematical tools are needed to process and invert the new data sets acquired directly in the Earth or using remote sensing from air or space.

For instance, we can find important issues regarding the life in an active Earth as earthquakes and eruption predictions. Although Geoscience is moving towards predictive capabilities of volcanic eruptions given the sensitive amount of data and better understanding of the causes, we have no idea why volcanic deformation culminates or not in an eruption. Understanding what determines the rates of magma accumulation in the chamber and what mechanisms make magmas eruptible could help to improve prediction capabilities. These mechanisms involve the inclusion of magmatic processes in the kinematic models actually used to integrate geophysical, geological and geodetic observations (e.g., ANDERSON and SEGALL 2013). By other side, understanding earthquakes and their hazards is a major challenge in Earth's sciences. Nowadays, it has become clear that the traditional statistical paradigm used to describe earthquake behavior (large earthquakes are spatially focused and temporally quasi-periodic) often fails, particularly within continents. Still little is known about the physics of faults where earthquakes occur or how faults form or why and how the earthquakes migrate between faults (e.g., KEILIS-BOROK et al. 2009). Again data integration and modeling for quantitative interpretation are required to understand such processes.

Other grand challenging questions in Earth Sciences are related to Earth's interior since our rock-sampling reach is limited to the upper tens of kilometers of the Earth's crust. Nowadays it is recognized that large-scale processes such as Plate Tectonics are driven by the nature of materials that make up the planet down to the smallest atomic scales, as thought for instance for the trigger of earthquakes (LAY and GARNERO 2011). The role of mantle plumes and their depth of origin are part of an intense debate. The subduction cycles or why subduction zones are developed are opened questions, also related with the Earth structure and geodynamics. New mathematical approaches that can be able of including more physical complexity into convection models are needed for the quantitative interpretation of geophysical, geochemical and geological data from a holistic perspective in terms of geodynamical relevant parameters as temperature, composition and rheology (e.g., CAMMARANO et al. 2011; AFONSO et al. 2015; FOULGER et al. 2015).

Some illustrative examples that are well understood by the general public are associated with climate and the habitability of the planet (HENDERSON-SELLERS and MCGUFFIE 1987, http://www.ipcc.ch). The mean global surface temperature of the Earth has risen since the beginning of the industrial age with the advent of CO_2 and other greenhouse gases emissions. The potentially serious consequences of global warming mark the need to understand, for instance, the fate of the Atlantic Meridional Overturning Circulation which reduction is likely to have strong implications for subtropical Atlantic temperatures and the position of the Intertropical Convergence Zone (e.g., SMEED et al. 2014). Geological proxies have revealed that the climate history of the planet is a combination of both variability and stability. Nevertheless, future climate projections depend on new mathematical advances to understand the thermodynamic and transport properties of the Earth's atmospheric-oceanic system. Many environmental issues and the presence and placement of many Earth's resources involve the role of fluid flow and transport. Landscape evolution and transport of environmental fluxes over the ground surface are scientific and social problems. It remains a challenge to include fine-scale features that are very important for modeling sediment transport (e.g., individual hill slopes, channel-river bank morphology) due to the strong nonlinearities of the sediment-transport laws (e.g., GARCIA-CASTELLANOS and JIMÉNEZ-MUNT 2015). The ability to asses and extract minerals, petroleum, natural gas and groundwater and to safely dispose of wastes depends on understanding flow of fluids. In hydrocarbon reservoirs and volcanic systems, the simulation of multiphase fluid flow is an important problem due to the large viscosity and density ratios involved (LONGO et al. 2012).

Focusing on the mathematical problems arising in the context of addressing such geoscientific challenges provides the opportunity to bring the expertise of geoscientists together with that of mathematicians to develop important insights and solve these challenges proving the enrichment and the link between theoretical and applied parts. To this end, the workshop "Mathematics and Geosciences: Global and Local Perspectives" was organized by the Institute of Mathematical Sciences (ICMAT) and the Institute of Geosciences (IGEO) under the patronage of the Spanish Council for Scientific Research (CSIC), Madrid Autonomous University (UAM), the Institute of Interdisciplinary Mathematics (IMI), Madrid Complutense University (UCM) and Technical University of Madrid (UPM). The workshop was held in Madrid ICMAT from 4th to 8th November 2013 to facilitate a fruitful interaction among a broad and geographically distributed group of mathematicians and geoscientists. The workshop was one of the year events of the Mathematics and the Planet Earth organized in Spain (MPE2013 2014). A first volume was published recently (Díaz et al. 2015).

The second volume of the Topical Issue includes some contributions presented at the meeting and others, related papers. It embraces 21 papers on different topics relating to Mathematics and Geosciences.

Holliday et al. apply a previously presented method for calculating probabilities for large events in systems such as earthquakes, typhoons, market crashes, electricity grid blackouts, floods, droughts, wars and conflicts, and landslides which can be unexpected and devastating, to the calculation of large earthquake probabilities in California-Nevada, USA. The method counts the number of small events since the last large event and then converts this count into a probability using a Weibull probability law. They consider a fixed geographic region and assumed that all earthquakes within that region, large magnitudes as well as small, were perfectly correlated. The model is extended to systems in which the events have a finite correlation length. They modify previous results by employing the correlation function for near mean field systems having long range interactions, an example of which is earthquakes and elastic interactions. Then they construct an application of the

method and show examples of computed earthquake probabilities.

The work carried out by Cho et al. explores the use of the Laplace–Fourier-domain full waveform inversion technique to deep-sea seismic data. It is a difficult attempt since the deep-sea layer reduces the amplitude of signals. To overcome this problem, and reduce the water layer's effect, they performed a downward continuation and built a macro-velocity model through refraction tomography which is used as an initial model for Laplace–Fourier inversion. This scheme is applied to both synthetic and field data from Sumatra. Limitations of this technique are discussed in the paper.

Khazaei et al. present a discrete element model, using Particle Flow Code, that allows direct modeling of stick–slip behavior in pre-existing weak planes such as joints, beddings, and faults. The model is used to simulate a biaxial sliding experiment from literature on a saw-cut specimen of Sierra granite with a single fault. They represent the fault by the smooth joint contact model. In addition, they develop an algorithm to record the stick–slip induced microseismic events along the fault. Once the results compared well with laboratory data, they conduct a parametric study to investigate the evolution of the model's behavior due to varying factors such as resolution of the model, particle elasticity, fault coefficient of friction, fault stiffness, and normal stress. The results show a decrease in shear strength of the fault in the models with smaller particles, smaller coefficient of friction of the fault, harder fault surroundings, softer faults, and smaller normal stress on the fault.

Edge detection is a useful tool in the interpretation of potential field data, and the existing edge detection filters are almost functions of first-order horizontal and vertical derivatives. Ma et al. propose step-edge detection filters to improve the resolution of edge detection results, which use the functions of different-order derivatives to accomplish the edge detection task. They demonstrate the proposed filters on synthetic potential field data, and the results show that the new methods can recognize the edges of the sources more precisely and clearly. They also discuss the application effect of different step-edge detection filters and apply the proposed filters to real potential field data.

Ghosh carry out a study in the transition zone of the Narmada-Son lineament (NSL) which is seismically active with various geological complexities. The area of investigation extends from longitude 80.25 to 81.50E and latitude 23.50N to 24.37N in the central part of the Indian continent. Different types of subsurface geological formations viz. alluvial, Gondwana, Deccan traps, Vindhyan, Mahakoshal, Granite and Gneisses groups exist in this area with varying geological ages. In this study area tectonic movement and crustal variation have been taken place during the past time and which might be origin of the variation of magnetic field. Magnetic anomaly suggests that the area has been highly disturbed which causes the Narmada-Son lineament trending in the ENE–WSW direction. Magnetic anomaly variation has been taken place due to the lithological variations subject to the changes in the geological contacts like thrusts and faults in this area. Shallow and deeper sources have been distinguished using frequency domain analysis by applying different filters. He studies the interpretation using total horizontal derivative, tilt angle derivative, horizontal tilt angle derivative and Cos (θ) derivative map to get source-edge locations.

The paper by Eshagh presents some integral formulae for recovering the sub-crustal stress from terrestrial gravimetric data. The formulation that is proposed follows Runcorn's theory, but from the practical point of view it allows the inclusion of high degree gravity model. He develops three novel methods to recover the Stress function (S) from terrestrial gravity anomalies: (1) direct integration with limited spectral kernel, (2) integral inversion with closed-form kernels and (3) integral inversion with limited spectral kernel. Finally, he applies them for modeling the sub-crustal stress in Iran and its surrounding countries showing that these integral methods are useful when the terrestrial gravity data of the area are used for estimating the stresses.

The Gravity Recovery And Climate Experiment (GRACE) mission has shown that it is possible to make detailed gravity measurements from space for climate dynamics and other purposes. To build the groundwork for a more advanced satellite-based gravity survey, the level of accuracy needed for precise estimation of fault slip in earthquakes must be estimated. Shultz et al. turn to numerical simulations of earthquake fault systems and use these to estimate gravity changes. The current generation of Virtual California (VC) simulates faults of any orientation, dip, and rake. They discuss these computations and the implications they have for accuracies needed for a dedicated gravity monitoring mission. Preliminary results are in agreement with previous results calculated from an older and simpler version of VC. Computed gravity changes are in the range of tens of μGal over distances up to a few hundred kilometers, near the detection threshold for GRACE.

Ali Shahrabi et al. provide an application of Gaussian clustering under the scheme of Expectation/Maximization algorithm to joint interpretation of seismic and magnetotelluric data to lithologic characterization of SehQanat anticline (Southwest Iran). The anticline is the most important oil trap of this area. Previous works, based only on seismic data, provide two lithological formations that belong to different geological times with a large interval and some formations in between with no history available. However, in this work the anticline is interpreted in six classes as different lithotypes which can also be related to particular stratigraphic units. The agreement of the found lithological model with markers of SQD-1 borehole located at Munir Block in the anticline underlines a suitable correlation of the seismic and MT models and performance of the Expectation/Maximization clustering.

The paper by Pavón-Carrasco et al. is focused on the reliability and fidelity of archaeomagnetic and volcanic records to recover the past evolution of the Earth's magnetic field. The authors compiled the palaeomagnetic data available for the last 400 years and compared palaeomagnetic data with historical predictions. They used the historical model GUFM1, based on a massive compilation of historical data from observations picked up by seamen in naval shipping and geomagnetic observatories, to provide an accurate vision of the directional geomagnetic field elements. The results show statistical agreement between archaeomagnetic data and directions given by the geomagnetic field model, concluding that the heated archeological materials are good recorders of the past Earth's magnetic field. In contrast, volcanic materials provide directions affected by an inclination shallowing. This systematic error is also observed when comparing recent magnetic records from lava flows with the International Reference Geomagnetic Field (IGRF) model. On average, the inclination error is around 3°, being systematically lower than the historical geomagnetic field model predictions. Although the mean flattening deviation is low, this error should be taken into account when accurate spatial and temporal evolution of the ancient geomagnetic field is analyzed.

Sánchez-Reales, Vigo and Trottini study, for the first time, variations in absolute surface geostrophic currents (SGC) using satellite data only. Their approach combines 18 years' altimetry data, which provide reliable measurements of absolute sea level (ASL), with a gravity field and steady-state ocean circulation explorer geoid model to obtain dynamic topography, and achieves unprecedented precision and accuracy. They overcome the main limitations of existing approaches based solely on altimetry data, and approximations based on in situ data. Features of annual variations of SGC are also addressed. As a result of their study they provide new absolute SGC climatology in the form of a 52-week data set of surface current fields, gridded at quarter degree longitude and latitude resolution and resolving spatial scales as short as 140 km.

Escapa et al. review and discuss the inconsistency of IAU2000 non-rigid earth nutation model (MHB). Given the complexity of the Earth rotation, it usually used a twofold approach for modeling his motion. Its long-term behavior is modeled by the precession theory whereas the theory of nutation is used to model the short-term behavior. The problem with such approach comes from considering different values of the parameters used to describe the motion in both theories. This lack of consistence might originate numerical differences that would be incompatible with nowadays accuracy requirements for Earth Orientation parameters predictions. Here, the authors discuss the effects of considering slightly different values of the dynamical ellipticity in both precession and nutation theories.

With increased geoid resolution provided by the gravity and steady-state ocean circulation explorer (GOCE) mission, the ocean's mean dynamic topography (MDT) can be now estimated with an accuracy not available prior to using geodetic methods. However, an altimetric-derived MDT still needs filtering to remove short wavelength noise unless integrated methods are used in which the three quantities are determined simultaneously using appropriate covariance functions. Sánchez-Reales, Andersen and Vigo study nonlinear anisotropic diffusive filtering applied to the ocean's MDT and a new approach based on edge-enhancing diffusion (EED) filtering is presented. EED filters enable controlling the direction and magnitude of the filtering, with subsequent enhancement of computations of the associated surface geostrophic currents (SGCs). Applying this method to a smooth MDT and to a noisy MDT, both for a region in the Northwestern they found that EED filtering preserves all the advantages that the Perona and Malik filter have over the standard linear isotropic Gaussian filters. Moreover, EED is shown to be more stable and less influenced by outliers. This suggests that the EED filtering strategy would be preferred given its capabilities in controlling/preserving the SGCs.

Galán de Sastre and Bermejo present in their paper a Lagrange-Galerkin hp-finite element method to calculate the numerical solution of a nonhydrostatic ocean model. This model is composed of the incompressible Navier–Stokes equations with Coriolis and buoyancy terms, two scalar advection–diffusion systems for temperature and salinity, and an equation of state for the density. To integrate the equations of the nonhydrostatic model, the authors propose a second-order projection method in combination with a Lagrange-Galerkin method for time discretization along the trajectories of the fluid particles and higher order hp-finite element method for the discretization in space of the differential operators. The Lagrange-Galerkin method yields a Stokes-like problem the solution of which is computed by a second-order rotational splitting scheme that separates the calculation of the velocity and pressure, the latter is decomposed into hydrostatic and nonhydrostatic components. They focus on the behavior of their method on density driven flows because they are relevant for many ocean phenomena such as water interchange through the Gibraltar's Strait.

The paper by Cea and Rodríguez develops coupled Hydrological–Hydraulic models that enable to analyze water movement in watershed as well as analyze the rainfall-runoff. They note that the rainfall in situ is principally responsible for damage in many cases, for example, in some Spanish Mediterranean regions as in the area around the township Alginet (Spain). The authors use the well-known Saint Venant equations on an unstructured mesh by finite volume to simulate the water transfer introduced upstream of the modeled area and they incorporate different hydrological situations calculating the run-off hydrograph cell level where they take into account four processes: precipitation, loss, transformation of excess rainfall in direct runoff and cost base. The results of their simulations show that the models can predict the water evolution and can be considered as a promising tool in the rainfall-runoff processes simulation.

Díaz and Gómez-Castro present a mathematical analysis study of the shape of chemical reactors, in particular for reactors designed for the treatment of wastewaters. They simplify the modeling by assuming a single chemical reaction with a monotone kinetics leading to a parabolic equation with a non-necessarily differentiable function. They assume that an ideal homogenization process was applied (by passing to the limit to zero on the porosity of the solid bed) so that the chemical reaction can be assumed as distributed over the reactor cylinder. Their main goal is to give a proper conceptual justification of why these reactors are wide and low, using natural techniques in homogenization in partial differential equations.

Cannavò and Palano provide a technical report of a new version of a previous software tool namely PlatE-Motion 2.0 (PEM 2.0). This tool, initially developed for easy-to-use file exchange with the GAMIT/GLOBK software package, allows inferring the Euler pole parameters by inverting the observed velocities at a set of sites located on a rigid block. The tool is open source and freely available for the scientific community.

The study of the fractal dimensions for the identification of bedrock lithology was carried out by Cámara et al. Their starting point is that Geographic Information System (GIS) technologies and the increasing availability and resolution of digital elevation data have greatly facilitated the delineation, quantification, and study of drainage networks. It is well known that drainage networks can exhibit different drainage patterns depending on the hydrogeological properties of the underlying materials. This study investigates the possibility of inferring geological information of the underlying material from fractal and linear parameters describing drainage networks automatically extracted from 5-m resolution LiDAR digital terrain model (DTM) data. According to the lithological information (scale 1:25,000), the study area is comprised of 30 homogeneous bedrock lithologies, the lithological map units (LMUs). Their results imply that the information included in a 5-m resolution LiDAR DTM and the appropriate techniques employed to manage it are the only inputs required to identify the underlying geological materials.

Uruk archeological site, which is located in Al-Muthanna Governorate southern Iraq, is studied by Al-Khersan et al. by integrated geophysical methods, ground penetration radar (GPR) and electric resistivity tomography (ERT) to image the historical buried structures. The GPR images show large radar attributes characterized by its continuous reflections having different widths. GPR attributes at shallower depth are mainly representing the upper part of Babylonian Houses that can often be found throughout the study area. In addition, radargrams characterized objects such as buried items, buried trenches and pits which were mainly concentrated near the surface. The ERT results show the presence of several anomalies at different depths generally having low resistivities. The map of the archeological anomalies distribution and 3D view of the foundations at the study area using GPR and ERT techniques clearly show the characteristics of the Babylonian remains. A contour map and 3D view of Uruk show that the archeological anomalies are concentrated mainly at the NE part of the district with higher values of wall height that range between 6 and 8 m and reach to more than 10 m. At the other directions, there are fewer walls with lower heights of 4–6 m and reach in some places the wall foot.

A new mathematical model for patchy landscapes in drylands is introduced in the paper by Kinast S., Ashkenazy Y., and Meron E. The model concerns the dynamics of biogenic soil crusts and their mutual interactions with vegetation growth. The identification of spatially uniform and spatially periodic solutions that represent different vegetation-crust states, and map them along the rainfall gradient are analyzed. A significant difference between the current and earlier models of patchy landscapes is found in the bistability range of vegetated and unvegetated states; the incorporation of crust dynamics shifts the onset of vegetation patterns to a higher precipitation value and increases the biomass amplitude. This new model may shed new light on the effects of biogenic crusts on the response of dryland ecosystems to rainfall variability, and may improve understanding of desertification processes.

The paper by San José Martínez et al. deals with the study of pore space soil structure of X-ray computed tomography CT images of soil columns. Their study uses mathematical morphology as source of a plethora of different mathematical techniques. They provide a guide to design the process from image analysis to the generation of synthetic models of soil structure to investigate key features of flow and transport phenomena in soil. In this work, they explore the ability of morphological functions built over Minkowski functionals with parallel sets of the pore space to characterize and quantify pore space geometry of columns of intact soil. These morphological functions seem to discriminate the effects on soil pore space geometry of contrasting management practices in a Mediterranean vineyard.

Zhou et al. perform a comparison between transverse electric-type (TE) fields and, for other hand, between transverse magnetic (TM) and transverse electromagnetic (TEM) fields by generating them from grounded-wire source and testing their distribution characteristics, detection depth, and sensitivity to anomalies, using both synthetic 1D model and two field surveys in China (two field examples are conducted to verify their comparisons of the TE, TE–TM, and combined TEM fields). The comparisons demonstrate that the detection depth of the TE–TM field is smaller than those of both the TE and combined TEM fields. Meanwhile, for electric field, the TE–TM response provides a better detection than the TEM one, but with an uneven distribution. Therefore, the TE–TM electric field requires well-designed arrangements of receiving positions when applied to real projects. For the magnetic field, the TEM response has the best detection capability compared to the TE and TE–TM ones, but is least sensitive to layer thickness and resistivity.

Acknowledgments

We thank the contributors to this and previous Topical Issue in Mathematics and Geosciences. We thank also Birkhauser personnel, without whom this volume would not have been possible. We especially appreciate the great and generous work carried out by many reviewers in the Mathematics and Geosciences intersection field. They have worked in most of the cases in the difficult intersection of two different fields, such as Earth Sciences and Mathematics. We would like to take this opportunity to thank the Instituto de Ciencias Matemáticas and the Institute of Geosciences for their financial and technical support in organizing the Mathematical and Geosciences Congress, and especially E. Fuentes, E. Frechilla, and A. Chacón from ICMAT. We are also grateful to CEI Campus Moncloa (UCM-UPM) and CEI UAM-CSIC for their support. The workshop was also partially supported with funds from research projects AYA2010-17448, MTM2011-26696, SEV-2011-0087, CGL2011-24790 and ESP2013-47780-C2-1-R of MINECO, (Spain) MTM2011-26119 of the DGISPI (Spain), the UCM Research Group MOMAT (Ref. 910480), and the ITN FIRST of the Seventh Framework Program of the European Community's (grant agreement number 238702). The editors also thank Dr. Renata Dmowska and Ganesh Priyanka for the help, suggestions, and support received during the editing of this topical issue.

REFERENCES

AFONSO, J.C., MOORKAMP, M., FULLEA, J., 2015. Imaging the Lithosphere and Upper Mantle (Chapter), In: MOORKAMP, M., KHAN, A., LINDE, N., LELIEVRE, P. (Eds.) Integrated Imaging of the Earth (AGU Book) (In press).

ANDERSON, K. and SEGALL, P., 2013. *Bayesian inversion of data from effusive volcanic eruptions using physics-based models: Application to Mount St. Helens 2004–2008.* J. Geophysical Res., *118*(5), 2017–2037, doi:10.1002/jgrb.50169.

CAMMARANO F., TACKLEY P. BOSCHI L., 2011. *Seismic, petrological and geodynamical constraints on thermal and compositional structure of the upper mantle: global thermo-chemical models.* Geophys. J. Int., *187*, 1301–1318.

Díaz, J.I, Fernández, J., Orive, R., Osete, M.L., 2015 (Editors). Pure and Applied Geophysics (Pageoph), Topical Issue "Mathematics and Geosciences. Global and Local perspectives: Volume I". 172/1, 1–196.

Foulger, G. R., Panza, G. F., Artemieva, I., Bastow, I. D. Cammarano, F., Doglioni, C., Evans, J. R., Hamilton, W. B., Lustrino, M., Thybo, H., Yanovskaya, T. B., 2015. *Teleseismic tomography: The challenges ahead.* In: EOS, Transactions of the Geophysical Union, Vol. in press.

Garcia-Castellanos, D., Jiménez-Munt, I., 2015. *Topographic evolution and climate aridification during continental collision: insights from computer simulations.* PLoS One, doi:10.1371/journal.pone.0132252.

Henderson-Sellers, A. and McGuffie, K., 1987. A Climate Modelling, John Wiley&Sons, Chichester, UK.

Keilis-Borok, V., A. Gabrielov, and A. Soloviev, Geo-complexity and earthquake prediction. In: Meyers R. (ed.) Encyclopedia of Complexity and Systems Science, Springer, New York, 2009, pp. 4178–4194. doi:10.1007/978-0-387-30440-3_246.

Lay, T. and Garnero, E. J., 2011. *Deep Mantle Seismic Modeling and Imaging.* Annu. Rev. Earth Planet. Sci., *39*, 91–123.

Longo A., M. Barsanti, A. Cassioli, P. Papale, 2012. *A finite element Galerkin/least-squares method for computation of multicomponent compressible-incompressible flows,* Computers & Fluids, *67*, 57–71, doi:10.1016/j.compfluid.2012.07.008.

McKenzie, D., 1966. *The viscosity of the lower mantle.* J. Geophys. Res., *71*, 3995–4010.

MPE2013, Mathematics of Planet Earth (2014). http://mpe2013.org/.

Smeed, D.A., G. D. McCarthy, S. A. Cunningham, E. Frajka-Williams, D. Rayner, W. E. Johns, C. S. Meinen, M. O. Baringer, B. I. Moat, A. Duchez and H. L. Bryden, 2014. *Observed decline of the Atlantic meridional overturning circulation 2004–2012,* Ocean Sci., 10, 29–38, doi:10.5194/os-10-29-2014.

Wegener, A., 1912. Die Entstehung der Kontinente. Geologische Rundschau, *3*, 276–292. doi:10.1007/BF02292896.

Pure Appl. Geophys. 173 (2016), 739–748
© 2014 Springer Basel
DOI 10.1007/s00024-014-0951-3

Pure and Applied Geophysics

Computing Earthquake Probabilities on Global Scales

James R. Holliday,[1] William R. Graves,[2] John B. Rundle,[3,4] and Donald L. Turcotte[5]

Abstract—Large devastating events in systems such as earthquakes, typhoons, market crashes, electricity grid blackouts, floods, droughts, wars and conflicts, and landslides can be unexpected and devastating. Events in many of these systems display frequency-size statistics that are power laws. Previously, we presented a new method for calculating probabilities for large events in systems such as these. This method counts the number of small events since the last large event and then converts this count into a probability by using a Weibull probability law. We applied this method to the calculation of large earthquake probabilities in California-Nevada, USA. In that study, we considered a fixed geographic region and assumed that all earthquakes within that region, large magnitudes as well as small, were perfectly correlated. In the present article, we extend this model to systems in which the events have a finite correlation length. We modify our previous results by employing the correlation function for near mean field systems having long-range interactions, an example of which is earthquakes and elastic interactions. We then construct an application of the method and show examples of computed earthquake probabilities.

1. Introduction

Many systems in nature and society, including earthquakes, typhoons, market crashes, electricity grid blackouts, floods, droughts, wars and conflicts, and landslides (Sachs *et al.* 2012; Scholz 2002; Taleb 2007; Turcotte 1997; Sornette 2009; Malamud *et al.* 2005), display power-law statistics for frequency and magnitude. Calculation of event probabilities in these systems is often of considerable importance to reduce the damage (physical, financial

and social) that may be associated with the largest events. These power laws appear to be stable, in the sense that including more events in the statistics over larger regions and longer periods of time improves the fit to a power law.

In these power-law systems, a fixed number of the frequent small events is associated with one large event. This property suggests a strategy for the computation of the infrequent large event probabilities. Let N small events correspond to one large event. Then counting the number of small events occurring since the last large event provides a measure of the "natural time" elapsed between large events (Rundle *et al.* 2012).

Rundle *et al.* (2012) used a Weibull probability law to convert the small event count to a probability of occurrence of the next large event during a future time interval Δt. The means of conversion from natural time to calendar time was the long-term average rate of small events.

The previous paper applied these ideas to earthquakes in California and Nevada, USA. An underlying assumption of the analysis presented there was that earthquakes within the analyzed region were perfectly correlated, but that events outside the region had no effect on events within the region. Thus, the region of California-Nevada was considered to be an isolated system having no communication with the external region.

In this article, we relax these assumptions and instead consider earthquake systems having a finite correlation length. In this more general model, all events within a finite designated region are affected by events outside the region. Or in other words, the events inside the region interact with external events with a finite external correlation length ξ. It is assumed that all events within the finite region are highly correlated. There is no longer a privileged

[1] Department of Physics, University of California, Davis, CA, USA.
[2] Open Hazards Group, Davis, CA, USA.
[3] Departments of Physics and Geology, University of California, Davis, CA, USA. E-mail: jbrundle@ucdavis.edu
[4] The Santa Fe Institute, Santa Fe, NM, USA.
[5] Department of Geology, University of California, Davis, CA, USA.

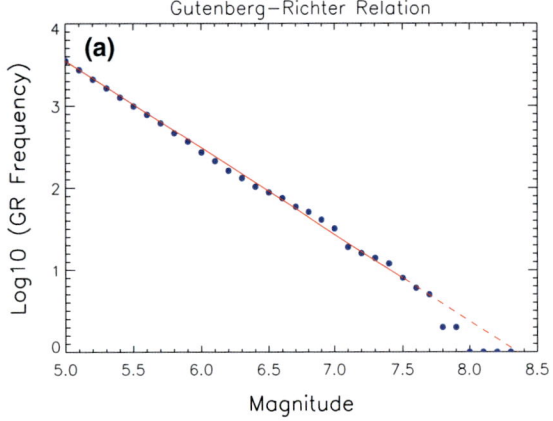

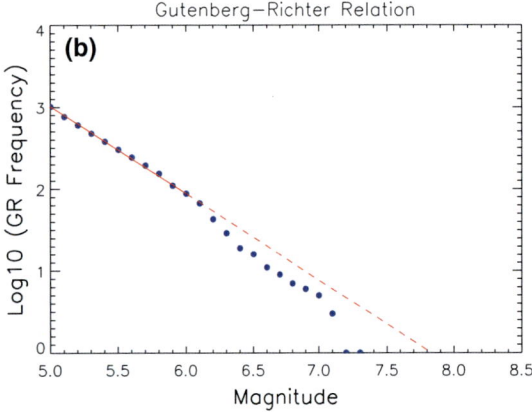

Figure 1

a Gutenberg-Richter magnitude frequency relation for the region encompassing a radius of 1,000 km around Tokyo, Japan, as determined by fitting events from the ANSS catalog. The time period for the earthquakes was from July 1961 through the last event prior to the 11 March 2011 M9.1 Tohoku earthquake. The largest earthquake in this sequence was the $m = 8.3$ Hokkaido earthquake of 25 September 2003. The line fit was over the magnitude interval from 5.0 to 7.5, yielding a GR b value of 1.03 ± 0.03. It can be observed that the GR relation was largely complete to the level of $m = 8.3$, although the time interval was not long enough to determine completeness to magnitudes of $m = 9$. As noted in the text, this type of scaling analysis of the Gutenberg-Richter relation applies to time periods that are at most of the order of the recurrence time for the largest earthquakes in the region. **b** Gutenberg-Richter magnitude frequency relation for the region encompassing a radius of 1,000 km around Tokyo, Japan, as determined by fitting events from the ANSS catalog. The time period for the earthquakes was from just after the $m = 7.7$ aftershock on 11 March 2011 at 06:25:50.30 until 19 April 2014. The largest earthquake in this sequence was the $m = 7.3$ earthquake of 7 December 2012. It can be observed that the GR relation is not complete in the sense that the data do not define a linear scaling relation over the entire magnitude range. In fact, there is an apparent deficit of earthquakes larger than 6.0, which eventually must be filled in at the high end if a linear GR relation is to be observed. Projecting the line to the largest event suggests that an event of magnitude $m = 7.8$ is possible. The GR scaling line was fit over the magnitude interval 5.0–6.0 and yields the same b value of 1.01 ± 0.01

location or preferred region; instead, all locations are treated equivalently.

We begin by briefly summarizing the main result of our previous model. We introduce a correlation function having a finite correlation length and apply it to the statistical analysis. We then construct an algorithm to compute large event probabilities worldwide and give several examples.

2. Motivation

Earthquakes are associated with a frequency-size distribution, the cumulative Gutenberg-Richter magnitude frequency law (SCHOLZ 2002). Earthquake magnitude is measured by modeling seismic waveforms to obtain the seismic moment, then converted to a moment magnitude (HANKS and KANAMORI 1979). Examples of the frequency-magnitude statistical distribution for the region around Japan are shown in Fig. 1a and b. Similar statistics are shown in Figs. 2 and 3 for California-Nevada-Baja Mexico (Fig. 2) and for the Pacific NW of the USA (Fig. 3). Data are taken from online catalogs (USGS 2014). Unless otherwise stated, magnitudes are computed according to the method as stated in the ANSS catalog. For large earthquakes, this is usually moment magnitude. Small earthquake magnitudes may be body wave, local, coda magnitude or some other method.

For Fig. 1a, b, we present cumulative magnitude-frequency statistics for events within a 1,000-km circle surrounding Tokyo for two time periods. In Fig. 1a, we plot the cumulative magnitude-frequency relation for the time period from 1961 to just prior to the $m = 9.1$, 11 March 2011 Tohoku earthquake. In Fig. 1b, we plot the cumulative magnitude-frequency relation for the time period from just after the $m = 7.7$ aftershock that occurred on 11 March 2011 about 39 min after the $m = 9.1$ mainshock.

As is clear from comparing Fig. 1a and b, smaller earthquakes lie along a scaling line, but a deficit in earthquakes larger than about $m > 6$ is apparent. Technically, the GR relation is exponential in cumulative frequency-magnitude, but power law in cumulative frequency-moment. Seismic moment is defined by HANKS and KANAMORI (1979) as a linear relation between magnitude and $\log_{10}$(moment). The

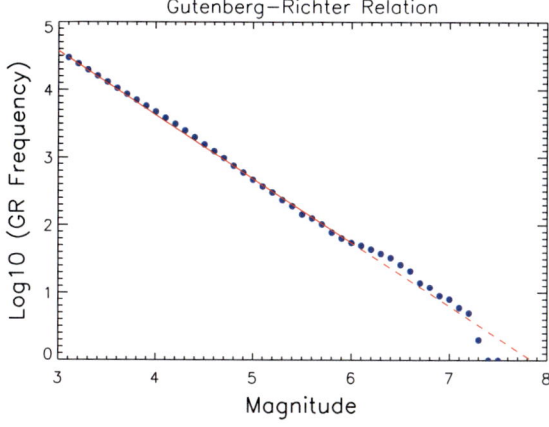

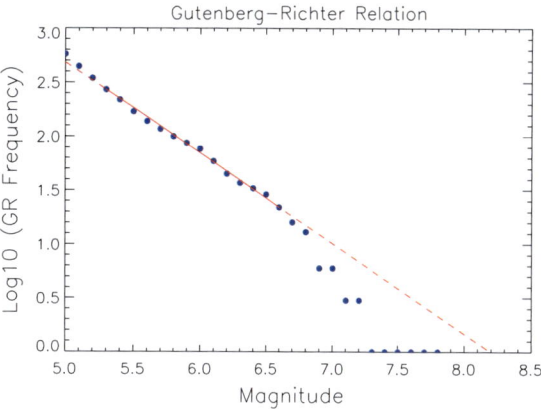

Figure 2

Gutenberg-Richter magnitude frequency relation for California-Nevada-Baja California (Mexico) between latitudes 29 and 42°N and between longitudes of −127 and −113°W. Data from 1900 to 19 April 2014, although there is only one event prior to 1932. From 1932 onward, it is generally considered that the catalog is complete at the low end down to magnitude 3.0. The largest event in this sequence of data is the 21 July 1952 $m = 7.5$ Kern County earthquake, although this event is generally considered by other sources to have a magnitude of 7.3. The most recent large event is the 4 April 2010 $m = 7.2$ El Major-Cucupah (Mexico) event. Data were fit by the straight line over the magnitude interval 3.0–6.0 and yielded a b value of 0.96 ± 0.01. It can be observed that the scaling line is largely complete at the high end to magnitudes of about $m \sim 7.3$. Projecting the scaling line at the high end suggests that a $m = 7.8$ earthquake would be expected if the GR statistics are to define a linear relation at the high end

Figure 3

Gutenberg-Richter magnitude frequency relation for the region encompassing a 1,000-km radius around Seattle, WA. The earliest event in this record drawn from the ANSS catalog is in 1912, although the data only become relatively complete at the low end beginning in about 1963. At the small magnitude end, data are probably only complete within this region, which encompasses much of the oceanic areas west of Oregon, Washington and British Columbia, at magnitudes above 5.0. The largest event in this sequence is the 27 October 2012 $m = 7.8$ Haida Gwaii earthquake. Data were fit by the straight line over the magnitude interval 5.0–6.6 and yielded a b value of 0.81 ± 0.02. It can be observed that the scaling line is not complete at the high end, and projecting the scaling line to one event suggests that a $m = 8.2$ earthquake is possible if the scaling relation is to be linear over its entire range. Note that this region encompasses the area of Cascadia, where a $m \sim 9$ earthquake is thought to have occurred on 26 January 1700 AD

Gutenberg-Richter distribution will ultimately fill in, so over time this deficit can be expected to eventually be removed by the occurrence of larger earthquakes to "fill in" the scaling relation. This observation is the basic rationale behind the NTW method as described in (RUNDLE et al. 2012). The method thus arises from the expectation that the filling-in of the Gutenberg-Richter distribution is accomplished by the occurrence of future large earthquakes. Converting this deficit to a forecast is accomplished by means of the Weibull probability law, which is frequently used to describe the statistics of failure and reliability in engineered systems (RUNDLE et al. 2012).

Similarly, Fig. 2 shows Gutenberg-Richter statistics for the region of California-Nevada-Baja Mexico (Fig. 2) between latitudes 29 and 42°N and between longitudes of −127 and −113°W. Data from 1900 to 19 April 2014, although there is only one event prior to 1932. Figure 3 shows Gutenberg-Richter statistics for the region of Northern California-Oregon-

Washington-British Columbia within 1,000 km of Seattle, WA. It can be seen that Fig. 2 is similar to Fig. 1a in that both appear to be relatively complete at the high magnitude end. Figures 1b and 3 are similar in that both appear to display a deficit of large earthquakes, because the frequency of high magnitude events falls well below the scaling line. Extrapolating the scaling line to a frequency of one event suggests that the 1,000 km radius regions around Tokyo and Seattle can only be made complete at the high magnitude end by the occurrence of future large earthquakes.

As a qualification, note that this type of scaling analysis of the Gutenberg-Richter relation applies to time periods that are at most of the order of the recurrence time for the largest earthquakes in the region. For California, the largest earthquakes are about $m \sim 8$, and we know that this recurrence time is about 200 years. For the Tohoku region in Japan, we know that the largest earthquakes are about $m \sim 9$, and we now know that this time period is about 1,000 years, the last such earthquake being the Jogan event in 869 AD.

In a later section, we apply the methods of the present article to compute the probability of earthquakes having magnitudes $m \geq 6$ within a 200-km radius area of the Pacific NW. This circular area is centered on the most recent major earthquake to occur within the Pacific NW, the magnitude $m = 6.6$ earthquake of 24 April 2014.

3. Basic Method

Here we summarize the basic method as described in (RUNDLE et al. 2012) as a prelude to further developments. We defined a seismically active region in California-Nevada between latitudes 29 and 42° north and longitudes -127 and $-113°$ west. It is well known that the Gutenberg-Richter magnitude-frequency relation holds for this region over time, as well as active subregions (RUNDLE et al. 1995, 1997; TIAMPO et al. 2003), and this can be seen in Fig. 2. For smaller regions, power law statistics improve with longer waiting times.

We computed the probability of a large earthquake m_L from observations of small earthquakes m_S, where $m_S < m_L$. (Note that subscripts S and L denote "small" and "large" magnitudes here and *do not* refer to surface wave or body wave magnitudes.) We used the fact that Gutenberg-Richter distribution implies that for every large earthquake $m \geq m_L$, there are N small earthquakes:

$$N = 10^{b(m_L - m_S)} \tag{1}$$

In California, we used $m_S = 3.0$ and $m_L = 6.0$ in (RUNDLE et al. 2012). Here b is the Gutenberg-Richter b value, typically about $b \approx 1$.

We assumed that all earthquakes in the region are correlated internally, but that there is no effect from earthquakes outside the region. We then computed the conditional probability that a large earthquake will occur a time interval Δt in the future, given that $n(t)$ small earthquakes have occurred during the time t since the last large earthquake:

$$P(\Delta t | t) = 1 - \exp\left\{ \left[\frac{n(t)}{N} \right]^{\beta} - \left[\frac{n(t) + \omega \Delta t}{N} \right]^{\beta} \right\} \tag{2}$$

Here ω is the time-averaged rate of small earthquakes having magnitude $m_S \leq m < m_L$. From (2), it can be seen that just after a large earthquake, the probability $P(\Delta t | 0)$ is not zero, but rather:

$$P(\Delta t | 0) = 1 - \exp\left\{ - \left[\frac{\omega \Delta t}{N} \right]^{\beta} \right\} \tag{3}$$

We can also define the time-averaged rate Ω of large earthquakes having $m \geq m_L$ by:

$$\Omega \equiv \frac{\omega}{N} = \omega \, 10^{-b(m_L - m_S)} \tag{4}$$

In our earlier paper, we used these ideas to compute and verify an earthquake probability time series for California-Nevada-Baja California (Mexico), assuming that this is an isolated region that does not interact with earthquakes occurring outside the region.

It will be seen later that Eq. (2) is of the type often called a "renewal" model or "characteristic earthquake" model, where the probability falls after a large event, then increases leading up to the next large event (KIREMIDJIAN and ANAGNOS 1984). The probability depends not only on the rate of activity, but is also conditioned on the (natural) time since the last large earthquake.

4. Finite Correlation Length

The model in (1)–(4) assumes that all events within the geographic region considered are perfectly correlated and completely uncorrelated with events outside the region. This is not a good assumption for many cases of interest, so here we generalize the model to conditions in which the correlation length ξ is finite.

4.1. Definitions

We first consider a local region around x_i. As a practical matter, we define a global coarse-grained grid of boxes or "pixels" with grid size 0.1°, indexing these so that the pixel is centered on point x_i. We then extend a circle outwards until it contains at least 300 small earthquakes and at least 5 large earthquakes having the target magnitude m_L (Fig. 3). This region will be used to determine the time averaged rate Ω_i of earthquakes having magnitude $m \geq m_L$. We find Ω_i either from direct observation or by extrapolating from smaller magnitude events as in (4).

For a seismically active location, $R(x_i, t)$ can be small, of the order of kilometers or tens of kilometers. For less active regions, $R(x_i, t)$ is limited by assumption to a radius of a maximum of 1,800 km. Since small earthquake seismicity tends to be highly clustered in space, it is not generally possible to use the same number of small earthquakes for each pixel. The pixel that is at the center of the large circle is the central pixel. The most recent of these five large earthquakes within the circle will be used as the reference time from which to begin counting small earthquakes.

We also consider a worldwide earthquake catalog, such as the ANSS catalog (ANSS http://earthquake. usgs.gov/), beginning at a time t_0 and continuing up to a present time t. Given the quality of the data, we choose $t_0 = 1980$. The fraction of small earthquakes $m_S \leq m < m_L$ contained within $R(x_i, t)$, which occurs specifically within the central pixel between times (t_0, t), will be referred to as $\rho(x_i, t) = \rho_i(t)$. The time-averaged rate of small earthquakes is denoted by ω_i, and the time-averaged rate of large earthquakes is denoted by Ω_i over the time interval (t_0, t). The Gutenberg-Richter b value of the small earthquakes occurring within $R(x_i, t)$ is denoted by b_i, and the number of small earthquakes since the last large earthquake within the circle is $n_i(t)$.

Given an optimal value of exponent β, we can then compute the exponential factor as (RUNDLE et al. 2012):

$$H_i(t, \Delta t, m_L) = \left(\frac{n_i(t)}{N} + \Omega_i \Delta t\right)^\beta - \left(\frac{n_i(t)}{N}\right)^\beta \quad (5)$$

If there were no other earthquakes outside $R(x_i, t)$, the probability of a large earthquake within Δt from t would be:

$$P_i(\Delta t|t) = 1 - \exp\{-H_i(t, \Delta t, m_L)\} \quad (6)$$

4.2. Effect of Distant Earthquakes

Earthquakes external to $R(x_i, t)$ can affect earthquakes inside $R(x_i, t)$ through stress transfer (RUNDLE

et al. 1997; STEIN 1999). While the process is complex, our model includes this effect in a statistically simple way. If an external large earthquake with $m \geq m_L$ occurs outside $R(x_i, t)$ the probability $P_i(\Delta t|t)$ within $R(x_i, t)$ may tend to decrease as stress is relieved in the region. However, if that same large earthquake increases the number of small earthquakes within $R(x_i, t)$, then the probability within $R(x_i, t)$ will tend to increase as a result. The factor by which $P_i(\Delta t|t)$ within $R(x_i, t)$ increases or decreases is assumed to be the static correlation function $C(r)$, where r is the distance outside $R(x_i, t)$. r is defined so that the distance of the large earthquake from the central pixel at x_i is $r + R(x_i, t)$.

In models of the earthquake process published in the recent past (KLEIN et al. 2000; GOLDENFELD 1992), it has been pointed out that earthquakes are spatially correlated with a correlation length roughly equal to the size of the largest earthquake in the region. The source dimensions of the largest earthquakes can vary from 300 to 500 km (1857, 1906 California; 2011 Tohoku) to occasionally as large as 1,600 km (2004 Sumatra-Andaman). We therefore adopt at this point the globally uniform value of $\xi = 400$ km, a reasonable value for most areas of the world.

We adopt the simplest form for the correlation function for a mean field system in $d = 2$ space dimensions (GOLDENFELD 1992):

$$C(r) = e^{-r/\xi} \quad (7)$$

We use the $d = 2$ dimensional correlation function since the distances between the large earthquakes are usually significantly larger than their depths. Future work (RUNDLE et al. 2014) will explore the effects of using the $d = 3$ correlation function.

When small earthquakes occur, we use (7) to adjust the count $n_i(t)$:

$$\begin{aligned} n_i(t+1) &\to n_i(t) + 1 \quad \text{for} \quad r \leq R(x_i, t) \\ n_i(t+1) &\to n_i(t) + e^{-r/\xi} \quad \text{for} \quad r > R(x_i, t) \end{aligned} \quad (8)$$

When large earthquakes occur, $n_i(t)$ is incremented as follows:

$$\begin{aligned} n_i(t+1) &\to 0 \quad \text{for} \quad r \leq R(x_i, t) \\ n_i(t+1) &\to n_i(t) \times (1 - e^{-r/\xi}) \quad \text{for} \quad r > R(x_i, t) \end{aligned} \quad (9)$$

In this model, a large more distant earthquake decreases the small $n_i(t)$ event count, but not by as much as a large event within the range $r \le R(x_i, t)$ for which $n_i(t + 1) \to 0$. Similarly, small earthquakes outside $R(x_i, t)$ increase $n_i(t)$, but not by as much as small earthquakes inside $R(x_i, t)$.

4.3. Correction Factors

Only a fraction $\rho_i(t)$ of the small earthquake activity within $R(x_i, t)$ is associated with the pixel at x_i. In addition, $H_i(t, \Delta t, m_L)$ must be consistent with the constraint of an observable, long time-averaged rate of large earthquakes within $R(x_i, t)$. Or to put it another way, the time average of $H_i(t, \Delta t, m_L)$ over (t_0, t) must be equal to the observed time average. So a time-dependent correction factor $f_i(t)$ for each pixel may be needed.

Let $n_L(x_i) \equiv \Omega_i \Delta t$ be the expected average number of large events $m \ge m_L$ during the forecast time interval Δt within $R(x_i, t)$. $n_L(x_i)$ can be determined by either extrapolation of the rate of small events to larger magnitudes or directly from the rate of large magnitude events.

We first compute the integrals:

$$I_i(t, \Delta t) \equiv \int_{t_0}^{t} H_i(t', \Delta t, m_L) dt' \qquad (10)$$

$$\langle I_i(t, \Delta t) \rangle \equiv \frac{1}{(t_0 - t)} \int_{t_0}^{t} H_i(t', \Delta t, m_L) dt' \qquad (11)$$

Then the normalization $f_i(t)$ is defined by:

$$f_i(t) = \frac{n_L(x_i)}{\langle I_i(t, \Delta t) \rangle} \qquad (12)$$

$f_i(t)$ provides the correction factor that makes the time average of $H_i(t, \Delta t, m_L)$ consistent with the number of large events $n_L(x_i)$ within $R(x_i, t)$.

We now define the contribution $h_i(t, \Delta t, m_L)$ of the pixel at x_i to $H_i(t, \Delta t, m_L)$ using $\rho_i(t)$:

$$h_i(t, \Delta t, m_L) \equiv \rho_i(t) n_L(x_i) \frac{I_i(t, \Delta t)}{\langle I_i(t, \Delta t) \rangle} \qquad (13)$$

The probability for a large earthquake occurring at the central pixel is:

$$P_i(\Delta t | t) = 1 - \exp\{-h_i(t, \Delta t, m_L)\} \qquad (14)$$

The probability for a large earthquake occurring in any of the subset $\{x_i\}$ of pixels is then:

$$P(\Delta t | t) = 1 - \exp\left\{-\sum h_i(t, \Delta t, m_L)\right\} \qquad (15)$$

In going from Eq. (14) to (15) we have assumed that $R(x_i, t)$ is large enough so that $h_i(t, \Delta t, m_L)$ does not depend sensitively on details of its size. We have conducted a series of trials with varying $R(x_i, t)$ that lend support to this assumption.

In addition, we have also assumed that the rate of large earthquake occurrence within any subset of pixels $\{x_i\}$ is proportional to the fraction of small earthquakes that occurs within $\{x_i\}$ relative to the total number that occurs within $R(x_i, t)$. This assumption is supported by results from previous studies (HOLLIDAY et al. 2005).

We also see that Eq. (13) can be rewritten as:

$$h_i(t, \Delta t, m_L) \equiv \rho_i(t) \, \Omega_i \Delta t \, \frac{I_i(t, \Delta t)}{\langle I_i(t, \Delta t) \rangle} \qquad (16)$$

The nature of Eq. (16) implies that the time-dependent probability $P_i(\Delta t | t)$ can be regarded as consisting of a perturbation $I_i(t, \Delta t)/\langle I_i(t, \Delta t) \rangle$ on the expected average number of large earthquakes $\Omega_i \, \Delta t$ during the future time interval Δt. The appropriate fraction $\rho_i(t)$ of the perturbation, based on the relative activity of small earthquakes in $R(x_i, t)$, is then assigned to the central pixel at x_i.

Alternatively, we may define:

$$\Omega'_i(t, \Delta t) \equiv \rho_i(t) \, \Omega_i \, \frac{I_i(t, \Delta t)}{\langle I_i(t, \Delta t) \rangle} \qquad (17)$$

so that:

$$h_i(t, \Delta t, m_L) \equiv \Omega'_i(t, \Delta t) \Delta t \qquad (18)$$

4.4. Data Issues

A major constraint on the method is variation in the quality of the data with location. Data quality is highest in the USA, specifically California. Here the data catalogs are generally complete (at least in southern California) from 1932 for magnitudes larger than about $m \ge 3.0$ (USGS 2014; LEE et al. 2002). These data were recorded on analog recorders and later digitized and added to the catalogs many years later. However, even though these early events may be included in the catalog, details of their source parameters such as locations and magnitude determinations were probably not as reliable as the data recorded in the modern era (since about 1990) by digital networks.

The global digital network has only been in place for a decade or two, and it has been upgraded with more stations over the years since its establishment in 1986 (SORNETTE 2009). A number of global locations are poorly covered even today by these stations, particularly in areas of Asia including China, India and the Middle East, such as Pakistan, Iran and Iraq. Since the method described here uses data back to 1980 to determine event counts, this lack of completeness in the small events of the global catalog is a serious problem.

A manifestation of this problem occurs if time-averaged rates of large events are extrapolated from counts of historic small earthquake rates. Since the global catalogs are missing many of the small earthquakes, counts will generally be too small in many regions of southern Asia and the Middle East. For earlier periods in these regions of the world, small earthquake counts may not be accurate, which will affect the accuracy of the NTW forecast.

On the other hand, global events larger than $m \geq 6.0$ during the last 30 years are almost always accurately represented in the catalog. Therefore, the time-averaged rates of large events are generally more accurate if the larger events $m \sim 6$ are used to determine time-averaged rates. This is the basic reason for defining $R(x_i, t)$ to include at least five large earthquakes in the region.

4.5. Verification and Validation

In RUNDLE et al. (2012) we discussed the issue of forecast verification and presented a selection of results. The methods we used have been extensively documented in the literature (JOLIFFE and STEPHENSON 2003; HSU and MURPHY 1986; KHARIN and ZWIERS 2003). They are based on the Briar Score (HSU and MURPHY 1986), leading to the Reliability/Attributes (R/A) test, and the receiver operating characteristic (ROC) test (KHARIN and ZWIERS 2003). In this article, we will not repeat this analysis, reserving this

discussion for a future paper (RUNDLE et al. 2014). Our backtesting yielded the best-fitting parameter $\beta = 1.4$ for California. Since the catalog data are best for California, we have adopted this value for global use.

5. Automated Computations

One of the challenges (RUNDLE et al. 2012) in research of this type is to present the information in usable form. Earthquake probabilities are dynamic, for example, earthquake aftershocks and triggered events are common phenomena. Thus, we require a platform that allows automated computation, updating and independent analysis in real time. For that reason, we have coded the methods described here onto open access online platforms (Open Hazards www.openhazards.com; Quakesim www.quakesim. org). A user interface (UI) includes a toolset that allows retrieval of probability information worldwide for display onscreen via the standard web browsers.

At about midnight East Coast time, catalog feeds are downloaded and combined to form the input data. These feeds are primarily the ANSS catalog and the USGS 30-day real-time feed (USGS 2014). Other data feeds may be used as appropriate. Since the ANSS catalog is not always updated in real time, and since our goal is to provide real-time probabilities, the catalog must be updated daily. When combining the catalogs, event IDs are checked to eliminate the possibility of including multiple listings of the same event.

The methods described in the foregoing discussion are then applied to compute the factors $h_i(t, \Delta t, m_L)$ at pixels of size 0.1°. These factors are then assigned as tags to each screen pixel and also used to construct a KML file for displaying the probabilities on screen. The factor $h_i(t, \Delta t, m_L)$ is nonzero only at pixels where seismicity exists, about 4 % of the earth's surface. As a result, NTW probability has sharp spatial boundaries. However, some of this sharpness can be due to errors in the location of the catalog data, and a method is needed to include these errors.

For that reason, a spatial smoothing or smearing-out operation is used, typically a Gaussian smoothing over radial distances of about 0.2° for all earthquakes occurring since $t_0 = 1980$. In addition, where recent activity has been low, but previous activity has been high, the time-averaged rate is not negligible. The probability displayed is therefore a combination or ensemble forecast consisting of 80 % NTW and 20 % smoothed seismicity forecast (HOLLIDAY *et al.* 2007). This combination of validated forecasts provides adequate spatial smoothing consistent with uncertainty in global earthquake locations.

The interested user is referred to the web site (Open Hazards www.openhazards.com) for application to other areas of interest.

6. Application

Probability time series for $m_L \geq 6$ earthquakes occurring within the next year, computed using Eq. (15), are shown in Fig. 5 for a circular area in the Pacific NW (Fig. 4). The circular area in Fig. 4 has a radius of 200 km. Red circles in Fig. 4 represent the earthquakes that have occurred since 1 January 2008. Table 1 shows earthquakes having magnitudes $m_L \geq 6$ that have occurred in the region. The largest event during that time interval was the $m_L = 7.8$

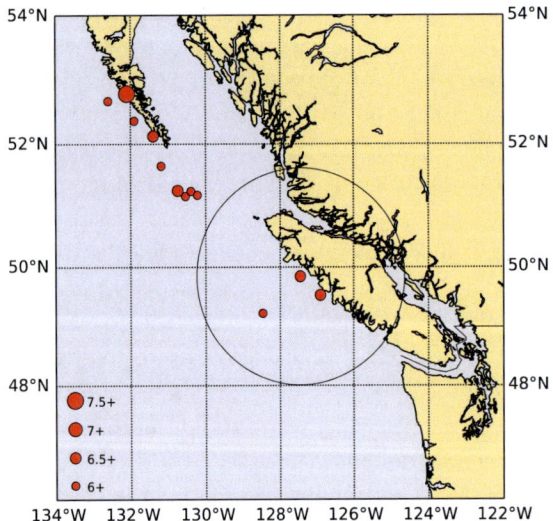

Figure 4

Location map of the Pacific Northwest and British Columbia. Large earthquake activities since 1 January 2008 are shown as *red circles*. Figure 5 shows earthquake probability computed by the methods in the text for a $m_L \geq 6$ earthquake occurring within the large circular region shown (radius 200 km). The circle is centered on the $m = 6.6$ earthquake that occurred on 24 April 2014 (UTC)

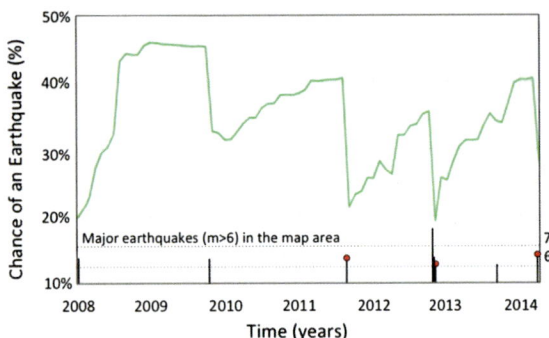

Figure 5

Earthquake probabilities through time since 1 January 2008, computed for events having $m > 6$ and plotted at monthly intervals. Probabilties are for events occurring within the large 200-km *radius circle* shown in Fig. 4 using the methods described in the text. Shown at the bottom are the earthquakes having $m > 6$ that occurred within the map region of Fig. 4 (see also Table 1). The three events with the *small red circles* are the three events that occurred within the 200-km radius region. It can be seen that probabilities decreased suddenly at the time of these three events, as well as for an earlier large earthquake that occurred in late 2009

Table 1

Major earthquakes in the Pacific Northwest between latitudes 46 and 54°N, between longitudes −134 and −122°W, and from 1 January 2008–19 April 2014

Date	Magnitude (m)	Latitude (°)	Longitude (°)	Depth (km)
2008/01/05	6.6	51.25	−130.75	15.0
2008/01/05	6.4	51.16	−130.54	10.0
2008/01/09	6.1	51.65	−131.18	10.0
2009/11/17	6.6	52.12	−131.40	17.0
2011/09/09	6.5	49.54	−126.89	22.0
2012/10/28	7.8	52.79	−132.10	14.0
2008/10/28	6.3	52.67	−132.60	9.0
2012/10/30	6.2	52.37	−131.90	9.0
2012/11/08	6.1	49.23	−128.48	13.7
2013/09/03	6.1	51.24	−130.40	2.7
2013/09/04	6.0	51.18	−130.22	9.9
2014/04/24	6.6	49.85	−127.44	11.4

Haida Gwaii earthquake of 27 October 2012, shown as the largest circle in Fig. 4 and listed in Table 1. The time-dependent nature of the forecast probabilities for $m_L \geq 6$ earthquakes within the selection circle can be seen in Fig. 5. Note that the probability time series is plotted at monthly intervals.

Sudden decreases in the forecast occur at times when earthquakes having $m_L \geq 6$ strike either within the 200-km circle or near it. The large events outside the circular region influence the probability of large events inside the circle via the influence of the exponential term in Eq. (9). Sudden increases in probability are generally due to aftershocks of the large earthquakes occurring in either the circle itself or nearby, influencing the probability via the exponential term in Eq. (8).

7. Summary

The procedure we discuss can be applied to other systems in which complex events with power law distributions interact in a space with a finite correlation length. Earthquakes are known to have a quadrupole stress increase/decrease pattern so the interaction is spatially complex (AKI and RICHARDS 1980). In addition, the correlation length is finite, as shown by aftershock patterns, for example (STEIN 1999; AKI and RICHARDS 1980).

It is also known that aftershocks occur preferentially in regions where stress has increased following the main shock. From Eq. (9), it can be seen that probability $P_i(\Delta t | t)$ for any location within a distance ξ of a large earthquake would be reduced following the occurrence of the earthquake. However, the regions where stress transfer leads to aftershocks would find an increase in $P_i(\Delta t | t)$. This increase would be due to the burst of aftershock activity and its effect on $n_i(t)$ through Eq. (8).

It can be seen that this simple statistical model accounts for a number of the important aspects of earthquake stress transfer and earthquake dynamics. We predict that the model also has applications to other systems having complex dynamics but with finite correlation lengths and power-law scaling of frequency-size event statistics. Probabilities for infrequent large events can be computed using the idea that the frequency-size distribution must be filled in over time.

We finally note that further evidence for the importance of finite correlation length in earthquake occurrence was given in (SYKES and JAUME' 1990; JAUME' and SYKES 1999). While these authors do not

argue for any particular correlation length, a value of $\sim 400\ km$ is not inconsistent with their results.

Acknowledgments

Research by JBR and JRH was performed with funding from NASA NNX12AM22G to the University of California, Davis.

REFERENCES

M.K. SACHS, M.R. YODER, D.L. TURCOTTE, J.B. RUNDLEand B.D. MALAMUD, Eur. Phys. J. Special Topics, 205, 167 (2012).

SCHOLZ, C.H., The Mechanics of Earthquakes and Faulting, Cambridge (2002).

N. TALEB, The Black Swan: The Impact of the Highly Improbable, 1st edn. (Random House, New York, 2007), ISBN 9781400063512.

D.L. TURCOTTE, Fractals and Chaos in Geology and Geophysics, 2nd edn. (Cambridge University Press, Cambridge, U.K., 1997), ISBN 0521561647.

D. SORNETTE, Int. J.Terraspace Sci. Eng., 2, 1 (2009).

B.D. MALAMUD, J.D.A. MILLINGTON, G.L.W. PERRY, Proc Nat Acad Sci USA 102(13), 4694 (2005).

J.B. RUNDLE, J.R. HOLLIDAY, W.R. GRAVES, D.L. TURCOTTE, K.F. TIAMPO AND W. KLEIN and W. KLEIN, Phys Rev E, 86, 021106 (2012).

T.C. HANKS and H. KANAMORI, J Geophys Res 84, 2348 (1979).

USGS (Accessed 4/18/2014): http://pubs.usgs.gov/fs/2011/3021/.

J.B. RUNDLE, W. KLEIN, S. GROSS, and D.L. TURCOTTE, Phys Rev Lett, 75, 1658 (1995).

J.B. RUNDLE, W. KLEIN, S. GROSS and C.D. FERGUSON, Phys Rev E, 56, 293 (1997).

K.F. TIAMPO, J.B. RUNDLE, W. KLEIN, J.S.S. MARTINS, and C.D. FERGUSON, Phys Rev Lett, 91, 238501(1–4) (2003).

A.S. KIREMIDJIAN and T. ANAGNOS, Bull. Seism Soc Am, 74, 739 (1984).

ANSS and related catalogs: http://earthquake.usgs.gov/.

R.S. STEIN, Nature, 402, 605 (1999).

W. KLEIN, C.D. FERGUSON, M. ANGHEL, J.B. RUNDLE and J.S.S. MARTINS, Geocomplexity and the Physics of Earthquakes (Monograph), ed. J.B. Rundle, D.L. Turcotte and W. Klein, Amer. Geophys. Un., Washington DC (2000).

N. GOLDENFELD, Lectures on Phase Transitions and the Renormalization Group, Addison Wesley, Reading MA (1992).

J.B. RUNDLE et al. (2014), to be published.

J.R. HOLLIDAY, K.Z. NANJO, K.F. TIAMPO, J.B. RUNDLE and D.L. TURCOTTE, Nonlin Proc Geophys, 12, 965 (2005).

J.R. HOLLIDAY, D.L. TURCOTTE AND JB RUNDLE and JB RUNDLE, BASS, an alternative to ETAS, Geophys Res Lett, 34 Art. No. L12303 doi:10.1029/2007GL029696 (2007).

W.H.K. LEE, H. KANAMORI, P.C. JENNINGS, and C. KISSLINGER, *International Handbook of Earthquake and Engineering Seismology*, Academic Press, San Diego, CA (2002).

I.T. JOLIFFE and D.B. STEPHENSON, *Forecast Verification: A Practitioners' Guide in Atmospheric Science*, John Wiley & Sons (2003).

W.R. HSU and A.H. MURPHY, Int J Forecasting, *2*, 285 (1986).

V.V. KHARIN and F.W. ZWIERS, J Climate, *16*, 4145 (2003).

Open Hazards: www.openhazards.com.

Quakesim: www.quakesim.org.

K. AKI and P.G. RICHARDS, *Quantitative Seismology, Theory and Methods, Vols. I and II*, WH Freeman, San Francisco (1980).

L.R. SYKES and S.C. JAUME', Nature, *348*, 595 (1990).

S.C. JAUME' and L.R. SYKES, PAGEOPH, *155*, 279 (1999).

(Received May 26, 2014, revised September 29, 2014, accepted October 1, 2014, Published online October 16, 2014)

Pure Appl. Geophys. 173 (2016), 749–773
© 2015 Springer Basel
DOI 10.1007/s00024-015-1125-7

| Pure and Applied Geophysics

Laplace–Fourier-Domain Full Waveform Inversion of Deep-Sea Seismic Data Acquired with Limited Offsets

Yongchae Cho,[1] Wansoo Ha,[2] Youngseo Kim,[3] Changsoo Shin,[4] Satish Singh,[5] and Eunjin Park[4]

Abstract—Laplace–Fourier-domain full waveform inversion is considered one of the most reliable schemes to alleviate the drawbacks of conventional frequency-domain inversion, such as local minima. Using a damped wavefield, we can reduce the possibility of converging to local minima and produce an accurate long-wavelength velocity model. Then, we can obtain final inversion results using high-frequency components and low damping coefficients. However, the imaging area is limited because this scheme uses a damped wavefield that makes the magnitudes of the gradient and residual small in deep areas. Generally, the imaging depth of Laplace–Fourier-domain full waveform inversion is half the streamer length. Thus, dealing with seismic data in the deep-sea layer is difficult. The deep-sea layer reduces the amplitude of signals and acts as an obstacle for computing an exact gradient image. To reduce the water layer's effect, we extrapolated the wavefield with a downward continuation and performed refraction tomography. Then, we performed Laplace–Fourier-domain full waveform inversion using the refraction tomography results as an initial model. After obtaining a final velocity model, we verified the inversion results using Kirchhoff migration. We presented common image gathers and a synthetic seismogram of Sumatra field data to prove the reliability of the velocity model obtained by Laplace–Fourier-domain full waveform inversion. Through the test, we concluded that Laplace–Fourier-domain full waveform inversion with refraction tomography of the downward-continued wavefield recovers the subsurface structures located at depth despite a relatively short streamer length compared to the water depth.

Key words: Full waveform inversion, Laplace–Fourier domain, downward continuation.

1. Introduction

When performing pre-stack depth migration, we must construct exact subsurface structures for the migration image with high resolution. A variety of research has been conducted to develop velocity-building schemes, such as inversion or reflection tomography. Reflection tomography (Chiu *et al.* 1986; Zhang and Toksoz 1998; Murphy and Gray 1999) has been the method of choice for recovering subsurface velocity structures since the introduction of a method that uses the travel-time of seismic reflections (Bishop *et al.* 1985). However, reflection tomography requires the reflectors to be defined through interpretation. Defining the exact reflection boundary is difficult and requires a great deal of time. In addition, if the defined geological structure is incorrect, we cannot expect to obtain a reliable velocity model. As an alternative to tomography, it is still discussed whether full waveform inversion (FWI) using the difference between the simulated data and field data to recover the subsurface parameters can be run in a reflection mode such as full waveform tomography (FWT). This idea has been greatly advanced by the development of the back-propagation algorithm (Lailly 1983; Tarantola 1984; Virieux and Operto 2009). Nevertheless, conventional frequency-domain FWI has weaknesses in recovering field data, such as the local minimum problem (Mora 1987). In addition, an absence of low-frequency components in the seismic data makes it difficult to obtain long-wavelength velocity models using conventional frequency-domain FWI. Increasing the offset during acquisition could be considered as an alternative because we can obtain long-wavelength signals that are important for building a

[1] Schlumberger Information Solutions, 136 Sejong-daero, Jung-gu, Seoul 100-768, Republic of Korea.
[2] Pukyong National University, 45 Yongso-ro, Nam-gu, Busan 608-737, Republic of Korea.
[3] Saudi Aramco, Dhahran, Saudi Arabia.
[4] Seoul National University, 1 Gwanak-ro, Gwnak-gu, Seoul 151-744, Republic of Korea. E-mail: ej0417@snu.ac.kr
[5] Institute de Physique du Globe de Paris, 1 Avenue Jussieu, 75238 Paris, France.

macro-velocity model through acquisition using wide apertures. However, performing additional seismic acquisition to obtain seismic data with increased offset requires a huge budget, and gaining a low-frequency signal from short-offset field data is difficult due to technical limitations (SYMES 2008).

SHIN and CHA (2008) suggested a Laplace-domain FWI using a damped wavefield to overcome the conventional limitations. The use of a damped wavefield eliminates the concern regarding the absence of a low-frequency signal. After multiplying the damping function over the field data, certain regions of the frequency are altered to a signal with low frequency, allowing us to obtain a long-wavelength signal from field data that have few or no long-wavelength components. Thus, despite the previously mentioned limitations, Laplace-domain inversion can recover the macro-velocity structure of field data (SHIN and HA 2008). Laplace-domain inversion is very sensitive to the noise that is present around the first-break signal; therefore, one must eliminate the noise that occurs early in the shot gather through muting (SHIN and CHA 2008; HA *et al.* 2010). After acquiring a macro-velocity model through Laplace-domain inversion, we can recover the high-resolution velocity model by combining it with the conventional frequency-domain FWI called Laplace–Fourier-domain FWI (SHIN and CHA 2009; SHIN *et al.* 2010; KAMEI *et al.* 2013). This scheme is similar to the method in which a long-wavelength velocity model is first constructed and then the short-wavelength components are used to recover the velocity structures, as suggested by PRATT and WORTHINGTON (1988).

Laplace- and Laplace–Fourier-domain inversions can handle seismic data with streamer lengths that are longer than twice the target imaging depth (HA *et al.* 2012b). However, we cannot expect to recover the exact subsurface structure in the case of field data with a deep-sea layer. When we consider deep-sea seismic data using Laplace-domain inversion, applying a damping function to seismic data significantly decreases the amplitude of the signal propagating through the deep area. In addition, this damped wavefield in the Laplace domain hinders the accurate computation of a gradient image. For this reason, the imaging depth of Laplace-domain inversion is generally less than half the streamer length (HA *et al.* 2012b).

In this study, we applied an extrapolated wavefield with a downward continuation to the field data to reduce the water layer's effect. This process made the construction of a macro-velocity model through refraction tomography easier. Once the macro-velocity model with downward-continued data was built, we performed Laplace–Fourier inversion to recover a more accurate velocity using high-frequency signals. To prove the reliability of the inversion results, we performed Kirchhoff depth migration based on the velocity model obtained through inversion. In the second section of this paper, we will briefly explain the theory of Laplace–Fourier inversion and downward continuation. We applied this scheme to both synthetic and field data. The results and related explanations were presented in the synthetic and field data parts. In the last section, we presented a brief comparison of the results and our expectations for future developments in this area.

2. Theory

2.1. FWI in the Laplace–Fourier Domain

The finite element equation of the wave equation with boundary conditions (CLAYTON and ENGQUIST 1977; REYNOLDS 1978) can be presented as

$$\mathbf{M\ddot{u}} + \mathbf{C\dot{u}} + \mathbf{Ku} = \mathbf{f}, \tag{1}$$

where $\mathbf{u}$ is a wavefield in the time domain and $\dot{\mathbf{u}}$ and $\ddot{\mathbf{u}}$ denote the first- and second-order time derivative of $\mathbf{u}$, respectively. $\mathbf{M}$ is a mass matrix, $\mathbf{C}$ is a damping matrix, $\mathbf{K}$ is a stiffness matrix and $\mathbf{f}$ is a source vector (Marfurt 1984). Taking the Laplace–Fourier transform of Eq. (1) yields

$$\mathbf{S\tilde{u}} = \mathbf{\tilde{f}}, \tag{2}$$

with $\mathbf{S} = \mathbf{M}s^2 + \mathbf{C}s + \mathbf{K}$,

$$\tilde{u}(s) = \int_0^\infty u(t)e^{-st}dt = \int_0^\infty u(t)e^{-i\omega t}e^{-\sigma t}dt, \tag{3}$$

where s, a complex number, is given by $\sigma + i\omega$; σ and ω represent the Laplace damping constant and angular frequency, respectively. $u(t)$ is a time-domain wavefield, and i is $\sqrt{-1}$. According to SHIN and CHA (2008), a zero-frequency component of the damped wavefield is used for the Laplace-domain FWI, where ω is zero and σ is a real number. In contrast, the term s is treated as a true complex number for Laplace–Fourier-domain FWI; this idea enables us to use both terms (σ, ω) for FWI. SHIN and CHA (2009) called the Laplace-transformed wavefield by a complex number s as the Laplace–Fourier-domain wavefield or the hybrid-domain wavefield.

The logarithmic objective function at a certain angular frequency (ω) can be presented as follows (SHIN and MIN 2006):

$$E(\mathbf{p}) = \frac{1}{2} \sum_{i=1}^{N_s} \sum_{j=1}^{N_r} \left(\ln \frac{|\tilde{u}_{ij}|}{|\tilde{d}_{ij}|} \right) \left(\ln \frac{|\tilde{u}_{ij}|}{|\tilde{d}_{ij}|} \right)^* \quad (4)$$

where $\mathbf{p}$ is the model parameter, $\tilde{u}_{ij}$ represents the modeled wavefields and $\tilde{d}_{ij}$ is the Laplace–Fourier-transformed data. N_s and N_r denote the number of sources and receivers, respectively. The asterisk (*) denotes a complex conjugate. The term $\ln \frac{|\tilde{u}_{ij}|}{|\tilde{d}_{ij}|}$ is termed the residual $\mathbf{r}$, and our aim is to set the direction of the gradient to reduce this residual. By taking the partial derivative of Eq. 4, the steepest descent direction with respect to the k-th model parameter can be obtained as follows:

$$\frac{\partial E(\mathbf{p})}{\partial p_k} = \mathbf{Re} \left[\sum_{i=1}^{N_s} \sum_{j=1}^{N_r} \frac{1}{\tilde{u}_{ij}} \frac{\partial \tilde{u}_{ij}}{\partial p_k} \left(\ln \frac{|\tilde{u}_{ij}|}{|\tilde{d}_{ij}|} \right)^* \right]. \quad (5)$$

To calculate the partial derivative wavefields, we used the differentiating form of the Laplace–Fourier-domain wave equations $\mathbf{S}\tilde{\mathbf{u}} = \tilde{\mathbf{f}}$ (MARFURT 1984). Taking the partial derivative of this equation with respect to a model parameter p_k yields the following:

$$\frac{\partial \mathbf{S}}{\partial p_k} \tilde{\mathbf{u}} + \mathbf{S} \frac{\partial \tilde{\mathbf{u}}}{\partial p_k} = 0, \quad (k = 1, 2, \ldots, m) \quad (6)$$

or

$$\frac{\partial \tilde{\mathbf{u}}}{\partial p_k} = \mathbf{S}^{-1} \mathbf{v}_k, \quad (k = 1, 2, \ldots, m) \quad (7)$$

where m is the number of parameters and

$$\mathbf{v}_k = -\frac{\partial \mathbf{S}}{\partial p_k} \tilde{\mathbf{u}}. \quad (k = 1, 2, \ldots, m). \quad (8)$$

In Eq. 8, $\mathbf{v}_k$ is the virtual source vector required to perturb the k-th model parameter (PRATT et al. 1998). By substituting Eq. 7 into Eq. 5, we can obtain the gradient of the objective function as follows:

$$\frac{\partial E(\mathbf{p})}{\partial p_k} = \mathbf{Re} \left[\left(\mathbf{S}^{-1} \mathbf{v}_k \right)^{\mathrm{T}} \mathbf{r} \right], \quad (9)$$

where

$$\mathbf{r} = \left[\frac{1}{\tilde{u}_1} \ln \frac{|\tilde{u}_1|}{|\tilde{d}_1|} \frac{1}{\tilde{u}_2} \ln \frac{|\tilde{u}_2|}{|\tilde{d}_2|} \cdots \frac{1}{\tilde{u}_{N_r}} \ln \frac{|\tilde{u}_{N_r}|}{|\tilde{d}_{N_r}|} 0 \cdots 0 \right]^{\mathrm{T}}. \quad (10)$$

The impedance matrix $\mathbf{S}$ has a symmetric form and satisfies the reciprocity condition. Thus, we can modify Eq. 5 as follows:

$$\frac{\partial E(\mathbf{p})}{\partial p_k} = \mathbf{Re} \left[(\mathbf{v}_k)^{\mathrm{T}} \mathbf{S}^{-1} \mathbf{r} \right] \quad (11)$$

Consequently, using the steepest descent method, the gradient of the objective function E can be computed by determining the zero-lag convolution between the back-propagated residual data in a time-reversed order using a two-way wave equation and the virtual source vector (PRATT et al. 1998).

We can enhance the efficiency by considering only the diagonal components of the pseudo-Hessian matrix (HA et al. 2012a)

$$\mathbf{p}_{l+1} = \mathbf{p}_l - \alpha \left(diag\mathbf{H}_p + \lambda \mathbf{I} \right)^{-1} \nabla_{\mathbf{p}} E(\mathbf{p}). \quad (12)$$

Equation 12 can be rewritten as

$$\mathbf{p}_{l+1} = \mathbf{p}_l - \alpha \sum_{i=1}^{N_f} \frac{\sum_{j=1}^{N_s} \mathbf{Re} \left[(\mathbf{v}_k)^{\mathrm{T}} \mathbf{S}^{-1} \mathbf{r} \right]}{\sum_{j=1}^{N_s} \mathbf{Re} \left[(\mathbf{v}_k)^{\mathrm{T}} (\mathbf{v}_k)^* \right] + \lambda}, \quad (13)$$

where $\mathbf{H}_p$ is the pseudo-Hessian matrix, l is the iteration number, and λ is a stabilizing factor. Based on Eq. 13, we can design the algorithm of the Laplace–Fourier inversion as shown in Fig. 1.

2.2. Downward Continuation

There are a couple of challenges for performing stable downward continuation: the effect of the sea floor's roughness and wrong estimation of the water

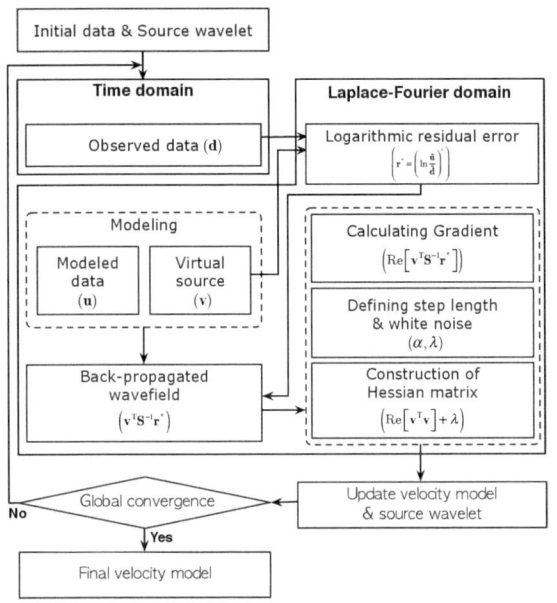

Figure 1
Workflow of Laplace–Fourier-domain full waveform inversion

velocity. Incorrect seawater depth data can cause artifacts near the water bottom, distorting the other aspects of the results. We calculated the seawater depth through a near-offset gather. After selecting the first arrival in the near-offset gather, the velocity of the water layer was fixed at 1498 m/s. Given these data, we can compute the water depth and use the fixed water velocity for downward continuation. However, the water velocity depends strongly on the location in deep-sea regions. Thus, a fixed water velocity may yield incomplete depth information during the calculation of seawater depth.

When we handle seismic data in the deep-sea layer, the first-break signal at the far-offset appears at a later time compared to that in a shallow-sea layer. In addition, most of the signals are composed of reflections that are not combined with refraction. However, it is imperative to consider refractions to build a macro-velocity model. If acquisition is re-performed with a streamer that is long enough to offset the effect of the deep-water layer, limited offset will not act as an obstacle for building a velocity model using a damped wavefield. Nevertheless, it is not easy to go back to the

seismic acquisition process. In this study, we eliminated the effect of water by extrapolating the wavefield using the downward continuation method. The principal objective of downward continuation is to transform the seismic wavefield $u(x_s, x_r, t, z_0)$ to the wavefield $u(x_{s'}, x_{r'}, t', z_0)$ at a depth z, in which z_0 indicates the location of the sea surface and z indicates the depth of the shots and receiver location following downward continuation. We presented Fourier-based wavefield extrapolation below, and this scheme can be implemented by multiplying a moving kernel to the transformed wave function as follows (CLEARBOUT 2009):

$$u(k_{xs}, k_{xr}, \omega, z) = u(k_{xs}, k_{xr}, \omega, z_0)e^{[i(\tau_s + \tau_r + \Delta x_s + \Delta x_r)]}$$

(14)

where $\tau_s = k_{zs}z$, $\tau_r = k_{zr}z$, $\Delta x_s = \frac{k_{xs}}{k_{zs}}z$ and $\Delta x_r = -\frac{k_{xr}}{k_{zr}}z$. At last, we can obtain the seismic wavefield $u(x_{s'}, x_{r'}, t', z)$ of the time-distance domain using a double inverse Fourier transform with the wavenumber (k_x, k_z) and angular frequency (ω). Finally, we can obtain the seismic wavefield $u(x_{s'}, x_{r'}, t', z)$ of the time-distance domain using a double inverse Fourier transform with the wavenumber (k_x, k_z) and angular frequency (ω). The downward continuation process was carried out in three steps. First, we performed (1) vertical extrapolation of the receiver gathers to the seafloor, then (2) sorted the extrapolated shot gathers into common receiver location gathers and (3) performed the downward continuation of the shot positions to the seafloor, sorting back into shot gathers. Figure 2a presents the impact of downward continuing of a 12.0-km-long streamer shot record from a WG2 profile to obtain source and receiver positions on the sea floor (GHOSAL et al. 2014). As shown in Fig. 2b, the direct-wave signals and the sea floor reflections disappear, and the region of refraction is enhanced. After applying a downward continuation, the reflection signals are distorted and the ability to use complete signals in several imaging schemes, such as frequency-domain FWI, is limited. However, the downward-continued data are much easier to interpret and effective to construct a macro-velocity model through refraction tomography.

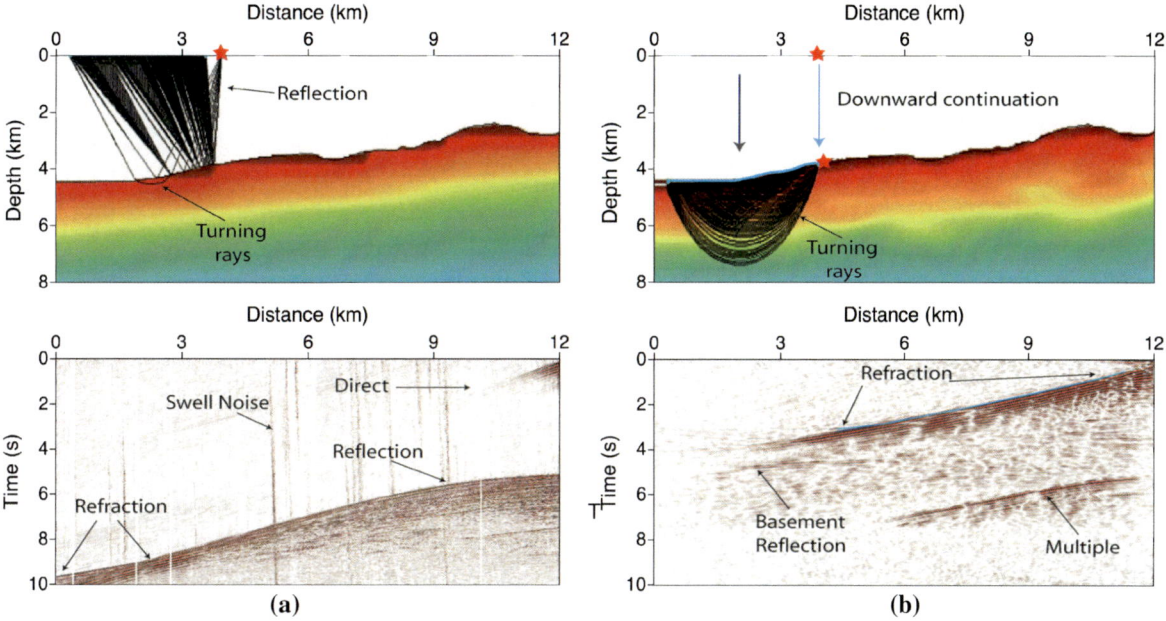

Figure 2
Downward-continued shot gather. The Sumatra dataset WG2 was used for the extrapolation. Through downward continuation, the locations of the shot and receivers are moved to the sea bottom. **a** A shot gather prior to performing the downward continuation implies a short area of refractions. **b** However, a shot gather exhibits extended refraction signals after downward continuation

2.3. Source Estimation for the Logarithmic Function

When we handle synthetic seismic data, we can obtain exact source wavelets. However, the exact source information is difficult to obtain when handling field data. In addition, one must recover an accurate source wavelet to reduce the differences between the modeled wavefield and observed data. SHIN and MIN (2006) demonstrated that a source wavelet can be estimated by separating the amplitude and phase of a complex source function. The objective function for the source wavelet can be presented as follows:

$$E = \frac{1}{2}\sum_{i=1}^{N_s}\sum_{j=1}^{N_r}\left[\left\{\ln\left(\frac{A^{\mathrm{src}}A_{ij}^{G}}{A_{ij}^{d}}\right)\right\}^2 + \theta^{\mathrm{src}} + \theta_{ij}^{G} - \theta_{ij}^{d}\right],$$
(15)

where N_s and N_r are the total number of shots and receivers, respectively. The values A and θ represent the amplitude and phase. The superscripts src, G, and d denote the source wavelet, Green's function and observed data, respectively. Hence, the modeled data and observed data can be written as follows:

$$u_{ij} = A^{\mathrm{src}}A_{ij}^{G}e^{i\left(\theta^{\mathrm{src}}+\theta_{ij}^{G}\right)}, \quad d_{ij} = A_{ij}^{d}e^{i\theta_{ij}^{d}}.$$
(16)

If we set the logarithmic amplitude of the source wavelet to $z = \ln A^{\mathrm{src}}$, the gradient and Hessian of the objective function to recover the amplitude and phase of the source are expressed as follows:

$$\frac{\partial E}{\partial z} = \sum_{i=1}^{N_s}\sum_{j=1}^{N_r}\mathbf{Re}\left[z + \ln\frac{A_{ij}^{G}}{A_{ij}^{d}}\right], \quad \frac{\partial^2 E}{\partial z^2} = \sum_{i=1}^{N_s}\sum_{j=1}^{N_r}1,$$
(17)

$$\frac{\partial E}{\partial \theta^{\mathrm{src}}} = \sum_{i=1}^{N_s}\sum_{j=1}^{N_r}\delta\theta_{ij}, \quad \frac{\partial^2 E}{\partial \theta^{\mathrm{src}2}} = \sum_{i=1}^{N_s}\sum_{j=1}^{N_r}1, \quad (18)$$

respectively. We can estimate the amplitude and phase of the source wavelet by updating it through l iterations as follows:

$$z(l+1) = z(l) - \frac{\sum_{i=1}^{N_s}\sum_{j=1}^{N_r}\mathbf{Re}\left[z + \ln\frac{A_{ij}^{G}}{A_{ij}^{d}}\right]}{\sum_{i=1}^{N_s}\sum_{j=1}^{N_r}1},$$
(19)

$$\theta_{\mathrm{src}}(l+1) \;=\; \theta_{\mathrm{src}}(l) - \frac{\sum_{i=1}^{N_s}\sum_{j=1}^{N_r} \theta^{\mathrm{src}} + \theta_{ij}^{G} - \theta_{ij}^{d}}{\sum_{i=1}^{N_s}\sum_{j=1}^{N_r} 1}$$

$$(20)$$

3. Synthetic Data Examples

We performed Laplace–Fourier-domain inversion on two sets of synthetic data to verify the reliability of the scheme to create a macro-velocity model in an ultra-deep-sea region. The first is the Marmousi model (VERSTEEG 1994), which implies several faults, folds and a trap structure. Though this model is a synthetic velocity model, it includes various types of geological structures. Thus, the Marmousi model is broadly used as a benchmark model.

In this chapter, we added a water layer to the original velocity model to generate both shallow and deep-sea velocity models. We applied water layer depths of 0.2 and 3 km to the shallow and deep-sea models. We obtained time-domain seismic data using a wave propagation-modeling scheme with both of the velocity models. We performed a Laplace–Fourier-domain inversion based on these seismic data. Then, we compared the result to the original velocity model to prove the reliability of the recovered velocity models.

3.1. Marmousi with Water Layer

We generated a synthetic velocity model based on the original Marmousi model with a shallow water layer. The depth of the water layer in the shallow model is 0.2 km, and its velocity model is presented in Fig. 3. The deep model consisted of water that is

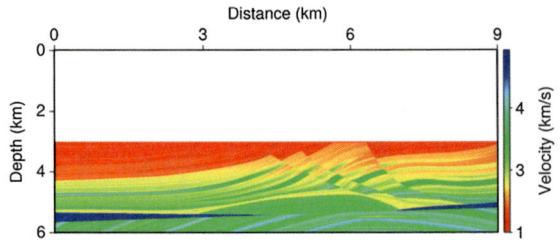

Figure 4
Marmousi velocity model with a 3.0-km-deep water layer

3.0 km deep, as shown in Fig. 4. The velocities ranged from 1.5 to 5.5 km/s. We used a wave propagation-modeling scheme in this model to obtain the seismic data. For wave propagation modeling in the time domain, we used a finite difference scheme with a fourth-order staggered grid (GRAVES 1996) and perfectly matched layer (PML) boundary conditions. The PML boundary exhibits high accuracy and a small noise reflector in half-space acoustic media. However, this form requires a first-derivative form in the spatial domain. Thus, a staggered grid must be applied that can calculate the pressure by separately computing the velocity and stress terms. We used 5-m grids to obtain the seismic data shown in Figs. 5 and 6. The distances of the shot and receiver intervals were 100 and 5 m, respectively. The number of shots was 151. We assumed the length of the streamer to be 6 km. The maximum frequency used in the model was 30 Hz, and we used the Ricker wavelet as a source function. The total recording time was 4 and 8 s for the shallow- and deep-sea models, respectively. The time-sampling interval was 4 ms. We used 0.5 ms in the modeling procedure and resampled the data at 4 ms for stable wavefields.

3.2. Laplace–Fourier-Domain FWI

We performed an inversion in the Laplace–Fourier domain based on the data presented in Figs. 5 and 6. We applied a finite element scheme to calculate the gradient and virtual source using a back-propagation algorithm. We set the maximum frequency as 15 Hz for the inversion procedure. The model grid size was 25 m × 25 m. The maximum and minimum velocities were set at 5.5 and 1.5 km/s, respectively. The stabilizing factor was 10^{-6}. We used a vertical, linearly increasing velocity model as

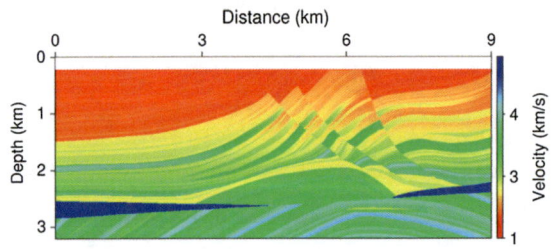

Figure 3
Marmousi velocity model with a shallow 0.2-km-deep water layer

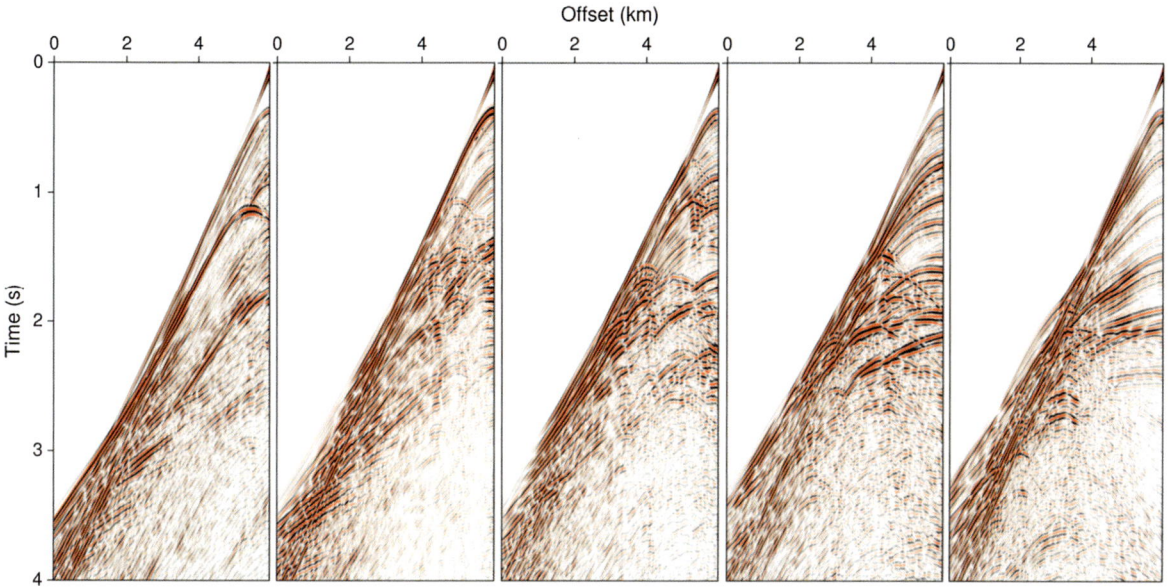

Figure 5
Seismic data obtained from the velocity model presented in Fig. 3. The locations of the shot from *left* to *right* are 5, 6, 7, 8, and 9 km from the *left* margin

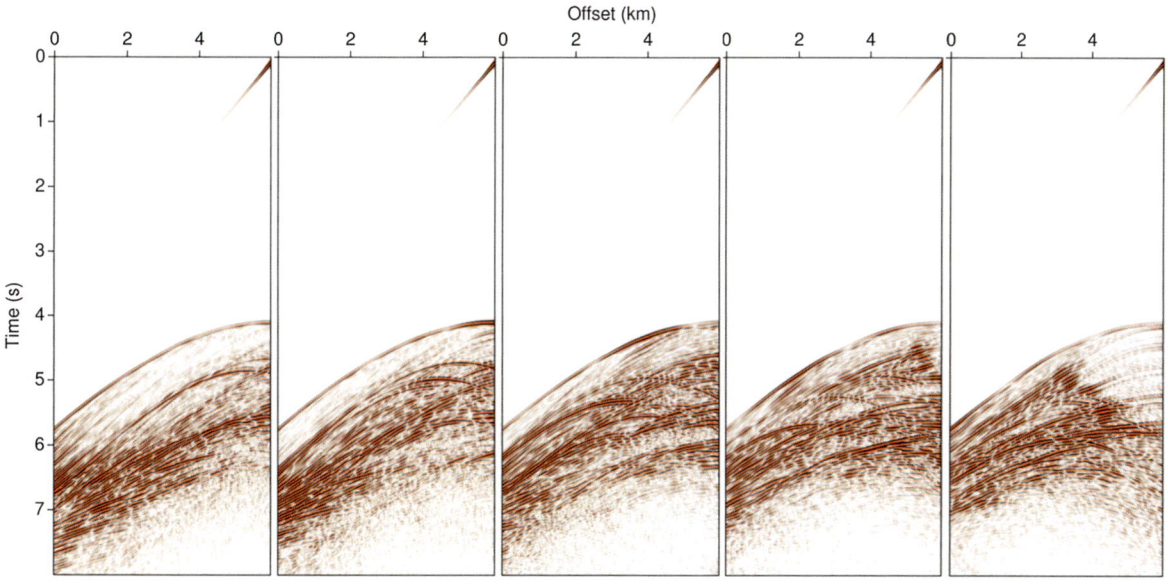

Figure 6
Seismic data obtained from the velocity model presented in Fig. 4. The locations of the shot from *left* to *right* are 5, 6, 7, 8, and 9 km from the *left* margin

a starting model, and the velocity ranged from 1.5 to 4.0 km/s. The ranges of the Laplace damping coefficient and the frequency component are shown in Table 1. We used clusters consisting of 64 cores for the inversion. Thus, we set 64 cores as 1 group to optimize the calculation process. The frequency components (ω, s) were selected in increasing order on the s axis, then moved 1 step right to the ω axis for

Table 1

Frequency loop used in the Laplace–Fourier inversion of the Marmousi model

	Laplace damping coefficients		Frequency components	
	Range	Interval	Range	Interval
1st part	1 ~ 13	2	0.26 ~ 6.00	0.25
2nd part	0.75		0.25 ~ 5.25	0.25
3rd part	0.50		0.25 ~ 5.25	0.25
4th part	0.15		0.25 ~ 5.25	0.25
5th part	0.10		1.00 ~ 15.0	0.25

the direction of the triangle (▼), as displayed in Fig. 7. The selected frequency component was added to the group until the number of components reached 64. When the group size of the components became 64, the group reformed by adding a new component as a replacement of the oldest component in the group. The cluster implemented this process until it covered all components of the (ω, s) field as presented in Fig. 7. There are myriad numbers of methods to compose the sequence of frequency components. Generally, we first considered damping coefficients s with low-frequency ω. This process enabled us to reduce the possibility of converging to a local minimum and build a macro-velocity model with long wavelength.

In Laplace–Fourier-domain FWI, we build a low-frequency velocity model using a Laplace damping coefficient. Meanwhile, the time-domain wavefield is transformed to the Laplace domain. Then, we can obtain a low-frequency component that did not exist in the original seismic data after applying a damping function to the wavefield (Ha and Shin 2012). Those low-frequency components that are generated during the damping process contribute to building a macro-velocity model in FWI. Thus, using a damped wavefield is helpful to calculate a long-wavelength velocity model, but a limitation also exists. The relatively shallow target depth of FWI is attributed to the damped wavefield. In other words, Laplace-domain FWI is generally considered appropriate to recover the velocity structure in the area that relates to half the streamer length, a consideration based not on algebraic proof, but from experiments (Ha *et al.* 2012a). If we use a damped wavefield, the value of the wavefield from the deeper region will be small, approximately zero. Thus, accurate computations of

the residual value in deeper regions are problematic, and the gradient value can be inaccurate. An inexact gradient results in a distorted subsurface structure. Therefore, when we deal with seismic data from an ultra-deep-sea region, we must consider the deep-water effect. We will demonstrate that the water layer may distort the subsurface structure from that determined using Laplace–Fourier-domain FWI.

We set the streamer length to a relatively short value compared to the water depth to adequately determine the effect of the water layer. The maximum depth of the original model was 3 km, and we fixed the streamer length at 6 km in the Marmousi model so the ray path can cover the entire velocity model. If we add 3 km of water depth to the original model, the total depth of the velocity model will be 6 km. According to the general recoverable area in the Laplace–Fourier domain, ~3 km of structures will be constructed. We will provide several numerical examples as confirmation of this idea.

We chose a vertical, linearly increasing velocity model as a starting model for the inversion. The velocity area of the initial model ranged from 1.5 to 4.5 km/s. The procedures for the inversion in each model are presented in Figs. 8 and 9. Figure 8 shows the recovery process of the model in shallow water. When we perform an inversion in the Laplace domain, the damping coefficient typically ranges from 1 to 15 s^{-1}. In this test, we applied a range of damping coefficients from 1 to 13 s^{-1}. With a damped wavefield, we can reduce the possibility of convergence to local minima, which can act as a major obstacle in conventional frequency-domain FWI. Based on the relatively large damping coefficients, we can obtain a macro-velocity model with a long wavelength, as shown in Figs. 8b and 9b. After

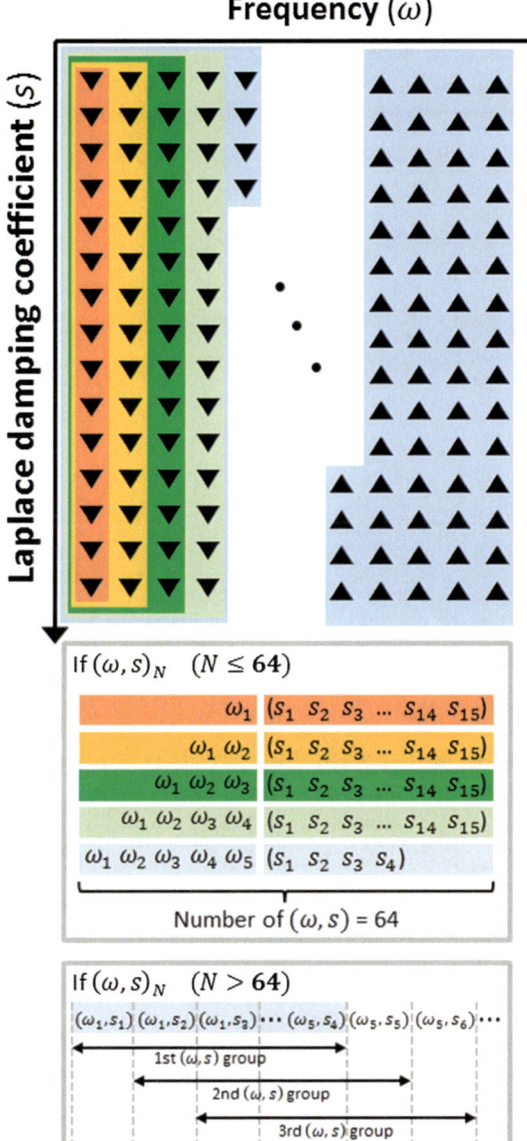

If $(\omega, s)_N$ $(N \leq 64)$

ω_1	$(s_1 \ s_2 \ s_3 \ \dots \ s_{14} \ s_{15})$
$\omega_1 \ \omega_2$	$(s_1 \ s_2 \ s_3 \ \dots \ s_{14} \ s_{15})$
$\omega_1 \ \omega_2 \ \omega_3$	$(s_1 \ s_2 \ s_3 \ \dots \ s_{14} \ s_{15})$
$\omega_1 \ \omega_2 \ \omega_3 \ \omega_4$	$(s_1 \ s_2 \ s_3 \ \dots \ s_{14} \ s_{15})$
$\omega_1 \ \omega_2 \ \omega_3 \ \omega_4 \ \omega_5$	$(s_1 \ s_2 \ s_3 \ s_4)$

Number of (ω, s) = 64

If $(\omega, s)_N$ $(N > 64)$

$(\omega_1, s_1)(\omega_1, s_2)(\omega_1, s_3) \cdots (\omega_5, s_4)(\omega_5, s_5)(\omega_5, s_6) \cdots$

1st (ω, s) group
2nd (ω, s) group
3rd (ω, s) group

Figure 7
Conceptual diagram of the frequency loop grouped into 64 components combined with the damping coefficients and frequencies. We set 64 cores as 1 group to optimize the calculation process. The selected frequency component is added to the group until the number of components reaches 64. When the group size of the components becomes 64, the group will reform by adding a new component as a replacement of the oldest component in the group. The cluster implements this process until it covers all the components in the (ω, s) field

constructing the macro-velocity model, we performed an inversion with a low damping coefficient and high-frequency component that is close to frequency-

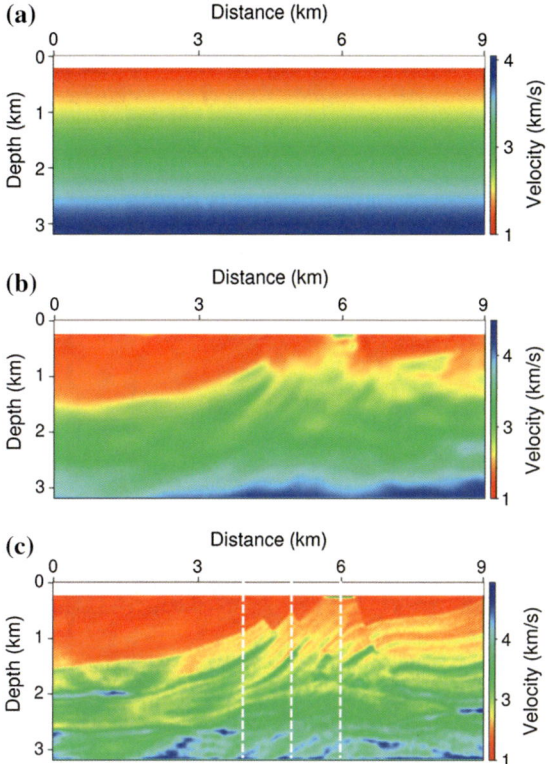

Figure 8
The velocity model of shallow water layer synthetic seismic data from the Laplace–Fourier-domain inversion. **a** The initial velocity model ranging from a velocity of 1.5–4.5 km/s, **b** an intermediate velocity model, and **c** the final inverted velocity model obtained through Laplace–Fourier-domain inversion. Without a thick water layer, we could recover the overall structure of the original model

domain inversion. We obtained the final velocity model shown in Figs. 8c and 9c after the 50th iteration. Because the length of the streamer was 6 km, we expected that a 3-km-deep subsurface structure could be reliable. As shown in Fig. 8c, we could recover the overall structure of the original model without a thick water layer. However, the velocity model presented in Fig. 9c is severely distorted compared to the original velocity model shown in Fig. 4. Because of the highly distorted subsurface structure, we need to consider a wavefield extrapolation technique to remove the water layer effect. The area at approximately 3 km is slightly recovered, but the entire velocity structure is far from the original velocity model. The deeper region of the model is severely distorted because of the inaccurate gradient that originates from the extremely small value of the damped wavefield.

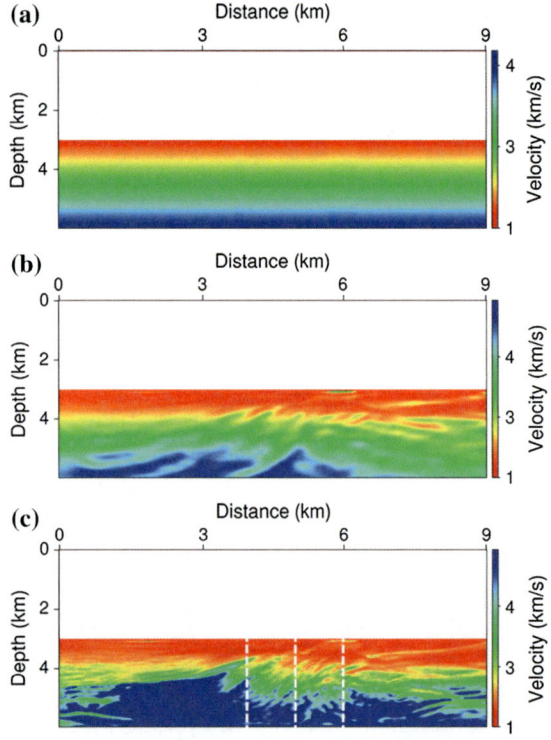

Figure 9

The velocity model of thick water layer synthetic seismic data from the Laplace–Fourier-domain inversion. **a** The initial velocity model with velocities ranging from 1.5–4.5 km/s, **b** an intermediate velocity model, and **c** the final inverted velocity model obtained through Laplace–Fourier-domain inversion. Because of the highly distorted subsurface structure, we need to remove the water layer's effect using the extrapolation technique

To compare the magnitudes of the velocities, we selected several target lines and plotted their velocities according to the depth. One of the principal objectives in the inversion test of the Marmousi model is the recovery of the trap structure and high-velocity area located below the faults. Thus, we plotted three lines (4, 5, and 6 km from the left) around the traps as presented on the graphs displayed in Fig. 10. The blue lines indicate the original velocity values according to the depth. The red and green lines indicate the velocities from shallow- and deep-water data, respectively. To make a comparison, we cut the sea water layer in the deep-sea model. As shown in the graph, the velocity recovered from the shallow water data has a similar tendency to that of the original model. However, the tendency of the green line from the deep-water data is far from that of

the original model when the depth is greater than 1.0 km, indicating 4.0 km depth in the deep-sea model. In Laplace–Fourier-domain FWI, signals from the far deeper part are damped too much because of their weak amplitudes, which might cause a couple of problems, such as convergence to the local minimum and non-uniqueness.

3.3. Recovering the Deep-Sea Marmousi Model

In FWI, it is imperative that the observed data d_{ij} and modeled data u_{ij} should be defined under the same wave propagation scheme as presented in Eq. 4. However, after wavefield extrapolation, the whole amplitude variation changed and the signals around the source are distorted. These problems make it difficult to reduce the residuals in FWI by distorting the observed data d_{ij}. In other words, the downward-continued wavefield cannot be directly matched with the wavefield that is calculated from a typical wave equation. Thus, we could not consider performing Laplace–Fourier-domain FWI directly on downward-continued data. To recover the velocity of the deep-sea Marmousi model, we first built a starting model using refraction tomography and then performed Laplace–Fourier-domain FWI on original seismic data. Using the downward-continued data presented in Fig. 11, we obtained the refraction tomography results shown in Fig. 12a. To perform refraction tomography, we employed the suppressed wave equation of travel-time (SWEET) method suggested by SHIN et al. (2002), who demonstrated that the SWEET method is equivalent to simultaneously solving the eikonal and transport equations. Using the SWEET algorithm, we calculated travel times and amplitudes from the solutions of the Laplace-transformed wave equation at a single Laplace frequency.

To use the refraction tomography results as a starting model for Laplace–Fourier-domain FWI, we extended the water layer as shown in Fig. 12b to fit to the size of the original data. The final inversion result after the 100th iteration is presented in Fig. 12c. To compare the magnitudes of the velocity models more exactly, we presented velocity profiles in Fig. 13 at the same locations as shown before. In the graph, the green and red lines are from conventional Laplace–

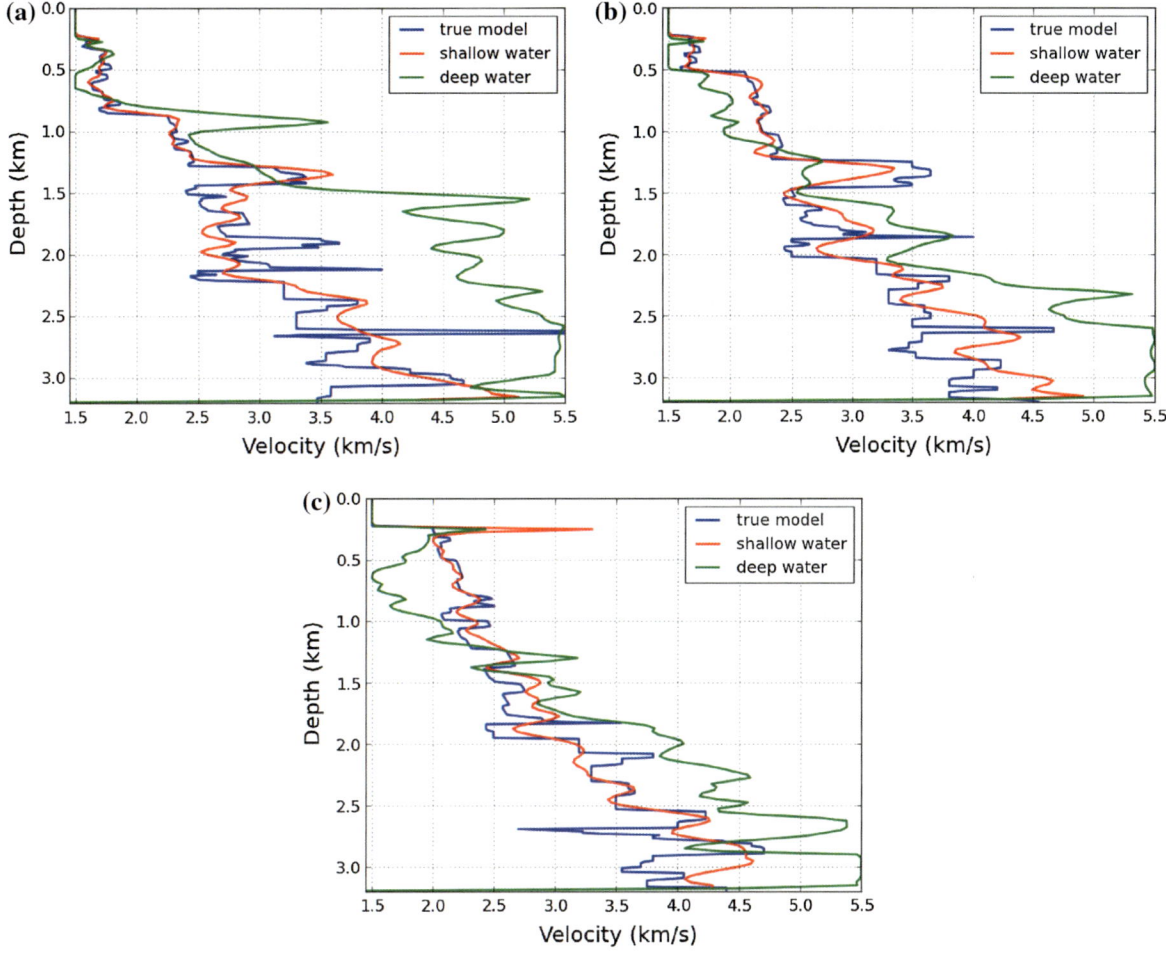

Figure 10

Comparison of the inversion results. The velocity profiles are **a** 4, **b** 5, and **c** 6 km from the *left*. As shown in the graph, the velocity values from the shallow water inversion data have similar tendencies to that of the original inversion. However, the tendency of the *green line* (the deep-sea data) differs from the original model at depths greater than 1.0 km. In Laplace–Fourier-domain FWI, the signals from the far deeper part are damped too much because of their weak amplitudes, which might cause a couple of problems: convergence to the local minimum and non-uniqueness

Fourier-domain FWI and refraction tomography combined with Laplace–Fourier-domain FWI. The red line is closer to the original model than the green one. Overall, a large portion of the geological structure is recovered, which is better than the image from only Laplace–Fourier inversion. Although the deepest area is still not accurate, we obtained a significantly higher quality velocity image compared to conventional Laplace–Fourier-domain FWI. Based on these results, we can expect that it is possible to apply downward continuation to enhance the imaging depth of field data with limited offset, which will be shown in the next chapter.

4. Field Data Examples

4.1. Sumatra Field Data WG2 Line

Sumatra (Indonesian: Sumatera) is the western-most of the Sunda Islands in western Indonesia and is the largest island located entirely within Indonesia (two larger islands, Borneo and New Guinea, are shared between Indonesia and other countries). The 2004 tsunami, formed by one of the largest earth-quakes (Mw = 9.3) in the last forty years, took more than 230,000 lives in this area. The earthquake initiated offshore Simeulue Island, SW of the tip of

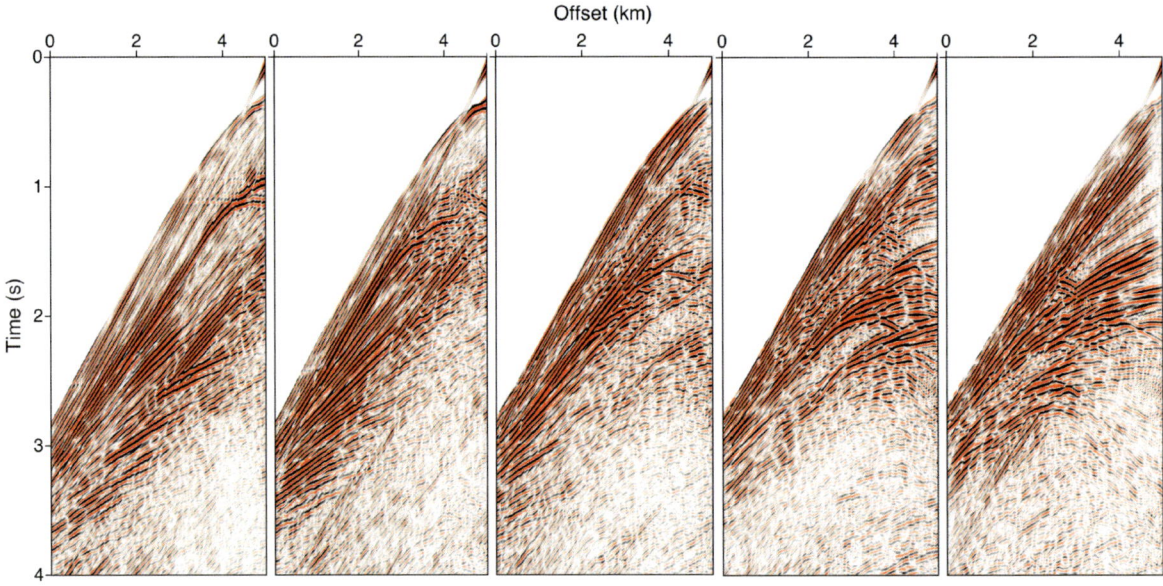

Figure 11
Shot gathers after downward continuation. The locations of the shot gather from *left* to *right* are 5, 6, 7, 8, and 9 km. Compared to the shot gather presented in Fig. 6, the signal appears in the early part

Sumatra, and ruptured over 1300 km of the plate boundary from northern Sumatra to the Andaman Islands (AMMON *et al.* 2005). Near the epicenter of the Sumatra–Andaman earthquake, the Indo-Australian oceanic plate subducts obliquely beneath the Sunda continental plate at a rate of 53 mm/year, decreasing to 43 mm/year at the latitude of the Andaman Islands (PRAWIRODIRDJO *et al.* 2000). Many studies have examined this area and the strong movement of the subsurface structure. As part of this study, seismic exploration was conducted by the Schlumberger WesternGeco company using the French Marion Dufresne and Geco Searcher vessels towing 8260 ($\sim$ 135.35 L) and 10170 ($\sim$ 163.87 L) cubic inch airgun array seismic sources, respectively (SINGH *et al.* 2008; CHAUHAN *et al.* 2009). The data acquisition area is presented in Fig. 14.

4.2. Data and Preprocessing

To verify the method for building velocity models in ultra-deep-sea regions, we selected an oceanic portion of the seismic data that has a water layer of $\sim$ 4.5 km. There were 1200 shots in the data set, and the shot interval was 50 m. The depth of the air-guns

was 10 m. The number of receivers was 958, and the interval was 12.5 m. The near-offset and far-offset were 237.5 and 12,000 m, respectively. The streamer depth was 15 m from sea level. The total recording time was 20 s, and the sampling rate was 2 ms. The water velocity was fixed at 1498 m/s. Figure 15 shows several shot gathers from the Sumatra data WG2 line.

The Sumatra field data have a considerably thick seawater layer of 4.5 km with strong noises between the direct waves and the first break. In addition, the amplitude of the direct wave in the far-offset region is too weak to locate exact time-points for muting. Thus, we eliminated the signals before the first break and considered only the reflections and refractions, not the direct wave. We then reconstructed the direct wave. To reconstruct the direct waves, we computed Green's function for half-space media, estimated the source wavelet from the original direct waves using the full Newton method in the frequency domain and then convolved the Green's function with the estimated source wavelet (KOO *et al.* 2011). After transforming the modeling results from the frequency domain to the time domain, we added the reconstructed direct wave to muted seismic data combined with reflections and refractions.

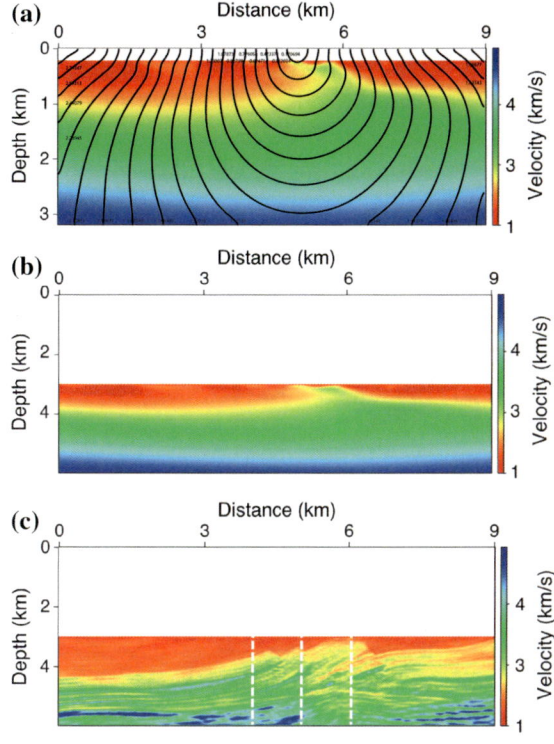

Figure 12
a Refraction tomography result with travel-time contour acquired from SWEET algorithm. **b** A velocity model with an extended water layer to fit the original deep-sea model. **c** Laplace–Fourier-domain inversion result using the tomography result as a starting model

To calculate the exact residual for the inversion, we must accurately define the source wavelet. In this study, we utilized a source estimation scheme using a direct wave. Figure 16 shows the spectrum of the source wavelet and the vertical source function observed during acquisition. To perform refraction tomography without the deep-water effect, we applied downward continuation on the field data. The entire procedure is the same as shown in the synthetic data test section. The downward-continued data shown in Fig. 17 were used to obtain a macro-velocity model for the starting model of Laplace–Fourier-domain FWI. After getting a long-wave-length velocity model through refraction tomography with downward-continued data, we

performed Laplace–Fourier-domain FWI with the original field data. These results will be shown in the next section.

4.3. FWI Results

We performed Laplace–Fourier-domain FWI using two different starting velocity models: a linearly increasing velocity model and a result from the refraction tomography.

First, we performed Laplace–Fourier inversion using the preprocessed original dataset. We used a linearly increasing model whose velocity varied from 1.9 km/s at the sea floor to 6.0 km/s at a depth of 11 km (Fig. 19a). For the frequency-domain modeling, we used the finite element scheme and applied a 25-m grid interval for the interpolation. The frequency loop used for the inversion is presented in Table 1. We also performed refraction tomography with a 50-m grid size using downward-continued data for the second inversion. After obtaining a macro-velocity model through refraction tomography, we performed Laplace–Fourier-domain FWI using the tomography results as a starting model. The inversion was implemented sequentially using 5 complex angular frequencies, with a complex portion fixed at 1.0 s^{-1} and a real portion with values ranging from 4 to 12 Hz at intervals of 2 Hz.

The Laplace–Fourier inversion results updated from the two initial models are presented in Figs. 18b and 19b. For a clear distinction of the sea floor, we eliminated the water layer with a velocity of 1.5 km/s from the results. The results from the linearly increasing starting model shown in Fig. 18 indicated that a thin low-velocity layer is located at a depth of ~6 km. In addition, an abnormal area of high velocity exists at a distance of ~35 km. This result is based on the starting model from tomography that has an inclined high-velocity basement below 8 km. On the other hand, Fig. 19 contains no abnormal high- or low-velocity structures around the shallow region of the structure. However, we observed a thin layer with a lower-velocity layer compared to that of the fringe area at a depth of ~9 km. As we compare

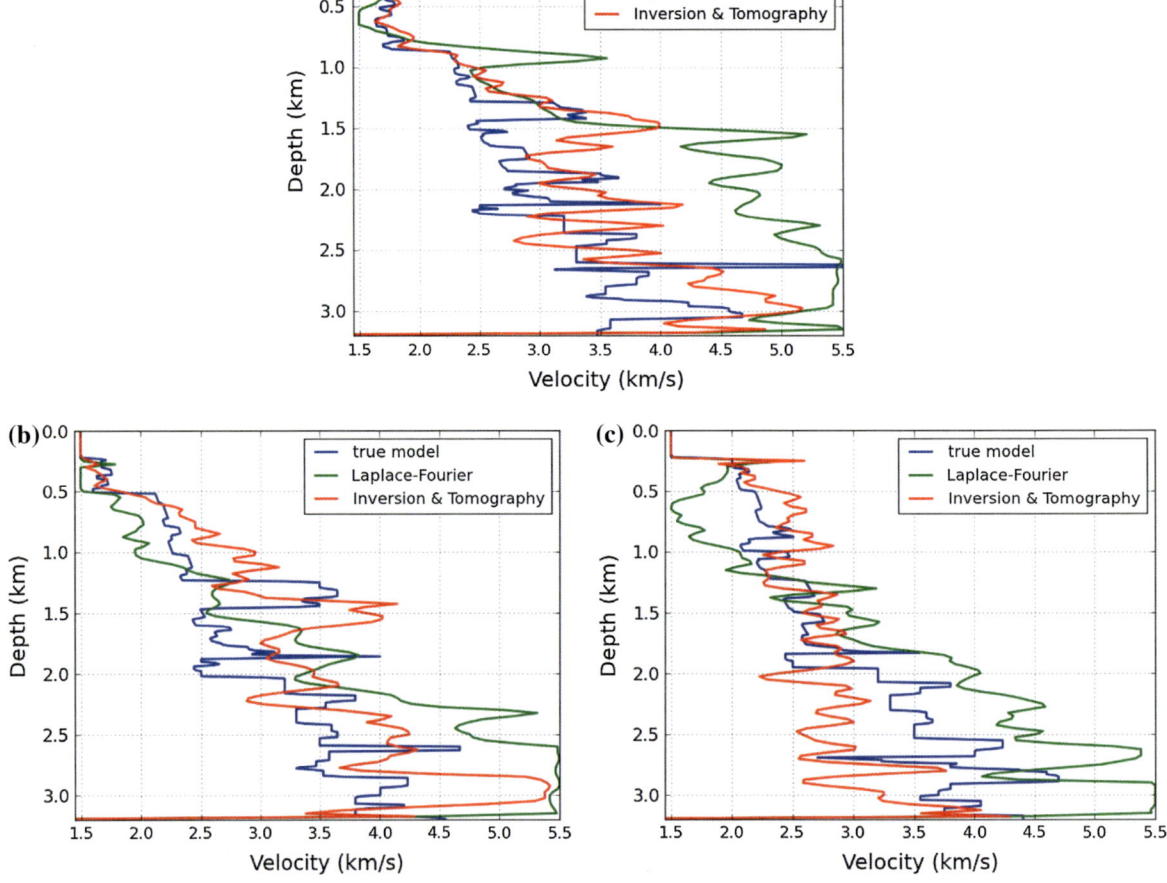

Figure 13

Comparison of the inversion results. We selected four lines around the traps from the *left* side at **a** 4, **b** 5, and **c** 6 km. Note that the accuracy of the result is significantly enhanced through the processes of tomography and Laplace–Fourier-domain inversion

these two results, we find that the overall tendencies in their subsurface structures are similar. However, several regions exhibit different structures, and we present the velocity-depth graph in Fig. 20 to compare these results more precisely. The profiles selected for the velocity comparison are shown in Figs. 18b and 19b. Both results have similar trends in their profiles at both ends; however, they have significantly different features at the other chosen profiles. To assess the reliability of the inversion results, we performed a Kirchhoff depth migration

with enough strength to accommodate layered media compared to reverse-time migration. The results of the depth migration are provided in next section.

4.4. Pre-Stack Depth Migration with FWI Model

We used the original data with a 2-ms time-sampling interval for the Kirchhoff depth migration. We interpolated each inversion result to a 12.5-m grid interval for the migration.

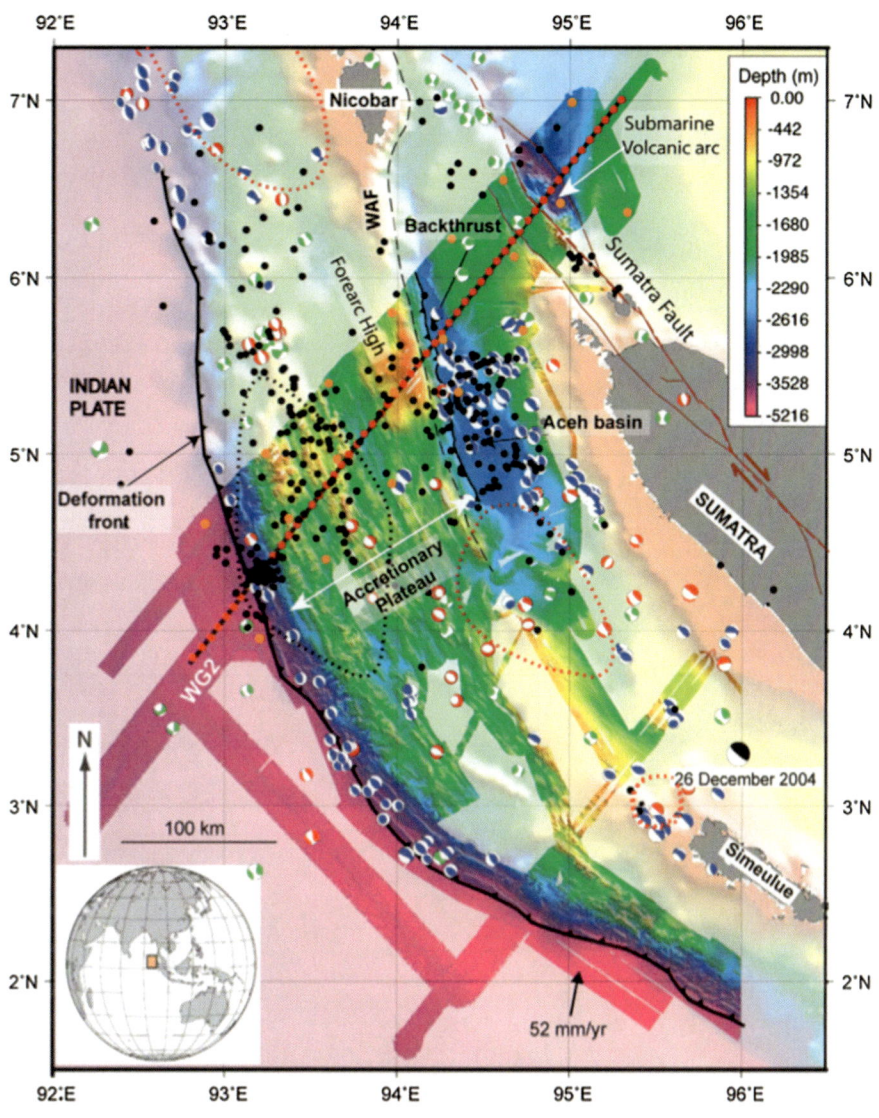

Figure 14

A map of the study area with geological information in Sumatra, Indonesia (SINGH *et al.* 2012). The *black line* is the WesternGeco WG2 seismic acquisition line. The *red dots* indicate OBS locations for a seismic refraction survey, and the *orange dots* are OBS locations for aftershock studies (SIBUET *et al.* 2007). The *red-dotted* contours represent the 10-m slip contour, and the *black-dotted* contours represent the 30-m slip contour from RHIE *et al.* (2007). *Blue beach balls* mean thrust. *Green* and *red* ones represent strike-slip and normal faulting mechanisms, respectively

We presented the Kirchhoff depth migration results in Figs. 21 and 22. Figure 21 is from the conventional Laplace–Fourier-domain FWI results. As we can see in the image, the basement around the 40 km distance is severely folded. However, Fig. 22, which is from a combination of Laplace–Fourier-domain FWI and tomography, shows a clear and flat basement structure. In addition, we can see clear fault structures in the overall area. We gathered common image points, which are presented in Figs. 23 and 24,

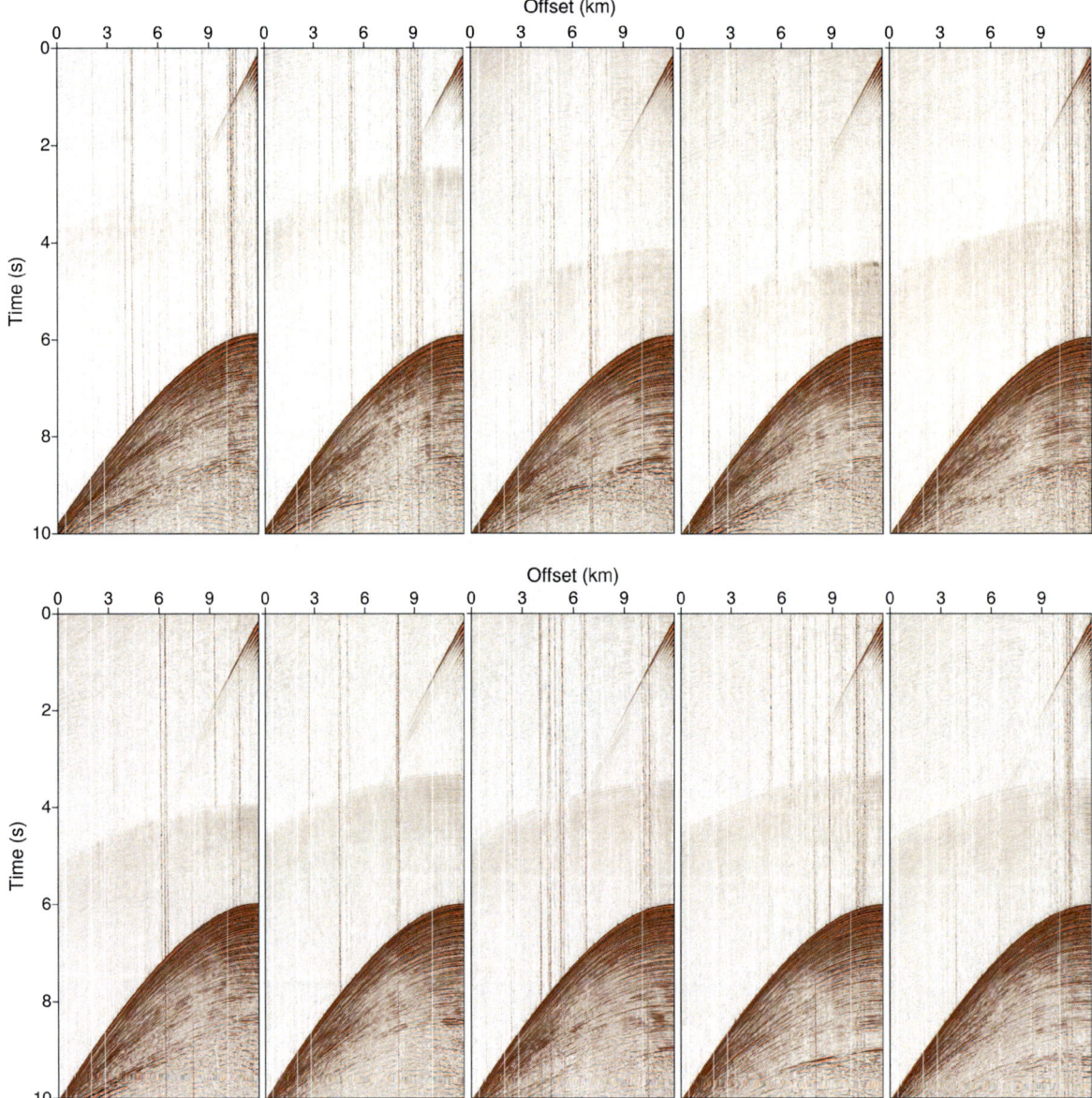

Figure 15
Shot gathers from the Sumatra data WG2 line

for exact analysis. The shape of the common image gather (CIG) from Laplace–Fourier inversion is curved downward, which means that the magnitude of the background velocity is higher than the exact value. Compared to the CIGs shown in Figs. 23 and 24 shows a clearer and flatter shape for the image points, which means the background velocity model is quite exact for depth migration. In addition, we displayed a shot gather of field data and the same shot gather in our synthetic data from our FWI result in

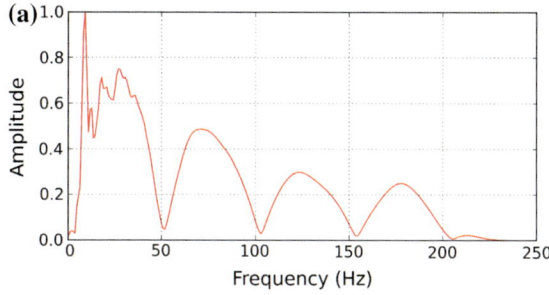

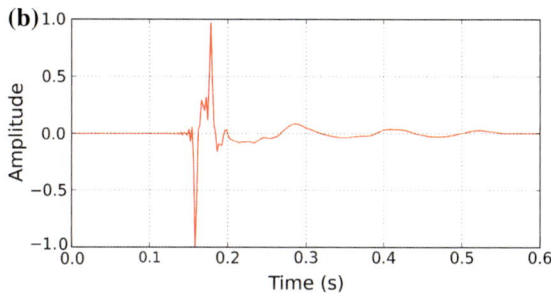

Figure 16

Spectrum of the source wavelet (**a**) and the source function acquired from air-guns (**b**)

Fig. 25 to add more confidence to the inversion results. Because there is no tangible variation in geological structure as presented in Fig. 22, most of the shot gathers look similar. Thus, we chose one set of shot gathers and extracted a seismic traces of the receiver at near-offset and far-offset in Fig. 25 (a) with that in (b) to compare the exact location of each reflection. Although the amplitudes of the traces do not exactly fit to each other in some parts, we found that most of the signals are well matched and show good results, especially around the time series from 9 to 10 s.

5. Conclusions

In this study, we attempted to build a reliable velocity model for deep-sea seismic data with a limited offset condition. For the first inversion method, we performed the well-known Laplace–Fourier-domain FWI to calculate the subsurface velocity structure. Using Laplace–Fourier-domain inversion for deep-sea data, however, has trouble inverting the velocity model because of the damped wavefield in the Laplace domain, which can yield inaccurate residuals and gradients. Therefore, we could not obtain an exact velocity structure of deep-water regions with this method.

As an alternative, we applied a wavefield extrapolation with downward continuation to eliminate the effect from the thick water layer. We then performed refraction tomography using the extrapolated seismic data. Afterward, Laplace–Fourier inversion was performed to increase the resolution of the tomography results. We determined that the refraction tomography with downward-continued data and the Laplace–Fourier-domain FWI can recover the subsurface structure located at considerable depths through experiments using both synthetic and field data. The velocity structure was successfully calculated for the synthetic data. When we applied our scheme to the field data to verify the inversion results, we performed a Kirchhoff depth migration based on the velocity model obtained through inversion and tomography. We identified common image gathers of the migration using tomography that is quite flat compared to that from the conventional Laplace–Fourier-domain FWI. In addition, we presented synthetic seismic data from the velocity model based on the proposed workflow to put more confidence on our results.

We obtained reliable results from both synthetic and field datasets. However, limitations also exist. Specifically, the downward continuation scheme used in this study increased the amplitude of the refractions, but distorted the reflections after extrapolation. Presumably, this effect is attributed to the endpoint effect associated with doing extrapolation in the Fourier domain. In addition, the two end regions of the data that correspond to the length of the streamer were unusable because of inconsistency in the number of extrapolated traces. Thus, a more stable extrapolation scheme is required for the application of downward-continued data to an FWI.

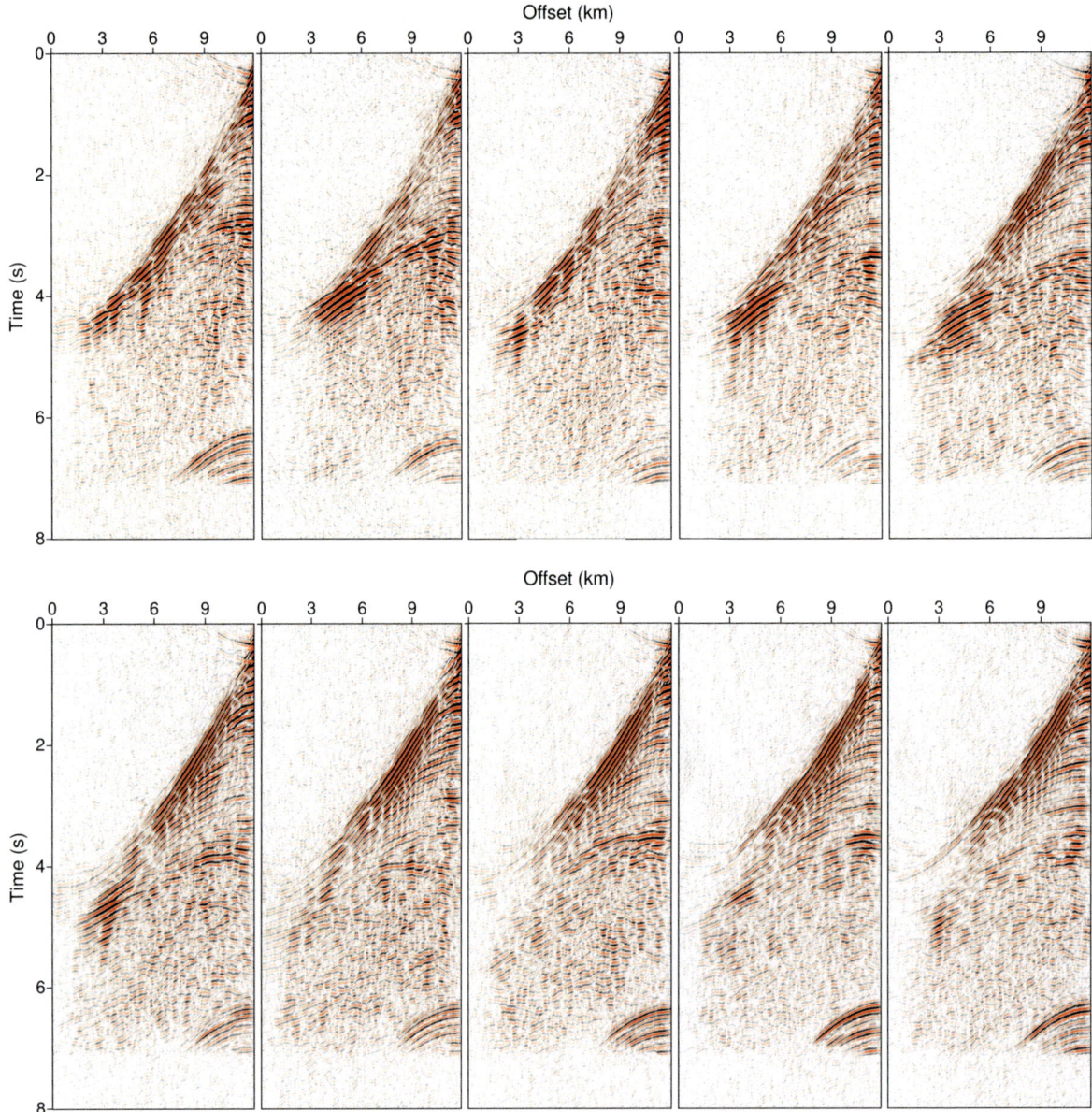

Figure 17
Shot profiles from the Sumatra data WG2 line presented in Fig. 15 following downward continuation

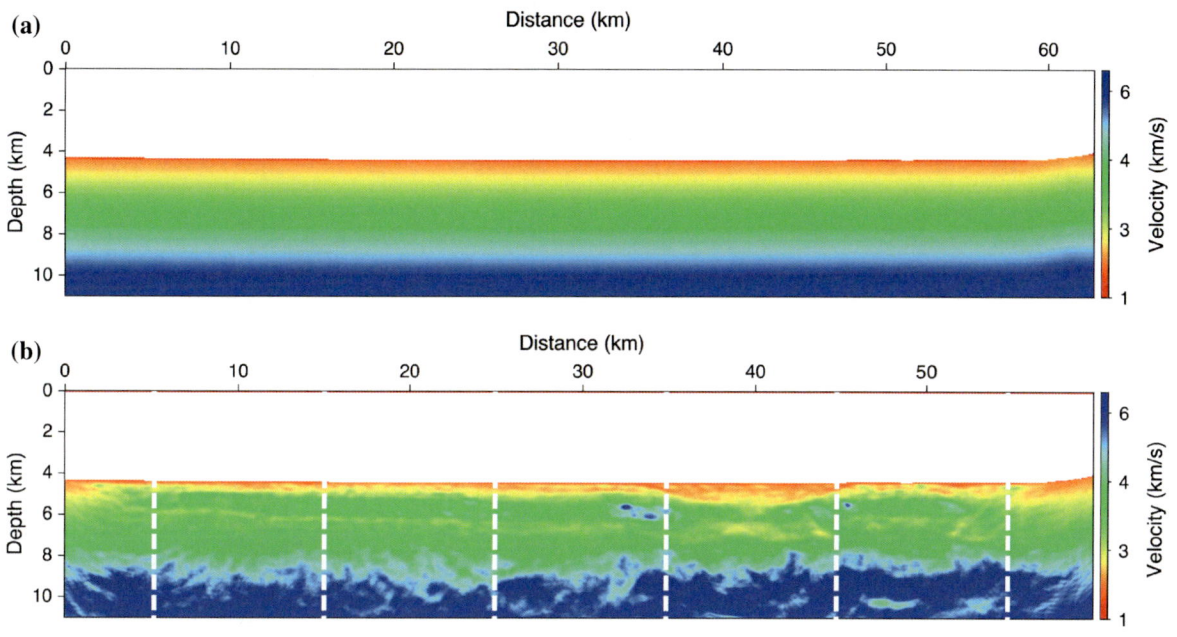

Figure 18
a Initial velocity model for the Laplace–Fourier-domain inversion and **b** the Laplace–Fourier-domain inversion result

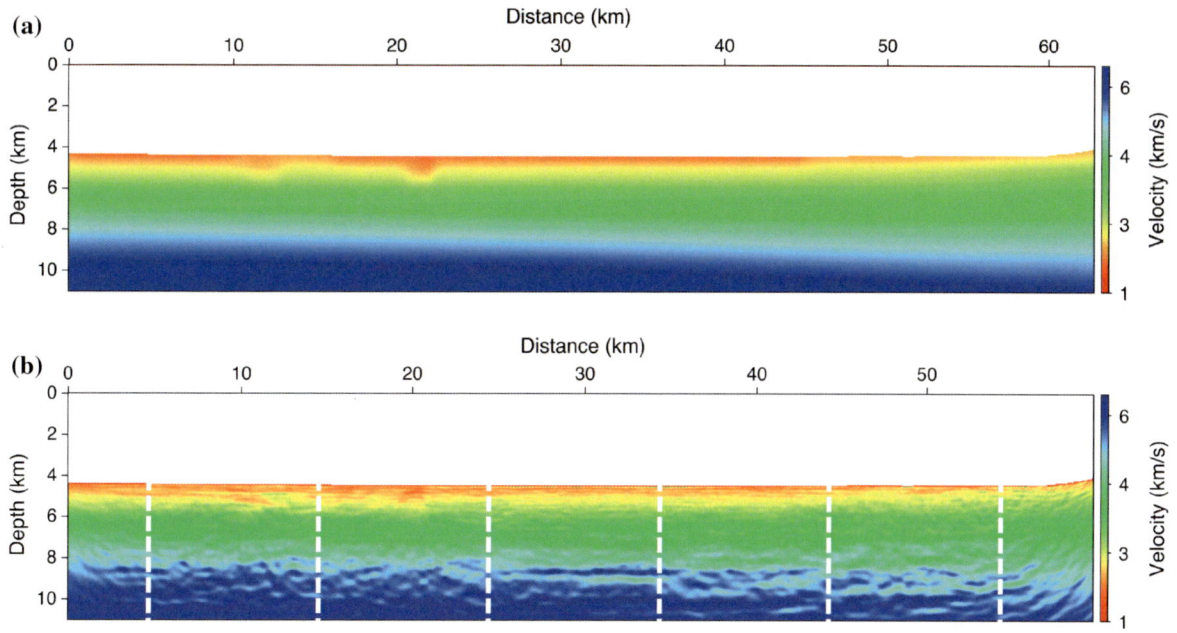

Figure 19
a Macro-velocity model obtained by refraction tomography using a downward-continued wavefield, and **b** the Laplace–Fourier inversion result

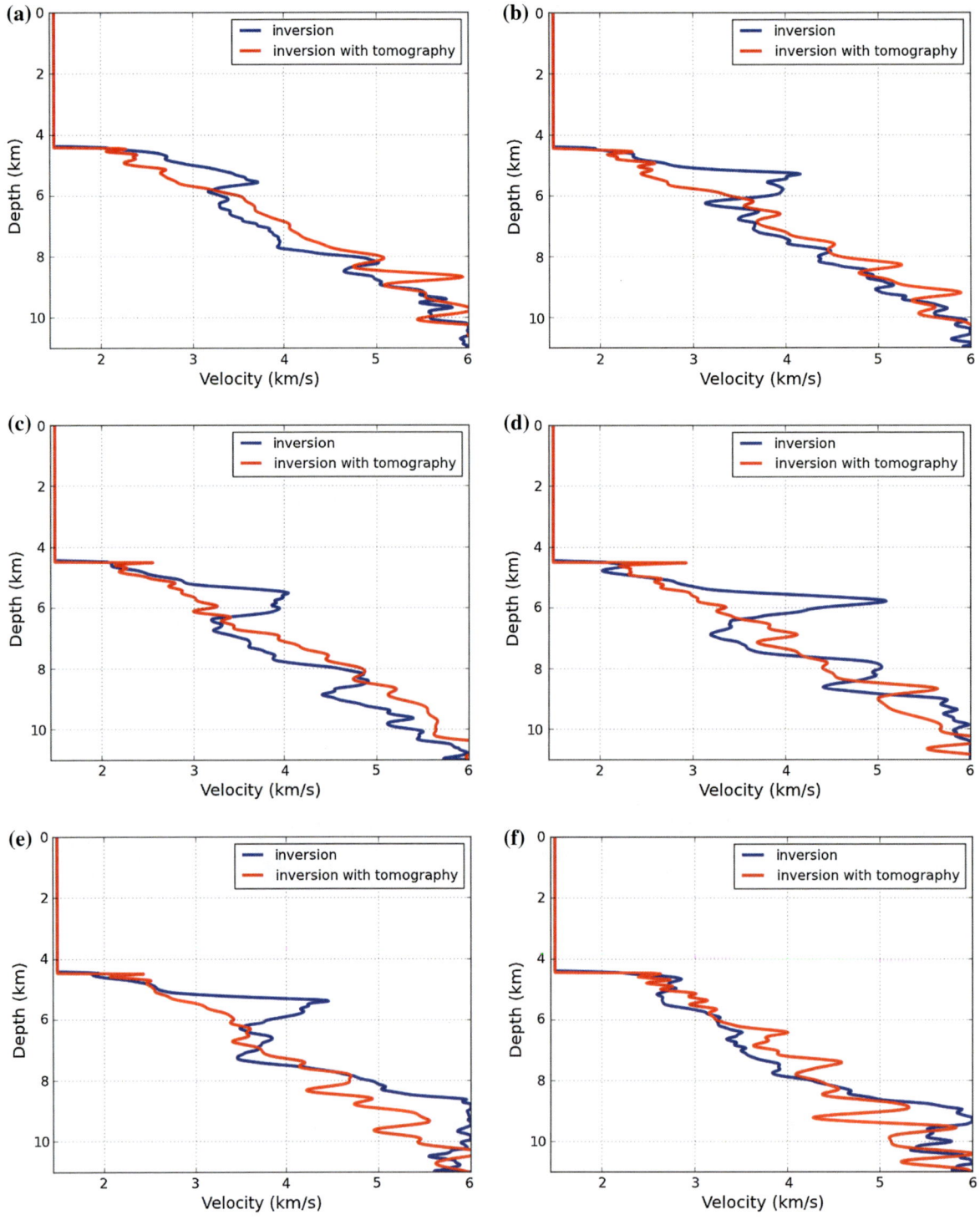

Figure 20
Comparison of the velocities corresponding to the *white-dotted lines* presented in Figs. 19b and 20b. We selected six lines at even intervals: **a** 5, **b** 15, **c** 25, **d** 35, **e** 45, and **f** 55 km from *left* to *right*. The *blue* and *red lines* indicate the Laplace–Fourier-domain inversion results and the inversion results combined with tomography, respectively

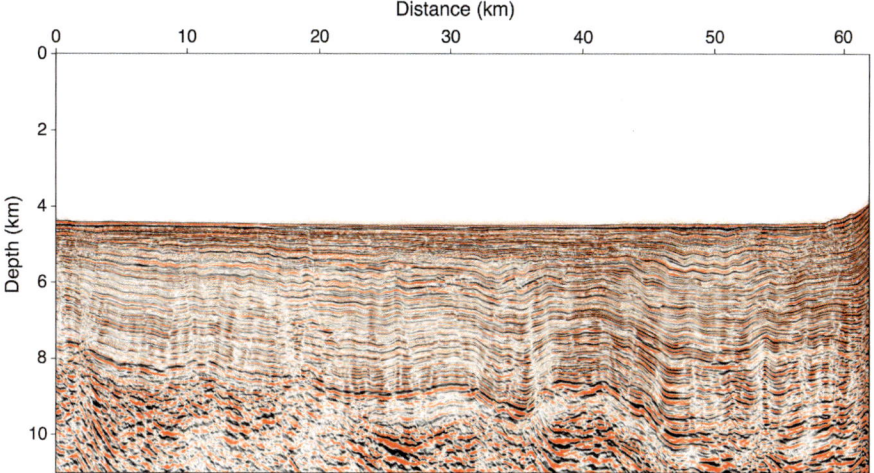

Figure 21
Kirchhoff depth migration results based on the velocity model obtained by conventional Laplace–Fourier-domain inversion

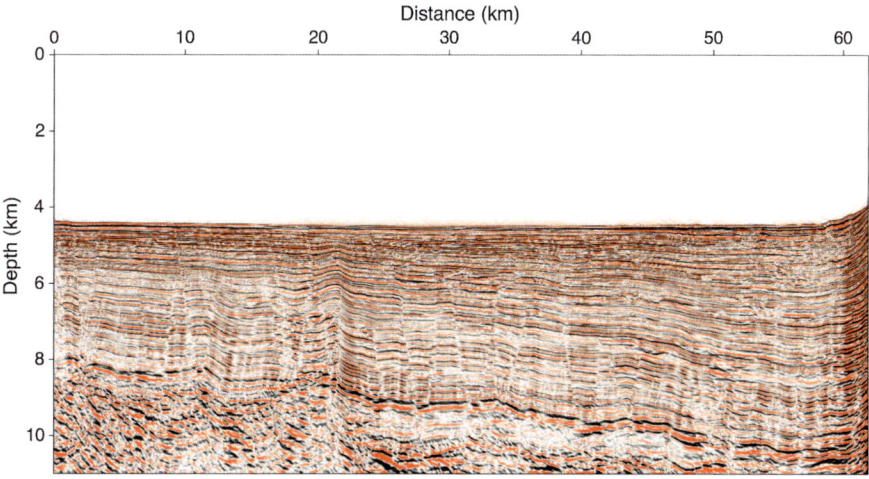

Figure 22
Kirchhoff depth migration results based on the velocity model obtained by a combination of tomography and Laplace–Fourier-domain inversion

Shot Number

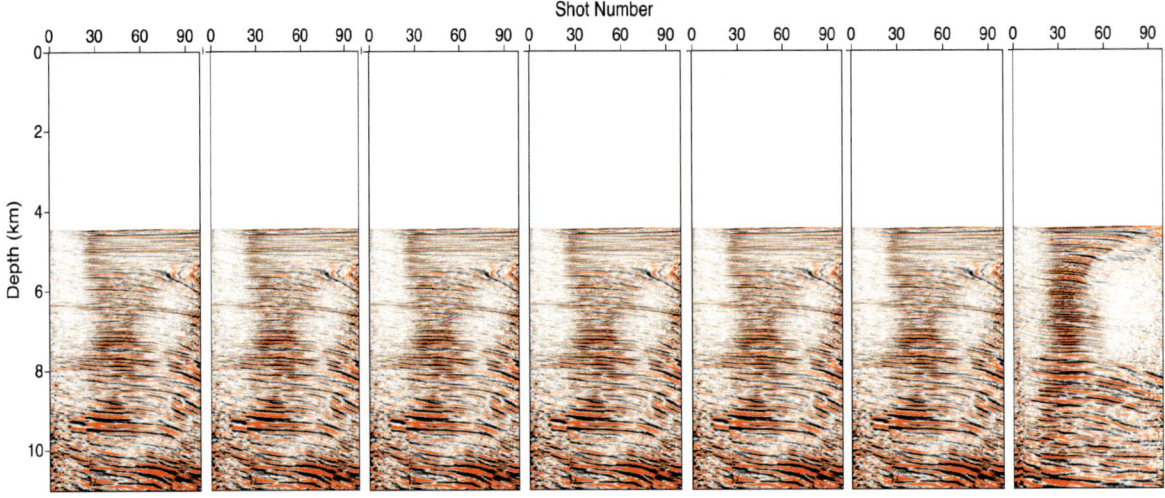

Figure 23
Common image gather from the Kirchhoff depth migration results presented in Fig. 21. The positions of the imaging points are 25, 31.25, 37.5, 43.75, 50, 56.25, and 62.5 km from *left* to *right*

Shot Number

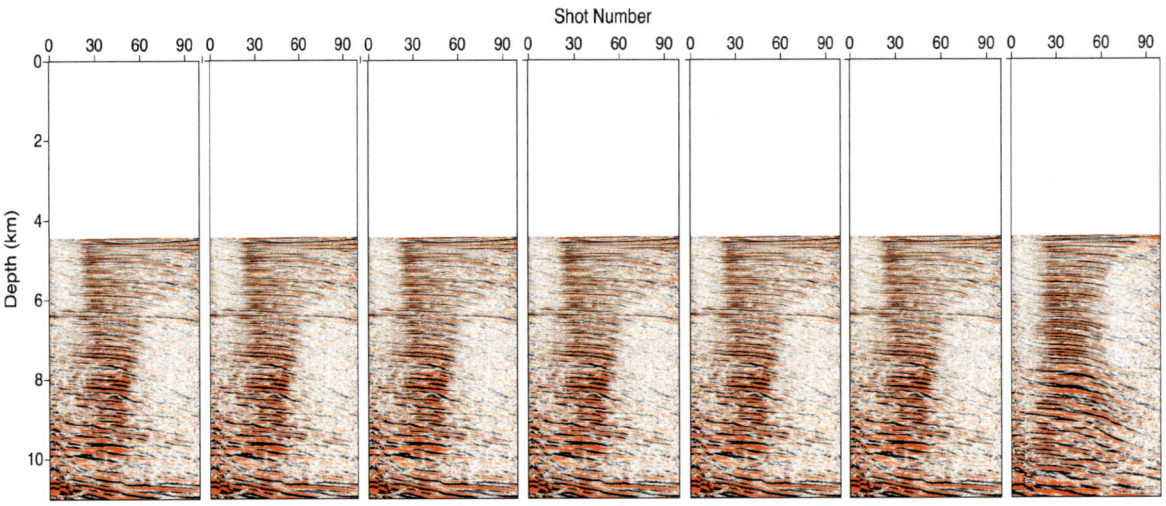

Figure 24
Common image gather from the Kirchhoff depth migration results presented in Fig. 22. The imaging points are positioned at 25, 31.25, 37.5, 43.75, 50, 56.25, and 62.5 km from *left* to *right*

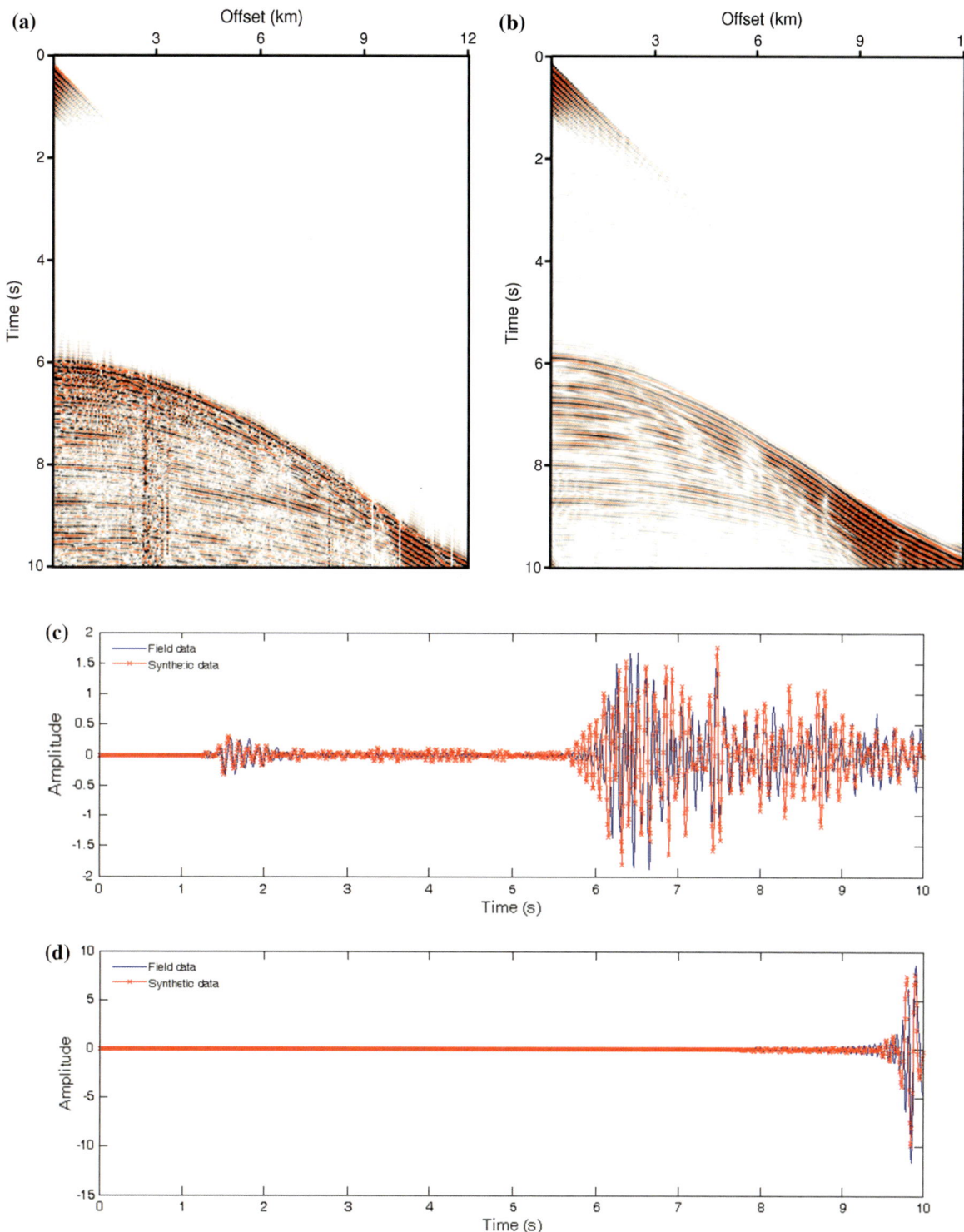

Figure 25
Comparison of the **a** field data and **b** synthetic seismogram, which is acquired from the final velocity model presented in Fig. 19b. Panels (**c**) and (**d**) compare the trace of the receiver at near-offset (**c**) and far-offset (**d**) in the panel (**a**) with that in panel (**b**)

Acknowledgments

This research was a part of the project titled 'The Study of Marine Geology and Geological Structure in the Korean Jurisdictional Seas', funded by the Ministry of Oceans and Fisheries, Korea, and was supported by the Energy Efficiency and Resources of the Korea Institute of Energy Technology Evaluation and Planning (KETEP) grant funded by the Korea government Ministry of Trade, Industry and Energy (No. 20132010201760). Also, we would like to extend our gratitude to Total oil company for providing a Sumatra seismic dataset to Seoul National University.

REFERENCES

AMMON, C. J., JI, C., THIO, H., ROBINSON, D., NI, S., HJORLEIFSDOT-TIR,V., KANAMORI, H., LAY, T., DAS, S., HELMBERGER, D., ICHINOSE, G., POLET, J. and WALD, D. (2005), *Rupture process of the 2004 Sumatra-Andaman earthquake*, Science *308*, 1133–1139.

BISHOP, T. N., BUBE, K. P., CUTLER, R. T., LANGAN, R. T., LOVE, P. L., REASNICK, J. R., SHUEY, R. T., SPINDLER, D. A., and WYLD, H. W. (1985), *Tomographic determination of velocity and depth in laterally varying media*, Geophysics *50*, 903–923.

CHAUHAN, A. P. S., SINGH, S. C., HANATO, N. D., CARTON, H., KLINGELHOEFER, F., DESSA, J. X., PERMANA, H., WHITE, N. J., GRAINDORGE, D., and the Sumatra OBS Scientific Team, (2009), *Seismic imaging of forearc backthrusts at northern Sumatra subduction zone*, Geophys. J. Int. *179*, 1772–1780.

CHIU, S. K. L., KANASEWICH, E. R., and PHADKE, S. (1986), *Three-dimensional determination of structure and velocity by seismic tomography*, Geophysics *51*, 1559–1571.

CLAERBOUT, J. and GREEN, I., Basic Earth Imaging – Madagascar edition (Stanford Exploration Project, 2009).

CLAYTON, R. and ENGQUIST, B. (1977), *Absorbing boundary conditions for acoustic and elastic wave equations*, Bull. Seismol. Soc. Am. *67*, 1529–1540.

GHOSAL, D., SINGH, S. C., and MARTIN, J. (2014), *Shallow subsurface morphotectonics of the NW Sumatra subduction system using an integrated seismic imaging technique*, Geophys. J. Int. *198*, 1818–1831.

GRAVES, W. (1996), *Simulating seismic wave propagation in 3d elastic media using staggered-grid finite differences*, Bull. Seismol. Soc. Am. *86*, 1091–1106.

HA, W., and SHIN, C. (2012), *Proof of the existence of both zero- and low- frequency information in a damped wavefield*, Journal of Applied Geophysics, *83*, 96–99.

HA, W., W. CHUNG, and C. SHIN, 2012, *Pseudo-hessian matrix for the logarithmic objective function in full waveform inversion*, J. Appl. Geophys. *21*, 201–214.

HA, W., CHUNG, W., PARK, E., and SHIN, C. (2012), *2-D acoustic Laplace-domain waveform inversion of marine field data*, Geophys. J. Int. *190*, 421–428.

HA, W., PYUN, S., YOO, J., and SHIN, C. (2010), *Acoustic full waveform inversion of synthetic land and marine data in the Laplace domain*, Geophys. Pros. 58, 1033–1047.

KAMEI, R., PRATT, R. G., and TSUJI, T. (2013), *On acoustic waveform tomography of wide-angle OBS data-strategies for pre-conditioning and inversion*, Geophys. J. Int. *194*, 1250–1280.

KOO, N., SHIN, C., MIN, D., PARK, K., and LEE, H. (2011), *Source estimation and direct wave reconstruction in Laplace-domain waveform inversion for deep-sea seismic data*, Geophys. J. Int. 187, 861–870.

LAILLY, P., *The seismic inverse problem as a sequence of before stack migrations* (SIAM, 1983).

MARFURT, K. (1984), *Accuracy of finite-difference and finite-element modeling of the scalar and elastic wave-equations*, Geophysics 49, 533–549.

MORA, P. (1987), *Nonlinear two-dimensional elastic inversion of multioffset seismic data*, Geophysics *52*, 1211–1228.

MURPHY, G. E., and GRAY, S. H. (1999), *Manual seismic reflection tomography*, Geophysics *64*, 1546–1552.

PRATT, R., and WORTHINGTON, M. (1988), *The application of diffraction tomography to cross-hole seismic data*, Geophysics *53*, 1284–1294.

PRATT, R., SHIN, C., and HICKS, G. (1998), *Gauss-newton and full newton methods in frequency-space seismic waveform inversion*, Geophys. J. Int. *133*, 341–362.

PRAWIRODIRDJO, L., BOCK, Y., GENRICH, J. F., PUNTODEWO, S., RAIS, J., SUBARYA, C., and SUTISNA, S. (2000), *One century of tectonic deformation along the Sumatran faults from triangulation and Global Positioning System surveys*, J. Geophys. Res. *105*, 28343–28361.

RAYNOLDS, A. C. (1978). *Boundary conditions for the numerical solution of wave propagation problems*, Geophysics *43*, 1099–1110.

RHIE, J., DREGER, D., BURGMANN, R., and ROMANOWICZ, B. (2007), *Slip of the 2004 Sumatra-Andaman earthquake from joint inversion of long-period global seismic waveforms and GPS static offsets*, Bull. Seismol. Soc. Am. *97*, S115–S127.

SIBUET J.-C., RANGIN, C., LE, P., SINGH, S., CATTANEO, A., GRAINDORGE, D., KLINGELHOEFER, F., LIN, J.-Y., MALOD, J.-A., MAURY, R., SCHNEIDER, J., SULTAN, N., UMBER, M., and YAMAGUCHI, H. (2007), *26th December 2004 great Sumatra-Andaman earthquake – seismogenic zone and active splay faults*, Earth Planet. Sci. Lett. *263*, 88–103.

SINGH, S. C., CHAUHAN, A., CALVERT, A. J., HANANTO, N. D., GHOSAL, D., RAI, A., and CARTON, H. (2012), *Seismic evidence of bending and unbending of subducting oceanic crust and the presence of mantle megathrust in the 2004 Great Sumatra earthquake rupture zone*, Earth Planet. Sci. Lett. *321*, 166–176.

SINGH, S. C., CARTON, H., TAPPONNIER, P., HANANTO, N. D., CHAUHAN, A. P. S., HARTOYO, D., BAYLY, M., MOELJOPRANOTO, S., BUNTING, T., CHRISTIE, P., LUBIS, H., and MARTIN, J. (2008), *Seismic evidence for broken oceanic crust in the 2004 Sumatra earthquake epicentral region*, Nat. Geosci. *11*, 777–781.

SHIN, C., MIN, D., MARFURT, K., LIM, H., YANG, D., CHA, Y., KO, S., YOON, K., HA, T., and HONG, S. (2002), *Traveltime and amplitude calculations using the damped wave solution*, Geophysics *67*, 1637–1647.

SHIN, C., and CHA, Y. (2008), *Waveform inversion in the Laplace domain*, Geophys. J. Int. *173*, 922–931.

SHIN, C., and CHA, Y. (2009), *Waveform inversion in the Laplace-Fourier domain*, Geophys. J. Int. *177*, 1067–1079.

SHIN, C., and HA, W. (2008), *A comparison between the behavior of objective functions for waveform inversion in the frequency and Laplace domains*, Geophysics *73*, VE119–VE133.

SHIN, C., KOO, N., CHA, Y., and PARK, K. (2010), *Sequentially ordered single-frequency 2d acoustic waveform inversion in the Laplace-Fourier domain*, Geophys. J. Int. *181*, 935–950.

SHIN, C., and MIN., D. (2006), *Waveform inversion using a logarithmic wavefield*, Geophysics *71*, R31–R42.

SYMES, W., (2008), *Migration velocity analysis and waveform inversion*, Geophys. Pros. *56*, 765–790.

TARANTOLA, A. (1984), *Inversion of seismic-reflection data in the acoustic approximation*, Geophysics *49*, 1259–1266.

VIRIEUX, J., and OPERTO, S. (2009), *An overview of full-waveform inversion in exploration geophysics*, Geophysics *74*, WCC1–WCC26.

VERSTEEG, R. (1994), *The Marmousi experience: Velocity model determination on a synthetic complex data set*, Leading Edge *13*, 927–936.

ZHANG, J., and TOKSOZ, M. N. (1998), *Nonlinear refraction traveltime tomography*, Geophysics *63*, 1726–1737.

(Received December 24, 2013, revised May 21, 2015, accepted June 11, 2015, Published online June 23, 2015)

Pure Appl. Geophys. 173 (2016), 775–794
© 2015 Springer Basel
DOI 10.1007/s00024-015-1036-7

| Pure and Applied Geophysics

Discrete Element Modeling of Stick-Slip Instability and Induced Microseismicity

CYRUS KHAZAEI,[1] JIM HAZZARD,[2] and RICK CHALATURNYK[1]

Abstract—Using Particle Flow Code, a discrete element model is presented in this paper that allows direct modeling of stick-slip behavior in pre-existing weak planes such as joints, beddings, and faults. The model is used to simulate a biaxial sliding experiment from literature on a saw-cut specimen of Sierra granite with a single fault. The fault is represented by the smooth-joint contact model. Also, an algorithm is developed to record the stick-slip induced microseismic events along the fault. Once the results compared well with laboratory data, a parametric study was conducted to investigate the evolution of the model's behavior due to varying factors such as resolution of the model, particle elasticity, fault coefficient of friction, fault stiffness, and normal stress. The results show a decrease in shear strength of the fault in the models with smaller particles, smaller coefficient of friction of the fault, harder fault surroundings, softer faults, and smaller normal stress on the fault. Also, a higher rate of displacement was observed for conditions resulting in smaller shear strength. An increase in *b*-values was observed by increasing the resolution or decreasing the normal stress on the fault, while *b*-values were not sensitive to changes in elasticity of the fault or its surrounding region. A larger number of recorded events were observed for the models with finer particles, smaller coefficient of friction of the fault, harder fault surroundings, harder fault, and smaller normal stress on the fault. The results suggest that it is possible for the two ends of a fault to be still while there are patches along the fault undergoing stick-slips. Such local stick-slips seem to provide a softer surrounding for their neighbor patches facilitating their subsequent stick-slips.

Key words: Stick-slip, Particle flow code (PFC), microseismic, fault instability.

1. Introduction

It is believed that the mechanism of fault instability involves multiple local on and off slips of patches referred to as "stick-slip" along the fault

(BRACE and BYERLEE 1966). Such small stick-slips may be recorded as "foreshocks" leading to "main shocks" and followed by "aftershocks" each releasing different levels of acoustic energy. Various aspects of fault instability have been already studied by laboratory experiments (BRUNE *et al.* 1993; BYERLEE and BRACE 1968; DIETERICH 1981; JULIAN *et al.* 1998; OHNAKA 1973) as well as numerical continuum models (DALGUER and DAY 2006; DAY *et al.* 2005; GALIS *et al.* 2008; XING *et al.* 2004) and numerical discontinuum models (FINCH *et al.* 2003; MORA and PLACE 1994; MORGAN 2004; PLACE *et al.* 2002).

Compared to continuum models, the discrete element method (DEM) has the capability to model geometrical heterogeneity by size distribution of particles as well as looking into the rupture process of rocks with more details. Using Particle Flow Code (PFC3D) v.5.0 in this research, a discrete representation of the fault is modeled, and release of acoustic energy due to its stick-slip instability is studied. For this purpose, a large scale laboratory experiment conducted on granite with a single fault originally reported by DIETERICH (1979) and recently repeated with microseismic recording by MCLASKEY and KILGORE (2013) was numerically simulated.

Traditionally, the approach for modeling microseismicity with PFC has been to consider each bond breakage (crack) as a single AE event with the further possibility of clustering the events to form more realistic magnitudes (HAZZARD 1998). This approach has been successfully used in modeling the intact rock problems where the events are believed to have a compressive induced crack nature (HAZZARD and YOUNG 2002, 2004; YOUNG and DEDECKER 2005; YOUNG *et al.* 2001; ZHAO and YOUNG 2011). However, in the present research, new routines have been developed for recording slip-induced microseismic

[1] NREF/CNRL Markin Bldg, Department of Civil and Environmental Engineering, University of Alberta, Edmonton, AB T6G 2W2, Canada. E-mail: khazaei@ualberta.ca; rjchalaturnyk@ualberta.ca
[2] Itasca Consulting Group Inc, Toronto, ON, Canada. E-mail: jhazzard@itascacg.com

events. The results have been compared with the experimental data. A parametric study has been conducted to study the effect of various factors on a fault's behavior.

This knowledge could also be useful for problems other than earthquake studies where there is likelihood of two planes sliding on each other such as in landslides (PENG and GOMBERG 2010), basal gliding of ice glaciers (JANSEN 2006; ROUX et al. 2008), and microseismic monitoring of a sedimentary rock mass in petroleum projects (FAIRHURST 2013; KRISTIANSEN et al. 2000).

2. Theory of Slip

The basics of how slip occurs in physics are briefly explained using Fig. 1 (left). As the applied driving force to the block, f_d, is increased, the resistive frictional force (solid red line) is also increased until at point a, f_d reaches the maximum static frictional resistance, the frictional resistance drops to lower values known as kinetic friction (or as simplified by the dotted blue curve), the spring unloads following a line with the slope equal to its stiffness (dashed green line), K, and the block starts to move.

The spring's unloading continues even after the applied force is equal to the kinetic friction at point c,

meaning deceleration of the block until the final stop at point e where the excess energy is dissipated (Δabc equals Δcde) (SCHOLZ 2002). In reality, once the motion eventually stops (point e), there has to be a "healing" mechanism for friction to regain its static value so that further slips can happen. This process of multiple slips and stops is called "stick-slip" and is believed to be the mechanism of earthquakes (BRACE and BYERLEE 1970). RABINOWICZ (1951, 1958) first suggested that if two surfaces are in contact in a stationary condition under load for time t, the coefficient of static friction increases approximately as log$t$. He also proposed that a minimum displacement of D_c, "critical slip distance", is required for transition from static friction state to kinetic friction state. With regard to Fig. 1 (left), the condition required for the onset of instability can be mathematically expressed by Eq. (1):

$$\left|\frac{\partial f_f}{\partial u}\right| = \frac{(\mu_s - \mu_k)N}{D_c} > K \qquad (1)$$

where $\left|\partial f_f / \partial u\right|$ is variations of resistive frictional force, f_f, for increments of displacement, u. The parameters μ_s and μ_k are static and kinetic coefficients of friction, respectively, N is the normal stress on the block, D_c is critical slip distance, and K is the stiffness of spring.

The physical implication of the stiffness of spring, K, in this model can be thought of as the

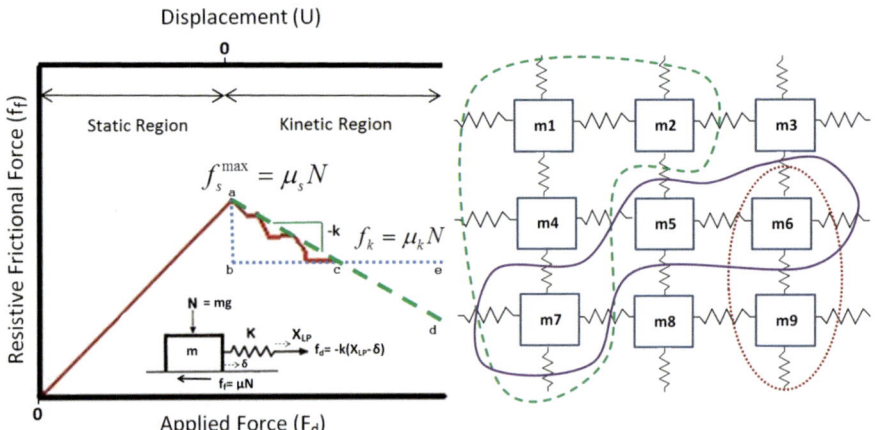

Figure 1

(Left) solid red line is variations of the resistive frictional force. The simplified change in friction is shown in dotted blue. The dashed green line shows variations of the driving force once the block starts to move until it stops. (Right) horizontal section of a more realistic model for the fault. The masses, stiffness of springs, and normal stress on each block are not necessarily equal due to heterogeneity. Three arbitrary patches are shown with dotted red, solid purple, and dashed green. The red and green patches can slide simultaneously

elastic conditions of the vicinity of a fault in nature or the stiffness of the loading machine in the laboratory. From Eq. (1), it is obvious that instability depends on the normal stress, elastic properties of the medium surrounding the fault, and roughness of the fault. The major microseismic events in the field are believed to be the result of stress drop followed by slip on fractures and consequently release of energy (DAMJANAC et al. 2010). However, if the stress drop is not large enough to impose excess energy to the system, i.e., after point "a" in Fig. 1 (left), the drop in resistive frictional force (solid red line) is such that it is placed above the unloading line (dashed green line), the slip would be aseismic. Therefore, in reality not every slip is the source of instability and considered seismic.

The assumption behind the model explained by Fig. 1 (left) is that the block can be regarded as a single mass point. PERSSON and TOSATTI (1999) suggest that for this assumption to be valid, the dimension of block, L_B, has to be smaller than a characteristic "elastic coherence length", ξ, otherwise, the block has to be discretized to smaller cells with the size, ξ, connected to each other by elastic springs. Therefore, it is necessary to take into account heterogeneity and nonuniform geometry of faults (NIELSEN et al. 1995, 2000). A more realistic model for the fault is shown in Fig. 1 (right). In this model, the mass, stiffness of springs, and normal stress on each block is different, and thus different patches may form and slide at different times or simultaneously.

3. Description of the Experiment

DIETERICH (1979) studied the scale dependence of the fault instability process through a large scale biaxial laboratory experiment. A large specimen of Sierra white granite with dimensions $1.5 \times 1.5 \times 0.419$ m was sawed diagonally at the quarry and roughened in the laboratory by lapping the two surfaces together with 30 grit silicon carbide abrasive to represent the fault with peak-to-trough surface roughness of 0.08 mm (DIETERICH 1981). He concluded that a minimum fault length was required so that confined shear instability could occur along a preexisting fault. Although strain gauges and velocity transducers were mounted on the model, no types of acoustic emission sensors were used during his experiment. The scalar seismic moments were later calculated from the general formula as:

$$M_0 = \mu DA \qquad (2)$$

where μ is the shear modulus, D is the average local seismic displacement and A is the area of the fault. A similar experiment on the same specimen was conducted by MCLASKEY et al. (2014) and MCLASKEY and KILGORE (2013) at the USGS, California, with 14 piezoelectric sensors recording the microseismic events during the test. This large model would allow the fault to be studied realistically so that some parts of it could slide while the rest was not slipping. In this way, individual stick-slips can occur during the loading with the possibility that they might trigger other slides. However, since this specimen had been used for 25 years resulting in many stick-slip events and cumulative slips without additional surface preparation, MCLASKEY et al. (2014) believed that the present surface was smoother than it was in 1981.

The test setup was a steel frame with four flat-jacks between the frame and the specimen as shown in Fig. 2. The model was loaded in σ_2 direction and unloaded in σ_1 direction by increasing and decreasing the pressures slowly along the 1 and 2 directions, respectively, with the same rate of 0.001 MPa/s. This way, the normal pressure on the fault was kept constant while the fault was shearing. They tried to model the earthquake cycle by loading, resetting and holding at four stages. Slip sensors were installed on top of the fault to measure relative slips from one side of the fault to the other. Piezoelectric sensors were also installed on top and bottom of the fault. The sensors would respond to the vertical component of motion in frequency ranges of ~ 100–1 MHz and were installed 200 mm off the fault. The onset of instability was observed at 3.7 MPa of shear stress. At this point, the fault would experience several small falls and rises in the shear stress each accompanied by a small displacement along the fault. These small displacements have been reported to range from 0.08 to 0.15 mm (G. MCLASKEY, personal communication, 2013). The

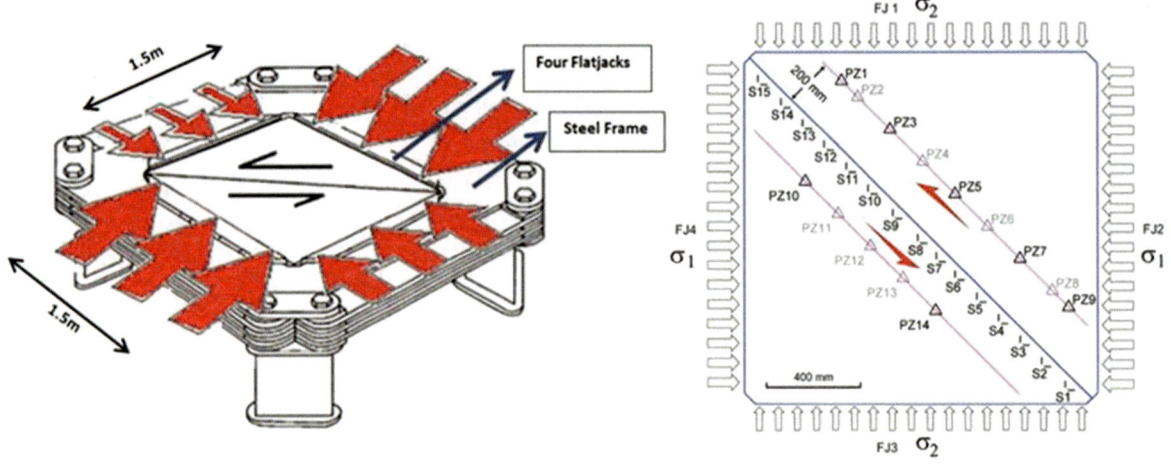

Figure 2
(*Left*) McLaskey's experimental setup. (*Right*) slip sensors and piezoelectric sensors. Piezoelectic sensors are mounted on the top and bottom of the sample. Flatjacks are marked FJ1 to FJ4. Strain gages are shown by S1 to S15. (McLASKEY and KILGORE 2013)

shear modulus and critical slip distance of the Sierra white granite were estimated to be around 20 GPa and 5 μm, respectively (G. McLASKEY, personal communication, 2013).

4. Numerical Model

In this research, Particle Flow Code (PFC3D) v.5.0 is used to model the experiment reported by McLASKEY and KILGORE (2013). PFC3D is an explicit implementation of the distinct element method (DEM) developed by Itasca, and despite continuum models, does not require mesh generation. In this code, it is assumed that the particles are rigid (non-deformable). In a PFC3D model, particles are bonded by models such as the parallel bond model and smooth joint model. Parallel bonds can transmit both forces and moments and can be envisioned as a short length beam or a "set" of elastic springs uniformly distributed over a circular cross-section at the contact point (POTYONDY and CUNDALL 2004).

The smooth joint model is used to simulate interfaces. The traditional way to model an interface is to change the properties such as strength of the bonds along the interface to those representing the real interface. The problem with this approach is bumpiness of the boundaries that can affect the behavior of the system. The effect of these bumpy

boundaries could become more important, when it comes to modeling joints or weak planes. A possible solution is to use smaller size particles along the interface which is not practical in large models. In order to solve this problem, another type of bond that simulates an "interface" regardless of the orientation of the particles along it is the "smooth joint contact model" (MAS IVARS et al. 2008). This model can be assigned to all the contacts between particles that lie on or along the opposite sides of the interface. This model overcomes the drawback of bumpy boundaries. The reason for calling it "smooth" is because of the constitutive behavior that allows particles bonded by this contact model to overlap and to "slide" on each other instead of moving around one another. Once a smooth joint model is created, the already existing contact or parallel bond models are deleted automatically for the contacts along interface. Figure 3 shows the behavior of smooth joint model compared to the usual parallel bond model.

In PFC3D, it is assumed that due to small time steps, a disturbance cannot propagate farther than neighbor particles, and therefore, velocities and accelerations can be considered constant during each cycle. Once a force is applied to the model, integrating twice the Law of Motion using the time step, velocities and then displacements are calculated that result in the updated position of each particle. The

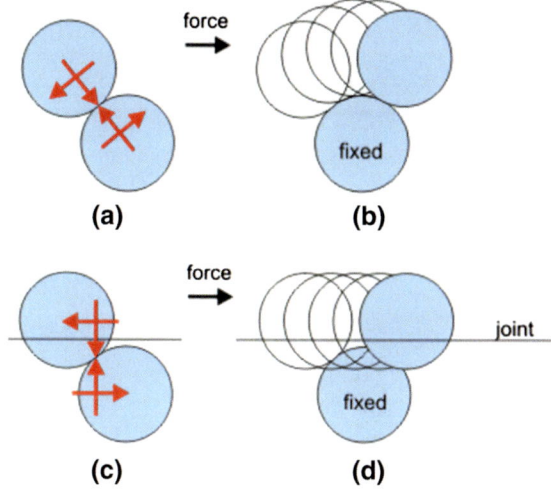

Figure 3
a Standard contact model with relative normal and tangential forces with regard to their orientation, **b** displacement of the particles bonded by standard contact model assuming the *bottom ball* is fixed, **c** normal and tangential forces on the particles whose contact is located along the smooth joint regarding their orientation with respect to the joint, **d** displacement of particles bonded by smooth joint model (MAS IVARS 2008)

updated positions are then used by the Force–Displacement Law to calculate the new forces, and so on. This process of cycling is continued until a predefined criterion is met.

4.1. Algorithm for Recording Slip-Induced Microseismicity

The numerical representation of the fault is composed of all the contact points whose at least one of the particles is located along the fault. Once a slip occurs at any of these contacts, if the normal stress of that contact is greater than 0.1 MPa, an "event" is created and the time at which slip started is recorded. The reason for this threshold can be explained using Fig. 1 (left). According to this figure, if the simple static-dynamic friction law (dotted blue curve) is used instead of the realistic frictional resistance curve (solid red), the rate of decrease in frictional resistance during slip will always be faster than the rate of decrease in driving force, and thus every slip will be seismic. It is known that low stiffness, high normal stress, and small D_c facilitate the unstable slip (DIETERICH 1979). Based on Eq. (1), assuming the frictional properties and the stiffness of

the surroundings are constant along the fault, normal stress will be the only controlling factor for the onset of instability, and thus a low normal stress results in the fault sliding stably. Therefore, it is necessary to set a minimum normal stress for the slips to be considered seismic. In practice, this threshold has the advantage of reducing the number of recorded events and faster computation time (HAZZARD and PETTITT 2013). However, it will be shown in Sect. 4.3.2 of this paper that the models presented in this study are not sensitive to this threshold.

According to laboratory stick-slip friction data reported by MCGARR (1994), the dynamic friction in this study has been assumed equal to 80 % of the static friction. Therefore, once a slip passes the normal stress criterion, the coefficient of friction of that contact is dropped to 80 % of its original value. Once the contact stops sliding, the coefficient of friction is readjusted to its static value. This is implemented to take into account the healing phenomenon (due to processes such as creep in the field or thermal mechanism in the lab causing the micro-asperities to weld together) that is believed to be a necessary component for generation of earthquakes in nature (MCLASKEY et al. 2012). For each seismic slip, the start and end times as well as displacement of the contact are recorded. Considering each slip as one event, results in all the magnitudes are assumed to be almost equal. A good practice is to assume a group of slips close in time and space forming a slip patch and clustering them to form a single event. Therefore, the events have been clustered if they had two conditions: they were within a space window and their duration of slip would overlap for at least one cycle. Another criterion has also been implemented for duration of each slip. Assuming the shear can propagate as slowly as 0.5 times the shear wave velocity of the material (HAZZARD and YOUNG 2000), the minimum duration of a slip to be considered seismic is calculated as:

$$T_{min} = \text{int}\left(\frac{\text{space\_window}}{0.5 \times \text{ae\_svel}}\right) \quad (3)$$

where T_{min} is the time required for the shear wave to propagate the space window, space_window is considered to be 0.42 m in this study, and ae_svel is the shear wave velocity equal to 2,700 m/s for Sierra granite as reported by McLaskey *et al.* (2014). Equation (3) shows that not only does the choice of space window affect the size of clusters and, therefore, distribution of event magnitudes, but also it affects the minimum lifetime of a slip to be considered seismic.

Therefore, choosing an appropriate space window can be considered part of the model's calibration. For a similar clustering algorithm, but for crack-induced events, Hazzard (1998) showed that the space window of five particle diameters would yield the best results in a 2D model, and above this value, the *b*-values were not dependent on the size of window, while for a 3D model a smaller space window would be more reasonable. Later in a study of unstable fault slip in Lac Du Bonnet granite, Hazzard *et al.* (2002) showed that three particle diameters would result in a realistic *b*-value. In the present model, considering the maximum diameter of the particles is 0.14 m, a space window equal to 0.42 m (three particle diameters) has been used. Comparison of the results with data reported by McLaskey *et al.* (2014) and McLaskey and Kilgore (2013) will be shown in the next sections and confirms that this is an appropriate choice.

At the end of the test, all the slips that occurred within the space window and with the durations overlapping at least one cycle are clustered together. The area of each clustered event is calculated assuming the event as an ellipse. The largest and smallest radii of the ellipse are calculated based on the farthest and closest particles to the center of the event, respectively.

In order to calculate the centroid of each event, the number of contacts forming that event has been used. However, it has to be emphasized that the number of "contacts" forming an event is not necessarily the same as the number of "slips" forming that event. As an example, the clustering process for three smooth joint contacts, A, B, and C, is schematically illustrated in Fig. 4.

In this figure, imagine contact A starts slipping from 0.10 to 0.15 s. At 0.13 s contact B starts slipping within the space window of contact A, and therefore, these two slips constitute one event. The contact B slips until 0.20 s. At 0.15 s contact C starts slipping until 0.16 s at which point contact A starts slipping

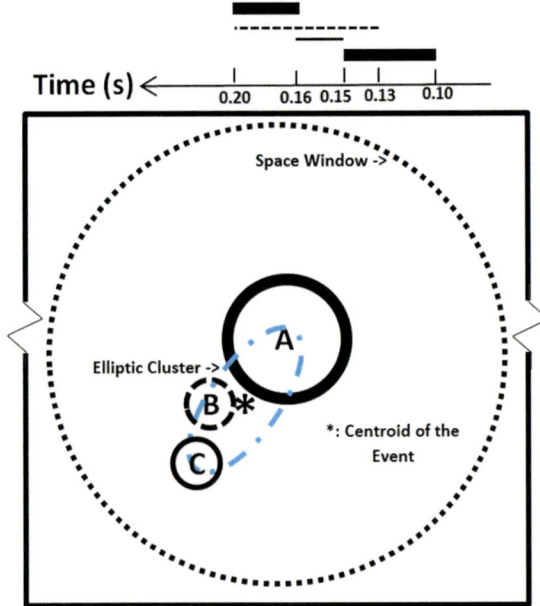

Figure 4

A section of the fault illustrating the clustering process for three contacts, A, B, and C. It's assumed that no other contacts within the space window slipped from 0.10 to 0.20 s. Size of the elliptic event is determined based on the closest and farthest contacts to the center of the event (i.e. B and A, respectively)

"again" until 0.20 s. No other contacts slipped within the space window of contact A from 0.10 to 0.20 s. In this example, although there are four slips forming one event, there are only three contacts involved, and therefore, the location of contact A is used once for calculating the centroid (although it slipped twice). The centroid is simply calculated as the average of the locations of the contacts forming the event. Average displacement is calculated as the sum of all displacements associated to all the slips, four slips in this example, forming one event divided by the number of contacts in that event/cluster. Having the area of the event and the average displacement of all the contacts forming the cluster, seismic scalar moment, M_0, is calculated using Eq. (2). Eventually, magnitudes are calculated using Eq. (4):

$$M_w = \frac{2}{3}\log M_0 - 6.0 \qquad (4)$$

The whole idea of clustering is justified considering the fact that in nature, most seismic events are made up of small ruptures and shearing of asperities (Scholz 2002).

4.2. Results

A discrete element model of the experiment reported by MCLASKEY *et al.* (2014) and MCLASKEY and KILGORE (2013) is made by PFC3D as shown in Fig. 5. There are six walls surrounding the model. The wall on the face is not shown in this figure to make the balls visible. The balls are bonded together by parallel bond model. In order to resemble the fault, smooth joint model has been installed in all the contacts of all the particles located along the diagonal fault extending from top right to the bottom left of the model.

There is no direct one-to-one correspondence between the micro-parameters of a discrete element model and macro parameters of the real rock. Therefore, the calibration process involves some trial and error attempt of varying the micro-parameters until the desired macro response is observed (ITASCA 1999). However, in the present model, in order to eliminate the effect of possible bond breakages (cracks) on the fault's behavior, the strength properties of parallel bonds were set to high values so that they would not control the results, and therefore, such calibration for the parallel bond properties was no longer necessary. The fault (smooth joint) parameters were chosen so that the expected stick-slip behavior would be observed at the onset of instability at about 3.7 MPa as reported by MCLASKEY and KILGORE (2013). In order to keep the normal stress on the fault constant at 5 MPa during the test, the top and bottom

walls were moving inwards while the right and left walls were moving outwards, all four with the same velocity. The two walls on the back and front were not moving during the test. However, it has to be kept in mind that due to complete symmetry in the fault's location and loading scheme of this test, as well as high strength properties of the parallel bonds, there will be practically no damage in any part of the model other than slip along the fault. The PFC3D parameters of the model are summarized in Table 1.

As the first pass to validate the model, the normal stress along the fault has been obtained during the test. In PFC3D, it is possible to record the stress and strain values in three different ways: wall-based, measurement sphere-based, and particle-based. The wall-based stresses are calculated as the sum of out-of-balance forces of all the particles in contact with the wall divided by the wall's area. Measurement spheres (or "circles" in 2D) are representative volumes in which an average value for the stress or strain is calculated. They can be defined at arbitrary places in the specimen. Particle-based measurements represent the value of stress at one particular point.

In order to ensure that the normal stress remains constant on the fault, the wall-based stresses (shown in dashed red and purple in Fig. 6) were monitored. Considering the whole model as one big element, the normal stress on the fault, however, is calculated using a 2D transformation matrix as shown in Eq. (5):

$$\sigma_n^{\text{fault}} = \frac{\sigma_x + \sigma_y}{2} - \frac{\sigma_x - \sigma_y}{2}\cos 2\theta - \tau_{xy}\sin 2\theta \quad (5)$$

where θ is 45°, τ_{xy} is assumed to be negligible and σ_x and σ_y are wall-based stresses along the x and y directions, respectively. As another measure of assurance, the normal stress on the fault is also determined by summing up all the normal forces of all the contacts with the smooth joint model assigned to them divided by their area. The wall-based and directly measured normal stresses along the fault, as well as the wall-based stresses along the x and y directions, are plotted in Fig. 6. As can be observed in this figure, if the onset of instability is assumed as the point at which the increase in shear stress stops (i.e. 3.7 MPa), this point corresponds to about 0.22 s past the start of the test. Despite the laboratory

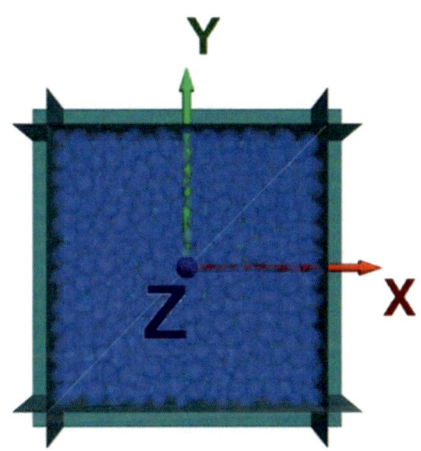

Figure 5
The PFC3D model. The fault extends from the *top right* to the *bottom left*

Table 1

Micro-parameters used in PFC3D model of Sierra granite

Ball parameters		Smooth joint parameters		
Average radius (m)	Young's modulus (GPa)	Normal stiffness (N/m^3)	Shear stiffness (N/m^3)	Coefficient of static friction
md_ravg	ba_Ec	sj_kn	sj_ks	sj_fric
0.056	2.1E10	2.1E10	2.1E10	1.05

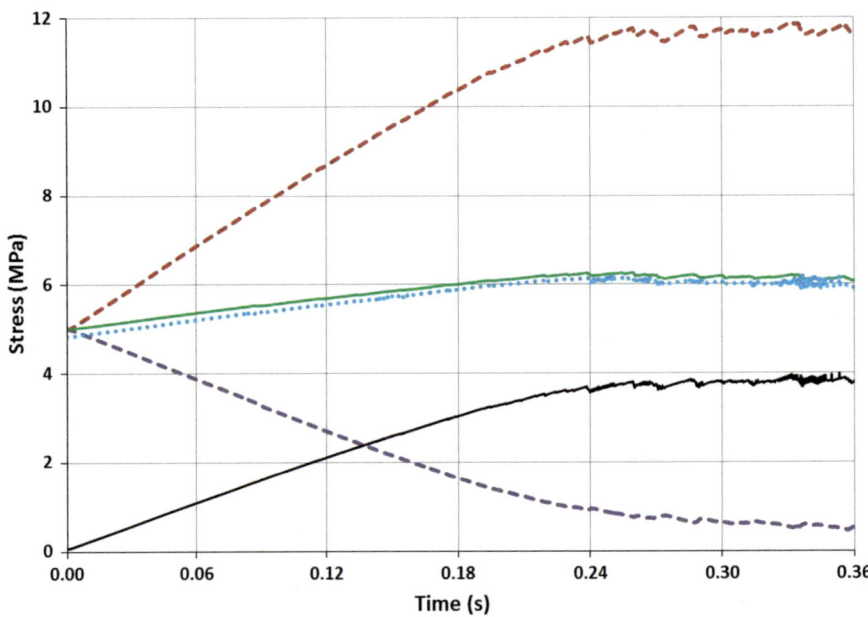

Figure 6

Stress measurements during the test. *Dotted blue curve* starting at 5 MPa is the direct measurement of normal stress from the particles along the fault. The *solid green curve* at 5 MPa is the normal stress along the fault determined from the wall based stresses by the transformation matrix. The *solid black curve* starting at 0 MPa is shear stress along the fault directly measured from the smooth joint contacts. The *two dashed symmetrical red* and *purple curves* are the wall-based stresses along the *x* and *y* directions, respectively

experiment which has practical limitations for how much the fault can slip, in this PFC3D model, there is really no criterion for when the test should stop. Therefore, the test was stopped once 0.1 % total displacement of the fault (i.e. 2 mm) was reached. Total displacement of the fault is calculated mathematically from the wall-based strains in the *x* and *y* directions.

As can be observed in Fig. 6, the normal stress directly measured from the balls is a little bit smaller than the wall-based measurement, which is as expected since the fault is modeled as a discontinuum surface. Other than that, the normal stress along the fault has been kept constant during the test. The

overall shear stress along the fault versus time as well as a magnified section of the final stick-slip behavior is plotted in Fig. 7.

As can be observed in this figure, the stick-slips have been successfully modeled by the numerical model. Prior to the onset of instability at 3.7 MPa, McLaskey and Kilgore (2013) observed a linearly increasing shear stress throughout the test. An interesting observation from this figure is that three stages in the slip behavior of the fault can be identified. The first stage is from 0 to about 3.2 MPa where the shear increases linearly with no major stick-slips. The second stage starts from 3.2 MPa until about 3.7 MPa with minor stick-slips while the

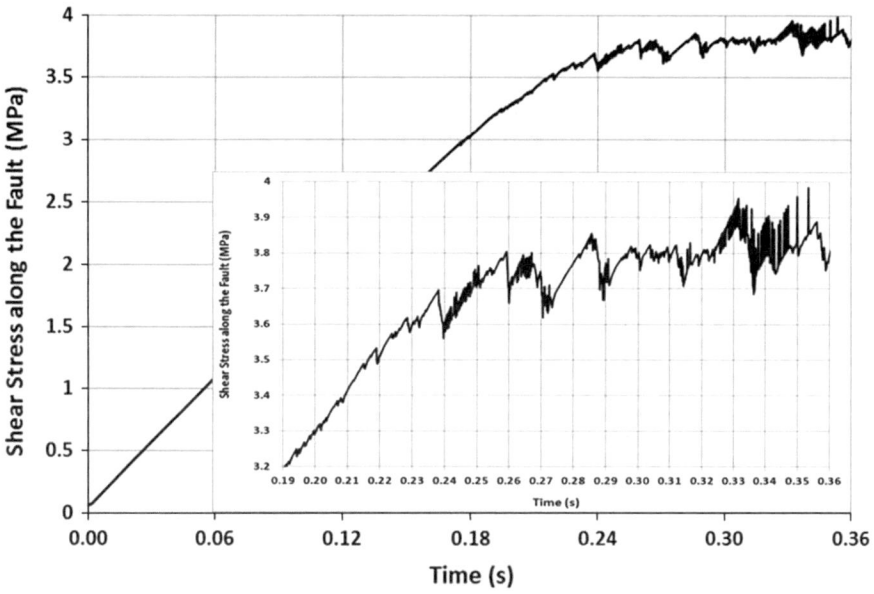

Figure 7

The average shear stress along the fault versus time. Shear force is a vector with three components. Average shear stress of all the contacts with smooth joint model along the x and y directions are calculated by summing up their respective forces divided by their areas. Then the average shear stress along the 45° fault is calculated by the stress transformation matrix

shear is still increasing, but non-linearly (i.e. the strength is still mobilizing). The final stage starts from 3.7 MPa where the falls and rises of shear stress become more significant and the overall trend of shear stress becomes almost constant while the fault is sliding.

Location of the slips at two sides of the fault recorded in the PFC3D model is shown in Fig. 8. As can be observed in this figure, the events are uniformly spread along the fault. This is in agreement with observation of McLASKEY et al. (2014).

In the PFC3D model, 223 slips have been recorded with the majority of magnitudes around −6 and some −7. As was previously mentioned, a reasonable approach would be to cluster the slips close in time and space to represent more realistic events. After clustering, the number of events is reduced to 131. As can be observed from the clusters shown in Fig. 9, the largest events have occurred at the central section of the fault which is in agreement with McLASKEY and KILGORE (2013). Because of difficulties in plotting ellipses, all the events in Fig. 9 are illustrated with circles with diameters equal to the major axis of ellipses; otherwise, as mentioned before, the area of each event has been calculated assuming it has an elliptic shape in general. It has to

be emphasized that the number of events in a PFC model is a function of model's resolution, and therefore, it is not reasonable to compare the number of PFC events directly with the real number of events recorded in the experiment.

In their experiment, McLASKEY and KILGORE (2013) reported that the normal stress along the fault was kept constant at 5 MPa. However, in their recent 2014 publication, the test was repeated with 4 and 6 MPa normal stresses along the fault. McLASKEY et al. (2014) were able to locate 16 and 32 events accurately for the test with 4 and 5 MPa normal stresses, respectively. In order to determine the mechanism of events, they performed a moment tensor inversion technique and reported that the majority of the events could be modeled by a double-couple mechanism resulting from frictional slips. The average displacement of each dynamic slip event (DSE) was reported to be about 50–150 μm occurring in about 3–5 ms with a source radii of about 3–6 mm. A few larger foreshocks $(M > -5.0)$ were not reported due to difficulty in analyzing them. In the PFC3D model, there are 98 smooth joint contacts forming the fault. The stick-slip behavior of a group of them during the test is shown in Fig. 10.

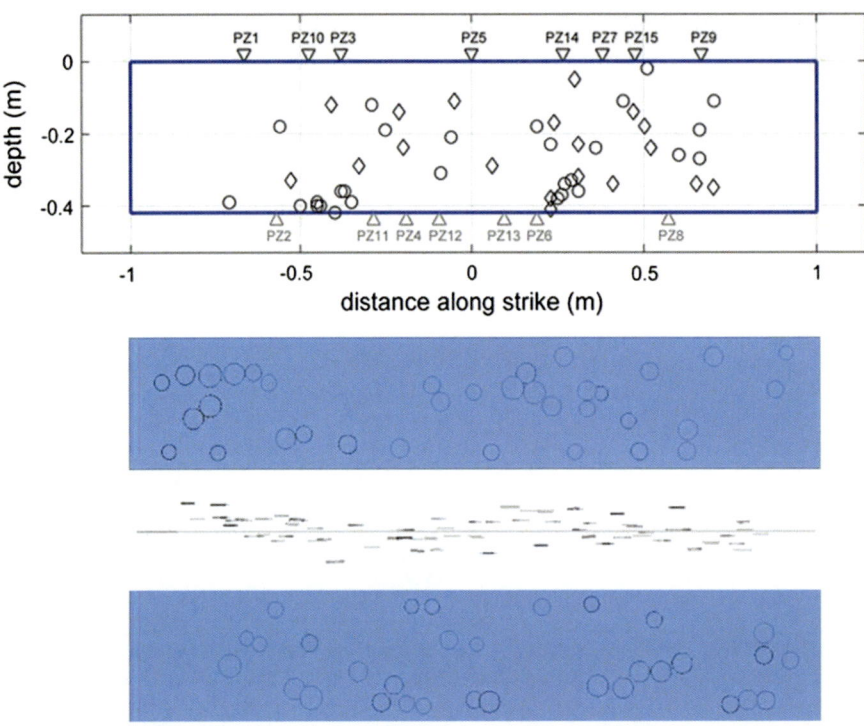

Figure 8

The *top figure* shows the location of events reported by (McLaskey *et al.* 2014). Foreshocks and aftershocks are shown by *circles* and *diamonds*, respectively. They have not reported the events near the fault ends. The *two figures* at the *bottom* with *blue background* are the location of slips in the PFC3D model at both sides of the fault. The *figure* between these two is a side view of the fault with the location of events around it. No clustering is shown in this figure

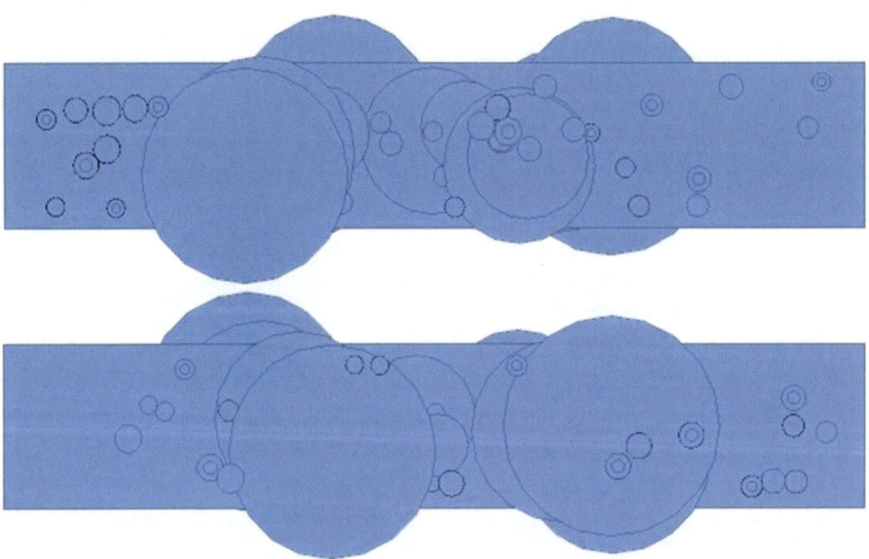

Figure 9

The clustered events in PFC3D model. A bigger radius shows a larger patch. *Concentric circles* represent consecutive slips on the same contact

Although a general trend of increase in the shear stress followed by stick slips is observed for the fault in Fig. 7, it can be observed from Fig. 10 that different contacts do not necessarily follow the same stick-slip pattern. In other words, for the small patches along the fault, not every drop in shear stress corresponds to a drop in the normal stress. This suggests the likelihood of the existence of another process responsible for generation of local stick-slips. It is in agreement with MCLASKEY and KILGORE (2013) that some aseismic slips can possibly trigger repeated earthquakes in their vicinity.

4.3. Parametric Study

A parametric study is presented in this section to investigate the effect of various parameters on the shear behavior of the fault as well as on generation of microseismic events.

4.3.1 Studying the Effect of Space Window's Size

A model with the same properties as mentioned in Table 1 is repeated with different space window sizes of 0.14, 0.28 and 0.42 m (equal to one, two, and three times the maximum diameter size of the balls,

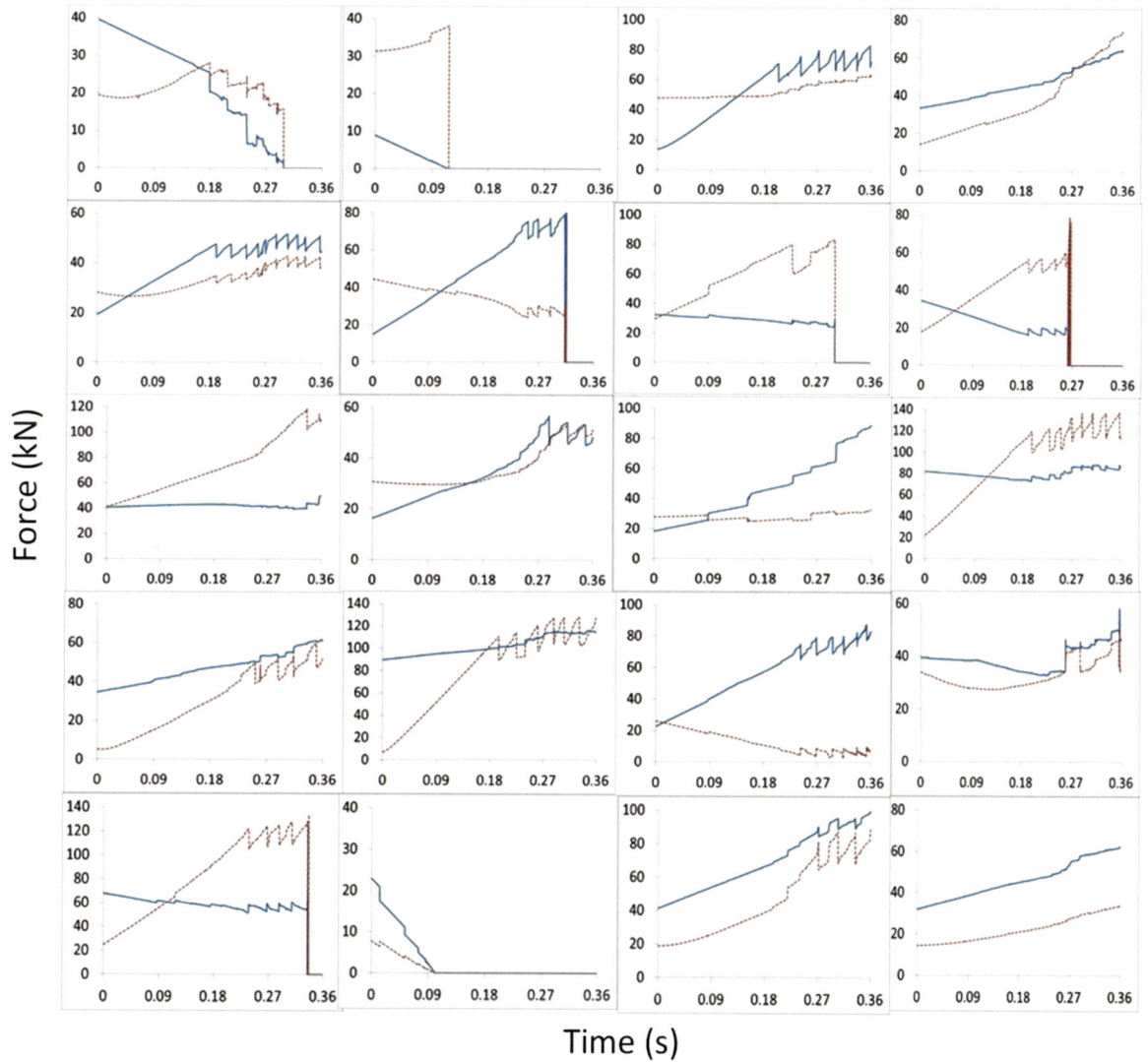

Figure 10
Variations of stick-slip behavior in different contacts along the fault. Normal and shear forces are shown with *solid blue* and *dashed red*, respectively

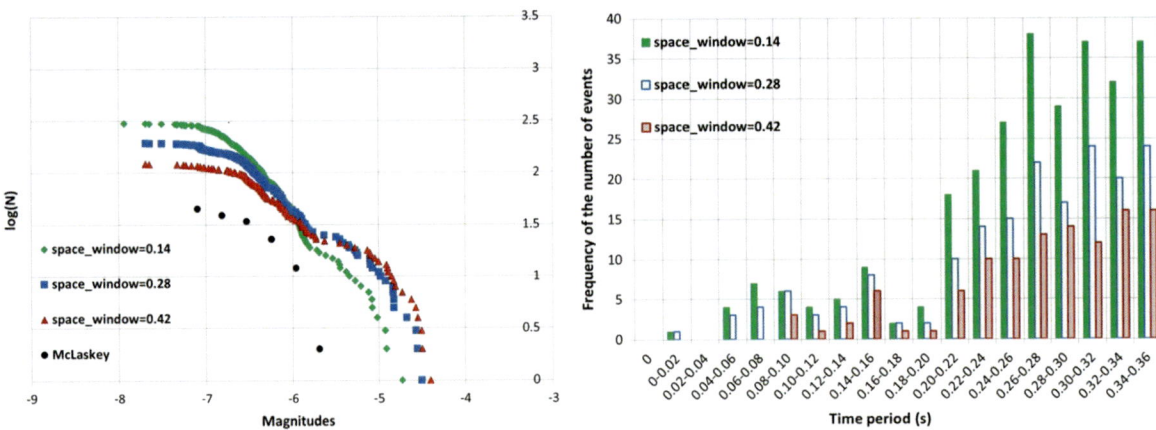

Figure 11
b-value plots as well as frequency of the number of events recorded during the tests for three different sizes of space windows

0.14 m). In order to calculate the magnitudes, the shear modulus of 2 MPa has been used for the slipping patch in all the tests. The *b*-value plots as well as variations in the number of events for the three tests are shown in Fig. 11.

As can be observed in this figure, an increase in the size of space window causes an increase in the appearance of magnitudes larger than −6 while for the smaller magnitudes, a reverse effect is observed. A comparison between the numerical *b*-value plots with the experiment is not accurate for three reasons: (a) McLaskey's data belong to the tests with 4 and 6 MPa normal stress (not 5 MPa). (b) The loading in this study was continuous, while in order to simulate the earthquake cycle, their experiments would include loading, resetting, and holding at four stages. (c) Some magnitudes greater than −5 have not been reported by them. However, for the sake of comparison and considering the *b*-values, regardless of *a*-values, the size of space window equal to three times the maximum diameter of the balls, 0.42 m, seems to provide the best match with reality, and therefore, it has been used for the other tests presented in the paper.

Also, a fewer number of clusters have been observed for greater sizes of space window which is as expected.

4.3.2 Studying the Effect of Normal Stress Threshold for Seismic Events

A model with the same properties as mentioned in Table 1 is repeated with three normal stress

thresholds of 0.0 MPa (i.e. no threshold), 0.1, and 3 MPa for the slips to be considered seismic (Fig. 12). As was mentioned in Sect. 4.1, although the coefficient of friction is decreased to 80 % of the static friction for the slips that pass the normal stress threshold, thus affecting the shear strength as well, no significant difference in shear-displacement and displacement–time plots was observed, and therefore, they are not included in the paper. However, as was expected, a lower threshold would result in a greater number of smaller events and also earlier appearance of events in the shear process. The *b*-values do not seem sensitive to this threshold.

In nature, the overall normal stress on the fault would depend on the in situ stresses as well as orientation of the fault, while for the local patches along the fault, heterogeneity would also play a role. Therefore, this threshold would be more important if faults with different resolutions and/or faults in different stress regimes were being compared. However, for the conditions tested in this section, the behavior of the model does not seem sensitive to the choice of this threshold, and thus 0.1 MPa has been used for the other tests presented in next sections.

4.3.3 Studying the Effect of Discretization (Size of Particles)

A model with the same properties as listed in Table 1 has been repeated with finer and coarser particles. The average radii of the particles in the three tests were 0.081 m (199 balls), 0.056 m (649 balls), and 0.029 m

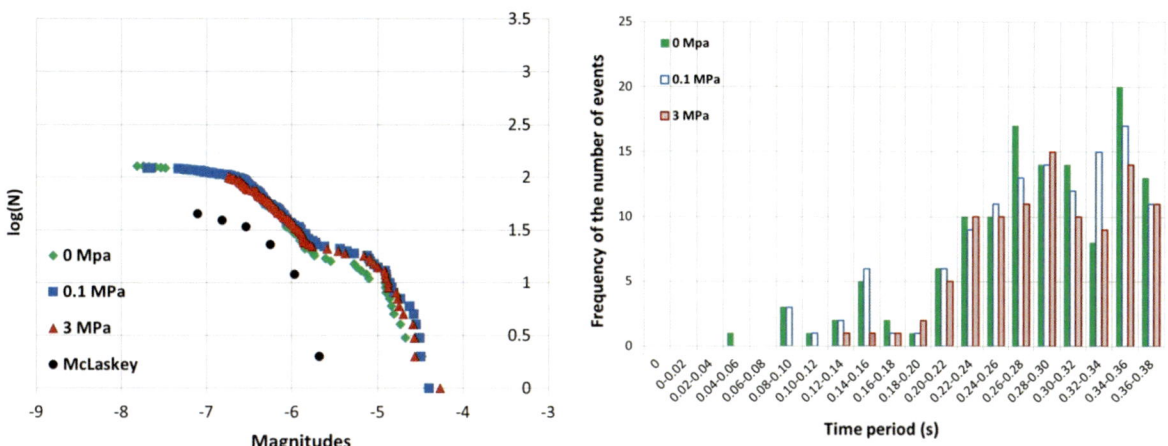

Figure 12
Variations of model behavior for different normal stress thresholds for the slips to be considered seismic

(5,164 balls). As shown in Fig. 13, coarser balls would result in higher strengths, greater stress drops during stick-slip, slower rate of displacements and smaller *b*-values. Although the events appear earlier in the finer model and the number of events are higher, the second stage in transitioning from elastic linear increase in shear strength to the final instability where the shear stress becomes constant, from 85 % to peak strength, is less obvious in the finer model. This could be observed both in the shear-displacement plots and in frequency of events versus time plots where the slope of getting to the peak number of events is very steep for the finest model (Fig. 13).

4.3.4 Studying the Effect of the Coefficient of Friction of the Fault (sj_fric)

A model with the same properties as mentioned in Table 1 is repeated with different coefficients of friction of the fault (1.05, 0.85 and 0.55) which is within the range of 0.6 to 1 as suggested by BYERLEE (1978). The results shown in Fig. 14 indicate a decrease in overall shear strength as well as an increase in the rate of displacements for smaller coefficients of friction. The number of magnitudes larger than −5.5 are not much sensitive to the fault's coefficient of friction while an increase in the frequency of events smaller than −5.5 is observed for smaller coefficients which is consistent with the greater number of events observed for these faults.

4.3.5 Studying the Effect of Particle Elasticity (ba_Ec)

A model with the same properties as mentioned in Table 1 is repeated with varying Young's moduli of the balls (2.1e9, 2.1e10, and 2.1e11). This would represent the stiffness of the medium surrounding the fault. As shown in Fig. 15, for the softest model (ba_Ec = 2.1e9), small stick-slip instability was observed as early as 0.1 mm displacements were reached. This behavior continued until about 7 mm displacements (at 3.7 MPa shear stress) that could be considered the onset of the second stage. The peak shear strength, 4.4 MPa, was reached after 9.6 mm displacements and 1.9 s after the start of the test. This is much longer compared to previous cases where the maximum shear strength was reached in a fraction of a second. The overall pattern of shear-displacement for the fault with the softest surrounding (i.e. ba_Ec = 2.1e9) included three stages in shear process similar to the model with Young's modulus of balls equal to 2.1e10, but only the portion until 2 mm displacement is shown in Fig. 15.

For the model with the surrounding region harder than the fault itself, a softening behavior is observed with almost no second stage in the shear process. With regard to the number of events, a softer surrounding has not resulted in more emissions for the same amount of displacement, and its only contribution has been to delay the process of displacements. An overall shift in the *a*-values

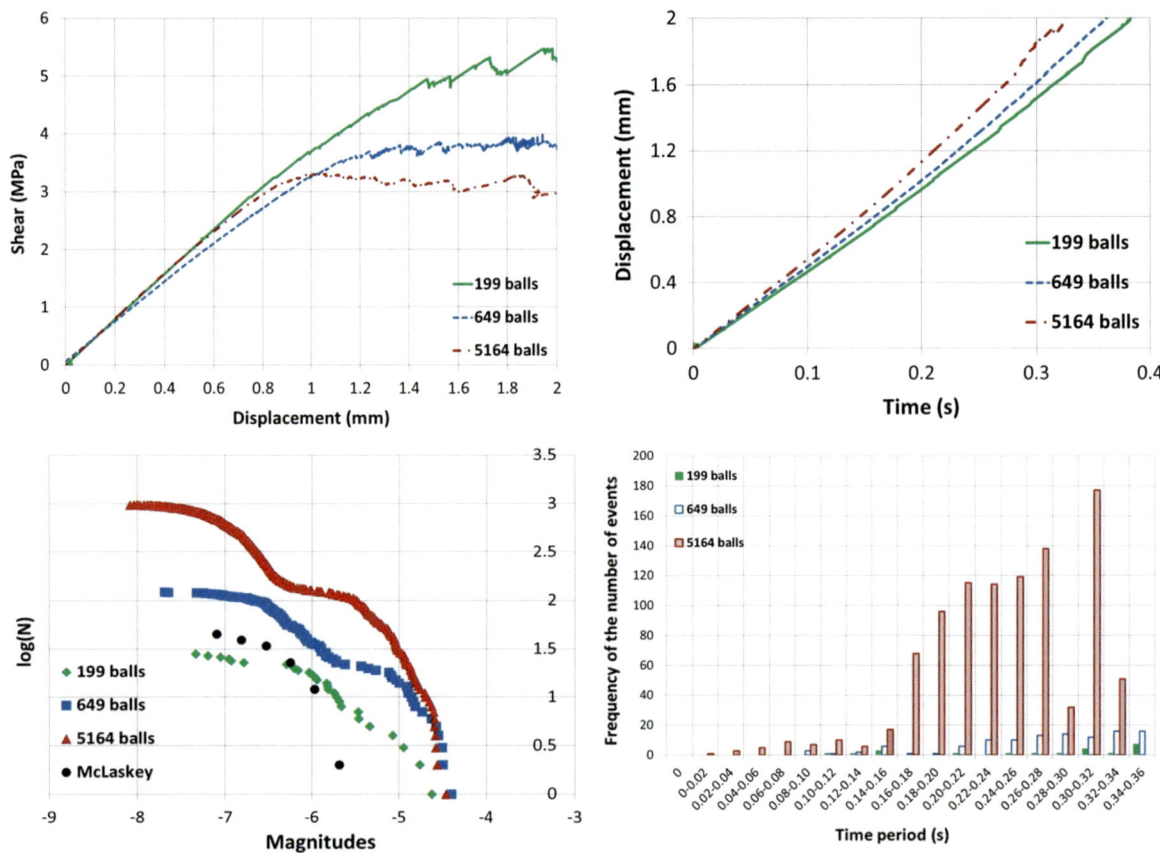

Figure 13
Variations of the model's behavior for different resolutions

towards larger magnitudes is also observed for the faults with softer surrounding while the *b*-values are not much sensitive.

4.3.6 Studying the Effect of the Fault's Elasticity (sj_kn and sj_ks)

A model with the same properties as mentioned in Table 1 is repeated with varying elastic properties of the smooth joint. As shown in Fig. 16, higher shear strengths are observed for harder faults. The rate of deformations and *b*-value plots, however, do not seem to be much affected. A lower number of events is observed for softer faults.

4.3.7 Studying the Effect of Normal Stress

A model with the same properties as mentioned in Table 1 is repeated with normal stresses along the

fault equal to 3, 5, and 7 MPa. The plots in Fig. 17 show an increase in the shear strength as well as a decrease in the rate of deformations for higher normal stresses. Also, for the same amount of deformation, higher normal stresses generate fewer emissions. An overall decrease in the *b*-value can be observed for higher normal stresses as well.

5. Discussion

As expected, the behavior of the numerical model was sensitive to the elastic properties of the medium surrounding the fault as well as the fault's properties and the size of particles. It is in agreement with the fundamental mechanism shown in Fig. 1. Sensitivity of the numerical model to the frictional parameters is also in agreement with the observation of MᴄLᴀsᴋᴇʏ *et al.* (2014) who believed that the recorded events

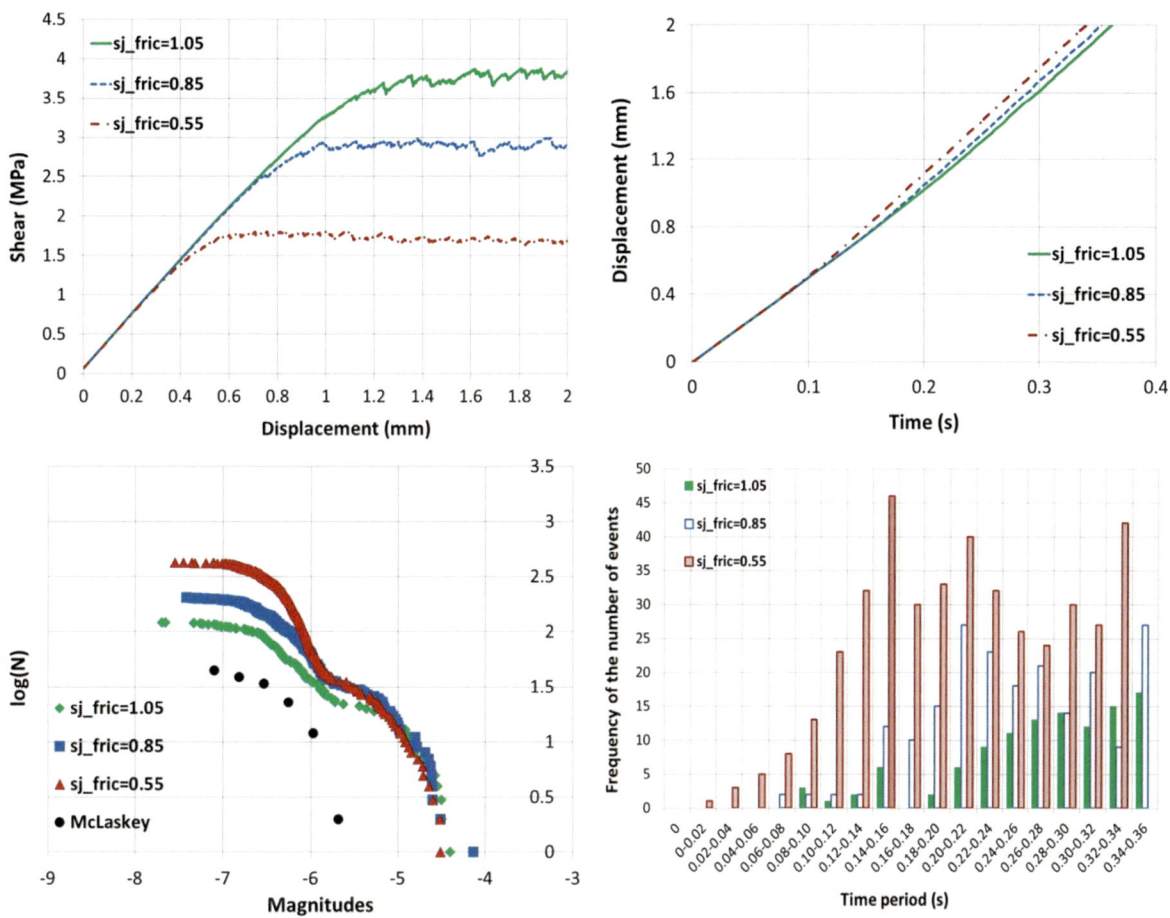

Figure 14
Variations of the model's behavior for different coefficients of friction

were not due to factors such as grain crushing or fracturing of the fresh rock. The majority of events were believed to have a double-couple mechanism indicating a shear dislocation along the fault. The ones with a non-double-couple mechanism were expected not to exceed 20 % of all the events (MCLASKEY *et al.* 2014).

They observed no specific difference between the focal mechanisms and magnitudes or even stress drop of the foreshocks and aftershocks. Both types of events were broadly distributed along the fault. This is consistent with the results of numerical modeling presented in this study. The results show that largest magnitudes appear mostly in the last stage of unstable stick-slip once the peak shear strength of the fault is already reached.

According to the model shown in Fig. 1 (left), whether or not a slip is seismic (or aseismic) would

depend on the elasticity of the medium (i.e. stiffness of spring in this model) as well as the amount of stress drop due to unloading the normal stress or frictional properties of the surface. In Fig. 15, it was shown that stick-slips would occur much earlier in the shear process for the fault with softer surrounding. This is in agreement with the fact that low stiffness facilitates unstable slips (DIETERICH 1979). Also, it has been suggested that local stick-slips could trigger or facilitate other slips in their vicinity (MCLASKEY and KILGORE 2013). A possible mechanism for this could be the fact that local slips (or in other words, local unloading normal stress conditions) provide a softened surrounding for their neighbor patches and thus, similar to what was shown in Fig. 15, facilitating their instability. The amount of how much of either seismic or aseismic slips would contribute to this could be the subject of another study.

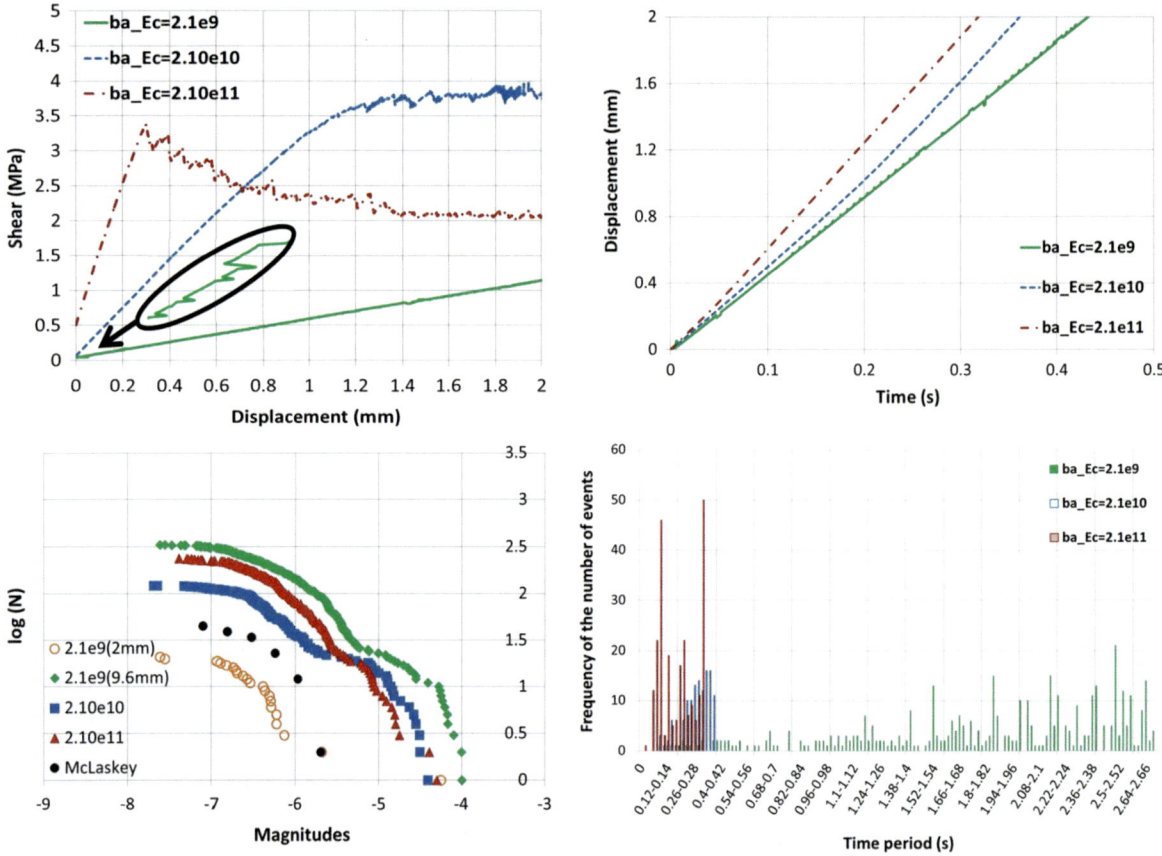

Figure 15
Variations of the model's behavior for different Young's moduli of the particles. Distribution of magnitudes for the softest model are plotted based on the events recorded until 2 and 9.6 mm displacements

6. Conclusion

This paper presented applicability of the discrete element modeling to reproduce stick-slips successfully on a large scale, which takes a lot of time and effort in laboratory to be studied. Once the model is calibrated with one set of experiments, it can be used to expand our knowledge to the cases that cannot be easily tested.

In this research, the microseismic release along preexisting faults during shearing was studied numerically. For this purpose, the experiment reported by McLaskey and Kilgore (2013) was modeled using PFC3D. PFC inherently allows modeling of stick-slips; however, an algorithm is also developed in this research to record the slip-induced acoustic emissions. Some advantages of the present model are as follows: (1) The three-dimensional and discrete

nature of model allows taking into account the geometrical heterogeneity and make the simulations more realistic. (2) Focusing on the fault's behavior by use of a smooth joint model allows eliminating the bond breakages affecting the results. (3) The algorithm developed for recording the stick-slip induced microseismic events.

Once the results compared well with real laboratory data, the model was then expanded upon to study the difference in shear and microseismic behavior in faults with different properties. A summary of the results is presented here:

1. A decrease in shear strength was observed for the models with smaller particles, smaller coefficient of friction of the fault, harder surroundings of the fault (higher Young's modulus for the particles), softer faults (smaller elasticity moduli of the fault), and

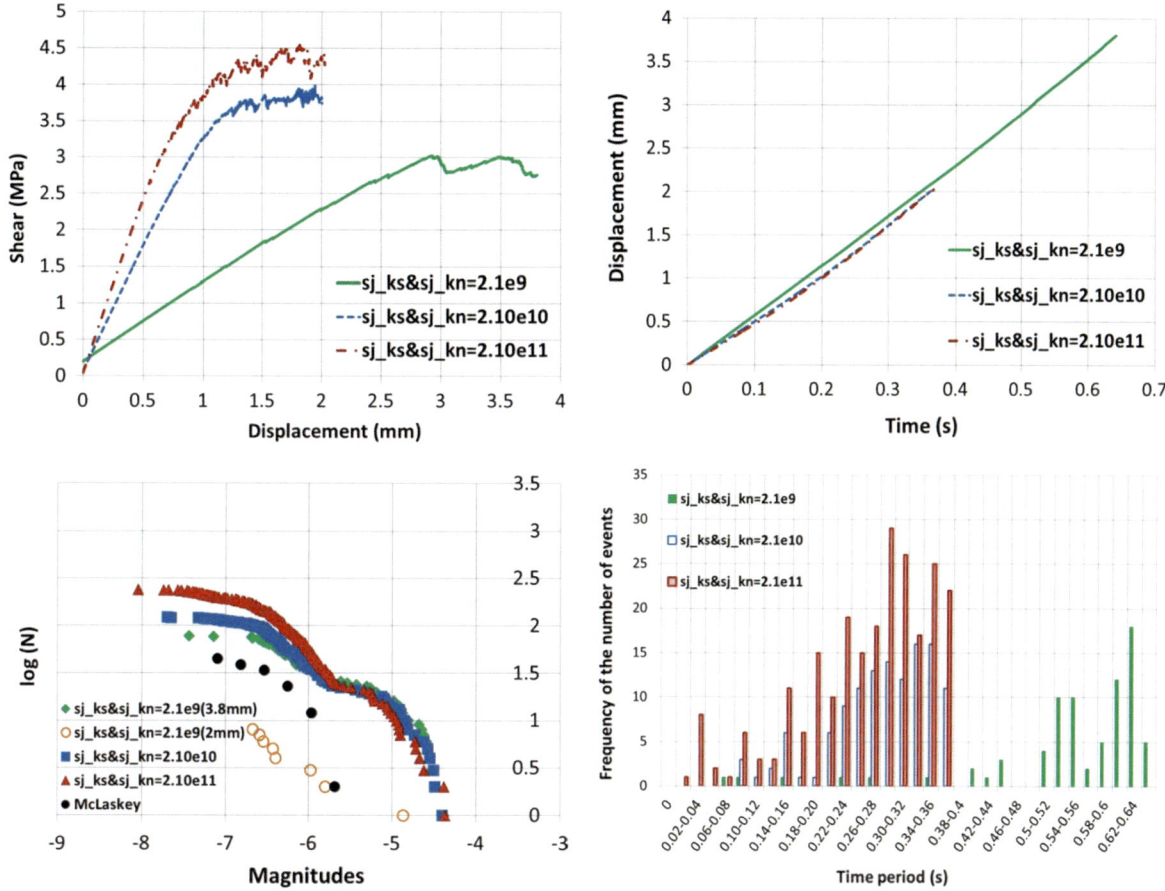

Figure 16
Variations of the model's behavior for different elastic properties of the fault. Distribution of magnitudes for the model with softest fault properties are plotted based on the events recorded until 2 and 3.8 mm displacements

smaller normal stress on the fault. Also, a softer behavior (i.e. decrease in the initial slope of shear-displacement curve) was observed for the softer faults as well as faults with softer surrounding.

2. A higher rate of displacements was observed for the faults with finer particles, smaller coefficient of friction, harder surrounding, and smaller normal stress. The fault's elastic properties did not seem to have much effect on the rate of displacements; however, a small increase in this rate was observed for the softest fault. A comparison between displacement–time and shear-displacement curves suggests that higher rate of displacements are observed for weaker conditions (i.e. conditions resulting in smaller peak strength) while this conclusion seems less obvious for the softer faults (Fig. 16).

3. With regard to the magnitudes, a greater size of space window resulted in an increase in the events larger than −6 while a reverse effect was observed for the smaller events. The b-values were not sensitive to the normal stress threshold of slips to be seismic. Increasing the resolution or decreasing the normal stress on the fault both caused an increase in the b-values. Decreasing the coefficient of friction did not have much effect on the magnitudes larger than −5.5 while for the smaller magnitudes, a decrease in this coefficient caused an increase in the number of events. Increasing the particles' elasticity caused a decrease in the a-values, but the b-values were not much affected. The b-values were not sensitive to the elastic properties of the fault.

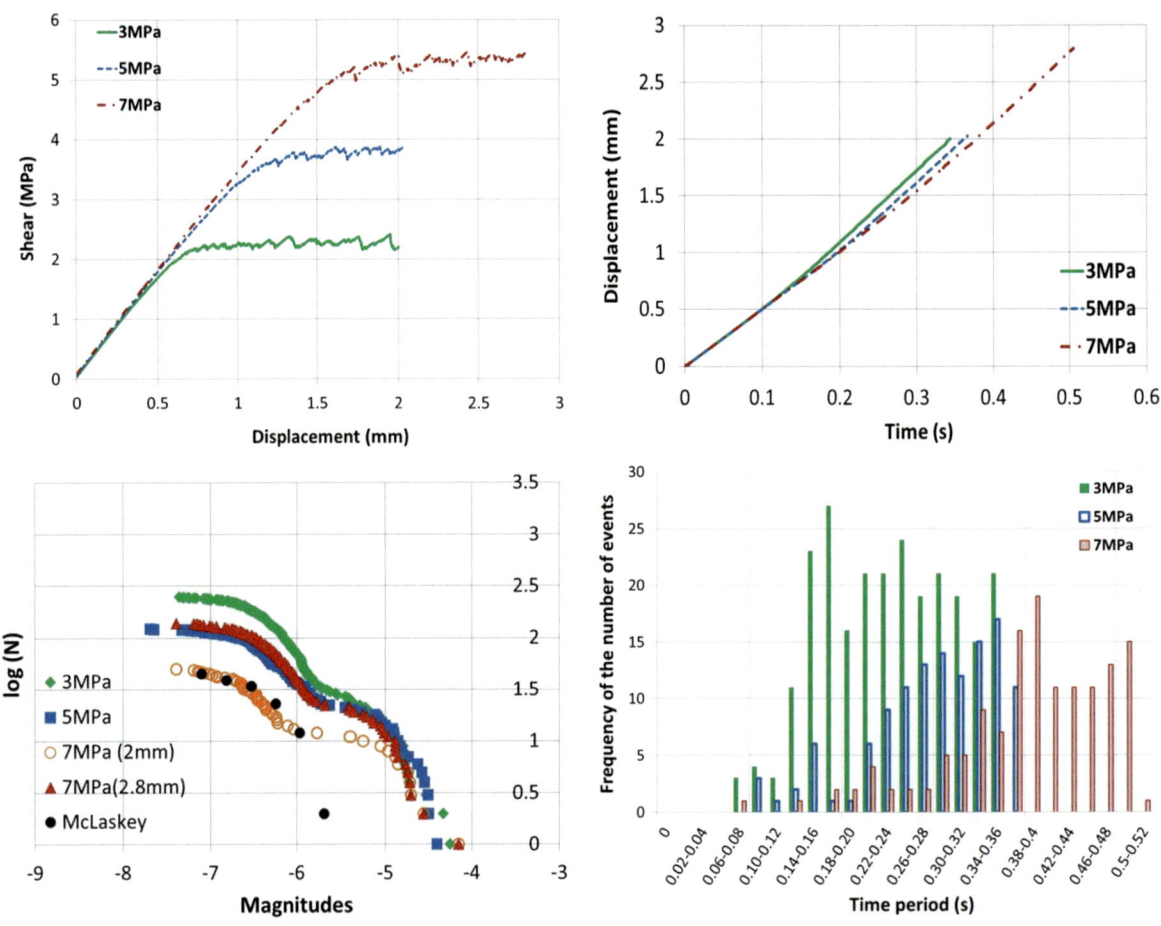

Figure 17
Variations of the model's behavior for different normal stresses. Distribution of magnitudes for the model with 7 MPa normal stress are plotted based on the events recorded until 2 and 2.8 mm displacements

4. A larger number of recorded events were observed for the models with smaller size of space window, smaller normal stress threshold, finer particles, smaller coefficient of friction, harder fault surrounding region, harder fault, and smaller normal stress on the fault. An obvious observation is that for the same amount of slip, the emissivity would depend on several factors affecting the release of microseismic energy. It was also suggested that there are three stages in the slip behavior of a fault: (1) linear increase in the shear stress until 85 % of the peak strength with no or very few stick-slips, (2) stable slip from 85 % of the peak shear strength until the maximum shear strength is reached, (3) unstable continuation of slip until for some reason the fault stops. An exception was

observed for the case with the fault's surrounding being harder than the fault itself (Fig. 15), where right after the first elastic stage in the shear process, the third stage started with a post-peak softening pattern. However, among these three stages, the last one which is unstable has the greatest number of stick-slips, and therefore, a comparison between the number of events versus time and shear versus displacement plots suggests that for the same amount of displacement, the conditions at which the third stage is reached earlier would be the most emissive ones.

The results suggest that in reality it is quite possible for the two ends of a fault to be still while there are patches along the fault undergoing stick-slips;

however, due to the geometry of the fault and loading scheme used in this research it was not investigated. Also, local stick-slips seem to provide a softer surrounding for their neighbor patches favoring their subsequent stick-slips.

With regard to the calibration of model for stick-slip behavior, the onset of instability (peak shear strength) as well as b-value seems to be the two most controlling parameters that need to be taken into account.

Acknowledgments

The authors wish to thank Dr. Sacha Emam for his guidance with PFC5.0. Also, the comments provided by Dr. Greg McLaksey about the details of his experiments as well as the comments and suggestions by an anonymous reviewer are gratefully acknowledged.

REFERENCES

BRACE, W. F., and BYERLEE, J. (1966), *Stick-slip as a mechanism for earthquakes*. Science, *153*(3739), 990–992.

BRACE, W. F., and BYERLEE, J. (1970), *California earthquakes: why only shallow focus?* Science (New York, N.Y.), *168*(3939), 1573–1575. doi:10.1126/science.168.3939.1573.

BRUNE, J. N., BROWN, S., and JOHNSON, P. A. (1993), *Rupture mechanism and interface separation in foam rubber models of earthquakes: a possible solution to the heat flow paradox and the paradox of large overthrusts*. Tectonophysics, *218*(1–3), 59–67. doi:10.1016/0040-1951(93)90259-M.

BYERLEE, J. (1978), *Friction of rocks*. Pure and Applied Geophysics PAGEOPH, *116*(4–5), 615–626. doi:10.1007/BF00876528.

BYERLEE, J., and BRACE, W. F. (1968), *Stick slip, stable sliding, and earthquakes—effect of rock type, pressure, strain rate, and stiffness*. Journal of Geophysical Research, *73*(18), 6031–6037.

DALGUER, L. A., and DAY, S. M. (2006), *Comparison of fault representation methods in finite difference simulations of dynamic rupture*. Bulletin of the Seismological Society of America, *96*(5), 1764–1778.

DAMJANAC, B., GIL, I., PIERCE, M., SANCHEZ, M., VAN AS, A., and McLENNAN, J. (2010), A New Approach to Hydraulic Fracturing Modeling In Naturally Fractured Reservoirs. *44th U.S. Rock Mechanics Symposium and 5th U.S.-Canada Rock Mechanics Symposium*. Retrieved from https://www.onepetro.org/conference-paper/ARMA-10-400.

DAY, S. M., DALGUER, L. A., LAPUSTA, N., and LIU, Y. (2005), *Comparison of finite difference and boundary integral solutions to three-dimensional spontaneous rupture*. Journal of Geophysical Research: Solid Earth (1978–2012), *110*(B12).

DIETERICH, J. H. (1979), *Modeling of rock friction: 1. Experimental results and constitutive equations*. Journal of Geophysical Research, *84*(B5), 2161. doi:10.1029/JB084iB05p02161.

DIETERICH, J. H. (1981), *Potential for geophysical experiments in large scale tests*. Geophysical Research Letters, *8*(7), 653–656. doi:10.1029/GL008i007p00653.

FAIRHURST, C. (2013), Fractures and Fracturing: Hydraulic Fracturing in Jointed Rock. *ISRM International Conference for Effective and Sustainable Hydraulic Fracturing*. Retrieved from https://www.onepetro.org/conference-paper/ISRM-ICHF-2013-012.

FINCH, E., HARDY, S., and GAWTHORPE, R. (2003), *Discrete element modelling of contractional fault-propagation folding above rigid basement fault blocks*. Journal of Structural Geology, *25*(4), 515–528. doi:10.1016/S0191-8141(02)00053-6.

GALIS, M., MOCZO, P., and KRISTEK, J. (2008), *A 3-D hybrid finite-difference—finite-element viscoelastic modelling of seismic wave motion*. Geophysical Journal International, *175*(1), 153–184.

HAZZARD, J. (1998), *Numerical modelling of acoustic emissions and dynamic rock behaviour. [electronic resource]*. University of Keele.

HAZZARD, J., COLLINS, D. S., PETTITT, W. S., and YOUNG, R. P. (2002), Simulation of unstable fault slip in granite using a bonded-particle model. In *The Mechanism of Induced Seismicity* (pp. 221–245). Springer.

HAZZARD, J., and PETTITT, W. S. (2013), Advances in Numerical Modeling of Microseismicity. In *47th US Rock Mechanics/Geomechanics Symposium*. American Rock Mechanics Association.

HAZZARD, J., and YOUNG, R. P. (2000), *Simulating acoustic emissions in bonded-particle models of rock*. International Journal of Rock Mechanics and Mining Sciences, *37*(5), 867–872. Retrieved from http://cat.inist.fr/?aModele=afficheN&cpsidt=1419981.

HAZZARD, J., and YOUNG, R. P. (2002), *Moment tensors and micromechanical models*. Tectonophysics, *356*(1), 181–197.

HAZZARD, J., and YOUNG, R. P. (2004), *Dynamic modelling of induced seismicity*. International Journal of Rock Mechanics and Mining Sciences, *41*(8), 1365–1376.

ITASCA, C. G. (1999). PFC 3D-User manual. *Itasca Consulting Group, Minneapolis*.

JANSEN, F. (2006), Numerical Simulation of Stick-slip Processes with Application to Seismology and Rock Glacier Dynamics.

JULIAN, B. R., MILLER, A. D., and FOULGER, G. R. (1998), *Non-double-couple earthquakes 1. Theory*. Reviews of Geophysics, *36*(4), 525. doi:10.1029/98RG00716.

KRISTIANSEN, T. G., BARKVED, O., and PATTILLO, P. D. (2000), Use of passive seismic monitoring in well and casing design in the compacting and subsiding Valhall field, North Sea. In *EURO-PEC: European petroleum conference* (pp. 231–241).

MAS IVARS, D. (2008), Bonded-Particle Model for the Deformation, Yield and Failure of Jointed Rock Masses. In *PhD Thesis*. Stockholm, Sweden.

MAS IVARS, D., POTYONDY, D. O., PIERCE, M., and CUNDALL, P. A. (2008), *The smooth-joint contact model*. Proceedings of WCCM8-ECCOMAS.

McGARR, A. (1994), *Some comparisons between mining-induced and laboratory earthquakes*. Pure and Applied Geophysics PAGEOPH, *142*(3–4), 467–489. doi:10.1007/BF00876051.

McLASKEY, G. C., and KILGORE, B. D. (2013), *Foreshocks during the nucleation of stick-slip instability*. Journal of Geophysical Research: Solid Earth, *118*(6), 2982–2997. doi:10.1002/jgrb.50232.

McLASKEY, G. C., KILGORE, B. D., LOCKNER, D. A., and BEELER, N. M. (2014), *Laboratory Generated M -6 Earthquakes*. Pure and Applied Geophysics. doi:10.1007/s00024-013-0772-9.

McLaskey, G. C., Thomas, A. M., Glaser, S. D., and Nadeau, R. M. (2012), *Fault healing promotes high-frequency earthquakes in laboratory experiments and on natural faults*. Nature, *491*(7422), 101–104. Retrieved from http://dx.doi.org/10.1038/nature11512.

Mora, P., and Place, D. (1994), *Simulation of the frictional stick-slip instability*. Pure and Applied Geophysics, *143*(1–3), 61–87.

Morgan, J. K. (2004), *Particle Dynamics Simulations of Rate- and State-dependent Frictional Sliding of Granular Fault Gouge*. Pure and Applied Geophysics, *161*(9–10). doi:10.1007/s00024-004-2537-y.

Nielsen, S., Carlson, J. M., and Olsen, K. B. (2000), *Influence of friction and fault geometry on earthquake rupture*. Journal of Geophysical Research, *105*(B3), 6069. doi:10.1029/1999JB900350.

Nielsen, S., Knopoff, L., and Tarantola, A. (1995), *Model of earthquake recurrence: role of elastic wave radiation, relaxation of friction, and inhomogeneity*. Journal of Geophysical Research: Solid Earth (1978–2012), *100*(B7), 12423–12430.

Ohnaka, M. (1973), *Experimental studies of stick-slip and their application to the earthquake source mechanism*. Journal of Physics of the Earth, *21*(3), 285–303.

Peng, Z., and Gomberg, J. (2010), *An integrated perspective of the continuum between earthquakes and slow-slip phenomena*. Nature Geoscience, *3*(9), 599–607. doi:10.1038/ngeo940.

Persson, B. N. J., and Tosatti, E. (1999), *Theory of friction: elastic coherence length and earthquake dynamics*. Solid State Communications, *109*(12), 739–744.

Place, D., Lombard, F., Mora, P., and Abe, S. (2002), Simulation of the micro-physics of rocks using LSMearth. In *Earthquake Processes: Physical Modelling, Numerical Simulation and Data Analysis Part I* (pp. 1911–1932). Springer.

Potyondy, D. O., and Cundall, P. A. (2004), *A bonded-particle model for rock*. International Journal of Rock Mechanics and Mining Sciences, *41*(8), 1329–1364. doi:10.1016/j.ijrmms.2004.09.011.

Rabinowicz, E. (1951), *The Nature of the Static and Kinetic Coefficients of Friction*. Journal of Applied Physics, *22*(11), 1373. doi:10.1063/1.1699869.

Rabinowicz, E. (1958), *The Intrinsic Variables affecting the Stick-Slip Process*. Proceedings of the Physical Society, *71*(4), 668–675. doi:10.1088/0370-1328/71/4/316.

Roux, P.-F., Marsan, D., Métaxian, J.-P., O'Brien, G., and Moreau, L. (2008), *Microseismic activity within a serac zone in an alpine glacier (Glacier d'Argentière, Mont Blanc, France)*. Journal of Glaciology, *54*(184), 157–168. doi:10.3189/002214308784409053.

Scholz, C. H. (2002), *The mechanics of earthquakes and faulting*. Cambridge university press.

Xing, H. L., Mora, P., and Makinouchi, A. (2004), Finite element analysis of fault bend influence on stick-slip instability along an intra-plate fault. In *Computational Earthquake Science Part I* (pp. 2091–2102). Springer.

Young, R. P., Collins, D. S., Hazzard, J., Pettitt, W., and Baker, C. (2001), *Use of acoustic emission and velocity methods for validation of micromechanical models at the URL* (p. 153). Toronto, Ont.: Ontario Power Generation, Nuclear Waste Management Division.

Young, R. P., and Dedecker, F. (2005), Seismic Validation of 3-D Thermo-Mechanical Models for the Prediction of the Rock Damage around Radioactive Waste Packages in Geological Repositories-SAFETI. *Final Report, European Commission Nuclear Science and Technology*.

Zhao, X., and Young, R. P. (2011), *Numerical modeling of seismicity induced by fluid injection in naturally fractured reservoirs*. Geophysics, *76*(6), WC167–WC180.

(Received May 27, 2014, revised December 31, 2014, accepted January 9, 2015, Published online February 1, 2015)

Pure Appl. Geophys. 173 (2016), 795–803
© 2015 Springer Basel
DOI 10.1007/s00024-015-1053-6

Pure and Applied Geophysics

Step-Edge Detection Filters for the Interpretation of Potential Field Data

Guoqing Ma,[1] Danian Huang,[1] and Cai Liu[1]

Abstract—Edge detection is a useful tool in the interpretation of potential field data, and the existing edge detection filters are almost functions of first-order horizontal and vertical derivatives. We propose step-edge detection filters to improve the resolution of edge detection results, which use the functions of different-order derivatives to accomplish the edge detection task. We demonstrate the proposed filters on synthetic potential field data, and the results show that the new methods can recognize the edges of the sources more precisely and clearly. We also discuss the application effect of different step-edge detection filters. Lastly, we apply the proposed filters to real potential field data, and the recognized edges of the stratigraphic markers are more precise and clear.

Key words: Step, edge detection, potential field, derivative.

1. Introduction

Edge detection of potential field data has been widely used as a significant tool of geophysical exploration technologies, which can delineate the horizontal locations of causative sources, and we can formulate the next plan based on the recognition results. Normally, various high-pass filters, such as horizontal derivative, vertical derivative and downward continuation, and so on, are used to recognize the edges of potential field data (Evjen 1936; Cordell 1979; Cordell and Grauch 1985; Hood and Teskey 1989; Roest *et al.* 1992; Hsu *et al.* 1996; Fedi and Florio 2001), but the edges of deeper sources recognized by them are fuzzy. In recent years, there have been many endeavors in the development of balanced edge detection filters, based on the ratio of first-order horizontal to vertical derivatives that can display the edges of shallow and deep bodies simultaneously, which is beneficial for finding weak anomalies among

strong anomalies. Miller and Singh (1994) proposed the tilt angle filter, which is the ratio of the vertical derivative to the horizontal derivative. Some authors also presented other formats of balanced edge detection filters (Verduzco *et al.* 2004; Wijns *et al.* 2005; Cooper and Cowan 2006, 2008; Ma and Li 2012; Ma 2013).

In our research, we find that the maxima of the total horizontal derivative and the zero of the vertical derivative correspond to the edges of the vertical fault, but the horizontal locations of the maxima of the total horizontal derivative and the zero of the vertical derivative are both bigger than the edges of the prism. We know that the shape of real geology models are closer to that of a prism, so the edges recognized by the existing balanced edge detection filters are bigger that the true horizontal locations. We prove that higher-order vertical derivatives can obtain more accurate edges, but the ratio of same higher-order derivatives will produce additional edges, so we present step-edge detection filters that use combinations of different-order derivatives to finish the edge detection task. We demonstrate the step-edge detection filters on synthetic and real potential field data, and the recognized edges are more precise and clear.

2. Step-Edge Detection Filters

The Theta map filter (Wijns *et al.* 2005) and the normalized horizontal tilt angle (TDX) filter (Cooper and Cowan 2006) are two common edge detection filters, and the expression of Theta map is

$$\text{Theta} = \frac{\sqrt{\left(\frac{\partial f}{\partial x}\right)^2 + \left(\frac{\partial f}{\partial y}\right)^2}}{\sqrt{\left(\frac{\partial f}{\partial x}\right)^2 + \left(\frac{\partial f}{\partial y}\right)^2 + \left(\frac{\partial f}{\partial z}\right)^2}} \quad (1)$$

[1] College of Geoexploration Science and Technology, Jilin University, Changchun, China. E-mail: maguoqing@jlu.edu.cn

where f represents the original potential field data. The normalized horizontal tilt angle (TDX) filter can be expressed as

$$TDX = \tan^{-1}\left(\frac{\sqrt{(\partial f/\partial x)^2 + (\partial f/\partial y)^2}}{|\partial f/\partial z|}\right) \quad (2)$$

The maxima of the above edge detection filters correspond to the edges of the sources (COOPER and COWAN 2006, 2008), and we can see that the Theta map and TDX filters get maxima when the first-order vertical derivative is zero. The expression of the first-order vertical derivative of a single vertical fault (BLAKELY 1995) is

$$\frac{\partial f}{\partial z} = 2G\rho\left(\text{atan}\frac{x}{h} - \text{atan}\frac{x}{H}\right) \quad (3)$$

where G is the gravitation constant, ρ is the residual density of the fault, and h and H are the top and bottom depths of the fault, respectively, and we can see that the zero of first-order vertical derivative corresponds to the edges of the fault. However, the fault is not a very common model in real geophysical exploration, while the prism is more common.

The first-order vertical derivative of a two-dimensional (2D) vertical prism can be expressed as

$$\frac{\partial f}{\partial z} = 2G\rho\left(\tan^{-1}\frac{H}{x+a} - \tan^{-1}\frac{h}{x+a}\right.$$
$$\left. - \tan^{-1}\frac{H}{x-a} + \tan^{-1}\frac{h}{x-a}\right) \quad (4)$$

where $2a$ is the horizontal width of the prism, and h and H are the top and bottom depths of the prism, respectively. The horizontal coordinates at which the first-order vertical derivative obtains zero are

$$x_{0,1} = \pm\sqrt{a^2 + Hh} \quad (5)$$

We can see that the horizontal coordinates are bigger than the true edges, so the edges recognized by the Theta map filter and the *TDX* filter are not very correct.

The first-order horizontal derivative of the 2D vertical prism is expressed as

$$\frac{\partial f}{\partial x} = G\rho\ln\frac{\left[(x+a)^2+H^2\right]\left[(x-a)^2+h^2\right]}{\left[(x+a)^2+h^2\right]\left[(x-a)^2+h^2\right]} \quad (6)$$

The horizontal coordinates at which the first-order horizontal derivative of potential field data obtains maxima are

$$x_m = \pm\sqrt{\frac{1}{6}\left\{\sqrt{[4a^2+(h^2+H^2)]^2+12h^2H^2} + 2a^2 - (h^2+H^2)\right\}} \quad (7)$$

The horizontal coordinates of the maxima of the first-order horizontal derivative are also bigger than the true edges, but are smaller than the zero coordinates of the first-order vertical derivative, so we use the maxima of the horizontal to recognize source edges that can obtain more accurate edges, but the total horizontal derivative cannot display the edges of the deeper bodies clearly.

The expression of the second-order vertical derivative of the 2D vertical prism is

$$\frac{\partial^2 f}{\partial z^2} = 2G\rho\left[\frac{x+a}{(x+a)^2+h^2} - \frac{x+a}{(x+a)^2+H^2}\right.$$
$$\left. - \frac{x-a}{(x-a)^2+h^2} + \frac{x-a}{(x-a)^2+H^2}\right] \quad (8)$$

The horizontal coordinates at which the second-order vertical derivative obtains zero are

$$x_{0,2} = \pm\sqrt{\frac{1}{6}\left\{\sqrt{[4a^2+(h^2+H^2)]^2+12h^2H^2} + 2a^2 - (h^2+H^2)\right\}} \quad (9)$$

The horizontal coordinates of the zero of the second-order vertical derivative are the same as the coordinates of the maxima of the total horizontal derivative. We initially considered using the ratio of the second-order derivatives to recognize the edges of the sources, and the second-order normalized horizontal tilt angle (TDX$_2$) filter can be given by

$$TDX_2 = \tan^{-1}\left(\frac{\sqrt{\left(\frac{\partial^2 f}{\partial z \partial x}\right)^2 + \left(\frac{\partial^2 f}{\partial z \partial y}\right)^2}}{\left|\frac{\partial^2 f}{\partial z^2}\right|}\right) \quad (10)$$

The maxima of the TDX$_2$ filter can be used to recognize the edges. In order to low the interference of noise, we use the following Laplace equation to compute the second order vertical derivative.

$$\frac{\partial^2 f}{\partial z^2} = -\left(\frac{\partial^2 f}{\partial x^2} + \frac{\partial^2 f}{\partial y^2}\right) \tag{11}$$

The second-order horizontal derivative can be computed in the space domain, which can lower the noise effect. Figure 1a shows the synthetic gravity anomaly of two prisms with depths of 15 and 20 m, and the white dotted lines represent the true edges. Figure 1b and c show the Theta map and TDX of the data in Fig. 1a. We can see that the recognized edges are significantly larger than the true edges. Figure 1d shows the second-order normalized horizontal tilt angle (TDX_2) of the data in Fig. 1a, computed using Eq. 10. Depending on the results, we can see that the edges recognized by the second-order normalized horizontal tilt angle filter produce low values surrounding the maximum values, and this will lead to a complicated interpretation.

We suggest using the ratio of the total horizontal derivative to the second-order vertical derivative to recognize the source edges depending on the distribution features of the different-order horizontal and vertical derivatives; this method is called the step-edge detection filter. The second-order vertical derivative is computed by Eq. 11, and the forms of the first-order step-edge detection (FSTE) filter are

$$FSTE_1 = \tan^{-1}\left(\frac{\sqrt{(\partial f/\partial x)^2 + (\partial f/\partial y)^2}}{k \times |\partial^2 f/\partial x^2 + \partial^2 f/\partial y^2|}\right) \tag{12}$$

$$FSTE_2 = \Re\left(\text{acos}\left[\frac{\left|\frac{\partial^2 f}{\partial z^2}\right| \cdot k}{\sqrt{\left(\frac{\partial f}{\partial x}\right)^2 + \left(\frac{\partial f}{\partial y}\right)^2}}\right]\right) \tag{13}$$

where $\partial f/\partial x$, $\partial f/\partial y$ and $\partial^2 f/\partial z^2$ are the derivatives of the data f, and k is a constant value, to make the equations have mathematical significance.

We also use the ratio of the second-order horizontal derivative to the third-order horizontal derivative to recognize the source edges, which is called a second-order step-edge detection (SSTE) filter. The expressions can be given by

$$SSTE_1 = \text{atan}\left[\frac{\sqrt{\left(\frac{\partial}{\partial x}\left(\frac{\partial^2 f}{\partial z^2}\right)\right)^2 + \left(\frac{\partial}{\partial y}\left(\frac{\partial^2 f}{\partial z^2}\right)\right)^2} \cdot p}{\left|\frac{\partial^2 f}{\partial z^2}\right|}\right] \tag{14}$$

$$SSTE_2 = \Re\left(\text{acos}\left[\frac{\left|\frac{\partial^2 f}{\partial z^2}\right|}{\sqrt{\left(\frac{\partial}{\partial x}\left(\frac{\partial^2 f}{\partial z^2}\right)\right)^2 + \left(\frac{\partial}{\partial y}\left(\frac{\partial^2 f}{\partial z^2}\right)\right)^2} \cdot p}\right]\right) \tag{15}$$

where p is a constant value, to make Eqs. 14 and 15 have the mathematical significance. Figure 1e and f show the first-order step-edge detection filters computed by Eqs. 12 and 13 with a k value of 50 m, respectively. Figure 1g and h show the second-order step-edge detection filters computed by Eqs. 14 and 15, with a p value of 1 m. We can see that the step-edge detection filters can display the edge more clearly and precisely.

Depending on the equations, we can see that the edges recognized by the step-edge detection filters are related to the parameters of k and p, and we use the synthetic gravity anomaly to test the application effect of different parameters. Figure 2a and b show the first-order step-edge detection filters computed by Eqs. 12 and 13 with a k value of 50 m, and Fig. 2c and d show the first-order step-edge detection filters computed by Eqs. 12 and 13 with a k value of 5 m. We can see that the edges recognized by the first-order edge detection filters are clearer with the increasing of k. Figure 2e and f show the FSTE of the data computed by Eqs. 12 and 13 with a k value of 200 m, and the recognized edges are not complete, so a very big value of k is not good for the application effect of the FSTE filters. We find that a better value of k is close to the following function, which is given by

$$k = \frac{\max\left|\frac{\partial f}{\partial z}\right|}{\max\left|\frac{\partial^2 f}{\partial z^2}\right|} \tag{16}$$

Figure 2g and h show the second-order step-edge detection filters computed by Eqs. 14 and 15, with a p value of 1 m, and Fig. 2i and j show the first-order step-edge detection filters computed by Eqs. 14 and 15, with a p value of 20 m. According to the results, we know that a smaller value of p can obtain clearer edges. Figure 2k and l show the SSTE of the data computed by Eqs. 14 and 15, with a p value of 0.1 m, and the edges are clear, but are not very complete, so a very small value of p is not good for the SSTE filters. We find that a better value of p is close to the following function.

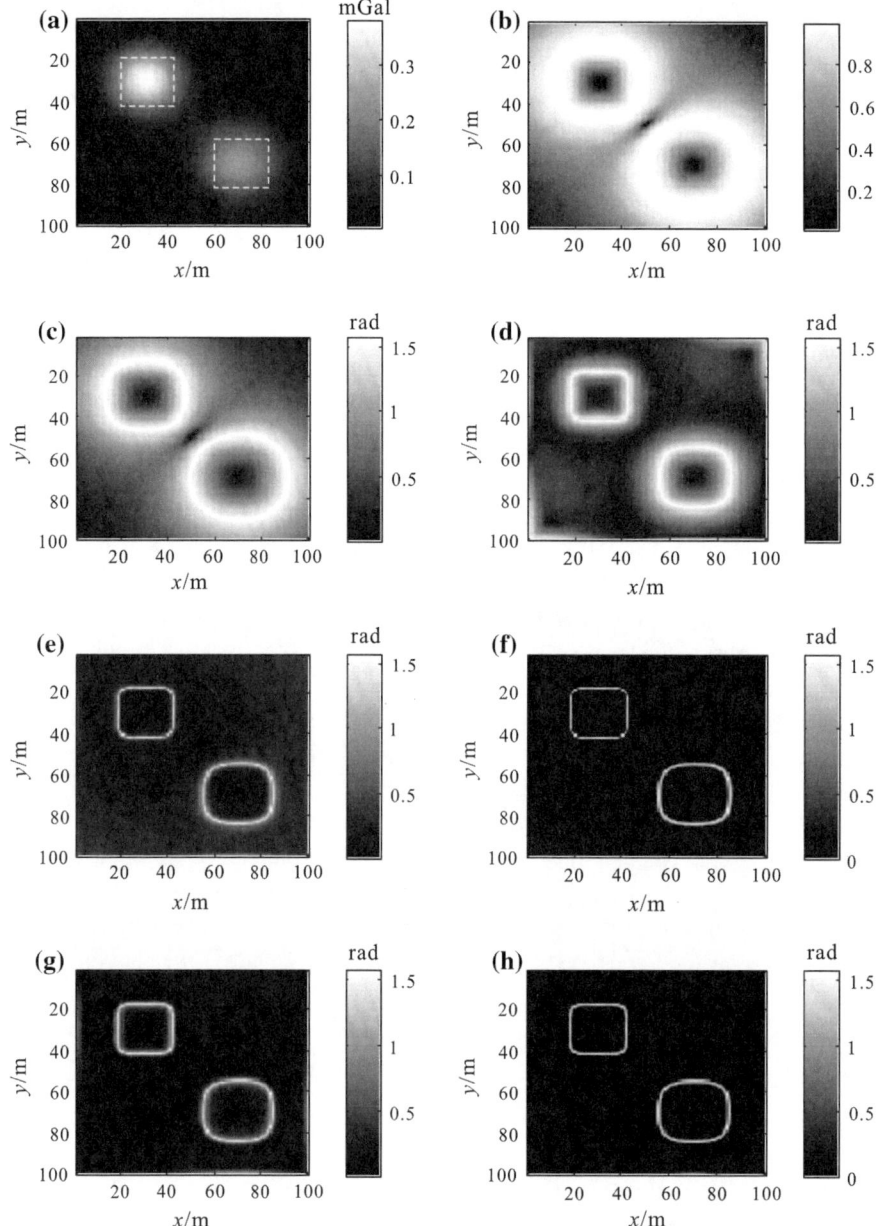

Figure 1

a Synthetic gravity anomaly generated by two prisms with depths of 15 and 20 m. **b** Theta map of the data in Fig. 1a. **c** TDX filter of the data in Fig. 1a; **d** TDX$_2$ filter of the data in Fig. 1a; **e** first-order step-edge detection filter of the data in Fig. 1a, with a k value of 50 m computed by Eq. 12; **f** first-order step-edge detection filter of the data in Fig. 1a, with a k value of 50 m computed by Eq. 13; **g** second-order step-edge detection filter of the data in Fig. 1a, with p of 1 m computed by Eq. 14; **e** second-order step-edge detection filter of the data in Fig. 1a, with p of 1 m computed by Eq. 15

$$p = \frac{\text{mean}\left|\frac{\partial f}{\partial z}\right|}{\text{max}\left|\frac{\partial^2 f}{\partial^2 z}\right|} \qquad (17)$$

The step-edge detection filters require the computation of higher-order derivatives, so we should test

the stability of the new methods. Figure 3a shows the data in Fig. 1a, adding Gaussian noise with a signal-to-noise ratio (SNR) of 50. Figure 3b and c show the theta map and TDX of the data in Fig. 3a, and d shows the second-order normalized horizontal tilt

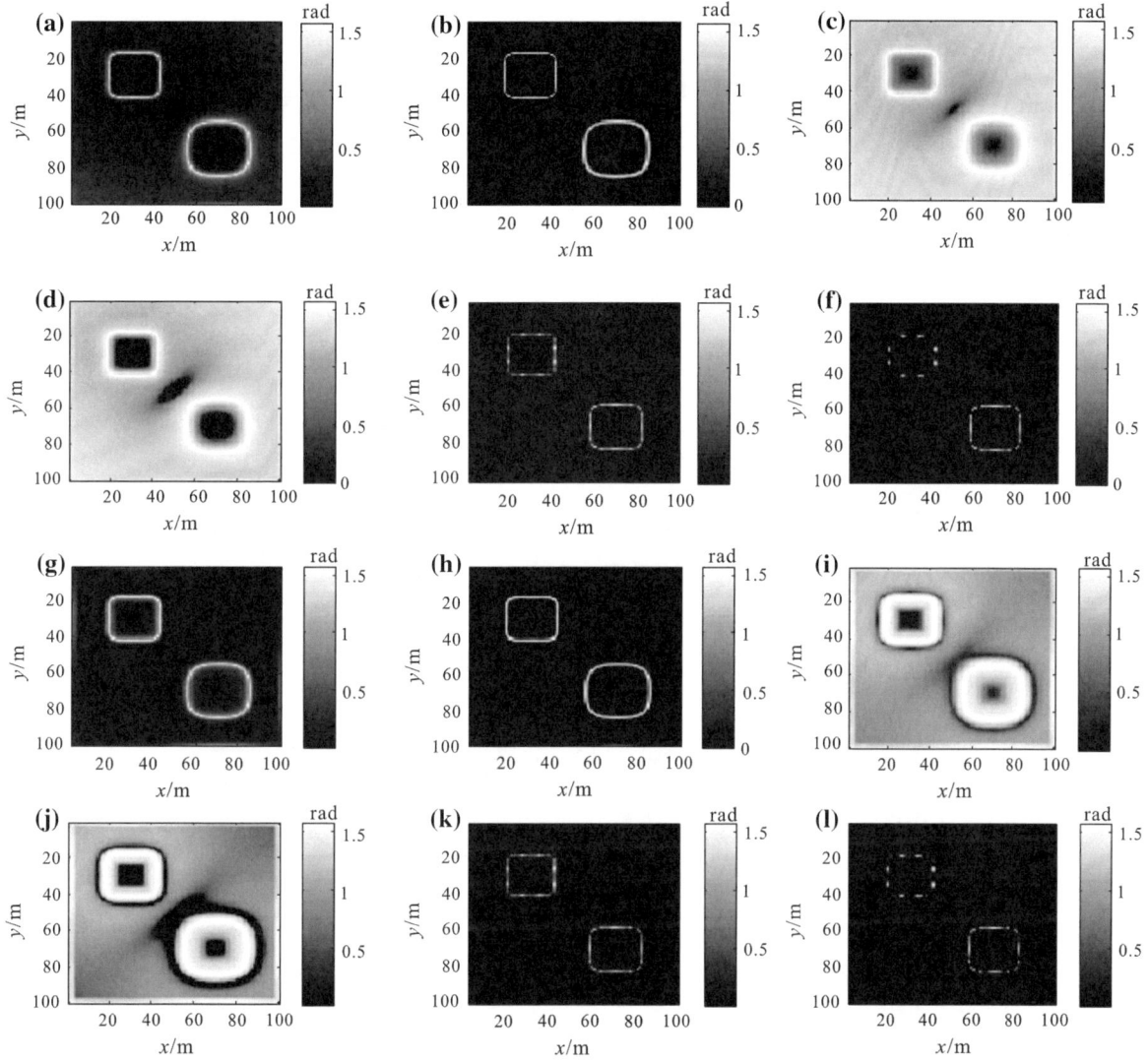

Figure 2

a First-order step-edge detection filter of the data in Fig. 1a, with a *k* value of 50 m computed by Eq. 12; **b** first-order step-edge detection filter of the data in Fig. 1a with a *k* value of 50 m computed by Eq. 13; **c** first-order step-edge detection filter of the data in Fig. 1a, with a *k* value of 5 m computed by Eq. 12; **d** first-order step-edge detection filter of the data in Fig. 1a, with a *k* value of 5 m computed by Eq. 13; **e** first-order step-edge detection filter of the data in Fig. 1a, with a *k* value of 200 m computed by Eq. 12; **f** first-order step-edge detection filter of the data in Fig. 1a, with a *k* value of 200 m computed by Eq. 13; **g** second-order step-edge detection filter of the data in Fig. 1a, with a *p* value of 1 m computed by Eq. 14; **h** second-order step-edge detection filter of the data in Fig. 1a, with a *p* value of 1 m computed by Eq. 15; **i** second-order step-edge detection filter of the data in Fig. 1a, with a *p* value of 20 m computed by Eq. 14; **j** second-order step-edge detection filter of the data in Fig. 1a, with a *p* value of 20 m computed by Eq. 15. **k** second-order step-edge detection filter of the data in Fig. 1a, with a *p* value of 0.1 m computed by Eq. 14; **l** second-order step-edge detection filter of the data in Fig. 1a, with a *p* value of 0.1 m computed by Eq. 15

angle. Figure 3e and f show the first-order step-edge detection filters computed by Eqs. 12 and 13, with a *k* value of $\frac{\max|\partial f/\partial z|}{\max|\partial^2 f/\partial^2 z|}$, respectively. Figure 3g and h show the second-order step-edge detection filters computed by Eqs. 14 and 15, with a *p* value of $\frac{\text{mean}|\partial f/\partial z|}{\max|\partial^2 f/\partial^2 z|}$. According to the edge detection results, we can see that the step-edge detection can still display the edges more clearly, and these filters are not dramatically disturbed by the noise, because of the computation of higher-order derivatives.

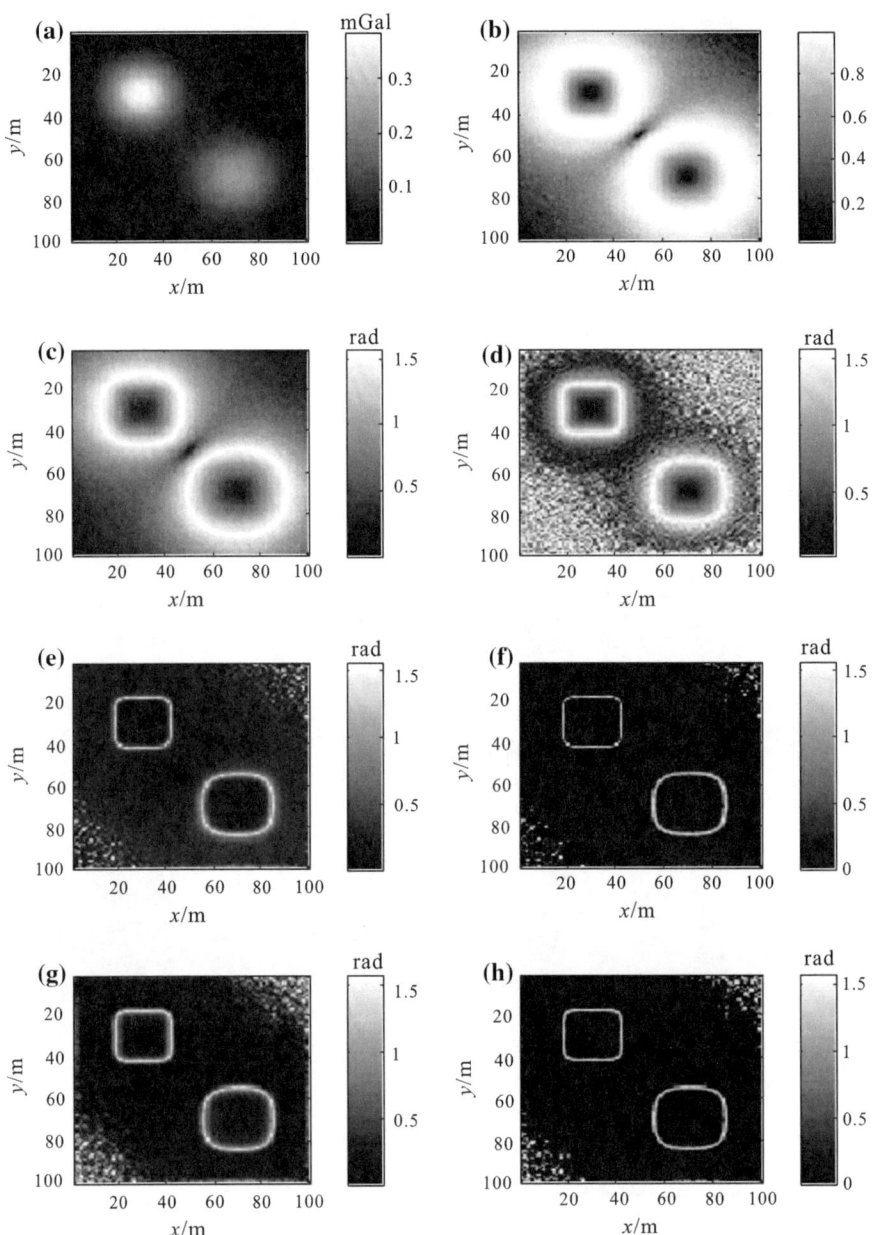

Figure 3

a Synthetic gravity anomaly in Fig. 1a, adding Gaussian noise. **b** Theta map of the data in Fig. 3a. **c** TDX filter of the data in Fig. 3a; **d** TDX$_2$ filter of the data in Fig. 3a; **e** first-order step-edge detection filter of the data in Fig. 3a, with a k value of $\frac{\max|\partial f/\partial z|}{\max|\partial^2 f/\partial^2 z|}$ computed by Eq. 12; **f** first-order step-edge detection filter of the data in Fig. 3a computed by Eq. 13; **g** second-order step-edge detection filter of the data in Fig. 3a, with a p value of $\frac{\mathrm{mean}|\partial f/\partial z|}{\max|\partial^2 f/\partial^2 z|}$ computed by Eq. 14; **h** second-order step-edge detection filter of the data in Fig. 3a computed by Eq. 15

In order to show the resolution of different step-edge detection filters, we test them on a complicated model. Figure 4a shows the synthetic gravity anomaly generated by three prisms with depths of 6, 15 and 15 m, and the white dotted lines represent the true horizontal locations. Figure 4b and c show the theta map and TDX of the data in Fig. 3a, and they can only show the edges of bigger bodies. Figure 4d shows the second-order normalized horizontal tilt angle filter of the data, and it can display the edges of

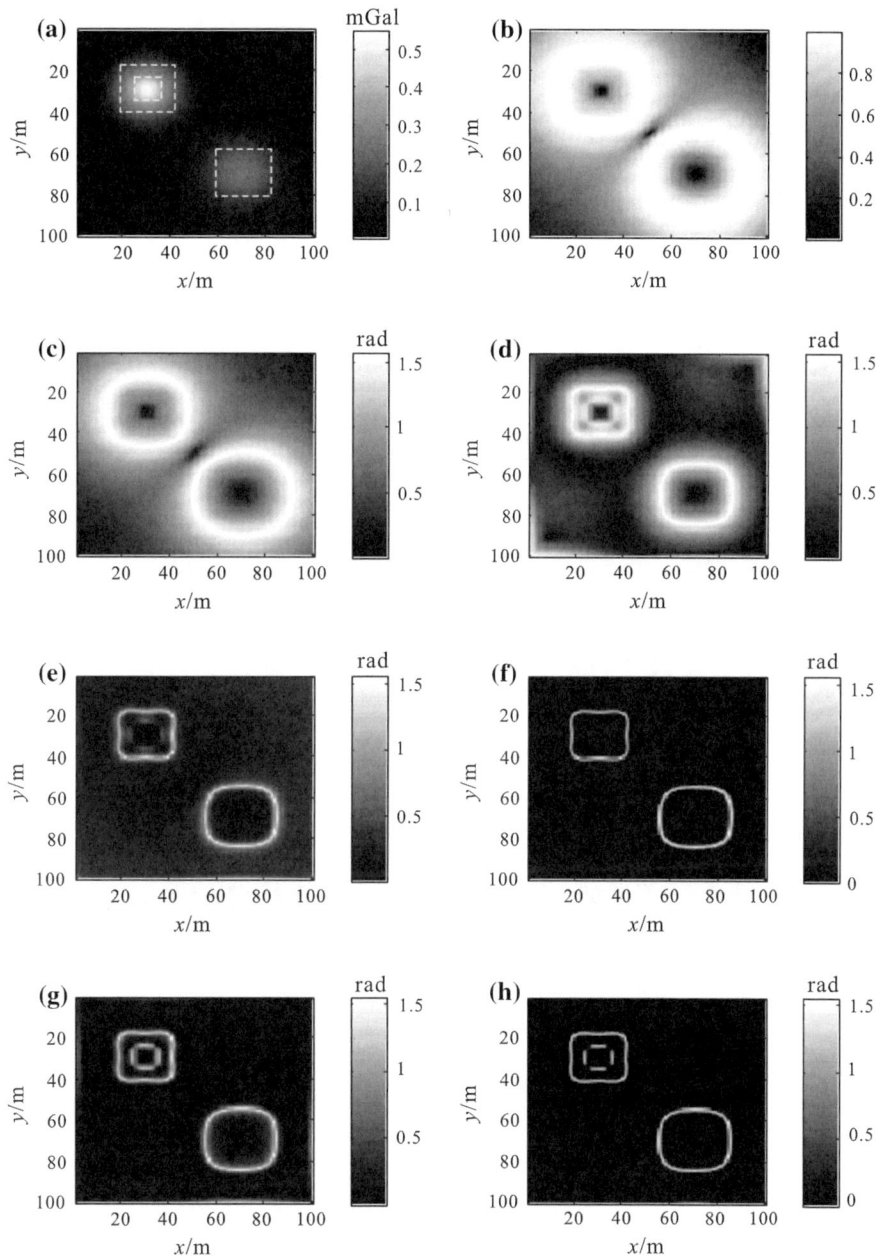

Figure 4

a Synthetic gravity anomaly generated by three prisms. **b** Theta map of the data in Fig. 4a. **c** TDX filter of the data in Fig. 4a; **d** TDX$_2$ filter of the data in Fig. 4a; **e** first-order step-edge detection filter of the data in Fig. 4a computed by Eq. 12; **f** first-order step-edge detection filter of the data in Fig. 4a computed by Eq. 13; **g** second-order step-edge detection filter of the data in Fig. 4a computed by Eq. 14; **e** second-order step-edge detection filter of the data in Fig. 4a computed by Eq. 15

the sources, but the edges are diffused and not clear. Figure 4e and f show the first-order step-edge detection filters computed by Eqs. 12 and 13, with a k value of $\frac{\max|\partial f/\partial z|}{\max|\partial^2 f/\partial^2 z|}$, respectively. Figure 3g and h show the second-order step-edge detection filters computed by Eqs. 14 and 15, with a p value of $\frac{\text{mean}|\partial f/\partial z|}{\max|\partial^2 f/\partial^2 z|}$. According to the edge detection results, we can see that the second-order step-edge detection filters have higher resolution, and can display the horizontal locations of smaller body.

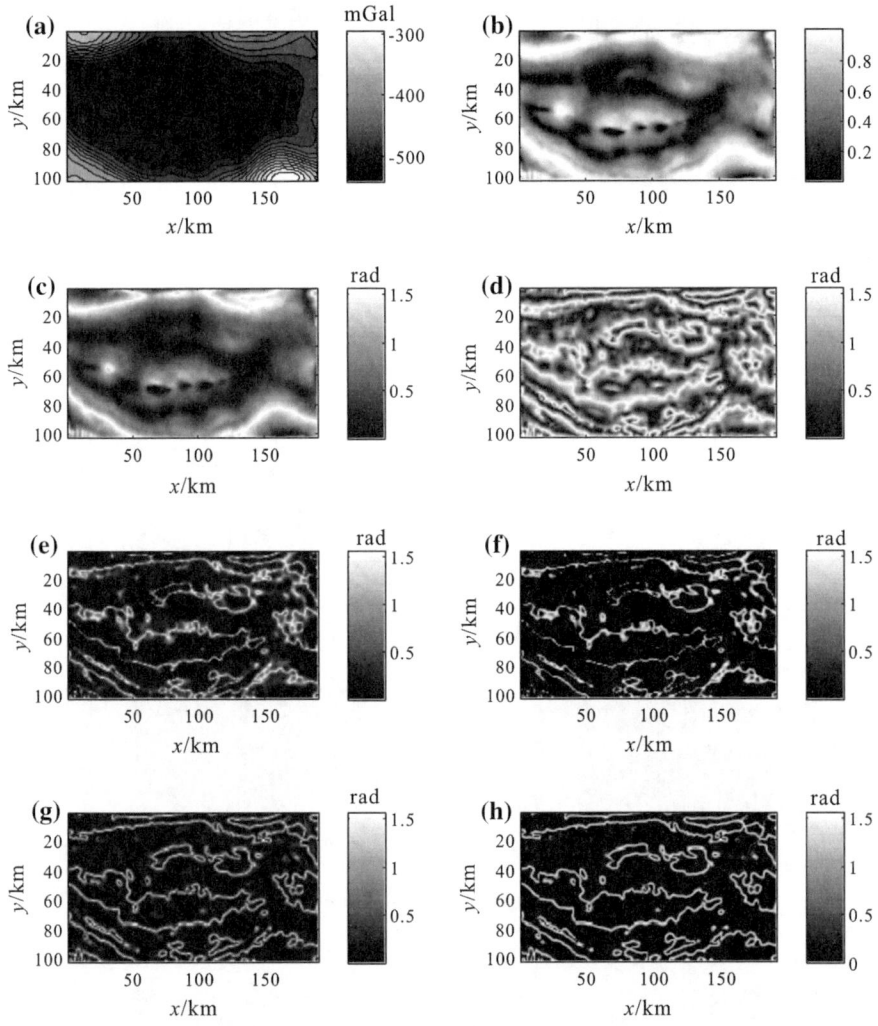

Figure 5

a Real gravity anomaly of Qinghai-Tibet plateau. **b** Theta map of the data in Fig. 5a. **c** TDX filter of the data in Fig. 5a; **d** TDX_2 filter of the data in Fig. 5a; **e** first-order step-edge detection filter of the data in Fig. 5a computed by Eq. 12; **f** first-order step-edge detection filter of the data in Fig. 5a computed by Eq. 12; **g** second-order step-edge detection filter of the data in Fig. 5a computed by Eq. 14; **h** second-order step-edge detection filter of the data in Fig. 5a computed by Eq. 15

It should also be noted that when we use the step-edge detection filters to interpret magnetic anomaly, we should reduce the magnetic anomaly to the North Pole, because the derivatives of magnetic anomaly are sensitive to magnetization direction.

3. Application to Real Potential Field Data

Qinghai–Tibet Plateau plays an important role in the interpretation of geophysical data, because of its high elevation and high seismicity. Figure 5a shows the Bouguer gravity anomaly of the whole Qinghai-Tibet plateau, southwest China, and the data is from th eChina Geological Survey. Figure 5b and c show the Theta map and TDX of the data in Fig. 5a, and they cannot display the edges of stratums clearly. Figure 5d shows the second-order normalized tilt angle filter of the data, and this method can recognize the edges, but they are diffused. Figure 5e and f shows the first-order step-edge detection filters with a k value of $\frac{\max|\partial f/\partial z|}{\max|\partial^2 f/\partial^2 z|}$ computed by Eqs. 13 and 14, respectively. We can see that the first step-edge detection filters can display the edge of stratums more clearly compared to the existing balanced edge detection filters.

Figure 5g and h show the second step-edge detection filters with a p value of $\dfrac{\text{mean}|\partial f/\partial z|}{\text{max}|\partial^2 f/\partial^2 z|}$ computed by Eqs. 15 and 16, respectively. Depending on the results, we can see that the second step-edge detection filters have higher resolution, and can show the features of structures more clearly.

4. Conclusions

In this paper, we find that higher-order derivatives can obtain more accurate edges, while the ratio of same-order derivatives will produce unwanted edges, so we present step-edge detection filters, which use combinations of different-order horizontal derivatives to finish the edge detection task, and we add parameters to make the step-edge detection filters have mathematical significance. We demonstrate the step-edge detection filters methods on synthetic and real potential field data, and they can display the edges more precisely and clearly, and the second-order step-edge detection filters have higher resolution. Because the step-edge detection filters do not need the computation of vertical derivatives, they are insensitive to noise. We also test the effect of different parameters on the application of edge detection, and the values of the parameters are related to the derivatives. We also apply the step-edge detection filters to interpret real gravity data, and the presented filters can obtain clearer edges of stratums.

Acknowledgments

We acknowledge the assistance of the associate editors and the reviewers in improving the paper, and this work is supported by the National Natural Science Foundation of China (Grant No. 41404089), the State Key Program of National Natural Science of China (Grant No. 41430322) and the China Postdoctoral Science Foundation (Grant No. 2014M550173).

References

BLAKELY, R. J., (1995), *Potential theory in gravity and magnetic applications*, Cambridge University Press.

COOPER,G. R. J. and COWAN, D. R., (2006), *Enhancing potential field data using filters based on the local phase*, Computers & Geosciences *32*, 1585–1591.

COOPER,G. R. J. and COWAN, D. R., (2008), *Edge enhancement of potential-field data using normalized statistics*, Geophysics *73*, H1–H4.

CORDELL, L., (1979), *Gravimetric expression of graben faulting in Santa Fe Country and the Espanola Basin*, New Mexico: New Mexico Geol. Soc. Guidebook, 30th Field Conf., 59–64.

CORDELL, L. and GRAUCH, V. J. S. (1985), *Mapping basement magnetization zones from aeromagnetic data in the San Juan basin*, New Mexico. in Hinzc. W. J. Ed. The utility of regional gravity and magnetic anomaly. Society of Exploration Geophysicists. 181–197.

EVJEN, H. M., (1936), *The place of the vertical gradient in gravitational interpretations*, Geophysics *1*, 127–136.

FEDI, M. and FLORIO, G., (2001), *Detection of potential fields source boundaries by enhanced horizontal derivative method*, Geophysical prospecting *49*, 40–58.

HOOD, P. J. and TESKEY, D. J., (1989), *Aeromagnetic gradiometer program of the Geological Survey of Canada*, Geophysics *54*, 1012–1022.

HSU, S., SIBUET, J. C. and SHYU, C. (1996), *High-resolution detection of geologic boundaries from potential field anomalies: an enhanced analytic signal technique*, Geophysics *61*, 373–386.

MA, G. and LI, L., (2012), *Edge detection in potential fields with the normalized total horizontal derivative*, Computers & Geosciences *41*, 83–87.

MA, G., (2013). *Edge detection of potential field data using improved local phase filter*, Exploration Geophysics *44*, 36–41.

MILLER, H.G. and SINGH, V., (1994), *Potential field tilt—a new concept for location of potential field sources*, Journal of Applied Geophysics *32*, 213–217.

ROEST, W. R., VERHOEF, J. and PILKINGTON, M., (1992), *Magnetic interpretation using the 3-D analytic signal*, Geophysics *57*, 116–125.

VERDUZCO, B., FAIRHEAD, J. D. and GREEN, C. M., (2004), *New insights into magnetic derivatives for structural mapping*, The Leading Edge *23*, 116–119.

WIJNS, C., PEREZ, C. and KOWALCZYK, P., (2005), *Theta map: edge detection in magnetic data*, Geophysics *70*, 39–43.

(Received September 16, 2014, revised January 13, 2015, accepted February 6, 2015, Published online February 26, 2015)

Reprinted from the journal

Pure Appl. Geophys. 173 (2016), 555–571
© 2015 Springer Basel
DOI 10.1007/s00024-015-1082-1

Pure and Applied Geophysics

Magnetic Data Interpretation for the Source-Edge Locations in Parts of the Tectonically Active Transition Zone of the Narmada-Son Lineament in Central India

G. K. Ghosh[1]

Abstract—The study has been carried out in the transition zone of the Narmada-Son lineament (NSL) which is seismically active with various geological complexities, upwarp movement of the mantle material into the crust through fault, fractures lamination and upwelling. NSL is one of the most prominent lineaments in central India after the Himalaya in the Indian geology. The area of investigation extends from longitude 80.25°E to 81.50°E and latitude 23.50°N to 24.37°N in the central part of the Indian continent. Different types of subsurface geological formations viz. alluvial, Gondwana, Deccan traps, Vindhyan, Mahakoshal, Granite and Gneisses groups exist in this area with varying geological ages. In this study area tectonic movement and crustal variation have been taken place during the past time and which might be reason for the variation of magnetic field. Magnetic anomaly suggests that the area has been highly disturbed which causes the Narmada-Son lineament trending in the ENE-WSW direction. Magnetic anomaly variation has been taken place due to the lithological variations subject to the changes in the geological contacts like thrusts and faults in this area. Shallow and deeper sources have been distinguished using frequency domain analysis by applying different filters. To enhance the magnetic data, various types of derivatives to identify the source-edge locations of the causative source bodies. The present study carried out the interpretation using total horizontal derivative, tilt angle derivative, horizontal tilt angle derivative and Cos (θ) derivative map to get source-edge locations. The results derived from various derivatives of magnetic data have been compared with the basement depth solutions calculated from 3D Euler deconvolution. It is suggested that total horizontal derivative, tilt angle derivative and Cos (θ) derivative are the most useful tools for identifying the multiple source edge locations of the causative bodies in this tectonically active and transition zone area. As this area is highly prone to hydrocarbon bearing zone, hence, the integrated interpretation could reliably image various thrusts and faults boundaries and the source edge locations with dip and strike orientation along with the basement lineation in encouraging exploration for better understanding of the geo-scientific data.

Key words: Narmada-Son lineament, magnetic data, tilt angle, horizontal tilt angle, total horizontal derivative.

Views expressed in this paper are that of author only and may not necessarily be of Oil India Limited (OIL).

[1] Oil India Limited, Duliajan, Assam, India 786602. E-mail: gk_ghosh@yahoo.com

1. Introduction

The area is highly disturbed and tectonically active and shows the variation of magnetic amplitude. The study area is located in the tectonically active transition zone area of the Narmada-Son lineament (NSL) near Barhi-Shahdol region and extends from longitude 80.25°E to 81.50°E and latitude 23.50°N to 24.37°N (Fig. 1). The Narmada-Son lineament (NSL) falls in the transition zone which is seismically active with various geological complexities like variation of seismic velocities, density, magnetic susceptibility, thermal variation, tectonic resettlement, etc. NSL is one of the most prominent lineaments prominently trending ENE–WSW in central India after the Himalaya in the Indian geology. The regional gravity and magnetic study have been brought out the nature of the hidden tectonic feature of the complex signature of the Narmada-Son lineament. The Central Indian Transition Zone (CITZ) constitutes an integral part of north central province (NCP) and is bounded by the Son Narmada North Fault (SNNF) in the north, a shear zone to the south of the Sausar Mobile Belt (SMB) grossly coinciding with Central Indian Shear Zone (CISZ), (Yedekar *et al.* 1990) in the south (Radhakrishna *et al.* 1988). It forms a distinct ENE–WSW trending of 120–150 km wide tectonic geological trend which divides the Indian continent into two tectano-magnetic/metamorphic provinces, namely northern (Bundelkhand) central provinces towards north and the southern (Deccan/Peninsular) crustal provinces comprising Bastar, Singhbum and Dharwar cratons in the south. The area with the Central Indian Transition Zone (CITZ) encompasses a major part of Central India with its diverse lithotectonic units of Archaean, Proterozoic and Phanerozoic in age (Fig. 2).

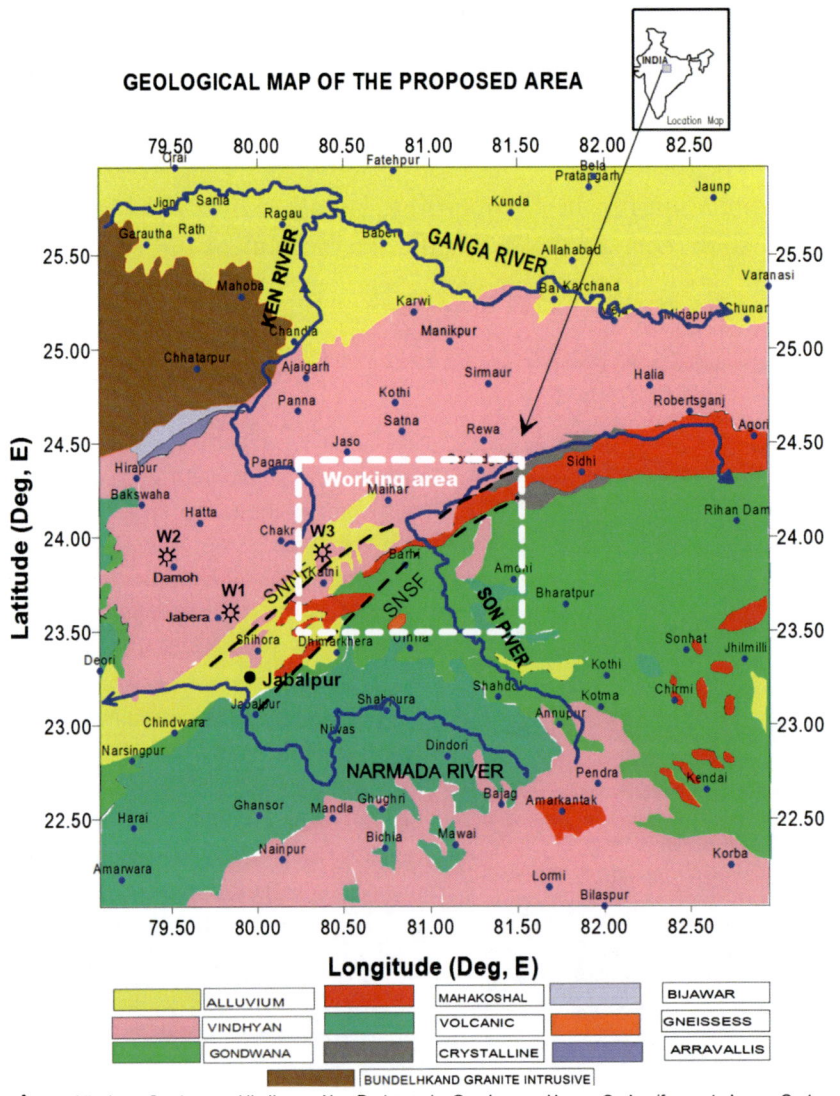

Figure 1
Simplified geology map of the study area and its surroundings (original map was after Geolological Survey of India 1993). The revised map was taken from GHOSH and SINGH (2011)

The CITZ comprises two sub-parallel structural domains viz. the Son-Narmada (SONA) sub-zone containing the Mahakoshal belt in the north and Sousar Mobile Belt (SMB) with associated basement gneisses, granitoids and granualites in the south which has been separated by wide interval of younger covered rocks including Gondwana sediments and Deccan traps. Major dislocation zones, such as regional faults, important ductile shear zones and prominent ENE-WSW trending lineaments are indication of significant crustal interaction in CITZ. The major structural features are including the Son-Narmada North Fault (SNNF), Son-Narmada South Fault (SNSF) and those in the vicinity including Balarampur-Tattapani fault, Tapti fault, Tan shear and the Central Indian Shear Zone (CISZ). Amongst these,

Narmada-Son lineament

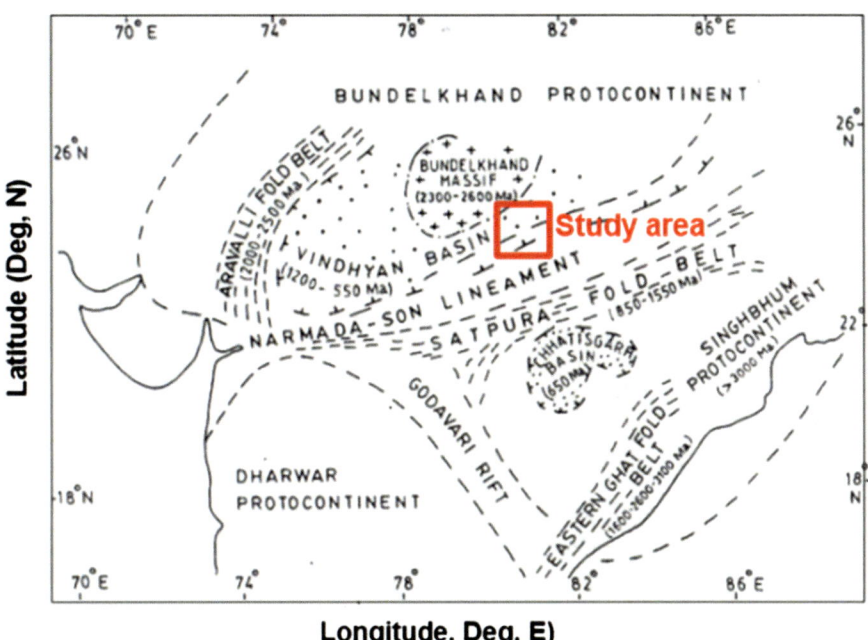

Figure 2
Simplified tectonic map of the northern Peninsular India (after NAQVI *et al.* 1974) shows the major proto continents around the Narmada-Son lineament

SNNF, SNSF and CISZ are significant in view of their deep crustal extension (KAILA *et al.* 1987, 1989; KUMAR *et al.* 2000). It has been observed that the collision of the Indian plate and the Eurasian plate have been caused to swell within the crust and horst structure behaves as suture zone in between the Bundelkhand protocontinent to the north and Dharwar protocontinent in the south.

In this present study, joint integrated approach has been carried using various derivatives of magnetic anomaly to understand the geological setting including thrust and fault boundaries and basement depth in the NSL region. Tilt angle derivative (TDR) (MILLER and SINGH 1994; VERDUZCO *et al.* 2004), horizontal tilt angle derivative (TDX) (COOPER *et al.* 2006) and the total horizontal derivative (THDR) (GRAUCH *et al.* 2001) are the most suitable techniques to determine the basement depth locations. WIJNS *et al.* (2005) studied the edge detection technique of the causative bodies using magnetic data where Cos (θ) map has been used to demarcate the source locations and edges. Although the gravity–magnetic and other geoscientific works have been carried out by different potential workers viz:

deep seismic sounding study in the Vindhyan basin in the NSL region (KAILA *et al.* 1989); nature of continental crust in the NSL region using gravity and deep seismic sounding (VERMA and BANERJEE 1992); crustal configuration (SINGH *et al.* 1995); 3-D structure and geodynamic evolution in the Narmada-Tapti region (SINGH 1998); crustal velocity study along the Hirapur-Mandla location (MURTY *et al.* 2004); 2-D and 3-D gravity modeling study along the Narmada-Tapti rift (BHATTACHARJI *et al.* 2004); subsurface imaging using electromagnetic data in the NSL region (RAO *et al.* 2004); study of Vindhyan basin at Jabera-Damoh structure in the central India (SRIVASTAVA *et al.* 2007); tectomagnetic study in the NSL (WAGHMARE *et al.* 2009); crustal density structure using gravity data (NAGESWARA *et al.* 2011) and Proterozoic sedimentation in the Vindhyan, Cuddapah and Pakhal basin (MISHRA 2011a, b). The results derived by them have not clearly indicated the proper source depth locations and horizontal thrust/fault boundaries. Further, GHOSH and SINGH (2011) have worked in this area using simultaneous gravity-magnetic data interpretation using 2.5 D modelling and 2D-3D spectral analysis where the derived results stand more encouraging. In continuation to this study, GHOSH and SINGH (2013a) carried out 3D Euler deconvolution (ED) technique using magnetic data which provided comparatively better source depth estimation. The detailed theoretical approaches for Euler depth estimation using potential field data are available in various texts (THOMPSON 1982; REID et al. 1990; YAGHOOBIAN et al. 1992 and STAVREV 1997). In continuation to study the source depth locations, the present study has been carried out using various derivative analysis [THDR, TILT, TDX and Cos (θ)] and the consequential results are being superimposed to the 3D Euler depth solutions derived by GHOSH and SINGH (2013b) for better correlation. Based on this analysis and the previous studies, a geological structural section has been prepared showing different thrust and fault locations. The present results are more transparent and vivid compared to the previous study.

2. Geological Setting of the Area

Different types of subsurface geological formations viz. alluvial, Gondwana, Deccan traps, Vindhyan,

Mahakoshal, Granite and Gneisses groups exist in this area with varying geological ages (ROY and BANDOPADHYAY 1998) (Fig. 1). A simplified stratigraphic sequence in part of the Narmada-Son lineament is shown in Table 1. Geologically the study area is separated into two parts: Vindhyan and Gondwana rocks, and these two rocks are separated by the upwarpment of Mahakoshal rocks. Both of these Vindhyan rocks (Upper Proterozoic) and Gondwana rocks (Upper Carboniferous) are exposed in the northern and southern parts, respectively. Vindhyan rocks are basically thick sequence of sandstones, shales and limestones and generally lie below the alluvium.

Past studies (PROJECT CRUMANSONATA 1995) suggest that the Mahakoshal rocks belong to nonmagnetic high intrusive rock situated in between the Vindhyan and Gondwana rocks. These high intrusive rocks are basically dyke and have been caused due to seismic and non-seismic activities during the past times. These intrusions have been taken place several times in the past decades due to the re-settlement and tectonic movement processes which might be the reason for the variation of magnetic field. Such type of intrusions were previously considered to have under a post orogenic extensional setting in the

Table 1

Simplified sequence stratigraphy in parts of the Narmada-Son lineament (NSL) area

Period/sub period	Age	Geological formations	Associated mineral deposit
Phanerozoic			
Recent to Pliestocene	1–18 M.Y	Alluvium, Laterite and Bauxite	Ground Water, Bauxite
Cretaceous to Palaeocene	64–97 M.Y.	Deccan trap Basalt	Quartz, Road Material, Limestone
Upper Carboniferous to Lower Cretaceous	97–320 M.Y.	Gondwana Supergroup	Coal deposit
Proterozoic			
Upper	570–900 M.Y.	Vindhyan Supergroup (Basic dykes, Son Valley)	Acid intrusive, limestone, Lead–Zinc deposit
Middle	900–1800 M.Y.	Synite and lamprophyre (Son Valley) and intrusive Granite	Equivalent granitoids, Limestone, dolomite, iron deposit
Lower	1600–2500 M.Y.	Aravalli Group, Bijawar Group and equivalent Granite of Mahakoshal Belt Mahakoshal Group (lower Agoni formation, middle Parsoi formation and the uppermost Dudhmaniya formations)	Manganese and cupper deposit
Archaean	Older than 2500 M.Y.	Bundelkhand Granite and gneisses with associated igneous bodies, unclassified gneisses and granitoids, Older Granite and Migmatites, Crystalline Basement	Gneisses, meta-volcanic and meta-sediments, Granite for cutting/polishing

Paleoproterozoic intrusions and also go through various tectonic disruptions in this transition zone.

Due to such kind of tectonic disturbances, the direction of flow of Narmada and Son river change opposite direction among each other; however, both the rivers have emerged from Amarkantak hill. It has been suggested that Narmada river's streams flow northward direction to westward direction and the Son river also flows from northwards direction to eastwards direction, but both the rivers' have the same source of origination as Amarkantak hill (QURESHY 1981; VALDIYA 1984). These changes of river's flow has been taken place due to the undulation of surface during the past time which might be the reason for the variation of magnetic field.

In several studies, the Mahakoshal Group is referred to as Bijawar Group (1600–2400 M. Y) (KUMAR et al. 2005; MATHUR 1995; KEDAR NARAIN 1962; PASCOE 1965; CHAKRABORTY et al. 1996). Different types of Bijawar rocks are observed with varying ages at Bundelkhand, Son and Central Narmada Valley. The Bijawar rocks which are observed at Son and Central Narmada valley are called Mahakoshal Groups and are younger than the Bijawar of Bundelkhand. Various younger alkaline intrusive rocks are observed at the Son Narmada valley which are absent at the Bijawar type area. It was premeditated that Bijawar lying over the Budenkhand granite has been caused for the development of fault boundaries along the Son valley. However, Mahakoshal rocks are more deformed, folded and distributed in comparison to the Bijawar type areas. These Mahakoshal rocks divide the NSL into two parts: Son Narmada North Fault (SNNF) and Son Narmada South Fault (SNSF).

The Vindhyan basin consists Proterozoic intracontinental basin which has been developed in the central part of the Indian shield together with several additional basins such as Cuddapah, Chattisgarh, etc. The strata have been exposed in the Son valley of the study area. Vindhyan rocks have been considered as shallow marine environment (VALDIYA et al. 1982; BHATTACHARYA and MORAD 1993). Recently, it has been assumed that Vindhyan sediments represent deeper part of the shelf and spread by fluvial process (CHAKRABORTY et al. 1996; CHAKRABORTY and BHATTARCHARYA 1996). Vindhyan sedimentation

initiates in an intra cratonic rift structure with a systematic signature of numerous rift phases (JOKHAN et al. 1996). It has been understood that the Vindhyan Basin, encompassing more than 5000 m thick sequence of sandstones, shales and limestones, extends into the Ganga valley in the north and northeast beneath the Tertiary sediment of the Himalayan foredeep. Vindhyan rocks are covered by Deccan volcanic in the south west direction.

Indian plate has been divided through first collision with Eurasian Plate in the north and with Burmese plate in the northeast. Since then, Bengal basin attained the status of convergent margin basin. The basin shows two distinct phases of development: Gondwana phase and Post–Gondwana phase. In the Gondwana phase, non-marine sediments have been deposited within a graben oriented in N–S direction. Marine wave covered almost the entire basin up to its western margin. A thick sedimentary prism has been deposited during Tertiary period.

The hydrocarbon exploration activity in and around the area is still under progress and few exploratory wells have been drilled. The striking of hydrocarbon in the Jabera area appears as a dome-shaped basin having different geological formations such as shale, sandstone, limestone and basaltic intrusive. The exploratory well data from this area suggest that the Gondwana rocks behave as a source rock (MATHUR 1995). Recently, Oil and Natural Gas Corporation Limited (ONGCL) has drilled three exploratory wells (Jabera-1, Damoh-1 and Kharkhari-1) in the Vindhyan Basin, particularly in the Jabera-Damoh location. The first well has been drilled up to 3597.7 m at Jabera; the second well has been drilled up to 3501 m depth at Damoh, where commercial hydrocarbon was not found and the same applies to the Kharkhari well. But during the testing phase at the well Jabera-1, a flow of non-commercial quantity of 2000–3000 m^3/day of gas has been confirmed (Directorate General of Hydrocarbon, DGH report; SAMAL and MITRA 2006). The two prominent faults viz; SNNF and SNSF have been elongated in the ENE–WSW direction (JAIN et al. 1995 and ACHARYA and ROY 2000). BHATTACHARYA and KOIDE (1978) stated that these two faults look similar as inward dipping funnel shaped normal fault as studied in the Red Sea area.

3. Magnetic Data Source and Theoretical Approach

The magnetic data acquired by Geological Survey of India (PROJECT CRUMANSONATA 1995) with the specified grid interval of 1.0 km using Scintrex Fluxgate (MFM-2) magnetometer. The accuracy of the data is 0.1 nT. The data has been processed further to reduce in mean sea level. Diurnal and International Geomagnetic Reference Field (I.F.R.F) corrections have been applied to the magnetic data. In this study, it has been briefly discussed total horizontal derivative (THDR), tilt angle derivative (TDR), horizontal tilt angle derivative (TDX), Cos (θ) map and Euler deconvolution to delineate the thrust/fault boundaries and source depth locations.

3.1. Total Horizontal Derivative (THDR)

A number of filters have been used to enhance the potential field data, such as downward continuation, vertical and horizontal derivatives to delineate the source edges In general, total horizontal derivative (THDR) is used for edge detection.

Total horizontal derivative (THDR) is defined by GRAUCH et al. (2001) as

$$\text{THDR} = \sqrt{(\partial T/\partial x)^2 + (\partial T/\partial y)^2}, \quad (1)$$

where $\partial T/\partial x$ and $\partial T/\partial y$ are the horizontal derivative in x and y directions and $\sqrt{(\partial T/\partial x)^2 + (\partial T/\partial y)^2}$ is total horizontal derivative of the magnetic field T. Many potential workers have been carried the magnetic data interpretation using THDR (FERREIRA et al. 2011, 2013; WANG et al. 2009).

3.2. Tilt Angle (TDR)

MILLER and SINGH (1994) have projected a new concept for identifying a potential field source using the tilt angle filter (TDR). Tilt angle is the ratio of the vertical derivative (VDR = $\partial T/\partial z$) to the absolute value of the total horizontal derivative (THDR) which improves large and small amplitude anomalies. However, for deeper causative bodies, the edges detected by tilt angles are blurred. Further, VERDUZCO et al. (2004) expressed tilt angle filter in the generalized way for both profile and gridded dataset.

$$TDR = \tan^{-1}\left(\frac{VDR}{THDR}\right) \quad (2)$$

Since the TDR is the ratio of the vertical and horizontal derivatives, the resulting amplitude functions are measures either in degrees or radians. TDR does not contain strength of geomagnetic field or the susceptibility of the source bodies. TDR limits to $\mp\pi/2$. THDR operates as an effective automatic gain control (AGC) filter over small and large amplitude total magnetic intensity anomalies converted to reduce to pole (RTP) and reflects as the tilt amplitude. The zero contour of the tilt anomaly or TDR is now located close to the boundary of the causative source bodies. The thrust/fault boundaries are identified from the zero contour map of the TDR and one can mark the lineament. The tilt angle is positive over the source bodies and negative outside the source bodies. Various potential works have interpreted the magnetic data using tilt angle analysis (SALEM et al. 2007, 2008, 2010; FERREIRA et al. 2013; LAHTI and KARINEN 2010; and FAIRHEAD et al. 2011 and SANTOS et al. 2012).

3.3. Horizontal Tilt Angle Derivative (TDX)

The horizontal tilt angle (TDX) has been introduced by COOPER and CROWN (2006) using THDR and absolute value of VDR:

$$TDX = \tan^{-1}\left(\frac{THDR}{|VDR|}\right) \quad (3)$$

TDX is varying with the angle $+\pi/2 > TDX > -\pi/2$ similar to tilt angle method. TDX responds to shallower and deeper bodies and also delineates the edges of the bodies. Both the methods TDX and TDR show a contrast variation along the sharper boundary. Various potential workers have carried out magnetic data analysis using TDX derivatives for estimating the source edge detection (FAIRHEAD and WILLIAMS 2006; CORREGGIO et al. 2012; PHILLIPS 2000).

3.4. Cos (θ)

The THDR delineates the edges of the largest amplitude anomaly but it is less impressive for the deeper bodies. WIJNS et al. (2005) studied the application of theta map for the detection of edge of the causative bodies using magnetic data. Cos (θ) is the ratio of THDR and normalized analytical signal |A| (NABIGHIAN 1972). The THDR effectively delineates the edges of the largest amplitude anomalous bodies; however, it is less impressive in the case of deeper bodies.

$$Cos(\theta) = \frac{THDR}{|A|}, \qquad (4)$$

where

$$A = \left| \sqrt{(\partial T/\partial x)^2 + (\partial T/\partial y)^2 + (\partial T/\partial z)^2} \right|, \quad (5)$$

where |A| is the amplitude of the 3D analytical signal. Various potential workers have interpreted magnetic data for better source edge locations and identification of magnetic boundaries (FAIRHEAD and WILLIAMS 2006; FAIRHEAD et al. 2007; REID 2007; PILKINGTON and KEATING 2010). A schematic diagram of the various derivatives with the geometric illustrations of the TDX, THDR and the Analytical signal, |A|, is shown in Fig. 3.

3.5. 3D Euler Deconvolution Method

The Euler deconvolution technique has been used for direct source depth estimation for magnetic source bodies. This is used for either gravity or

magnetic data with the help of homogeneity equation. THOMPSON (1982) has expressed as in Eq. (6)

$$(x - x_0)\frac{\partial f}{\partial x} + (y - y_0)\frac{\partial f}{\partial y} + (z - z_0)\frac{\partial f}{\partial z} = -N(f - B), \qquad (6)$$

where, x, y and z represent the coordinates of the points; x_0, y_0 and z_0 represent the coordinates of the sources as a function of $f(x, y, z)$. B is called as "background" term, describing the constant contribution of the regional field. The x axis point denotes north, y axis point denotes east and z axis denotes vertically downward in the Euler equation or Euler's homogeneity equation which is expressed in Eq. (7):

$$x\frac{\partial f}{\partial x} + y\frac{\partial f}{\partial y} + z\frac{\partial f}{\partial z} = Nf, \qquad (7)$$

where the function $f(x, y, z)$ has a common function that can be represented as in Eq. (8):

$$f(x, y, z) = \frac{Gr}{r^N}, \qquad (8)$$

where N is the structural index (SI),

Gr = a function which is not dependent on x, y, z

$$r = (x^2 + y^2 + z^2)^{1/2} \qquad (9)$$

THOMPSON (1982) has correlated the Structural index (SI) with various types of magnetic source bodies. Euler depth solution provides a numbers of solutions for depth as much the number of grid points using structural index, window-source distance and depth tolerance with window size. The other points are also considered for getting the solutions for depth variation, depth uncertainty, horizontal uncertainty and offset variations. These points are considered for finalizing and deriving a better solution. Euler solutions with higher root mean square (rms) misfit are considered and a predetermined threshold are rejected. Various solutions will be generated as such the number of grid points. However, those solutions are filtered with limiting criteria as mentioned above. Structural index can be defined as the measure of the rate of change of potential field function with respect to distance. The derived solutions may indicate spurious solution; however, utilizing the boundary limiting condition (depth tolerance limit, horizontal

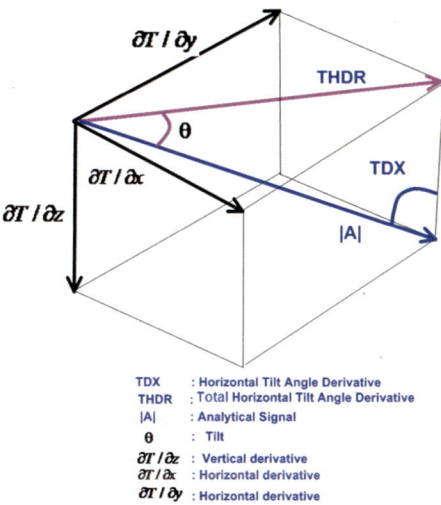

TDX	: Horizontal Tilt Angle Derivative
THDR	: Total Horizontal Tilt Angle Derivative
|A|	: Analytical Signal
θ	: Tilt
$\partial T/\partial z$	: Vertical derivative
$\partial T/\partial x$	: Horizontal derivative
$\partial T/\partial y$	: Horizontal derivative

Figure 3

Schematic diagram of the various derivatives with the geometric illustration of the TDX, THDR and Analytical signal, |A|, of magnetic data

offset and location uncertainty) accurate relevant valid solutions can be accepted. Different types of structural indices have been published by various authors (REID *et al.* 1990, 2014a, b; YAGHOOBIAN *et al.* 1992; STAVREV 1997; BARBOSA *et al.* 1999; MELO *et al.* 2013; SILVA and BARBOSA 2003; SILVA *et al.* 2001; REID and THURSTON 2014).

The depth estimate using 3D Euler deconvolution technique has been carried out by GHOSH and SINGH (2013b) using magnetic data in this area. In their study, various combination of structural index (SI = 0, 1, 2, 3 and 0.5) and window sizes (WS = 5 × 5, 10 × 10 and 15 × 15) has been worked out to calculate the 3D Euler depth solutions. Structural index for magnetic bodies vary for different sources SI = 0–0.5 for contact type bodies: SI = 1 for sill, dyke and sheet; SI = 2 for horizontal pipe or cylinder/vertical pipe and SI = 3 for spherical source bodies. However, the best results are estimated using SI = 0 and WS = 5 × 5. Window size = 5 × 5 has been considered because

most of the valid depth solutions are varying up to 4 km.

3.6. Qualitative Analysis of Magnetic Data

Magnetic data suggest that the magnetic anomaly trends are varying ENE-WSW direction in the study area (Fig. 4). The Higher magnetic field 1100 nT is observed in the northern part and lower magnetic anomaly −1100 nT is observed at the southern part. The extreme northern part shows higher magnetic anomaly at Maihar (1000 nT) and at Sarra (700 nT). The southern part of the area also shows lower value at Raipur (−1000 nT) and at Karwa (−700 nT) with undulation. Broad anomaly suggests regional features; however, strong anomaly suggests various shallow depth intrusions. The central part of the study area suggests that the anomaly oriented ENE to WSW direction due to the existence of Mahakoshal belts (PROJECT CRUMANSONATA 1995). The different

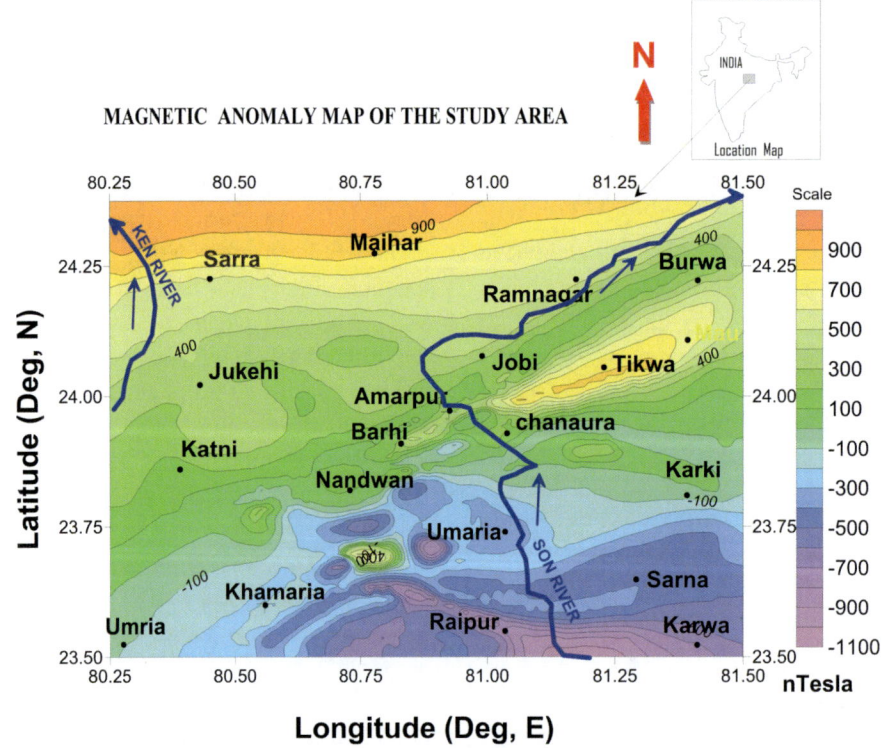

Figure 4
Magnetic anomaly map shows ENE–WSW direction. Higher magnetic anomaly observed in the northern part and lower magnetic anomaly observed at the southern part

type of Mahakoshal rocks are lower Agoni formation, middle Parsoi formation and the uppermost Dudhmaniya formations are in the lower Proterozoic period (Roy and Devarajan 2000).

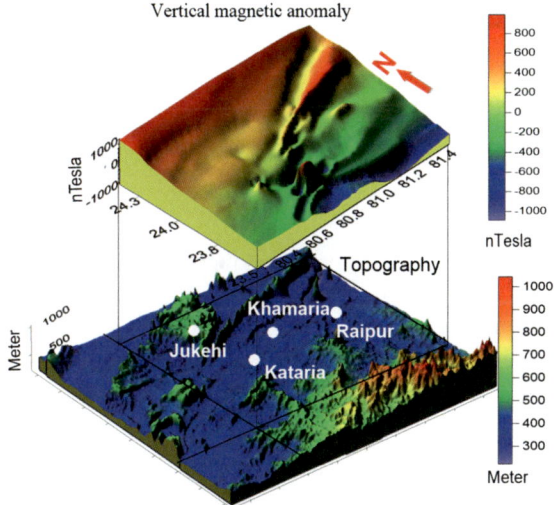

Figure 5
Surface topography correlated with magnetic data of the study area

The surface topography superimposed with the magnetic anomaly map (Fig. 5). The subsurface rocks are supposed to be the higher susceptibility and highly magnetized rocks. It has been observed that average magnetic susceptibility of Mahakoshal rocks vary with the range $10,304 \times 10^{-6}$ to $99,271 \times 10^{-6}$ in SI unit (Project Crumansonata 1995). The Mahakoshal rocks have high-density rock and also metamorphosed and rich in chemically with manganese and copper. It has been observed that in the Gondwana region, the magnetic anomaly changes due to the presence of buried magnetic material probably volcanic intrusions and variation of anomaly observed at Umaria and Sarna areas.

4. Data Interpretation

Attempt has been made to study the source edge detection (SED) technique for calculating the approximate edges of the magnetized source bodies and geological boundaries with the help of various derivatives from gridded magnetic data. A map has

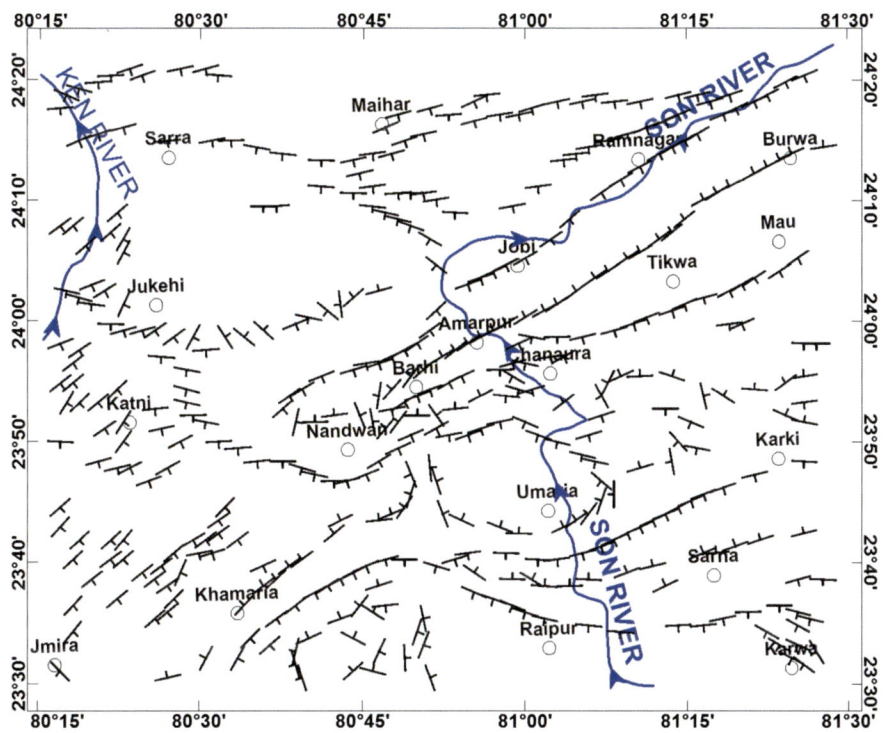

Figure 6
Map shows various dip and strike directions

Reprinted from the journal

been prepared which demonstrates the various dip and strike locations (Fig. 6). In practice, strike direction has been estimated in each grid points. Vertical gradient has been calculated in terms of frequency domain. The horizontal derivatives are computed in the direction perpendicular to the strike using least-square methodology (THURSTON and BROWN 1994; THURSTON and SMITH 1997; CORDELL and GRAUCH 1982).

Tilt angle derivative is normalized phase derivative of first order used for the mapping of structural edges for strappingly and feebly magnetized bodies. The Tilt angle derivative (TDR) map is superimposed on source edge location map along with Euler depth solution. It is suggested that TDR shows zero contour value at the contact bodies (VERDUZCO et al. 2004). Hence the zero crossing of the tilt delineates the borders or the contact of magnetic source bodies.

The magnetic anomalies with negative and positive signatures along with positive signature of horizontal gradients of magnetic anomaly have been considered for processing. In each grid point the dip and strike are calculated. Figure 6 shows the dips of the source bodies representing a symbol "⊥". The right hand is the dip direction when standing in the strike direction. The symbol plotted in the map shows the long axis

represented as the strike direction of the edge (contact) and the down gradient directions is the dip direction. There are four possible combinations of dip and strike directions which are varying from $0 \leq$ strike ≤ 90, $90 \leq$ strike ≤ 180, $180 \leq$ strike ≤ 270 and $270 \leq$ strike ≤ 360. Alternatively, Strike measures in degrees ($0°–360°$) in the clockwise direction from the coordinate axis Y denoted as north. Dip also measures in degree ($0°–180°$) relative to perpendicular to the strike (BLAKELY and SIMPSON 1986).

It has been observed that the results derived by the source edges in this study are well correlated with the 3D Euler source depth locations (GHOSH and SINGH 2013b) (Fig. 7). In this study the 3D Euler source depth locations are superimposed with the dip and strike direction (Fig. 8). The edges are clearly oriented in the ENE–WSW direction extending from the Burwa to Nandwan. The edges are oriented approximately in east–west direction in the northern part of the area surrounding Maihar. Source edges are oriented in the east–west direction in the Raipur-Karwa area. It has been observed that in many places 3D Euler depth solutions are lacking; however, the source edges are mostly populated all over the study area. It has been studied from the above two methods that source edge detection methods are more accurate

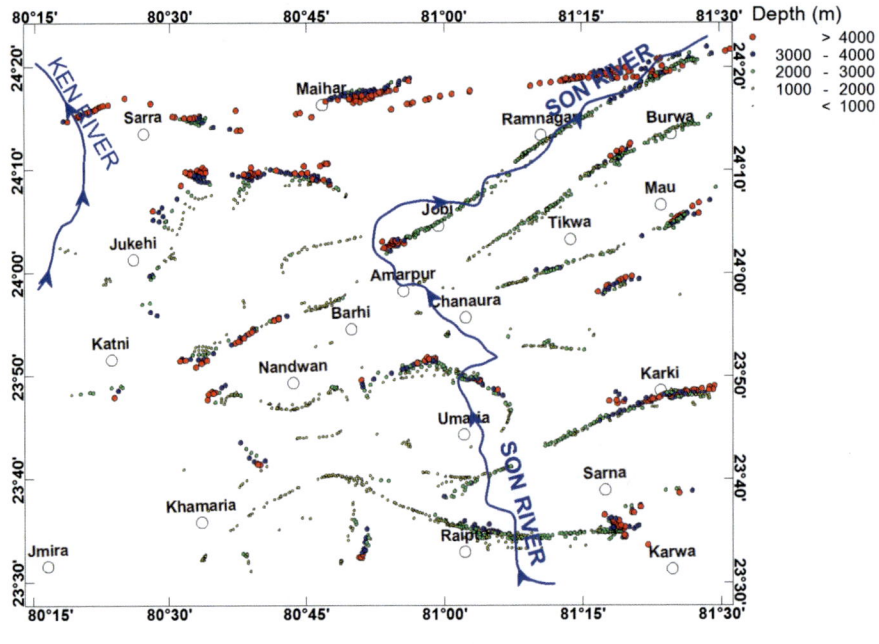

Figure 7
3D Euler source depth solution map (after GHOSH and SINGH 2013b)

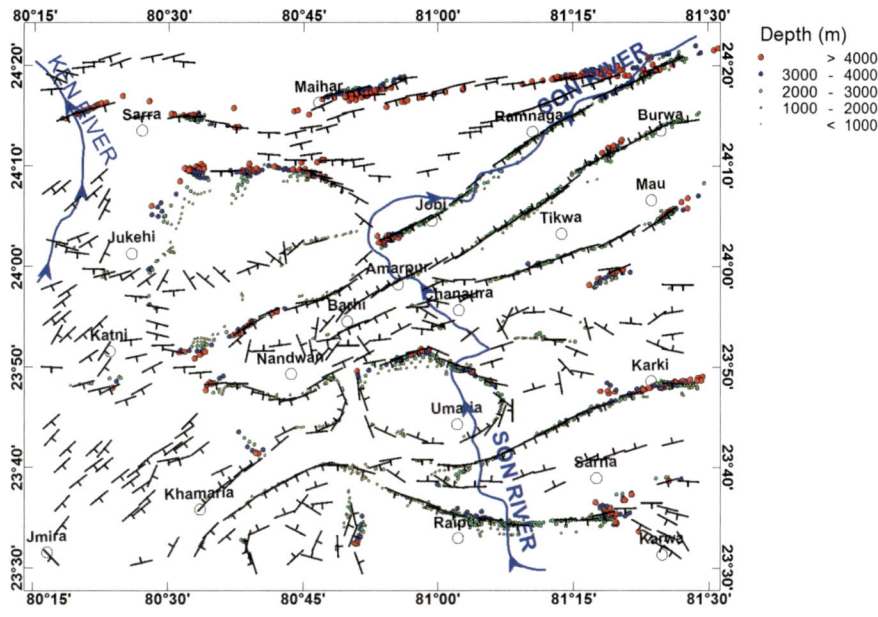

Figure 8
3D Euler depth solution (GHOSH and SINGH 2013b) superimposed with the results of dip and strike direction

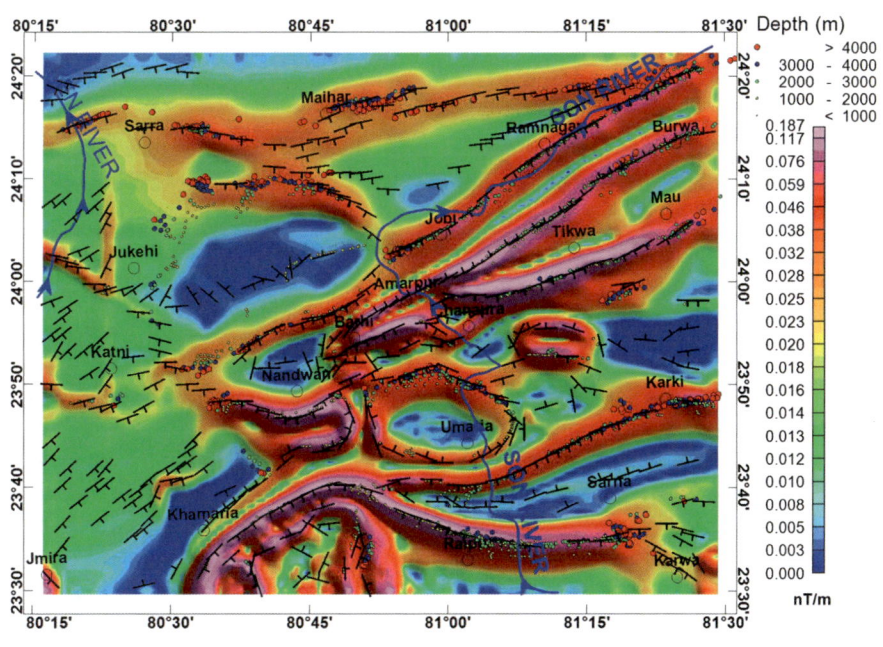

Figure 9
Total horizontal derivative (THDR) map superimposed with 3D Euler depth solution and dip and strike directions

and efficient. Euler deconvolution technique emphasizes the effect for getting deeper source locations (Fig. 7).

Source edge location map superimposed on the THDR map and 3D Euler depth solution (Fig. 9). The THDR lineament is oriented in the ENE–WSW direction and extends up to Nandwan and Khamaria. The source edges are well located in the THDR map where the maximum peak values are located at the edges of the source boundary of the magnetic bodies as reported from the field studies.

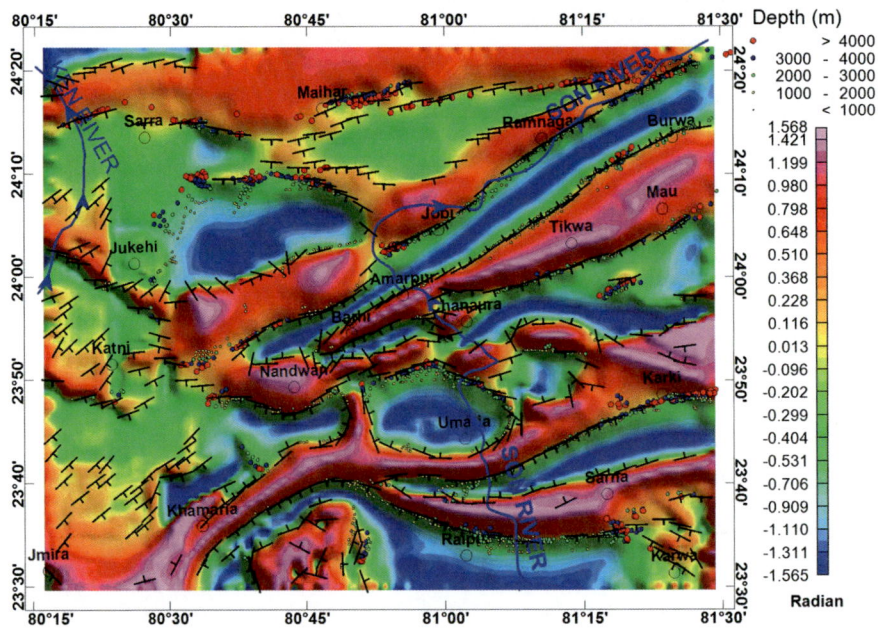

Figure 10
The Tilt angle derivative (TDR) maps superimposed with 3D Euler source depth solution and dip and strike directions

Tilt angle derivative is normalized phase derivative of first order used for the mapping of structural edges for strappingly and feebly magnetized bodies. The Tilt angle derivative (TDR) map is superimposed on source edge location map along with Euler depth solution (Fig. 10). It is suggested that TDR shows zero contour value in the scale bar which are supposed to be the contact bodies. Hence the zero crossing of the tilt delineates the borders or the contact of magnetic source bodies.

Both TDX and TDR derivative are widely used for identifying the source boundaries; however, TDX derivatives represent more sharper boundaries compared to the TDR. The zero crossing values in the TDX map delineate the borders of the magnetic source bodies. The horizontal tilt derivative (TDX) map is superimposed on the source edge location map with 3D Euler depth solution (Fig. 11). The strike and dip directions are also indicated.

The Cos (θ) map is clearly demarcating the source locations and edges with maximum positive values and is one of the most direct interpretation tools for identifying the source edge locations. The Cos (θ) map is superimposed with the existing source edge location map along with 3D Euler depth solution

(Fig. 12). The strike and dip directions are also indicated.

However, no single method is a best tool for source boundary identification; an integrated approach is always extremely compassionate for superior correlation and assessment to provide judicial more realistic information in this area. With the help of above studies including THDR, TDR, TDX, SED and Cos (θ) analysis, the derived results are strongly correlating to each other solutions. The Cos (θ) map covers entirely a better source edge location then projected other derivatives.

The basement depth information carried by GHOSH and SINGH (2011) using simultaneous gravity-magnetic 2.5D profile modeling and by GHOSH and SINGH (2013a) using 2D Euler and Werner depth solution stated the basement depth information along the few profiles only. Further, GHOSH and SINGH (2013b) carried out 3-dimensional Euler source depth investigation in the same area and provided a better depth understanding. However, the present investigation for source boundary estimations study using THDR, TDR, TDX and Cos (θ) provides a holistic view throughout the area. The above studies are summarized and a composite map has been generated after

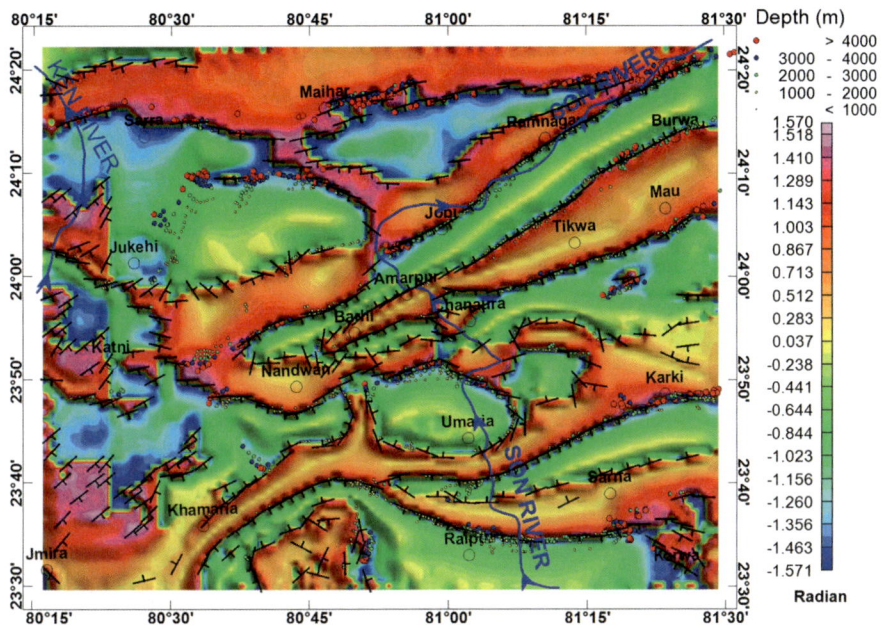

Figure 11
The horizontal tilt derivative (TDX) superimposed with 3D Euler depth solution and dip and strike directions

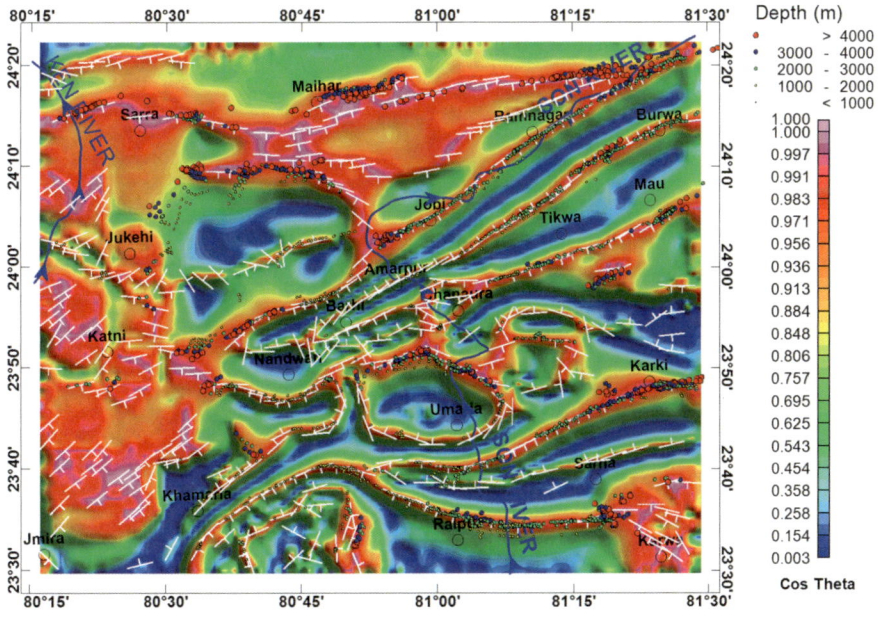

Figure 12
Cos (θ) map superimposed with 3D Euler depth solution and dip and strike location

integrating the various derived results (Fig. 13). As the study area is situated in the complex transition zone and is broadly prospective for hydrocarbon exploration, this integrated approach is more helpful for detailed investigation (SESHUNARAYANA *et al.* 2008; MUKHOPADHYAY *et al.* 2001a, b; JOKHAN 2005).

The recent tectonic activities suggest that the study area and its surrounding have recognized a

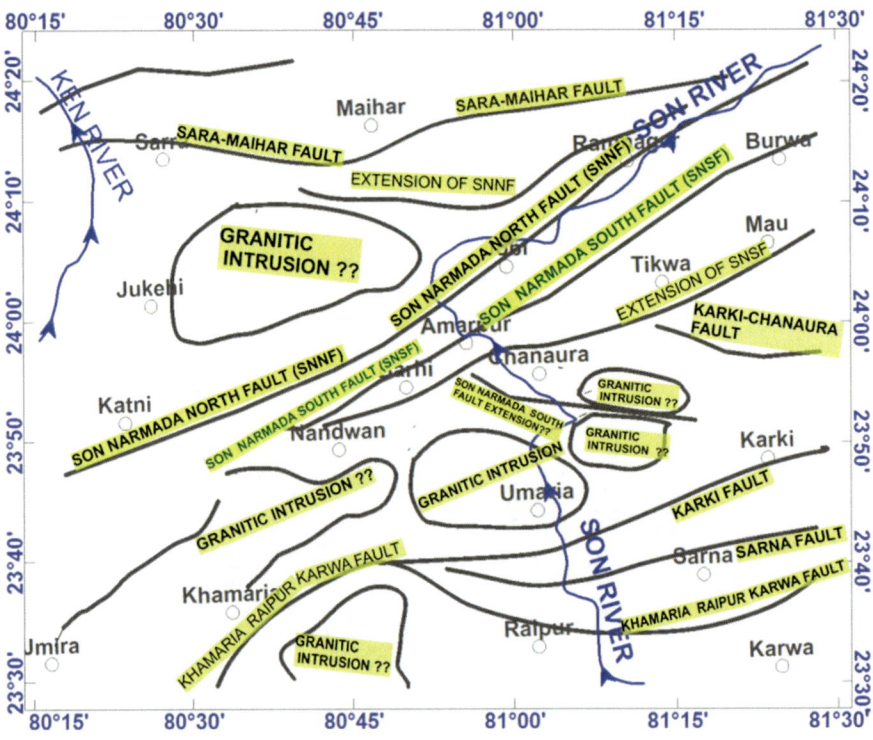

Figure 13
Interpreted geological section with various significant thrust and fault boundaries. Different probable granitic intrusions are also *marked*

hydrocarbon bearing field and large reservoir of coal bed methane gas. The various joint operators are presently engaged for exploration of hydrocarbon in this area. ONGCL struck a well recently at Damoh district of Madhya Pradesh (Fig. 1) in early 2012 and also reported initial discovery of gas in all the other four wells in the Damoh-Jabera-Katni nomination block. Geologically it was assumed that the region stands dated back to Proterozoic era, approximately 200 million years before the first Dinosaur. It was the time when the earth was on a stir and the region suffered a series of volcanic eruptions in the Vindhyan basin. The Vindhyan basins are concealed under nearly 2 km hard covered, 400-million-year-old rock.

The interpretation suggested various probable intrusions across the SNNF and SNSF. However, SNNF and SNSF have been seen as various extensions and marked as shown in the Fig. 13. The SNNF and SNSF are oriented in the ENE-WSW direction. The northern part of the study area signifies the dominated Northern fault that is referred as Sara-Maihar Fault. Similarly, southern fault is

situated in the southern end, named as Khamaria-Raipur-Karwa Fault. Mahakoshal rocks intruded in between SNNF and SNSF.

GHOSH and SINGH (2013b) studied the source edge location using 3D Euler deconvolution and carried out geological interpretation and show various faults; however, the locations are scattered and there are certain gaps in that interpretation. However, in the present study, the interpretation of various derivative analysis and the thrust-fault locations show very prominent and clear results. The THDR and cos (θ) map interpretation demarcates the boundary locations. Previous study shows faults are discontinuous; however, the present study clarifies that the northern fault (Sara-Maihar fault) and southern fault (Khamaria-Raipur-Karwa Fault) are continuous and shows the appropriate locations. The granitic intrusions are probably marked in the earlier methods which were unclear; however, this study provides the probable locations of granitic intrusions. It is understood that the present study provides a detailed interpretation showing thrust/fault boundaries using magnetic data.

5. Summary and Conclusion

The study area is situated in this transition zone which is seismically active and is one of the most prospective areas for hydrocarbon exploration. Narmada-Son lineament is one of the most geologically active prominent lineaments across the central India after Himalaya. Higher magnetic anomalies (1100 nT) are observed at the northern part and lower magnetic anomalies (−1100 nT) are observed at the southern part. Son Narmada North Fault (SNNF) and Son Narmada South Faults (SNSF) are the two major significant faults and have divided the study area after the collision of the Indian plate and the Eurasian plate. This causes swell in the crust and generates horst structure in between the Bundelkhand protocontinent to the north and Dharwar protocontinent in the south. Mahakoshal rocks belong to nonmagnetic high intrusive rock in between the Vindhyan and Gondwana rocks. This study has been carried out using joint interpretation using total horizontal derivative (THDR), tilt angle derivative (TDR), horizontal tilt angle derivative (TDX) and Cos (θ) using magnetic data. The results suggest noticeable source edge locations and structural boundaries which are very well correlated and in good agreement with the results obtained by GHOSH and SINGH (2013b). The above study suggests that THDR, TDR, TDX, SED and Cos (θ) are independent of inclination of the magnetic field. The patterns of the total magnetic field are similar to THDR, TDR and TDX which are oriented in the ENE–WSW direction. It is noted that TDR has its zero contour value indicated nearer to the boundary edges. However, the TDX derivative is more sharper compared to the TDR derivative at the maxima centered over the body edges. This area belongs to high potential in terms of coal bed methane, more prospective for hydrocarbon exploration and extensive coal mining operations; hence the derived results carried here through integrated studies will help to study the area for better understanding.

REFERENCES

ACHARYA, S. K. and ROY, A. (2000), *Tectonothermal history of the central Indian tectonics zone and reactivation of major faults/ shear zone,* J Geol Soc India, 239–256.

BARBOSA, V. C. F., SILVA, J. B. C., and MEDEIROS, W. E., (1999), *Stability analysis and improvement of structural index estimation in Euler deconvolution:* Geophysics, 64, 48–60, doi:10.1190/1. 1444529.

BHATTYACHARJEE, S. and KOIDE, H. (1978), *The origin and evaluation of rifts and rift valley structures: a mechanistic interpretation in Tectonic and Geophysics of continental rifts,* pp 29–37 (eds I. B. Ramberg and E. R. Neumann) (D'Reidal Publishing Company, England).

BHATTACHARYA, A. and MORAD, S. (1993), Proterozoic *braided ephemeral fluvial deposits: An example from the Dhandraul sandstone formation of the Kaimur group, Son Valley, Central India.* Sediment. Geol., *84,* 101–114.

BHATTACHARJI, S., SHARMA, R. and CHATTERJEE, N. (2004), *Two- and three- dimensional gravity modeling along western continental margin and intraplate Narmada-Tapti rifts: Its relevance to Deccan flood basalt volcanism,* Proc. Indian Acad. Sci. (Earth Planet. Science.), *113,* 4, 771–784.

BLAKELY, R.J. and SIMPSON, R.W. (1986), Approximating *edges of source bodies from magnetic or gravity anomalies.* Geophysics, *51,* 7, 494–1498.

CHAKRABORTY, C. and BHATTACHARYA, A. (1996), *Fan – Delta sedimentation in a foreland moat: Deoland formation, Vindhyan supergroup, Son Valley. In:* A. Bhattacharya (Ed.) Recent advances in Vindhyan Geology. Mem. Geol. Soc. India, no. *36,* 27–48.

CHAKRABORTY, P. P., BANERJEE, S., DAS., N. G., SARKAR, S. and BOSE, P. K. (1996), *Volcaniclastics and their sedimentological bearing in Proterozoic Kaimur and Rewa groups in central India.* In: A. Bhattacharya (Ed.), Recent advances in Vindhyan Geology, Mem. Geol. Soc. of India, *36,* 59–76.

CORDELL, L, and GRAUCH. V. J. S. (1982), *Mapping basement magnetization zones from aeromagnetic data in the San Juan Basin; New Mexico*: Presented at the 52nd Ann. Internat. Mtg., Sot. Explor. Geophys., Dallas; abstracts and biographies, 246–247.

CORDELL, L, and GRAUCH. V. J. S. (1985), *Mapping basement magnetization zones from aeromagnetic data in the San Juan basin, New Mexico.* in Hinze. W. J., Ed.. The utility of regional gravity and magnetic anomaly maps: Sot. Explor. Geophys. 181–197.

COOPER, G. R. J., and D. R. COWAN, (2006), *Enhancing potential field data using filters based on the local phase,* Computers and Geosciences, *32,* 1585–1591.

CORAGGIO, F., BERNARDELLI, P and GABBRIELLINI, G. (2012), *Structural reconstruction using potential field data in hydrocarbon exploration, SEG Las Vegas 2012, Annual Meeting.* doi:10.1190/ segam2012-0983.1.

DIRECTORATE GENERAL of HYDROCARBONS (DGH), (Under Ministry of Petroleum & Natural Gas, Govt. of India. (Web site http:// www.dghindia.org).

FAIRHEAD, J. D., A. SALEM, L. CASCONE, M. HAMMIL, S. MASTERTON, and E. SAMSON, (2011), *New developments of the magnetic tilt-depth methodto improve structural mapping of sedimentary basins,* Geophysical Prospecting, *59,* 1072–1086, doi:10.1111/j. 1365-2465-2478.2011.01001.x.

FAIRHEAD, J.D and WILLIAMS, S.E. (2006), *Evaluating Normalized Magnetic Derivatives for Structural Mapping,* SEG 2006 New Orleans Extended Abstract.

FAIRHEAD, D., WILLIAMS, S. and SALEM, A.B. (2007), *Structural mapping from high resolution aeromagnetic data in Namibia using normalized derivatives*, EGM 2007 International Workshop, Italy, April 15–18.

FERREIRA, FRANCISCO J. F.; de SOUZA, JEFERSON; de B. e S. BONGIOLO, ALESSANDRA; de CASTRO, LUÍS G, (2013), *Enhancement of the total horizontal gradient of magnetic anomalies using the tilt angle,* Geophysics, *78*, 3, J33–J41.

FERREIRA, F. J. F., L. G. de CASTRO, A. B. S. BONGIOLO, J. de SOUZA, and M. A. T. ROMEIRO, (2011), Enhancement of the total horizontal gradient ofmagnetic anomalies using tilt derivatives: Part II—Application to real data, 81st Annual International Meeting, SEG, Expanded Abstracts, 887–891.

FRANCISCO, J. F. F., JEFERSON, D. S, ALESSANDRA, D. B. and LUÍS, G.D.C. (2013), *Enhancement of the total horizontal gradient of magnetic anomalies using the tilt angle,* Geophysics, *78*, 3, J33–J41. doi:10.1190/geo2011-0441.1.

GHOSH, G.K. and SINGH, C.L. (2011), *Shallow crustal configuration of the Narmada–Son lineament transition zone near the Sahdol–Katni area of Central India using simultaneous gravity and magnetic observations,* Pure Applied Geophysics, 168, *5*, 845–860. doi:10.1007/s00024-010-0174-1.

GHOSH, G.K. and SINGH, C.L. (2013a), *Intrusion and upliftment of Mahakoshal Rocks between Vindhyan and Gondwana in Narmada Son Lineament, central India,* Journal Geological Society of India, *81*, 556–564. doi:10.1007/s12594-013-0071-1.

GHOSH, G.K. and SINGH, C.L. (2013b), *Crustal thickness mapping in Raipur-Katni area of Narmada Son Lineament in central India derived from 3-d Euler deconvolution of magnetic data,* Journal of earth system and Science, *122*, 5, 1399–1410. doi:10.1007/s12040-013-0352-5.

GRAUCH, V. J. S., M. N. HUDSON, and S. A. MINOR, (2001), *Aeromagnetic expression of faults that offset basin fill, Albuquerque basin, New Mexico,* Geophysics, *66*, 707–720.

JAIN, S. C, NAIR, K. K.K and YEDEKAR, D. B. (1995), *Tectonic evolution of the Son-Narmada-Tapti lineament zone,* Project Crumansonata, Special Publication. *10*, 333–371.

JOKHAN, RAM, (2005), *Hydrocarbon exploration in onland frontier basins of India- Perspectives and challenges,* Journal of the Palaeontological Society of India, *50*, 1, 1–16.

JOKHAN, RAM., SHUKLA, S. N., PARMANIK, A, G., VARMA, B. K., GYANESH, CHANDRA., and MURTY, M. S. N, (1996), *Recent investigations in the Vindhyan basin:Implications for the basin tectonics;* In: *Recent advances in Vindhyan geology* (ed.) Ajit Bhattacharya, Geol. Soc. India Memoir, *36*, 267–286.

KAILA, K. L., MURTY, P. R. K., MALL, D. M., DIXIT, M. M. and SARKAR, D. (1987), *Deep Seismic Soundings along Hirapur–Mandla profile: Central India.* Geophys. J. R. Astron. Soc., *89*, 399–404.

KAILA, K. L., MURTTHY, P. R. K., MALL, D. M. and DIXIT, M. M. (1989), *The evolution of the Vindhyan basin vis a vis the Narmada—Son lineament, Central India, from Deep Seismic Soundings,* Tectonophysics, *162*, 277–289.

KEDAR, NARAIN. (1962), *Systematic mapping in parts of Dudhi and Robertsganj tehsils of Mirzapur district. U.P. unpublished report,* Geological Survey of India, Progress Report.

KUMAR, P., TIWARI, H. C. and KHANDEKAR, G. (2000), An *anomalous high-velocity layer at shallow crustal depths in the Narmada zone, India.* Geophysics, J. Int., *142*, 95–107.

KUMAR, S., SCHIDLOWSKI, M. and JOACHIMSKI, M. (2005), *Carbon isotope Stratigraphy of the Palaeo-Neoproterozoic Vindhyan Supergroup, Central India: Implications for Basin Evolution and intrabasinal correlation,* Journal of the Palaeotological Society of India, *50*, 65–81.

LAHTI, I., and T. KARINEN. (2010), *Tilt derivative multiscale edges of magnetic data,* The Leading Edge, 29, 24–29, doi:10.1190/1.3284049.

MATHUR, P. (1995), *Precambrian emerging frontier of future exploration. Proceeding of International Petroleum conference,* Petrotech, *95*, II, 401–410.

MELO, F. F., BARBOSA, V. C. F., UIEDA, L., OLIVEIRA V. C. Jr., and SILVA, J. B. C. (2013), *Estimating the nature and the horizontal and vertical positions of 3D magnetic sources using Euler deconvolution:* Geophysics, *78*, 6, J87–J98, doi:10.1190/geo2012-0515.1.

MILLER, H. G. and SINGH, V. (1994), *Potential Field Tilt—a new concept for location of potential field sources,* Journal of Applied Geophysics, *32*, 213–217.

MISHRA, D. C. (2011a), *Long hiatus in Proterozoic Sedimentation in India: Vindhyan, Cuddapah and Pakhal Basins-A Plate Tectonics Model,* Journal of Geological Society of India. *77*, 1, 17–25. doi:10.1007/s12594-011-0004-9.

MISHRA, D. C. (2011b), *A Unified Model of Neoarchean-Proterozoic Convergence and Rifting of Indian Cratons: Geophysical Constraints,* International Journal of Geosciences, 610–630. doi:10.4236/ijg.2011.24063.

MUKHOPADHYAY, A., ADHIKARI, S., ROY, S.P., and BHATTACHARYYA, S., (2001a), *Eustacy, Climate, Tectonics, Sedimentary Environment and the Formation of Permian Coal measures in the Sohagpur coalfield, Madhya Pradesh, India,* Geological Survey of India, Spl. Pub. No. *54*: pp 305–320.

MUKHOPADHYAY, A., ROY, S.P., and ADHIKARI, S., (2001b), *Reactivation of Faults—Dulhara Fault of Sohagpur coalfield, An Example.* Geological Survey of India Spl Pub. No. *54*: pp 157–162.

MURTY, A. S. N., TEWARI, H. C. and REDDY, P. R. (2004), *2-D crustal velocity structure along Hirapur–Mandla profile in Central India: an update,* Pure and Applied Geophysics *161*, 165–184.

NABIGHIAN, M.N, (1972), *The analytic signal of two-dimensional magnetic bodies with polygonal cross-section: its properties and use for automated anomaly interpretation, 37*(3), 507–517. doi:10.1190/1.1440276.

NAGESWARA, RAO. B., KUMAR N., SINGH, A.P., PRABHAKAR, RAO. M. R. K., MALL, D. M. and SINGH, B. (2011), *Crustal density structure across the Central Indian Shear Zone from gravity data,* Journal of Asian Earth Sciences, *42*, 3, 341–353. doi:10.1016/j.jseaes.2011.04.023.

NAQVI, S.M, DIVAKAR RAO, V., and NARAYAN, H. (1974), *Archaean protocontinental growth of the Indian Shield and the antiquity of the rift valleys,* Precamb. Res, *1*, 345–398.

PASCOE, E. H. (1965), A manual of geology of India and Burma, Vol. I. Govt. of India publication, Calcutta.

PHILLIPS, J. D. (2000), *Locating magnetic contacts: a comparison of the horizontal gradient, analytic signal, and local wavenumber methods,* SEG Expanded Abstracts.

PILKINGTON, M. and KEATING, P. (2010), *Geologic applications of magnetic data and using enhancements for contact mapping,* EGM 2010 International Workshop, Capri, Italy, April 11–14, 2010.

PROJECT CRUMANSONATA (1995), Geoscientific studies of the Son-Narmada-Tapti lineament, pp. 371.

QURESHY, M. N. (1981), *Gravity anomalies, isostasy and crust mantle relations in the Deccan Trap and contiguous regions, India. In: Deccan volcanism and related basalt provinces in other parts in the world,* (eds. K. V. Subha Rao and R.N. Sukheswala) Memorial Geol. Soc. of India, *3,* 184–197.

RADHAKRISHNA, B. P. and RAMAKRISHNAN, M. (1988), Archaean Proterozoic boundary in India, Jour. Geol. Soc. of India, *32,* 263–278.

RAO, C. K., OGAWA Y., GOKARN, S. G. and GUPTA, G. (2004), Electromagnetic *imaging of magma across the Narmada Son lineament, central India.* Earth Planets Space, *56,* 229–238.

REID, A. (2007), *Semi-Automated Methods of Potential Field Interpretation—Innovations, and Recent and Future Developments,* EGM 2007 International Workshop. Capri, Italy, April 15–18.

REID, A. B, ALLSOP, J. M., GRANSER, H., MILLET, A, J, and SOMERTON, I. W. (1990), *Magnetic interpretation in three dimensions using Euler deconvolution,* Geophysics, *55,* 80–91.

REID, A.B, and THURSTON, J.B. (2014), *The structural index in gravity and magnetic interpretation: Errors, uses, and abuses,* Geophysics, *79.* J61–J66. doi:10.1190/GEO2013-0235.1.

REID, A, B., EBBING, J. O, and SUSAN, S. J. (2014), *Avoidable Euler Errors—the use and abuse of Euler deconvolution applied to potential fields,* Geophysical Prospecting, *62,* 1162–1168, doi:10.1111/1365-2478.12119.

ROY, A. and BANDYOPADHYAY, B. K. (1998), *Tectonic significance of ultramafic and associated rocks near Tal in the Mahakoshal belt, Sidhi district, M.P.,* Journal Geological Society of India, *32,* 397–410.

ROY, A. and DEVRAJAN, M. K. (2000), *A reappraisal of the stratigraphy and tectonics of the Palaeo-Proterozoic Mahakoshal supra crustal belt, central India,* Geol. Survey of India, Spl. Publ, *57,* 79–97.

SALEM, A., S. WILLIAMS, J. D. FAIRHEAD, D. RAVAT and R. SMITH, (2007), *Tilt-depth method; A simple depth estimation method using first-order magnetic derivatives,* SEG The Leading Edge, v. *26/12,* 1502–1505.

SALEM, A., WILLIAMS, S., FAIRHEAD, J. D., SMITH, R. and RAVAT, D.J, (2008), *Interpretation of magnetic data using tilt-angle derivatives,* Geophysics, *73,* L1–L10.

SALEM, A., S. WILLIAMS, E. SAMSON, D. FAIRHEAD, D. RAVAT, and R. J. BLAKELY, (2010), *Sedimentary basins reconnaissance using the magnetic tilt-depth method,* Exploration Geophysics, *41,* 198–209.

SAMAL, J. K. and MITRA, D. S. (2006), Field study of shear fractures—its tectonic significance and possible application in hydrocarbon exploration—an example from Vindhyan basin. 6th International Conference & Exposition on Petroleum Geophysics "Kolkata 2006" 35–42.

SANTOS, D. F., J. B. C. SILVA, V. C. F. BARBOSA, and L. F. S. BRAGA, (2012), *Deeppass-An aeromagnetic data filter to enhance deep features in marginal basins:* Geophysics, *77,* 3, J15–J22, doi:10.1190/geo2011-0146.1.

SESHUNARAYANA, T., VISWAJA, D., SRINIVAS, K. N. S. S. S., RAO, S. P., TRUPTI, S., KISHORE, P.P. (2008), Application of High Resolution Seismic Survey in CBM Exploration—A Case study, Sohagpur West Block, Madhya Pradesh, 7th international conference and exposition on petroleum geophysics, Society of Petroleum Geophysicist, Hyderabad.

SILVA, J. B. C., and BARBOSA, V. C. F., (2003), *3D Euler deconvolution: Theoretical basis for automatically selecting good solutions:* Geophysics, *68,* 1962–1968, doi:10.1190/1.1635050.

SILVA, J. B. C., BARBOSA, V. C. F., and MEDEIROS, W. E., (2001), *Scattering, symmetry, and bias analysis of source-position estimates in Euler deconvolution and its practical implications:* Geophysics, *66,* 1149–1156, doi:10.1190/1.1487062.

SINGH, A. P. (1998), *3-D structure and geodynamic evolution of accreted igneous layer in the Narmada-Tapti region (India),* Journal of Geodynamics 25, 129–141.

SINGH, A. P and MEISSNER, R. (1995), *Crustal configuration of the Narmada-Tapti region (India) from gravity studies,* Journal of Geodynamics, *20,* 111–127.

SRIVASTAVA, R. P., VEDANTI, N. and DIMRI, V. P. (2007), *Optimum design of a gravity survey network and its application to delineate the Jabera–Damoh structure in the Vindhyan Basin, Central India,* Pure and Applied Geophysics, *64,* 1–14, doi:10.1007/s00024-007-0252-1.

STAVREV, P. Y. (1997), *Euler deconvolution using differential similarity transformations of gravity or magnetic anomalies,* Geophysical Prospecting, *45,* 207–246.

THOMPSON, D. T. (1982), EULDPH: *A new technique for making computer-assisted depth estimates from magnetic data,* Geophysics, *47,* 31–37.

THURSTON and BROWN, (1994), *Automated source-edge location with a new variable pass-band horizontal-gradient operator:* Geophysics, *59,* 546–554.

THURSTON, J. B. and R. S. SMITH, (1997), *Automatic conversion of magnetic data to depth, dip, and susceptibility contrast using the SPI (TM) method.* Geophysics, *62,* 807–813.

VALDIYA, K. S. (1984), Aspects of tectonics focus on south-central Asia, Tata McGraw–Hill, New Delhi, p. 310.

VALDIYA, K.S., BHATIA, S.B. and GAUR, V.K. (1982), Geology of the Vindhyanchal. Hindustan Publishing Corporation, New Delhi, 231p.

VERDUZCO, B., J. D. FAIRHEAD, C. M. GREEN, and C. MACKENZIE, (2004), *New insights into magnetic derivatives.* The Leading Edge, *22,* 116–119, doi:10.1190/1.1651454.

VERMA, R. K. and BANERJEE, P. (1992), *Nature of continental crust along the Narmada-Son lineament inferred from gravity and deep seismic sounding data,* Tectonophysics, *202,* 375–397.

WAGHMARE, S. Y., PIMPRIKAR, S. D., GAWALI, P. B., CARLO, L. and PATIL, A.G. (2009), *Tectonomagnetic study in the seismoactive area of Narmada–Son lineament, central India, Preliminary results on repeat field observations.* Journal of Earth System and Science, *118,* 3, 261–272.

WANG, W., PAN, YU, and QIU Z, (2009), *A new edge recognition technology based on the normalized vertical derivative of the total horizontal derivative for potential field data,* Applied Geophysics, *6,* 3, 226–233, doi:10.1007/s11770-009-0026-x.

WIJNS, C., C. PEREZ, and P. KOWALCZYK, (2005), *Theta Map: Edge detection in magnetic data,* Geophysics, *70,* L39–L43.

YAGHOOBIAN, A. BOUSTEAD, G. A and DOBUSH, T. M. (1992), Object delineation using Euler's Homogeneity Equation. Location and Depth Determination of Buried Ferro-Metallic Bodies, Proceedings of SAGEEP 92, San Diego, California, 613–632.

YEDEKAR, D. B., JAIN, S. C., NAIR, K. K. K. and DUTTA, K. K. (1990), The Central Indian collision suture. In: Precambrian of Central India, Geol. Surv. India, Spl. Pub. no.28, 1–43.

(Received October 29, 2014, accepted April 7, 2015, Published online April 23, 2015)

Pure Appl. Geophys. 173 (2016), 805–825
© 2015 Springer Basel
DOI 10.1007/s00024-015-1107-9

| Pure and Applied Geophysics

Integral Approaches to Determine Sub-Crustal Stress from Terrestrial Gravimetric Data

MEHDI ESHAGH[1]

Abstract—The spherical harmonic expressions of the horizontal sub-crustal stress components induced by the mantle convection are convergent only to low degrees. In this paper, we use the method of stress (S) function with numerical differentiation and present a formula for determining the degree of convergence from the mean Moho depth. We found that for the global mean Moho depth, 23 km, this convergence degree is 622 and for Iran, 35 km, it is 372. Also, three methods are developed and applied for computing the sub-crustal stress, (1) direct integration with a spectral kernel limited up to the degree of convergence, (2) integral inversion with a kernel having closed-form formula without any frequency limit, and (3) solving an integral equation with limited spectral kernel to the convergence degree. The second method has no divergence problem and its kernel function is well behaving so that the system of equations from which the S function is determined is stable, and no regularisation is needed to solve it. It should be noted that for using this method the resolution of the recovery should be higher than $0.5° × 0.5°$, otherwise the recovered S function and correspondingly the stress components will have smaller magnitude than those derived from the other two methods. Our numerical studies for stress recovery in Iran and its surrounding areas show that the methods, which use the limited spectral kernels to the convergence degree, deliver consistent results to that of the spherical harmonic expansion.

Key words: Inversion, Integral equations, Integral method, Moho spectra, Stress function with numerical differentiation, Stress, Gravity.

1. Introduction

The gravity field of the Earth is a signature of the Earth's interior structures and it can be used for studying sub-lithospheric stress due to the mantle convection. The main challenging issue is how to extract such information from the gravimetric measurements. Modelling such stresses is important for geophysical and geographical interpretation of the

seismicity, volcanism, kimberlite magmatism, ore concentration, and tectonic and magnetic features (LIU 1977). The gravimetric data can play a significant role for modelling the stress in addition to the seismic data. In the following subsections, we divide the introduction into three parts. In the first part, the studies about the relation between the mantle convection and gravity field are presented and in the next two subsections, some studies on the relations between the tectonic and stress with gravity and geoid are mentioned, respectively.

1.1. Mantle Convection, Gravity and Geoid

Numerous studies have been done about the mantle convection modelling using gravimetric data. For such a purpose, the Earth is assumed to be constructed by different layers with special properties. The geoscientists tried to find the best fit to the gravity field and/or the geoid by changing the properties of the layers. However, in most cases, a forwards modelling method is used, but RUNCORN (1962, 1963, 1964, 1967) investigated the mantle convection currents and their determination from satellite gravity measurements. MCKENZIE (1967) used gravity anomalies for studying the heat flow and concluded that the long wavelength harmonics of the external gravity field cannot be supported by the strength of the lithosphere. WALTZER (1970) presented a simple and symmetric model for the convection current in the Earth's mantle by assuming that the mantle becomes more and more homogenous downward. Marsh and Marsh (1976) investigated a two-dimensional mantle convection model based on the global gravity anomalies and LUX et al. (1979) studied the movements of the lithospheric plate due to the mantle convection. RUNCORN (1980) discussed the mechanism of plate tectonic based on a three-

[1] Department of Engineering Science, University West, Trollhättan, Sweden. E-mail: mehdi.eshagh@hv.se

mantle convection pattern and mentioned that the geoid determined by satellite observations is the direct consequence of convective upwelling and downward currents in the mantle. HUANG and FU (1982) studied the convection pattern and force sources mechanism of the tectonic in China. HAGER (1983) studied global isostatic geoid anomalies for plate and boundary layer models of the lithosphere. RICHARDS and HAGER (1984) combined different types of data for improving the structure of the viscosity of the mantle. FORTE and PELTIER (1987) investigated the plate kinematics and mantle convection. FU (1989) investigated the mantle convection based on the thermal-convection model and concluded that the absolute motion of the rigid plates are correlated with the mantle thermal convection and associated with the geoid. In addition, the motion of the plates does not completely reflect the mantle flows below, and they are caused by coupling between the flowing viscous mantle and the rigid lithosphere. NATAF (1991) considered the relation between mantle convection and plate and hotspots and DAVIES and RICHARDS (1992) modelled this process. MONNEREAU and QUERE (2001) developed further the convection modelling by considering spherical shells. FU et al. (2003) presented a new convection models constrained by seismic tomography data.

In these studies, the long wavelength structure of the geoid and gravity field are used to study the mantle convection currents. Today, based on the advances in technology, different satellite missions, e.g., Chalenging MiniSatellite Payload (CHAMP) (REIGBER et al. 2004), the Gravity Recovery and Climate Experiment (GRACE) (Tapley et al. 2005) and the Gravity field and steady-state Ocean Circulation Explorer (GOCE) (FLOBERGHAGEN et al. 2011) have been developed and used for obtaining a precise and reliable long wavelength structure for the gravity field, which is useful for such studies.

1.2. Plate Tectonic, Gravity and Geoid

HAGER and O'CONNELL (1981) presented a simple global model of plate dynamics. RICARD et al. (1988) presented another model relating the geoid and global plate motions. HAGER and RICHARDS (1989) investigated the long wavelength variations of the Earth's geoid and

presented physical models and dynamical implications for them. JACOBY and SEIDLER (1981) studied the relation between the plate kinematics and the gravity field. SOURIAU and SOURIAU (1983) studied the global tectonic using geoid model, which derived by the gravimetric data. The global tectonic is the subject, which has been studied by the gravity field data and their changes in time as the lithosphere, containing the tectonic plates, is floating on the convective mantle changing the gravity field of the Earth. Such convective flows are the main reasons for the movements of the tectonic plates and the stresses in the crust. Therefore, these issues can also be studied by the gravimetric data.

1.3. Stress, Gravity and Geoid

RUNCORN'S (1967) simplified the Navier–Stokes equation to find the direct connection between the geoid with the sub-crustal stress. He presented this field in terms of spherical harmonics of the Earth's gravity field. ARTYUSHKOV (1973) mentioned that the stress in the crust is caused by crustal thickness inhomogeneities, which are somehow presented in the Earth's gravity field. Based on Runcorn's theory, LIU (1977) presented the convection pattern and stress system under the African plate and LIU (1978) did a similar study for the sub-crustal stress under Asia. Later on LIU (1979) presented a theory about the sub-crustal stress concentration and its relation with seismogenic model for the Tangshan earthquake. LIU (1980, 1985) studied the convection generated stress field and the intra-plate volcanism. DAHLEN (1982) investigated the isostatic geoid anomalies on the sphere, based on the HAXBY and TURCOTTE (1978) results, and mentioned that the stress in the crust influences the long wavelength structure of geoid, which is in agreement with what RUNCORN (1967) mentioned. McNUTT (1980) implemented the regional gravity field for studying the stress in the crust and upper mantle and stated that the stresses implied in the regional compensation scheme are an order of magnitude larger than those corresponding to local one. RUNCORN'S (1967) method gives the sub-crustal stress towards the north and the east and he assumed that there is no vertical stress in his model. However, his formulae are used to model the crustal or lithospheric stress as boundary conditions for solving the boundary-value problem of elasticity for the Earth. In this case, the tensor of stress, with six independent elements, can be computed for the crust or lithosphere. Fu and Huang (1983) used the satellite-derived gravitational harmonics for modelling the global stress fields in the lithosphere based on an elastic earth model and solutions of the elastic equations in spherical regions by Love (1944). Ricard *et al.* (1984) investigated the connection between the lithospheric stress and geoid height. Fu (1986) considered that the mantle is isoviscous, Newtonian liquid shell with a uniform distribution of heat sources, and tried to consider the boundary between the core and mantle in formulating the stress. Fu and Huang (1990) presented a global stress pattern constrained on deep mantle flow and tectonic features. Fu and Huang (1992) investigated the drag force caused by the mantle flow and the force system along plate boundaries to form the stress field in the lithosphere.

Pick and Charvatova-Jakubkova (1988) modified the Runcorn's formulae to reduce the contribution of the far-zone gravity anomaly and geoid height for local applications. Pick (1994) presented closed-form formulae for the kernel of the integral involving the gravity anomaly and geoid height to model the stress below crust. Naliboff *et al.* (2012) investigated the relation between the lithospheric thickness and density structure on the Earth stress field. Eshagh (2014a) developed the Runcorn (1967) formulae in such a way that the satellite gradiometry data can be used for determining the sub-crustal stress and Eshagh (2015) found the mathematical model between the sub-crustal stress and a gravimetrically-determined Moho model by Vening Meinesz–Moritz theory (Sjöberg 2009). Later on, Eshagh and Tenzer (2015) applied this method for sub-crustal stress determination, in some places of the world having special geophysical properties, and interpreted the results. Tenzer *et al.* (2015) studied the stress in subduction zones in Taiwan based on the gravimetric data.

1.4. In this Study

In this paper, we developed Runcorn's (1967) theory further and assume that it is valid. We present some integral formulae for recovering the sub-crustal stress from terrestrial gravimetric data, as Runcorn's formulae only allow the use of the Earth's gravity

models. On the other hand, his formula is not convergent and a high degree gravity model cannot be used. Eshagh (2014a) proposed the method S with numerical differentiation to make formulae convergent to higher degrees. S is a function, the derivatives of which towards the north and the east are the shear stress components. Theoretically, this method is the same with Runcorn's formulae but practically better. The convergence degree of S is dependent on the thickness of the outer layer in the two-layered Earth model and we present a formula to find it from this value. We develop three novel methods to recover S from terrestrial gravity anomalies: (1) direct integration with limited spectral kernel, (2) integral inversion with closed-form kernels and (3) integral inversion with limited spectral kernel. We will apply them for modelling the sub-crustal stress in Iran and its surrounding countries these integral methods are useful when the terrestrial gravity data of the area are used for estimating the stresses.

2. Sub-Crustal Shear Stress Components

The simplified Navier–Stokes equation by Runcorn (1967) is:

$$S_x\mathbf{e}_\theta + S_y\mathbf{e}_\lambda = \kappa \sum_{n=2}^{N} v_n \sum_{m=-n}^{n} T_{nm}X_{nm}^2(\theta,\lambda) \quad (1a)$$

where N is the degree of convergence and

$$\kappa = \frac{Mg}{4\pi(R-D)^2}, \quad v_n = \frac{1}{s^{n+1}}\frac{2n+1}{n+1} \text{ and } s = \frac{R-D}{R} \quad (1b)$$

and D stands for the mean Moho depth, M the mass of the Earth, g the gravity attraction, R is the mean radius of the Earth's sphere. $\mathbf{e}_\theta$ and $\mathbf{e}_\lambda$ are the unit vectors pointing towards the north and the east and S_x and S_y are the shear stress components in these directions. θ and λ stand for the co-latitude and longitude, respectively. $X_{nm}^2(\theta,\lambda)$ is the vector spherical harmonic of degree n and order m

$$X_{nm}^2(\theta,\lambda) = \frac{\partial X_{nm}^1(\theta,\lambda)}{\partial\theta}\mathbf{e}_\theta + \frac{\partial X_{nm}^1(\theta,\lambda)}{\sin\theta\partial\lambda}\mathbf{e}_\lambda \quad (1c)$$

with the following orthogonality property (e.g., Martinec 2003):

$$\iint_{\sigma} X_{nm}^{i}(\theta, \lambda).X_{n'm'}^{j}(\theta, \lambda)\mathrm{d}\sigma = 4\pi \begin{cases} \delta_{ij}\delta_{nn'}\delta_{mm'} \\ n(n+1)\delta_{ij}\delta_{nn'}\delta_{mm'} \end{cases}$$

$$(1d)$$

where δ stands for Kronecker's delta and σ is the unit sphere and $\mathrm{d}\sigma$ the surface integration element. $X_{nm}^{1}(\theta, \lambda)$ is the fully-normalised spherical harmonics. '.' stands for the inner product operator. According to Eq. (1d) the spherical harmonic coefficients of the disturbing potential (T) are derived by:

$$T_{nm} = \frac{1}{4\pi} \iint_{\sigma} T'X_{nm}^{1}(\theta', \lambda')\mathrm{d}\sigma \qquad (1e)$$

and the prime over T, θ and λ means that these parameters are related to the integration points, where T stands for disturbing potential and T_{nm} its spherical harmonic coefficients.

RUNCORN (1967) used a two-layered model for the Earth to solve the Navier–Stokes equation. He mentioned that the radial components of the velocity of tectonic movements must vanish at the boundary between the layers; this means that only tangential or shear stress components can be obtained in his solution. In order to obtain the full tensor of stress with 6 independent elements, the boundary-value problem of elasticity should be solved to model the stress inside the crust. To do so, Runcorn's formulae, Eq. (1a), are used as the boundary conditions to that problem. In fact, such a boundary-value problem of elasticity with Runcorn's stress formulae are used to propagate the sub-crustal stress into the crust; see e.g., FU and HUANG (1983, 1990) or LIU (1985) for more details.

The series (1a) is not convergent due to the presence of v_n, as $R - D < R$ and, therefore, the ratio $R/(R - D)$ is always larger than 1 and when it gets a power of $n + 1$, the ratio increases to infinity by increasing n, then the higher frequencies are amplified unboundedly and the series diverges. Limiting the series to a specific degree is an idea for controlling this divergence. PICK (1994) considered this ratio equals to 1 for developing some integrals with kernels having closed-form formulae. However, this assumption may not be realistic as amplification of high frequencies is an inevitable issue in such an inversion process, like the RUNCORN (1967) solution.

As we observe, Eq. (1a) involves the derivatives of the spherical harmonics, but in the case of considering the scalar spherical harmonics, the following function is obtained:

$$S = \kappa \sum_{n=2}^{N} v_n \sum_{m=-n}^{n} T_{nm}X_{nm}^{1}(\theta, \lambda) \qquad (1f)$$

This function contains v_n and is asymptotically convergent (MORITZ 1980) to higher degrees than Eq. (1a), which means that the series is convergent to a certain degree and after that it diverges. ESHAGH (2014a) named Eq. (1f), the S function, which has a scalar form and involves the spherical harmonics. Both of the equations contain v_n but Eq. (1a) involves the derivatives of the scalar spherical harmonics as well, which make the series even more divergent. Once S is determined, we can compute its derivatives towards the north and the east numerically as the stress components are the slopes of S in these directions. In the following, we present 2 propositions, presenting integral formulae, for recovering S from local gravimetric data.

Proposition 1 The direct integral formula for computing S from gravity anomaly is:

$$S = \frac{g}{16\pi^2 s^2 G} \iint_{\sigma} \sum_{n=2}^{N} \frac{(2n+1)^2}{n^2-1} \frac{1}{s^{n+1}} P_n(\cos\psi)\Delta g'\mathrm{d}\sigma$$

where $\Delta g'$ is the gravity anomaly at the integration points and $P_n(\cos\psi)$ stands for the Legendre polynomial of degree n and ψ is the geocentric spherical angle between the computation and integration points. G is the Newtonian gravitational constant.

Proof We can rewrite Eq. (1f) in the following spectral form:

$$S_n = \kappa v_n T_n \quad for \, n < N \qquad (2a)$$

where T_n and S_n are the Laplace coefficients of T and S. T_n has the following well-known relation with Laplace coefficient of gravity anomaly (Δg_n) (HEISKANEN and MORITZ 1967, P. 89):

$$\Delta g_n = \frac{n-1}{R}T_n \qquad (2b)$$

Considering Eq. (2b), Eq. (2a) becomes:

$$S_n = \kappa v_n \frac{R}{n-1}\Delta g_n \qquad (2c)$$

On the other hand, Δg_n is (cf. HEISKANEN and MORITZ 1967, p. 30):

$$\Delta g_n = \frac{2n+1}{4\pi} \iint\limits_{\sigma} \Delta g' P_n(\cos\psi)\mathrm{d}\sigma \qquad (2d)$$

Substituting Eq. (2d) into Eq. (2c), considering Eq. (1b) and further simplifications of the result yields:

$$S_n = \frac{gR^2}{16\pi^2 s^2 G} \frac{(2n+1)^2}{n^2-1} \frac{1}{s^{n+1}} \iint\limits_{\sigma} \Delta g' P_n(\cos\psi)\mathrm{d}\sigma. \qquad (2e)$$

Having taken a summation from both sides of Eq. (2e) from degree 2 to N, the proposition is proved. Q.E.D.

Note that in Eq. (2a) T_n should be free of any unit so that S is derived in Pascal (kg/ms^2). Therefore, we have to multiply T_n by R/GM. This is the reason of disappearance of M in the formula (2e). One issue of the integral of Proposition 1 is that it is not convergent due to appearance of s^{n+1} in the denominator of its kernel. This factor is always smaller than 1 and causes that $1/s^{n+1}$ grows up unboundedly when n increases. If n goes to ∞ the kernel diverges and finding a closed-form formula for that is not possible, but we can always use its spectral form to the degree of convergence (N) in practice. Similar integral formula to that of Proposition 1, was presented by PICK (1994) for recovering the stress components. However, PICK (1994) assumed that $s = 1$ so that he could find a closed-form formula for the kernel function needed for integration. He obtained the following integral for S:

$$S = \frac{g}{16\pi^2 s^2 G} \iint\limits_{\sigma} \sum_{n=2}^{\infty} \frac{(2n+1)^2}{n^2-1} P_n(\cos\psi)\Delta g'\mathrm{d}\sigma \qquad (2f)$$

and showed that

$$\sum_{n=2}^{\infty} \frac{(2n+1)^2}{n^2-1} P_n(\cos\psi) = \frac{2}{\sin(\psi/2)} + 1 - 9\sin(\psi/2)$$
$$- \frac{33}{4}\cos\psi - \frac{1}{2}(1+9\cos\psi)\ln(1+\sin(\psi/2))$$
$$+ \frac{1}{2}(1-2\cos\psi)\ln(\sin(\psi/2)) \qquad (2g)$$

This kernel is singular at the computation point, but it is very well behaving (ESHAGH 2011), which means

that it has its largest value at the computation point and decreases fast and uniformly to zero. In fact the integral (2f) tries to recover the high frequencies of the stress field from the gravity anomaly. The reason is that the assumption of $s = 1$ means that the mean Moho depth is at the sea level, i.e., $D = 0$. In other words, the sub-crustal stress is generated at sea level and not below the crust. Although mathematically, this assumption is very helpful for finding the closed-form formula (2g), but it makes the model unrealistic. Having s^2 in the denominator of the integral (2f) just amplifies the stress derived at sea level by a factor of 1.005 for $D = 23$ km, which is very small comparing to R. The signal amplification due to the inversion should be done through all frequencies and multiplication of such a factor does not continue the computed stress downward to the mean Moho depth level. It is recommended using the spectral kernel presented in Proposition 1 to the limited degree N.

Proposition 2 The integral equation for computing S from gravity anomaly is:

$$\frac{s^2 G}{Rg} \iint\limits_{\sigma} K(\psi,s)S'\mathrm{d}\sigma = \Delta g$$

where S' stands for S at the integration points and

$$K(\psi,s) = \frac{s^2(\cos\psi - 2s)}{l^3} + 3\frac{s^3(s-\cos\psi)^2}{l^5} - \frac{s}{l} + s \text{ and}$$
$$l = \sqrt{1+s^2-2s\cos\psi}.$$

Proof After solving Eq. (2c) for Δg_n, we obtain:

$$\frac{S_n}{\kappa v_n} \frac{(n-1)}{R} = \Delta g_n \qquad (3a)$$

According to Eqs. (1b), (3a) changes to:

$$\frac{S_n 4\pi(R-D)^2}{Mg} s^{n+1} \frac{(n+1)}{(2n+1)} \frac{(n-1)}{R} = \Delta g_n \qquad (3b)$$

On the other hand, the Laplace harmonics of S is (cf. HEISKANEN and MORITZ 1967, p. 30):

$$S_n = \frac{2n+1}{4\pi} \iint\limits_{\sigma} S' P_n(\cos\psi)\mathrm{d}\sigma \qquad (3c)$$

Substitution of Eq. (3c) into Eq. (3b) and simplification of the result yields:

$$\frac{s^2 G}{Rg} s^{n+1} \left(n^2 - 1\right) \iint_\sigma S' P_n(\cos\psi) d\sigma = \Delta g_n$$

By taking the summation from degree 2 to ∞ we obtain the following integral formula:

$$\frac{s^2 G}{Rg} \iint_\sigma K(\psi, s) S' d\sigma = \Delta g \tag{3d}$$

where

$$K(\psi, s) = \sum_{n=2}^{\infty} \left(n^2 - 1\right) s^{n+1} P_n(\cos\psi). \tag{3e}$$

Note that $K(\psi, s)$ is not divergent and mathematically contains all frequencies. To find its closed-form formula, according to the Legendre expansion of l^{-1}, we have (cf. HEISKANEN and MORITZ 1967, p. 33, Eq. 1–80):

$$l^{-1} = \sum_{n=0}^{\infty} s^n P_n(\cos\psi) \tag{3f}$$

where $l = \sqrt{1 + s^2 - 2s\cos\psi}$. Therefore, the integral (3f) has two separate terms having the following closed-form expressions:

$$\sum_{n=0}^{\infty} n^2 s^{n+1} P_n(\cos\psi) = s^2 \frac{\partial}{\partial s}\left(s\frac{\partial l^{-1}}{\partial s}\right)$$
$$= \frac{s^2(\cos\psi - 2s)}{l^3}$$
$$+ 3\frac{s^3(s - \cos\psi)^2}{l^5} \tag{4a}$$

$$\frac{s}{l} = \sum_{n=0}^{\infty} s^{n+1} P_n(\cos\psi) \tag{4b}$$

By considering Eqs. (4a) and (4b) and removing the zero- and first-degree terms, because the gravity anomaly does not contain these degrees, the kernel function presented in the proposition is derived. Q.E.D.

3. Numerical Considerations and Applications

In the first part of our numerical studies, the convergence of the original formulae of Runcorn and S is investigated. After that, the behaviours of the kernels of the integral approach and the integral inversion method are then presented and discussed. S will be recovered from the real gravity anomalies of Iran and its surrounding areas by our integral approaches. Finally, the sub-crustal stress components are computed and their magnitudes are presented based on each method. The maps of the recovered stresses are compared with the distribution of volcanic areas, the active seismic points of World Stress Map database (HEIDBACH et al. 2008) and the tectonic boundaries.

3.1. Convergence and its Relation with Mean Moho Depth

As already mentioned the idea of using S is to have a convergent series to higher degrees than that of Runcorn. By squaring and adding both formulae of Runcorn, Eq. (1a), and taking the global average, the surface integration over the unit sphere according to Eq. (1d), we obtain:

$$S_H = \frac{1}{4\pi} \iint_\sigma \left(S_x^2 + S_y^2\right) d\sigma$$
$$= \kappa^2 \sum_{n=2}^{\infty} \frac{n(n+1)}{s^{2n+2}} \left(\frac{2n+1}{n+1}\right)^2 c_n \tag{5a}$$

where $c_n = \sum_{m=-n}^{n} T_{nm}^2$.

After squaring S, Eq. (2f), and taking the global average of the result based on Eq. (1d), we derive:

$$S_S = \frac{1}{4\pi} \iint_\sigma S^2 d\sigma = \kappa^2 \sum_{n=2}^{\infty} \frac{1}{s^{2n+2}} \left(\frac{2n+1}{n+1}\right)^2 c_n. \tag{5b}$$

It is obvious that neither of Eq. (5a) nor Eq. (5b) is convergent due to involving s^{2n+2} in their denominators, but finding a maximum degree at which the series is convergent is possible (MORITZ 1980) numerically. The signal spectra of each series can be plotted to see if they are smooth and uniformly decreasing through all degrees. If the signal spectra start growing up uniformly after a certain degree, the series will diverge. This degree, which we call it the convergence degree, is lower for Eq. (5a) than Eq. (5b) because of including the coefficient $n(n+1)$. In other words, Eq. (5a) contains two amplifying factors of s^{2n+2} and $n(n+1)$, whilst

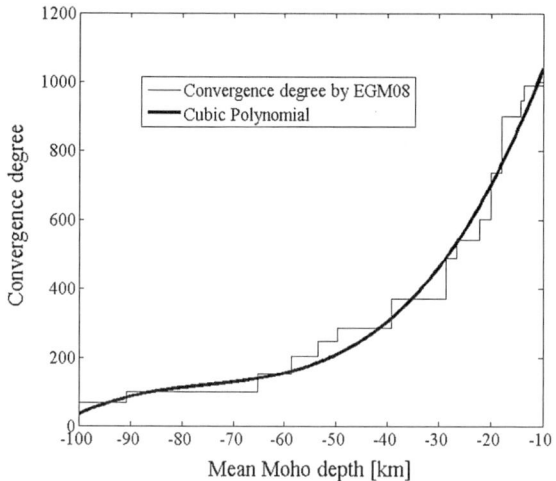

Figure 1
Convergence degree and mean Moho depth

3.2. Behaviour of Kernels of Integral Formulae

The behaviours of the kernels of the presented integral formulae are important as they show how significant the contribution of the quantity being integrated is to the result. In the case of using the integral presented in Proposition 1, the spectral form of the kernel should be used to the degree of convergence. Here, we selected two mean Moho depths of 23 and 35 km for presenting the kernel in Fig. 2a. Our goal for considering these two depths is to show that when D is smaller, or in other words, it is closer to mean sea level, higher frequencies of the stress components are achievable and the contribution of far-zone anomalies are less significant. In the case of considering $D = 23$ km, the kernel oscillates more than for $D = 35$ km and it has larger value at the computation point, $\psi = 0°$ than the other case. This means that higher frequencies of S are recoverable by such a kernel. Another issue is that both kernels have closer values to zero for $\psi > 5°$. This means that the integration area should be larger by $5°$ from each side of the desired area to reduce the effect of spatial truncation error of the integral.

Eq. (5b) only includes the former. Here, EGM08 (PAVLIS et al. 2012) to degree and order 1000 and the normal gravity field of the Geodetic Reference System 80 (GRS80) (MORITZ 2000) are used to generate c_n and plot the signal spectra of S. Another important issue is that s is also dependent on the mean Moho depth (D), and when D is large $1/s^{2n+2}$ is large as well and the convergence degree is lower. ESHAGH (2015) showed that this degree is about 25 for Eq. (5a) and about 600 for Eq. (5b) for when $D = 23$ km.

In order to find the relation between D and the convergence degree of Eq. (5b), we tested different values of D, from 10 to 100 km, and generated the signal spectra of S in each case. We found the degree of the spectrum having the smallest power and selected that as the convergence degree. Figure 1 shows the plot of convergence degrees derived from EGM08 for different values of D. We have tested different polynomials and found out that a cubic polynomial has the best fit to the plot. The following equation shows the mathematical relation between the convergence degree and D:

$$N = \text{fix}\left(-0.003D^3 + 0.67D^2 - 52D + 1500\right) \quad (5c)$$

where fix(.) is the rounding operator to have the smaller integer number for the degree of convergence. For example, when $D = 35$ km, $N = 372$ and when $D = 23$ km, $N = 622$.

In Fig. 2b we plotted the kernel of the integral formula presented in Proposition 2, which should be inverted to recover S. The kernel is well-behaving and suitable for the integral inversion of gravity anomaly to S and has a large value around $\psi = 0°$ meaning that higher frequencies can be recovered when the mean Moho is smaller. In addition, the kernel decays fast which means that system of equations, organised by discretising the integral formula, should be more stable numerically. Figure 2c shows the same kernel, but generated by its spectral form limited to maximum degree 372 as we already estimated that the formula is convergent to this degree for a Moho depth of $D = 35$ km. In fact, the plot of the kernel for $D = 23$ and 35 km coincided each other and therefore we presented one of them. This kernel has a large value at $\psi = 0°$, oscillates largely with an asymptotic decrease. As the plot shows, the kernel has non-zero values even up to $\psi = 20°$. We can expect to have a very ill-conditioned system of equations, after discretising the integral formula, for

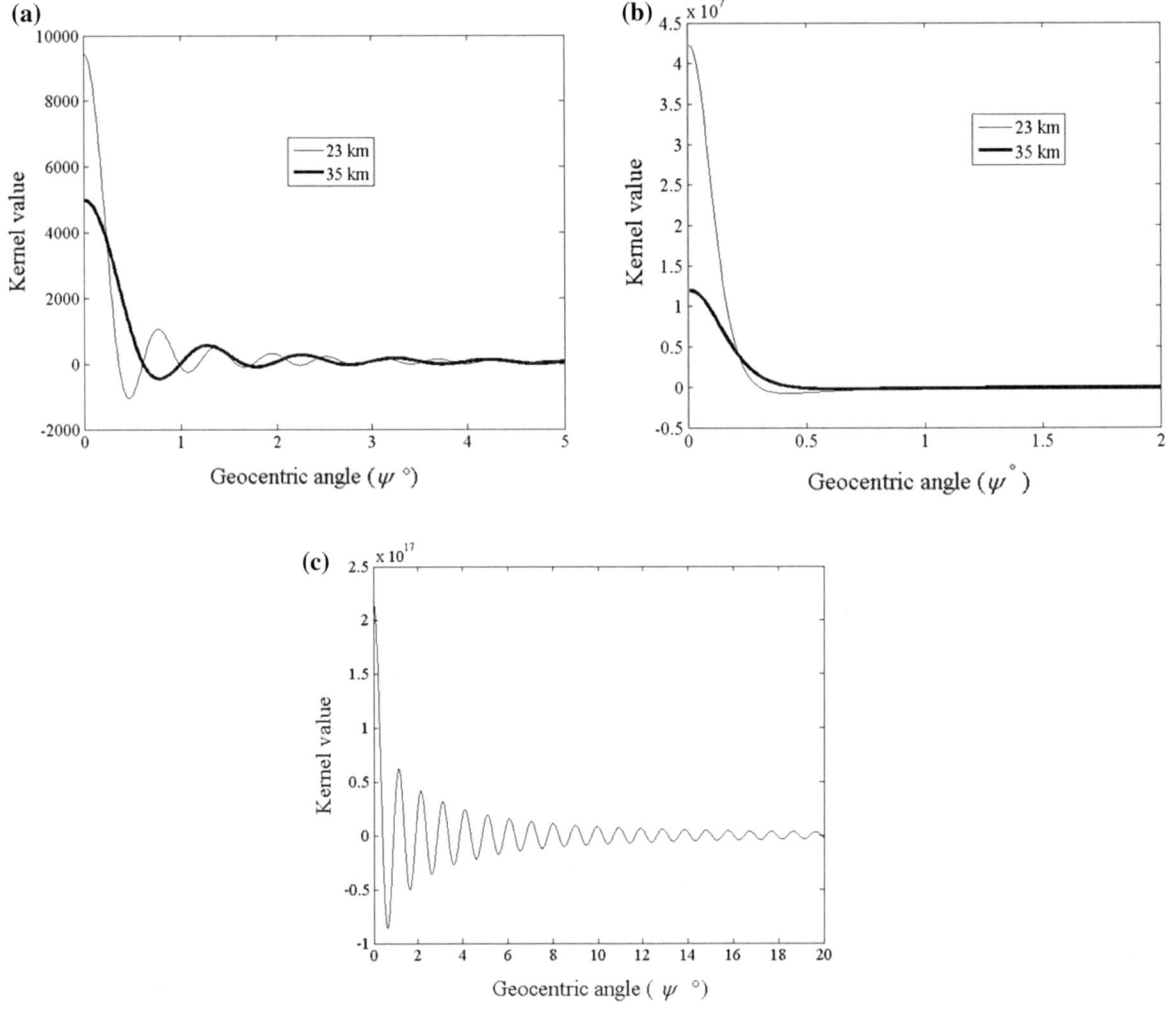

Figure 2
Kernel of **a** integral of Proposition 1, **b** integral of Proposition 2, **c** integral of Proposition 2 to degree 372

solving S by such a kernel especially when the resolution of recovery is high.

3.3. Sub-Crustal Shear Stress in the Study Area

The purpose of this section is the real application of the presented methods for the determination of the sub-crustal stress in the area. First, the geographical location of the country and the data are introduced, later on, the problem of recovering S by methods of spherical harmonic expression with EGM08, the direct integration of gravity anomalies with limited spectral kernel, Proposition 1, integral inversion of the anomalies by the closed-form formula of its

kernel, Proposition 2 and the spectral kernel are investigated. A discussion will be presented about the differences between the direct integration and integral inversion methods. In the last subsection, the problem of determining the sub-crustal stress components by numerical differentiation of the recovered S is presented and the results are compared with other geophysical information in the area.

3.3.1 The Study Area and Data

Here, an area limited between the latitudes 19°N and 46°N and the longitudes 39°E and 69°E comprising Iran and some parts of its neighbouring countries is

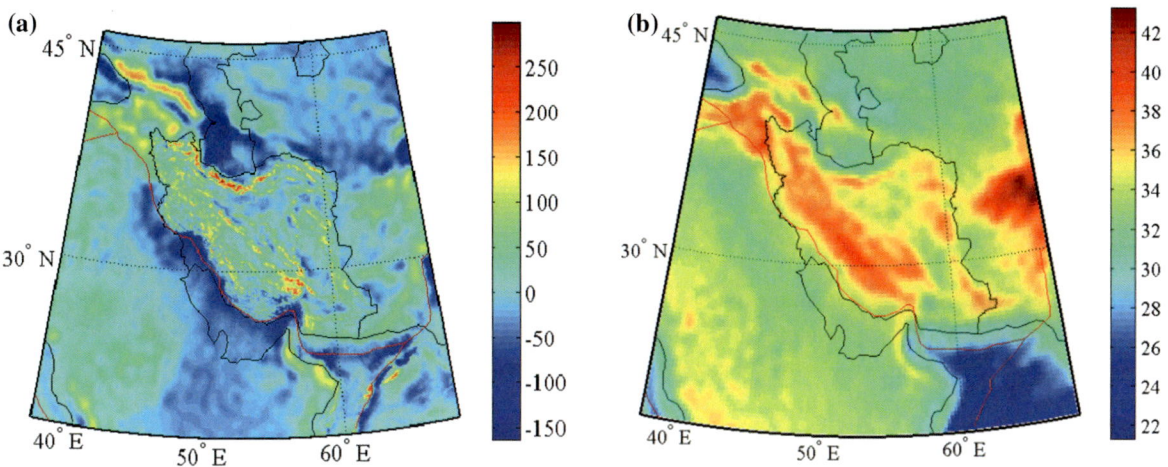

Figure 3
Map of **a** gravity anomalies (mGal), and **b** Moho depths (km)

selected as the test area. Iran has experienced many earthquakes so far, and the tectonic boundary between Arabian and Eurasian plates is extended from its south-east to the north-west part. In fact, this tectonic boundary partly follows the coastal line of the Persian Gulf, shown by the red line in Fig. 3. The area has been covered by 26125 gravimetric measurements, amongst which 9566 data belongs to the international gravimetric bureau (BGI), 8949 of them collected by the national cartographic centre (NCC) of Iran and 7610 are ship-borne gathered mainly in the Persian Gulf. Figure 3a shows the map of the free-air gravity anomalies with a resolution of $5' \times 5'$ at the surface of the reference ellipsoid. As can be seen, the largest values are seen around the northern and southern parts of the territory. Small values of the anomalies are seen along the tectonic boundary showing a mass deficiency there. The gravity anomalies of the area range from -150 to 300 mGal. Figure 3b is the map of the Moho depths computed gravimetrically based on the Vening Meinesz–Moritz theory (Sjöberg 2009) from EGM08 to degree and order 360 with a global average density contrast of 600 kg/m^3. Although gravity anomalies are large in the northern areas, but we observe the deepest part of the Moho under the Zagros Mountains, in the south-west part. The reason is that for computing the gravimetric model of Moho the Bougeur gravity anomalies of EGM08 to degree and order 360 have been used and the effect of topographic masses are

removed from them. The spherical harmonic coefficients of the Bouguer gravity anomalies are computed based on the topographic height of DTM2006 (Pavlis *et al.* 2006) and a constant density of 2.67 gr/cm^3 to degree and order 360 and applied for Moho modelling over the area. The maximum, mean, minimum, and standard deviation of the gravimetric Moho are 39.9, 34.9, 29.6 and 1.5 km, respectively. The corresponding ones for the CRUST1.0 (Laske *et al.* 2013) are 49.6, 37.5, 8.5 and 7.6 km. Eshagh (2014b) mentioned in the case of considering a mean depth of 35 km, the gravimetrically-determined Moho model is close to the seismic Moho model of CRUST1.0. This was the reason for considering $D = 35$ km for computing the kernels of the integral formula in the previous section. Consequently, according to Eq. (5c) for $D = 35$ km the convergence degree is 372, this means that not more than this degree should be used for generating S either from EGM08 or the terrestrial gravity anomalies by the spectral kernel of the integral formula.

3.3.2 Recovery of S Function

In this section, the free air gravity anomalies of the area and EGM08 is used to recover S by three methods of spherical harmonics, Eq. (1f), integral formula presented in Proposition 1 and inverting the integral of Proposition 2. Two resolutions of $0.25° \times 0.25°$ and $0.5° \times 0.5°$ are considered to see the importance of discretisation of the integrals in the recovered S. Resolution is an important factor in numerical integration especially for integral inversion, because the system of equations from which S is derived is more unstable when the resolution of S is higher.

The integral of Proposition 2 is discretised according to the resolution of recovery. A system of equations is organised in such a way that the S values are its unknowns and the right hand side of the equation is the vector of gravity anomalies. In fact, the integral equation is numerically solved and the gravity anomalies are inverted to the S function values by solving the system. The stability of the system is highly related to the resolution of the discretisation and the behaviour of its kernel function.

Now, we perform a closed-loop simulation study to see which method is better than another. Table 1 summarises the statistics of the differences between S generated from EGM08 directly by Eq. (1f) and the one generated based on our integral methods from the gravity anomalies generated from EGM08. The idea is to see which integral method delivers closer S to that of EGM08 directly. According to the table, the smallest root mean squared error (RMS) of the differences is related to the direct integration method with the limited spectral kernel to degree of convergence (DIR). Both inversion methods with and without closed-form kernels (INV + CLF) and (INV+SPC+REG) have RMSs of 0.49 and 0.50 MPa when the resolution of the recovery is $0.5° \times 0.5°$. In the case of considering higher resolution, again, the direct method is the best with an RMS of 0.40 MPa.

Figure 4a is the map of S determined by EGM08 to a maximum degree of 372 using Eq. (1f). Large and positive values of S are seen in the western part of the area. Figure 4b is the map of S derived by the direct integral formula, presented in Proposition 1. Here, the real gravity anomalies with a resolution of $0.5° \times 0.5°$ are integrated and the kernel is generated to degree 372. This map very well shows that S is not as smooth as that is obtained directly from EGM08. Since the maximum degree of the kernel is 372, then it removes the higher frequencies of the gravity anomalies as well. This is due to the orthogonality property of the Legendre polynomials in the kernel and in the Laplace coefficients of the anomaly. In Fig. 2a, we can see that the kernel value decreases, more or less, to zero for $\psi > 5°$. Therefore, the values of S recovered for the marginal areas are not very reliable, but the map is very similar to that is obtained from EGM08. However, detailed information are visible in Fig. 4b than Fig. 4a as the gravity anomalies, generated from an existing global model like EGM08 to degree and order 372, are smoother than the terrestrial anomalies. Definitely, such models present the global structure of the gravity field and S and for recovering regional information of S the terrestrial anomalies should be used. This is the main reason of obtaining such a detailed information from the terrestrial data.

Figure 4c illustrates the map of S determined using the integral inversion method and the integral of Proposition 2. Here, we use the closed-form formula of the kernel function for our inversion purpose. The values of S are smaller in this map comparing to those in Fig. 4a and b. Another issue is

Table 1

Statistics of differences between recovered S by spherical harmonic series and EGM08 to degree 372, Eq. 1f, and direct integration with spectral kernel to degree 372 (DIR), Proposition 1, Solving integral equation with closed-form kernel of Proposition 2 (INV+CLF) and with regularisation (INV+CLF+REG) and solving the integral equation of Proposition 2 but with spectral kernel to degree 372 (INV+SPC)

	$0.5° \times 0.5°$					$0.25° \times 0.25°$				
	Max	Mean	Min	Std	RMS	Max	Mean	Min	Std	RMS
DIR	1.03	0.12	−0.95	0.35	0.38	1.08	0.12	−1.11	0.38	0.40
INV+CLF	1.00	−0.13	−1.27	0.48	0.49	1.10	−0.01	−1.19	0.41	0.41
INV+CLF+REG	–	–	–	–	–	0.98	−0.15	−1.46	0.51	0.53
INV+SPC+REG	0.95	−0.15	−1.13	0.48	0.50	–	–	–	–	–

Unit: 1 MPa

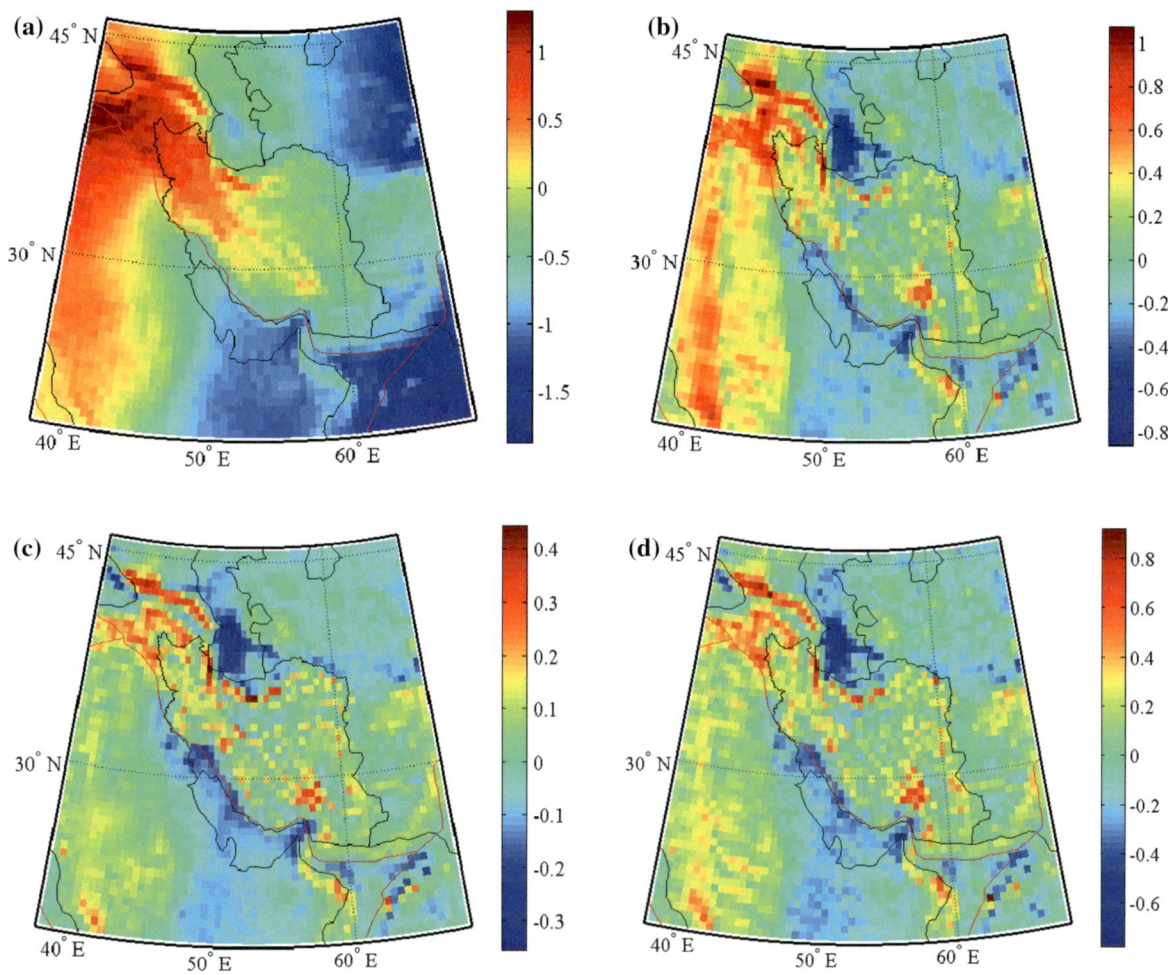

Figure 4

Maps of S with a resolution of $0.5° \times 0.5°$ derived by **a** spherical harmonic series and EGM08 to degree 372, Eq. (1f), **b** direct integration with spectral kernel to degree 372, Proposition 1, **c** Solving integral equation with closed-form kernel of Proposition 2, **d** solving the integral equation of Proposition 2 but with spectral kernel to degree 372. Unit: 1 MPa

that the system of equations, obtained after discretisation of the integral, is not ill-conditioned and we did not need to stabilise the system by any regularisation method. The condition number, the ratio of maximum and minimum eigenvalues, became 2.52! We refer to Fig. 2b showing the behaviour of the kernel of this integral formula to explain the reason. The kernel has very small values for $\psi > 0.5°$. This means that value of the kernel for the far-zone S is very small and the kernel practically removes them from the integration process. In other words, the coefficient matrix of the system will have more or less a block diagonal structure. Consequently, the system of equations, organised by discretising such

an integral, is not ill-conditioned. Why is S smaller than those obtained by the direct integration and EGM08? Again, this is due to the kernel function, as it uses S in the integration domain and not all of the gravity anomalies over the study area influence the solution due to the structure of the coefficient matrix. This causes that the recovered signal of S becomes weak. Now, we perform another test using the spectral form of the kernel, Eq. (3e), to degree 372 and carry out the inversion based on that. In this case, the system of equations is ill conditioned with a condition number of 4.93×10^{16}. The system is linear and we do not have to linearise it and, therefore, no iteration is required to solve it.

However, in order to control the instability of the system we used Tikhonov regularisation (TIKHONOV 1963) due to its simplicity. The principle of this method is to add a small positive number to the diagonal elements of the system of normal equations for stabilisation purpose. The main question is how to estimate this value in practice and numerous methods have been proposed for such a purpose but we selected the generalised cross validation (GCV) method (WAHBA 1976) for this goal as we could get the best result in this study by that. We used Regularisation Tools (HANSEN 2007) of MATLAB to solve the system by the Tikhonov regularisation and estimating its regularisation parameter. Figure 4d shows the recovered S and has a similar pattern as Fig. 4a and b but with larger values. In order to explain the reason of the instability of the system of equations, we refer to Fig. 2c, which shows the behaviour of the limited spectral form of the kernel to degree 372. The value of the kernel decreases with increasing ψ, but does not decay to zero even to $\psi = 20°$. However, the kernel has large values for $\psi = 0°$, which is a positive issue. An integral with such a kernel uses S all over the area and by inverting it, all gravity anomalies contributes to the recovery process of S as well. Since, the kernel does not decay fast to zero, the contribution of far-zone S remains significant, and the elements of the coefficient matrix of the system becomes large and more or less at the same order. This causes that the coefficient matrix has a large condition number and becomes unstable. Regularisation is a necessary issue for solving such a system of equations.

The use of EGM08 or any other gravity field model with the spherical harmonic series is always possible for stress determination. However, our idea was just to use the local gravity data for such purpose rather than a global model. Local data can present the properties of the region much better than a global model. Here, we showed that both spectral kernels to a limited degree of 372 are applicable for recovering S, but the integral inversion with the closed-form kernel cannot yield S with a comparable values with those obtained from the spectral kernels. The reason was that the kernel value is close to zero for $\psi > 0.5°$. Now, the question is if by increasing the resolution we can obtain better results. To answer this question, we

increased the resolution up to $0.25° \times 0.25°$ and discretised all of the integral formulae. Figure 5a shows the map of S generated from EGM08 to degree 372 with this resolution and Fig. 5b is the corresponding one from real gravity anomalies by direct integral formula with the limited spectral kernel to degree 372. There are many artefacts in the map, which mean that considering a high resolution is not very meaningful when the kernel is limited to degree 372. Now, the integral equation of Proposition 2 is discretised and inverted without any regularisation because the system is already stable with a condition number of 13.08; see Fig. 5c. More details of S are seen in this map, but its magnitude is considerably large. This is because of the fact that the gravity anomalies contain all frequencies and the kernel, having the closed-form formula, uses all frequencies of the anomalies and amplifies them through the inversion process. If we solve the system by the truncated singular value decomposition (TSVD) method with L-curve for estimating the truncation number, the map in Fig. 5d is obtained. In this method, the spectra, eigenvalues of the coefficients matrix of the system is analysed and the small ones, which contain the high frequencies of the solution, are removed gradually from the system. This causes that the solution becomes smoother and smoother by such a process. In our present case that the system is stable, we have the divergence problem; in other words, by considering high resolution of recovery we observe the divergence problem again. In fact, by removing the small spectra, we remove the diverging frequencies locally and we estimated this truncation number of these values when we have the balance between the error and size of the estimates by the L-curve. In other words, by using the L-curve method we can control this divergence problem and estimate how many small eigenvalues should be removed for this purpose. One may use another filtering method as well but since we organised a system of equation by discretising the integral equation we preferred to use TSVD method. We emphasise that the system of equations is not ill conditioned and the applied regularisation process is just for smoothing the solution further. Now, the values of S are closer to those determined by the direct integration with the spectral kernel, see Fig. 5d.

Here, we use the real gravity anomalies and the integral formulae to recover S. Table 2 presents the

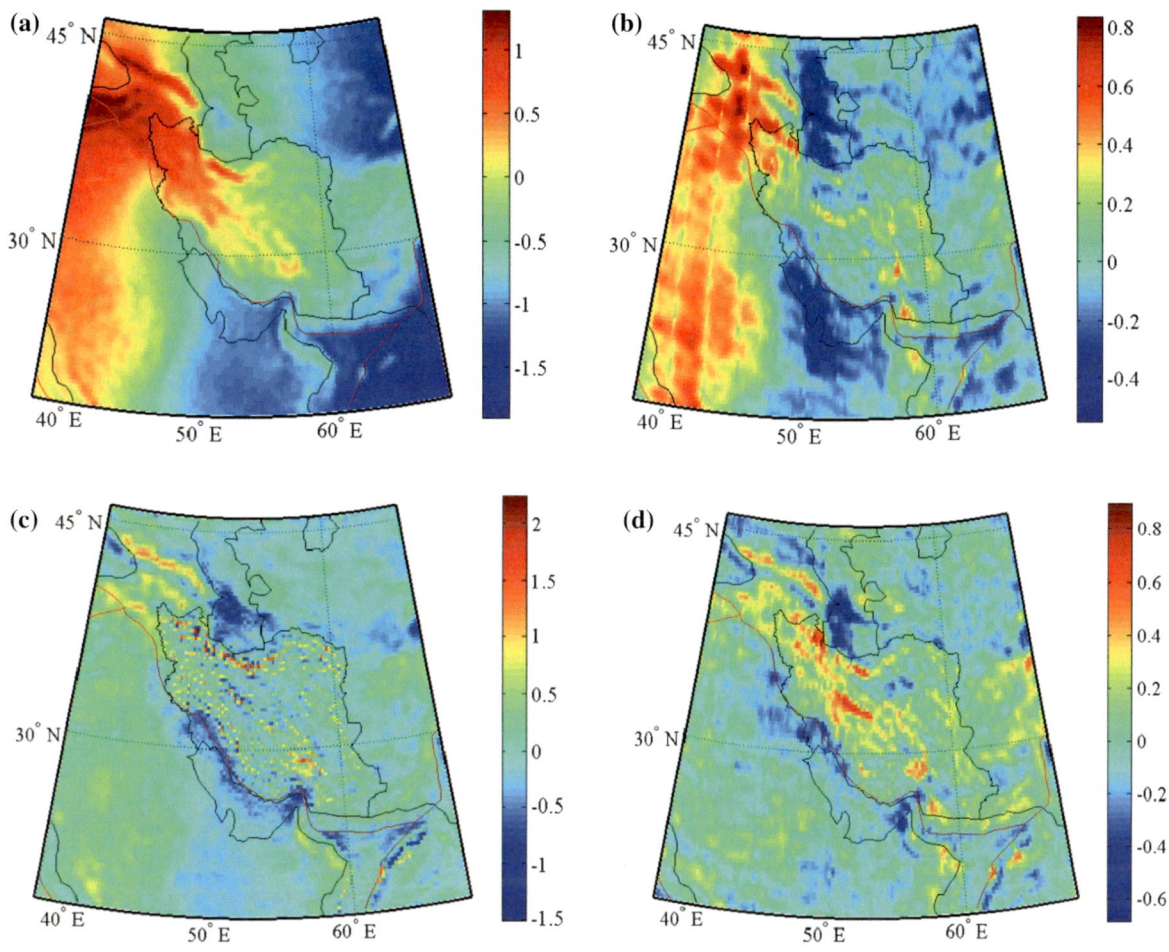

Figure 5

Maps of S with a resolution of $0.25° \times 0.25°$ derived by **a** spherical harmonic series and EGM08 to degree 372, Eq. (1f), **b** direct integration with spectral kernel to degree 372, Proposition 1, **c** solving integral equation with closed-form kernel of Proposition 2, **d** solving the integral equation of Proposition 2 but with spectral kernel to degree 372. Unit: 1 MPa

statistics of the recovered S based on the methods of spherical harmonic series with EGM08 to degree and order 372, direct integration (DIR) and inversion (INV+SPC) with spectral kernels limited to degree 372. Furthermore, integral inversion with the closed-form kernel (INV+CLF) and its regularised one (INV+CLF+REG). As already shown, having a higher resolution than $0.5° \times 0.5°$ for the gravity anomalies does not lead to recovering better results, and the recovered S based on the direct integration is close to that based on EGM08. Consequently, considering this resolution is not very meaningful as the kernel is limited to degree 372 and is not able to recover higher frequencies. The values of S range from -0.35 to 0.44 MPa when the integral inversion

with the closed-form kernel is used and in the case of using the spectral form of the kernel to degree 372, the values range from -0.78 to 0.92 MPa, which are closer to those obtained from the direct integration method. Since the system of equations organised based on the closed-form kernel was stable, no regularisation was applied for solving it. Nevertheless, the regularised solution by truncated singular value decomposition (TSVD) and L-curve provides similar results to that of the direct integration method. We should mention that no significant difference was seen between the regularised solutions by Tikhonov regularisation and the solution without regularisation due to stability of the system. However, the system, organised based on the limited spectral kernel for the

Table 2

Statistics of S derived by spherical harmonic series and EGM08 to degree 372, Eq. 1f, direct integration with spectral kernel to degree 372 (DIR), Proposition 1, Solving integral equation with closed-form kernel of Proposition 2 (INV+CLF) and with regularisation (INV+CLF+REG) and solving the integral equation of Proposition 2 but with spectral kernel to degree 372 (INV+SPC)

	$0.5° \times 0.5°$				$0.25° \times 0.25°$			
	Max	Mean	Min	Std	Max	Mean	Min	Std
EGM08	1.29	−0.36	−1.88	0.67	1.29	−0.36	−1.91	0.67
DIR	1.07	0.09	−0.86	0.27	0.83	0.07	−0.54	0.24
INV+CLF	0.44	0.01	−0.35	0.10	2.24	0.03	−1.51	0.32
INV+CLF+REG	–	–	–	–	0.90	0.00	−0.69	0.16
INV+SPC+REG	0.92	0.02	−0.78	0.21	–	–	–	–

Unit: 1 MPa

resolution $0.25° \times 0.25°$, was very unstable and we could not regularise it to find a smooth solution.

3.3.3 A Discussion about the Recovered S Functions

We used the spectral form of the kernel limited to degree 372 for the inversion process and observed that the system of equations becomes very unstable. Nevertheless, the system, organised based on the closed-form kernel, is stable and no regularisation is needed for the solution. As Fig. 5c shows the magnitudes of S reach to about 2 MPa whilst the one derived from EGM08 to degree and order 2160 to some values about 60 MPa, as the gravitational signal is strongly amplified after degree 372. We have done this test just to show how the convergence problem influences the results. However, Fig. 5c, shows that even if we consider the real gravity anomalies containing all frequencies of the gravity field and invert the integral of Proposition 2, the recovered S is still in the order of what obtained from EGM08. However, by regularisation we can smooth this solution further.

The map of S determined by the direct integration with spectral kernel, Proposition 1, is more similar to that derived from EGM08 by spherical harmonic series (1f) than the other methods. The reason is that the mathematical formulae of both methods are principally the same. In fact, the integral of Proposition 1 is the transferred form of Eq. (1f) to the spatial domain. If we look at the spherical harmonic series (1f) and the integral formula (2e), we observe v_n in both of them. Let us rewrite Eq. (1f) in the following form:

$$S_n = \frac{g}{4\pi G s^2 R} v_n T_n \tag{5d}$$

According to Eqs. (2b) and (2d) we can write

$$S_n = \frac{g}{4\pi G s^2 R} v_n \frac{R}{n-1} \frac{2n+1}{4\pi} \iint_\sigma \Delta g' P_n(\cos \psi) d\sigma \tag{5e}$$

which is the same with Eq. (2d). Therefore, we can expect to see a similar result from Eq. (1f) and the integral presented in Proposition 1. However, the solution based on the integral inversion is different. In order to show the difference, we write Eq. (3b) in the following form:

$$\frac{4\pi s^2}{g} \frac{(n-1)}{v_n} S_n = \Delta g_n \tag{5f}$$

which is also principally the same with Eq. (5e), but in Eq. (5f) v_n is in the denominator and diminish the high frequencies of S. By considering Eq. (3c), we can see that v_n is more or less disappeared in the formula and simplified to $(n+1)s^{n+1}$ and since $s < 1$ it makes the mathematical model convergent. In fact, by organising an integral formula from Eq. (5f), the high frequencies of S are diminished based on $(n^2 - 1)s^{n+1}$ in the integration domain. In other words, the signal regularisation is done locally and by inverting the integral, the high frequencies of gravity anomalies are amplified locally or the local frequencies are amplified. This is different from the solutions based on the spherical harmonics (1f) and Eq. (2e) as in both of them the global frequencies are amplified by v_n during converting the gravity field to S.

Table 3

Statistics of sub-crustal stress components derived from simulated gravity anomalies by EGM08 to degree 372, Eq. (1f), direct integration with spectral kernel to degree 372 (DIR), Proposition 1, solving integral equation with closed-form kernel of Proposition 2 (INV+CLF) and with regularisation (INV+CLF+REG) and solving the integral equation of Proposition 2 but with spectral kernel to degree 372 (INV+SPC) and with regularisation (INV+SPC+REG)

	$0.5° \times 0.5°$					$0.25° \times 0.25°$				
	Max	Mean	Min	Std	RMS	Max	Mean	Min	Std	RMS
DIR										
δS_x	24.51	0.30	−32.70	5.17	5.18	42.68	0.26	−62.73	11.36	11.37
δS_y	45.55	3.058	−30.73	7.48	8.08	92.58	3.15	−68.43	13.34	13.71
INV+CLF										
δS_x	30.68	0.31	−33.06	6.97	6.97	190.25	0.56	−79.63	12.35	12.36
δS_y	69.39	3.66	−51.77	9.57	10.25	81.82	3.35	−75.80	9.15	9.74
INV+CLF+REG										
δS_x	–	–	–	–	–	74.47	0.27	−80.24	13.40	13.40
δS_y	–	–	–	–	–	107.07	3.75	−82.53	15.08	15.54
INV+SPC+REG										
δS_x	29.20	0.31	−31.73	5.42	5.42	–	–	–	–	–
δS_y	55.60	3.69	−45.44	8.15	8.94	–	–	–	–	–

Unit: 1 MPa

3.3.4 Generation of Sub-Crustal Stress from Recovered S Functions

Now, the recovered values of S are converted to the sub-crustal stress components towards the north and the east. This process is done based on the resolution of S and using:

$$S_x = \frac{\Delta S_\theta}{\Delta \theta} \text{ and } S_y = \frac{\Delta S_\lambda}{\sin \theta \Delta \lambda} \qquad (6a)$$

where ΔS_θ and ΔS_λ are the differences between the values of S in two neighbouring cells towards the north and the east, $\Delta \theta$ and $\Delta \lambda$ are the difference between each two cells towards the corresponding directions, respectively and they must be in the radian unit. The magnitude of the sub-crustal stress (S_T) is computed by:

$$S_T = \sqrt{S_x^2 + S_y^2}. \qquad (6b)$$

In order to investigate the quality of each method, a closed-loop simulation was performed. We generated the gravity anomalies and the stress components using EGM08 to degree and order 372. After that, the gravity anomalies and the three integral methods were used to recover the stress components. We expect that the generated components and corresponding ones by EGM08 are equal. The differences

are considered as a criterion for judging about the quality of each method. Table 3 presents the statistics of these differences over the study area. It shows that in case of using the spectral kernels the quality of the stress components recovery is better than the cases where the closed-form formula is used. This is normal, as the mathematical models are more or less consistent to the spherical harmonic series, which is limited to the convergence degree N. The direct method is slightly better than the inversion method with regularisation as the RMS of δS_x is 5.18 MPa for the former and 5.42 MPa for the latter. The corresponding values are 8.08 MPa and 8.94 MPa for δS_y. Inversion with the closed-form formula is slightly worse as the RMS of δS_x is 6.97 MPa and for δS_y 10.25 MPa as this method does not limit the frequencies of the solution and during the inversion process signal is considerably amplified. The advantage of this method is that it is considerably faster than those methods, which use the spectral forms of the kernels. In case of increasing the resolution of the recovery to $0.25° \times 0.25°$ the errors become larger. The inversion of the integral formula with spectral form of the kernel is not possible even with regularisation. However, as we showed earlier that the spherical harmonic expansion of the S is asymptotically convergent to degree and order 372, which is somehow corresponding to a resolution of $0.5° \times 0.5°$. This means that by increasing the resolution of computations the solution is not improved as there is no frequency to be considered when the gravity anomalies are computed from EGM08 to degree and order 372. On the other hand, the series of S is divergent for higher degrees and orders and considering higher resolution means recovering higher frequencies, which do not exist. Therefore, it is natural to see that increasing the resolution just adds extra numerical errors in the solutions. One can see from the table that the northwards stress component S_x is recovered with lower error than S_y. Another issue that should be

explained here is that for using the direct integral method with the spectral kernel we need to have the gravity anomalies in grid forms. This means that all of the gravity data should be interpolated prior to using, but the inversion methods do not need the gridded data and we can simply use them in any distribution. The important issue here is that the data should have a good coverage all over the study area.

Now, we present the maps of S_T generated from the recovered S by each method. Figure 6a is the map of S_T generated from spherical harmonic series and EGM08 to degree and order 372. Large values of S_T are seen along the tectonic boundary between Arabian and Eurasian plates, and in the northern

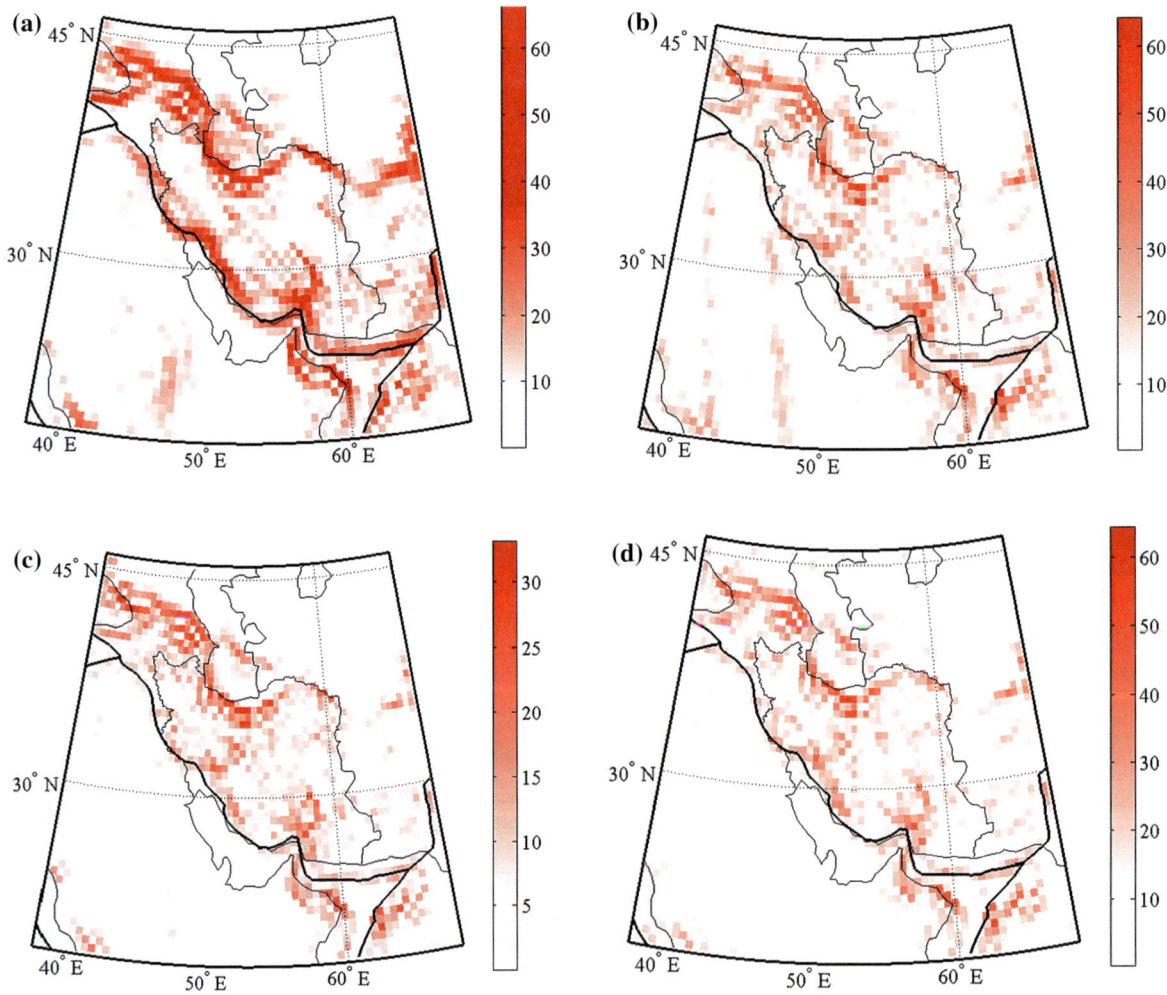

Figure 6

Maps of sub-crustal stress S_T with a resolution of $0.5° \times 0.5°$ derived by **a** spherical harmonic series and EGM08 to degree 372, Eq. (1f), **b** direct integration with spectral kernel to degree 372, Proposition 1, **c** solving integral equation with closed-form kernel of Proposition 2, **d** solving the integral equation of Proposition 2 but with spectral kernel to degree 372. Unit: 1 MPa

volcanic part of Iran. Figure 6b is a similar map to that of Fig. 6a, but determined from S recovered from direct integration with the limited kernel to degree 372. It does not show the stress along the tectonic boundary. In the case of using the gravity anomalies generated from EGM08 to degree 372 the stress along this boundary is more visible than the case where to degree 2160, which has a similar map with that of the real anomalies. S_T of this map is not as smooth as S_T of Fig. 6a because the terrestrial gravity anomalies have been used for its generation.

Figures 6c and d are maps of S_T derived from the integral inversion, Proposition 2, with closed-form

and spectral kernels, respectively. These maps have very similar interpretations and the only difference is that Fig. 6c shows smaller values for S_T than the rest of them.

Figure 7a is the map of S_T generated from EGM08 to degree and order 372 with a resolution of $0.25° \times 0.25°$ and very well shows that S_T follows the tectonic boundary shown by the thick black line. S_T derived from the direct integral methods, Fig. 6b, is not smooth and shows more details, but mixed with numerical errors. Figure 6c is the map of S_T recovered from integral inversion of gravity anomalies using the formula presented in Proposition 2 without any regularisation. A large stress exists along the

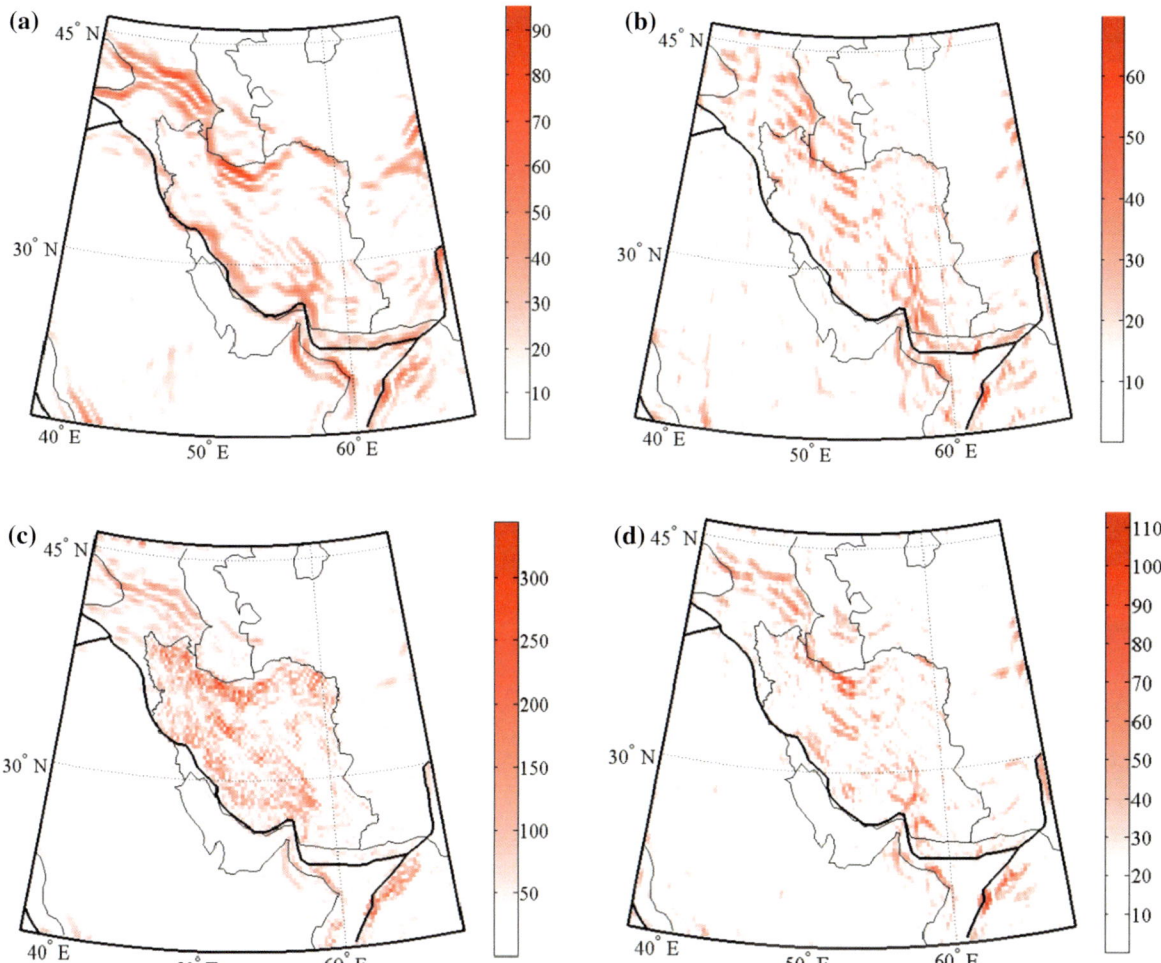

Figure 7
Maps of sub-crustal stress S_T with a resolution of $0.25° \times 0.25°$ derived by **a** spherical harmonic series and EGM08 to degree 372, Eq. (1f), **b** direct integration with spectral kernel to degree 372, Proposition 1, **c** solving integral equation with closed-form kernel of Proposition 2, **d** solving the integral equation of Proposition 2 with an extra regularisation. Unit: 1 MPa

Table 4

Statistics of sub-crustal stress components derived by spherical harmonic series and EGM08 to degree 372, Eq. (1f), direct integration with spectral kernel to degree 372 (DIR), Proposition 1, solving integral equation with closed-form kernel of Proposition 2 (INV+CLF) and with regularisation (INV+CLF+REG) and solving the integral equation of Proposition 2 but with spectral kernel to degree 372 (INV+SPC) and with regularisation (INV+SPC+REG)

	$0.5° \times 0.5°$				$0.25° \times 0.25°$			
	Max	Mean	Min	Std	Max	Mean	Min	Std
EGM08								
S_x	57.35	−0.91	−50.79	11.31	90.47	−0.91	−76.17	14.44
S_y	40.08	−2.94	−43.80	7.77	65.65	−2.94	−52.29	9.53
DIR								
S_x	62.35	0.02	−58.60	11.24	46.98	0.00	−47.11	10.13
S_y	48.30	−0.67	−63.03	9.91	45.43	−0.61	−52.58	8.76
INV+CLF								
S_x	31.82	0.02	−30.55	5.68	264.15	0.29	−244.06	34.44
S_y	22.91	−0.11	−29.46	4.31	214.67	−0.44	−314.08	26.18
INV+CLF+REG								
S_x	–	–	–	–	92.65	0.00	−85.97	15.46
S_y	–	–	–	–	67.43	−0.10	−101.93	11.33
INV+SPC+REG								
S_x	64.37	0.04	−54.04	10.85	–	–	–	–
S_y	38.07	−0.12	−52.32	7.89	–	–	–	–

Unit: 1 MPa

tectonic boundary in the south-east under the oceanic lithosphere. This map is different from those have been presented so far, because it contains all possible frequencies of the sub-crustal stress and there is no truncation of frequencies in that. If we smooth this map based on TSVD method with L-curve we obtain the map of Fig. 7d, which is very similar to those already derived by the other methods.

Now, the real data of the area is used. Table 4 shows the statistics of the computed sub-crustal stress components S_x and S_y. These components have been derived by numerical differentiation of the recovered S by Eq. (5a). Generally, with increasing the resolution of S the values of the stress components increased except the case where the direct integration method with the spectral kernel is used. According to Table 1, the values of S derived from DIR are already smaller as well when the resolution is $0.25° \times 0.25°$. We observe that when the resolution is $0.5° \times 0.5°$ the values of the components are small comparing those derived from other methods, whilst they are larger for the resolution $0.25° \times 0.25°$. Note that in neither of these resolutions, the system of equations from which S is estimated is ill-conditioned. This means that the system is stable and there is no

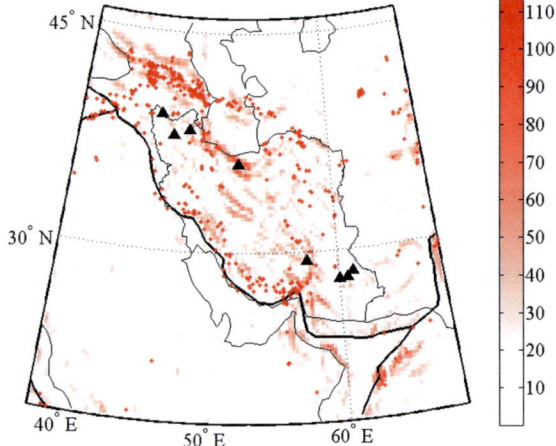

Figure 8

Distribution of stress points of WSM and volcanos of Iran on the background of sub-crustal stress computed from gravity anomalies of Iran by inverting integral of Proposition 2 with an extra regularisation. Unit: 1 MPa

frequency amplification due to the numerical problems. The values that we observe in Table 3, derived based on Proposition 2 and spherical harmonic series (1f) with EGM08, are more or less at the same level because in all of them higher frequencies than 372 are removed from the solution. Nevertheless, there is no frequency limit for inverting the integral presented in Proposition 2.

For a better comparison, we show the distribution of the active seismic points of the study area determined in the project World Stress Maps (WSM) (HEIDBACH *et al.* 2008), as well as 8 volcanos of Iran to compare them with our results. The seismic points and these volcanos are shown in Fig. 8 on the background of the sub-crustal stress field determined by inverting the integral of Proposition 2 after smoothing by TSVD. The red points are the seismically active points determined by HEIDBACH *et al.* (2008) and the black triangles show the position of the volcanos. The seismic points of WSM have good agreement with sub-crustal stresses especially those along the tectonic boundary between the Arabian and Eurasian plates.

4. Conclusions and Recommendations

- We have developed a simple formula showing the relation between the mean Moho depth and the convergence degree of the S function and based on that we found that the convergence degree for the global mean Moho depth, 23 km, is 622 and for the mean Moho of Iran, 35 km, is 372. This means that for generating the sub-crustal stress by the spherical harmonic series a global gravity model to this degree is enough for Iran.
- Our direct integral approach with its spectrally-limited kernel, presented in Proposition 1, to degree 372 showed that the recovered S function has a similar pattern as the one derived from EGM08. This is because this approach is more or less the same with the spherical harmonic method but in the spatial domain and the kernel filter out the higher frequencies of the gravity data.
- The closed-form formula of the integral of Proposition 2 is very well-behaving and gets close to zero after a geocentric angles of 0.5°. This causes that the system of equations, organised after discretisation of this integral, becomes well-conditioned but the magnitudes of the derived S are underestimated. However, it is recommended considering higher resolution for the S and use an extra regularisation to smooth the overestimated S.

- The use of spectral kernel for the integral of Proposition 2 is recommended if the resolution of the recovering process of S is not high. Because the system of equations is hard to stabilise for high resolutions as the kernel is not well-behaving and the effect of far-zone S remains significant even to a geocentric angle of 20°.
- The recovered stress in Iran and surrounding areas has a good agreement with the tectonic boundary and the volcanic areas in the northern parts as well as the stress points of WSM.

Acknowledgments

The author is thankful to Professor Lars E. Sjöberg at Royal Institute of Technology (KTH) in Stockholm, Sweden, for introducing the subject. Mr. Mohsen Romeshkani and the National Cartographic centre of Iran are appreciated for providing the gravimetric data. Professor Carla Braitenberg and Professor Robert Tenzer are appreciated for their helps and advices. The two unknown reviewers are also appreciated for their useful comments.

REFERENCES

ARTYUSHKOV E.V. (1973) *Stresses in the lithosphere caused by crustal thickness inhomogenities*, J Geophys. Res., *78*, 32, 7675–7708.

DAHLEN F.A. (1982) *Isostatic geoid anomalies on a sphere*, J Geophys. Res. *87*, B5, 3943–3947.

DAVIES G. F. and RICHARDS M.A. (1992) *Mantle convection*, J Geol., *100*, 151–206.

ESHAGH M. (2011) *The effect of spatial truncation error on integral inversion of satellite gravity gradiometry data*, Adv. Space Res., *47*, 1238–1247.

ESHAGH M. (2014a) *From satellite gradiometry data to subcrustal stress due to mantle convection*, Pure Appl. Geophys., *171*, 2391-2406.

ESHAGH M. (2014b) *Determination of Moho discontinuity from satellite gradiometry data: linear approach*, GRIB. *1*(2):1–13.

ESHAGH M. (2015) *On the relation between the Moho depth and sub-crustal stress due to mantle convection*, J Geophys. Eng., *12*, 1742–2140.

ESHAGH M. and TENZER R. (2015) *Sub-crustal stress determined using gravity and crust structure models*. Compt. Geosci. *19*, 115–125.

FLOBERGHAGEN R., FEHRINGER M., LAMARRE D., MUZI D., FROMM-KNECHT B., STEIGER C., PIÑEIRO J. and da COSTA A (2011) *Mission design, operation and exploitation of the gravity field and steady-state ocean circulation explorer mission*. J Geod 85:749–758.

FORTE A.M. and PELTIER W.R. (1987) *Surface Plate kinematics and mantle convection*, Geodynamics series, Composition, Structure and dynamics of the lithosphere-Asthenosphere system, *16*, 125–136.

FU R. and HUANG P. (1983) *The global stress field in the lithosphere obtained from satellite gravitational harmonics*, Phys. Earth planet. Int., *31*, 269–276.

FU R. (1986) *A numerical study of the effects of boundary conditions on mantle convection models constrained to fit the low degree geoid coefficients*, Phys. Earth planet. Int., *44*, 257–263.

FU R. (1989) *Plate motions, Earth's geoid anomalies and mantle convection*, Geophysical monography series, International union of geodesy and geophysics and American Geophysical Union, *49*, 47–54.

FU R. and HUANG J. (1990) *Global stress pattern constrained on deep mantle flow and tectonic features*, Phys. Earth planet. Int., *60*, 314–323.

FU R. and HUANG J. (1992) *Deep mantle flow, global tectonic pattern and the background of seismic stress field in China*, Acta Seism. Sin., *5*,2,-271–281.

FU R., HUANG J., DONG, S. CHANG X. (2003) *New mantle convection models constrained by seismic tomography data*, Chinese J Geophys., *46*, 6, 1106–113.

HAGER B.H. (1983) *global isostatic geoid anomalies for plate and boundary layer models of the lithosphere*, Earth Planet. Sci. Lett., *63*, 97–109.

HAGER B.H. and RICHARDS M.A. (1989) *Long wavelength variations in the Earth's geoid: physical models and dynamical implications*, Phil. Trans. Soc. Lond. A *328*, 309–327.

HAGER B.H. and O'CONNELL R.J. (1981) *A simple global model of plate dynamics and mantle convection*, J Geophys. Res. *86*, B6, 4843–4867.

HANSEN P. C. (2007) *Regularization Tools version 4.0 for Matlab 7.3*, Numerical Algorithms, *46*: p. 189–194.

HAXBY W.F. and TURCOTTE D.L. (1978) *On isostatic geoid anomalies*, J Geophys. Res. *83*, B11, 5473–5478.

HEISKANEN W.A. and MORITZ H. (1967) *Physical Geodesy*, *W.H.Freeman anc company*, San Francisco and London.

HEIDBACH O., TINGAY M., BARTH A., REINECKER J., KURFEß D. and MÜLLER B. (2008) *The World Stress Map database release 2008* doi:10.1594/GFZ.WSM.Rel2008.

HUANG P. and FU R. (1982) *The mantle convection pattern and force sources mechanism of recent tectonic movements in China*, Phys. Earth planet. Int., *28*, 261–268.

JACOBY W.R. and SEIDLER E. (1981) *Plate tectonic and the gravity field*, Tectonophys., *74*, 155–167.

LASKE G., MASTERS G., MA Z. and PASYANOS M. (2013) *Update on CRUST1.0—A 1-degree Global Model of Earth's Crust*. In EGU General Assembly Conference Abstracts *15*, 2658.

LIU H.S. (1977) *Convection pattern and stress system under the African plate*, Phys. Earth planet. Int., *15*, 60–68.

LIU H.S. (1978) *Mantle convection pattern and subcrustal stress under Asia*, Phys. Earth planet. Int., *16*, 247–256.

LIU H.S. (1979) *Convection-generated stress concentration and seismogenic models of the Tangshan Earthquake*, Phys. Earth planet. Int., *19*, 307–318.

LIU H.S. (1980) *Convection generated stress field and intra-plate volcanism*, Tectonophys., *65*, 225–244.

LIU H.S. (1985) *Geodynamical basis for crustal deformation under the Tibetan Plateau*, Phys. Earth planet. Int., *40*, 43–60.

LOVE A.E.H. (1944) *A treatise on the mathematical theory of elasticity*, Dover Publication, New York, pp. 249.

LUX R.A., DAVIS, G.F. and THOMAS J.H. (1979) *Moving lithospheric plates and mantle convection*, Geophys J R. astr Soc. *57*, 209–228.

MARTINEC Z. (2003) *Green's function solution to spherical gradiometric boundary-value problems*, J. Geod. *77*, 41–49.

MARCH B.D. and MARSH J.G. (1976) *On global gravity anomalies and two-scale mantle convection*, J Geophys. Res. *81*, 29, 5267–5280.

MCKENZIE D. P. (1967) *Some remarks on heat flow and gravity anomalies*, J Geophys. Res. *72*, 24, 6261–6273.

MCNUTT M. (1980) *Implication of regional gravity for state of stress in the Earth's crust and upper mantle*, J Geophys. Res. *85*, No. B11, 6377–6396.

MONNEREAU M. and QUERE S. (2001) *Spherical shell models of mantle convection with tectonic plates*, EPSL, *184*, 575–587.

MORITZ H. (1980) *Advanced Physical Geodesy*, Wichmann.

MORITZ H. (2000) *Geodetic reference system 1980*. J Geod. *74*, 128-133.

NALIBOFF J. B., LITHGOW-BERTELLONI C., RUFF L. J. and de KOKER N. (2012) *The effects of lithospheric thickness and density structure on Earth's stress field*, Geophys. J. Int. (2012) *188*, 1–17.

NATAF H.C. (1991) *Mantle convection, plates and hotspots*, Tectonophys., *187*, 361–371.

PAVLIS N.K., HOLMES S.A., KENYON S.C., Factor J.K. *(2012) The development and evaluation of the Earth gravitational model 2008 (EGM2008)*. J. Geophys. Res.-Sol. Ea. *117*: B04406.

PAVLIS N.K., FACTOR J. and HOLMES S. (2006) *Terrain-related gravimetric quantities computed for the next EGM, presented at the 1st International symposium of the International gravity service 2006*, Istanbul, Turkey.

PICK M. (1994) *The geoid and tectonic forces, In Geoid and its Geophysical interpretations EDS*. P. VANICEK and N. CHRISTOU, CRC Press, p. 386.

PICK M. and CHARVATOVA-JAKUBKOVA I. (1988) *Modification of the Runcorn's equations on the convection flows*, Stud. Geophys. Geod. *32*:47–53.

REIGBER Ch., JOCHMANN H., WÜNSCH J., PETROVIC S., SCHWINTZER P., BARTHELMES F., NEUMAYER K.-H., KÖNIG R., FÖRSTE Ch., BALMINO G., BIANCALE R., LEMOINE J.-M., LOYER S. and PEROSANZ F. (2004) *Earth Gravity Field and Seasonal Variability from CHAMP*. In: REIGBER, Ch., LÜHR, H., SCHWINTZER, P., WICKERT, J. (eds.), Earth Observation with CHAMP—Results from Three Years in Orbit, Springer, Berlin, 25–30.

RICARD Y., FLEITOUT L. and FROIDEVAUX C. (1984) *Geoid heights and lithospheric stresses for a dynamic Earth*, Ann. Geophys, *2*,3, 267–286.

RICARD Y., FROIDEVAUX C. and LEITOUT L. (1988) *Global plte mostion and the geoid: a physical model*, Geophys. J, *93*, 477–484.

RICHARDS M.A. and HAGER B.H. (1984) *Geoid anomalies in a dynamics Earth*, J Geophys. Res., *89*, B7, 5987–6002.

RUNCORN S.K. (1962) *Convection currents in the Earth's mantle*, Nature *195*, 1248–1249.

RUNCORN S.K. (1963) *Satellite gravity measurements and convection in the mantle*, Nature, *200*, 628–630.

RUNCORN S. K. (1964) *Satellite gravity measurements and laminar viscous flow model of the Earth mantle*, J. Geophys. Res., *69*, 20, 4389–4394.

RUNCORN S. K. (1967) *Flow in the mantle inferred from the low degree harmonics of the geopotential*, Geophys. J. R. astr. Soc. *14*, 375–384.

RUNCORN S. K. (1980) *Mechanism of plate tectonics: mantle convection currents, plumes, gravity sliding or expansion?* Tectonophys., *63*, 297–307.

SJÖBERG L.E., (2009) *Solving Vening Meinesz-Moritz Inverse Problem in Isostasy*, Geophys. J. Int. *179*, 1527–1536.

SOURIAU M. and SOURIAU A. (1983) *Global tectonics and the geoid*, Phys. Earth planet. Int., *33*, 126–136.

TAPLEY B., RIES J. BETTADPUR S., CHAMBERS D., CHENG M., CONDI F., GUNTER B., KANG Z., NAGEL P., PASTOR R., PEKKER T., POOLE S. and WANG F. (2005) *GGM02-An improved Earth gravity field model from GRACE.* J Geod, Vol. *79*, 467-478.

TENZER R., ESHAGH M. and GU X. (2015) *Subduction generated subcrustal stress in Taiwan.* Terr. Atm. Oceanic Sci. (accepted).

TIKHONOV A. N. (1963) *Solution of incorrectly formulated problems and regularization method, Soviet Math.* Dokl., *4*: 1035–1038, English translation of Dokl. Akad. Nauk. SSSR, *151*:501–504.

WAHBA G. (1976) *A survey of some smoothing problems and the methods of generalized cross-validation for solving them.* In Proceedings of the conference on the applications of Statistics, held at Dayton, Ohio, June 14–17, 1976, ed. By P.R. Krishnaiah.

WALTZER U. (1970) *Convection currents in the Earth's mantle and the spherical harmonic development of the topography of the Earth,* Zentralinstitut Physsik der Erde, Bereich Im Jena, Burgweg 11 m Report No. *74*, 73–92.

(Received December 13, 2014, revised May 5, 2015, accepted May 15, 2015, Published online May 29, 2015)

Pure Appl. Geophys. 173 (2016), 827–838
© 2014 Springer Basel
DOI 10.1007/s00024-014-0926-4

▐ **Pure and Applied Geophysics**

Simulating Gravity Changes in Topologically Realistic Driven Earthquake Fault Systems: First Results

KASEY W. SCHULTZ,[1] MICHAEL K. SACHS,[1] ERIC M. HEIEN,[2,3] JOHN B. RUNDLE,[1,2] DON L. TURCOTTE,[2] and ANDREA DONNELLAN[4]

Abstract—Currently, GPS and InSAR measurements are used to monitor deformation produced by slip on earthquake faults. It has been suggested that another method to accomplish many of the same objectives would be through satellite-based gravity measurements. The Gravity Recovery and Climate Experiment (GRACE) mission has shown that it is possible to make detailed gravity measurements from space for climate dynamics and other purposes. To build the groundwork for a more advanced satellite-based gravity survey, we must estimate the level of accuracy needed for precise estimation of fault slip in earthquakes. We turn to numerical simulations of earthquake fault systems and use these to estimate gravity changes. The current generation of Virtual California (VC) simulates faults of any orientation, dip, and rake. In this work, we discuss these computations and the implications they have for accuracies needed for a dedicated gravity monitoring mission. Preliminary results are in agreement with previous results calculated from an older and simpler version of VC. Computed gravity changes are in the range of tens of μGal over distances up to a few hundred kilometers, near the detection threshold for GRACE.

Key words: Numerical simulation, co-seismic gravity changes, virtual california, earthquakes, statistics.

1. Introduction

Current methods for predicting seismic events rely on present day seismicity records, surface deformation from GPS instruments, and surface deformation from synthetic aperture radar interferometry (InSAR) mapping. These measurements are all made at the surface, leaving us to infer the processes and dynamics at depth. However, gravity changes at the surface are a direct measure of stress and strain integrated over all depths of the underlying medium.

The experimental and theoretical foundation for using surface gravity changes to monitor fault systems is already well established. RUNDLE (1978) considered the question of gravity changes from dilation sources and thrust faults, and found that gravity changes in these cases were free-air anomaly (dilation) and Bouguer anomaly (thrust fault). WALSH and RICE (1979) computed these by a different method and found the same result. OKUBO (1992) listed gravity and potential Green's functions for all possible sources for the general case. Okubo also showed that dilatational gravity change (due to compression) corresponds directly to subsurface density change from seismic activity.

The Gravity Recovery and Climate Experiment (GRACE) satellites launched in 2002, and was designed to map gravitational field changes accurate to several μGal with a spatial resolution of a few hundred kilometers and a temporal resolution of around a month. SUN and OKUBO (2004) studied the errors in GRACE data and showed that the satellites should be able to detect gravity changes from large earthquakes with magnitudes of about 7.5 or larger. Since its launch in 2002, GRACE data have been used to map the two-dimensional distribution of co-seismic gravity changes for the 2004 Sumatra–Andaman earthquake ($M = 9.0$–9.3) (HAN *et al.* 2006), for the 2010 Central Chile earthquake ($M = 8.8$) (HAN *et al.* 2010; HEKI and MATSUO 2010), and for the 2011 Tohoku-Oki earthquake ($M = 9.0$). The

---
[1] Department of Physics, University of California-Davis, One Shields Ave, Davis, CA 95616, USA. E-mail: kwschultz@ucdavis.edu
[2] Department of Earth and Planetary Sciences, University of California-Davis, Davis, USA.
[3] Computational Infrastructure for Geodynamics, University of California-Davis, Davis, USA.
[4] Jet Propulsion Laboratory, California Institute of Technology, 4800 Oak Grove Dr, Pasadena, CA 91109, USA.

GRACE data in these studies showed co-seismic gravity changes of magnitude 5–10 μGal.

HAYES et al. (2006) took the Okubo Green's functions and evaluated them with an earlier version of Virtual California. This early version used a much simpler San Andreas fault system, including only strike-slip motion. Now, Virtual California has an updated infrastructure that enables the fast simulation of tens of thousands of years of simulated seismic history on arbitrarily complex fault networks, as well as enabling the evaluation of gravity Green's functions across the entire fault network.

2. Virtual California

Virtual California (VC) is a boundary element code that simulates fault systems based on stress interactions between fault elements. VC is designed to explore the long-term statistical behavior of topologically complex fault networks (RUNDLE 1988a, b; RUNDLE et al. 2006a b). The most recent version of VC is a highly streamlined and sophisticated tool that simulates earthquakes in a high performance computing environment (SACHS et al. 2012; HEIEN and SACHS 2012). VC can quickly simulate many thousands of years of seismic history on fault networks with any physically realizable geometry.

VC consists of three major components: a fault model, a set of quasi-static elastic interactions (Green's functions), and an event model. Despite the suggestive name, the only component of Virtual California specific to California is the fault model. This model can be changed to any arbitrary fault geometry and function properly the simulation physics and event model.

It is worth summarizing the major simplifications of Virtual California. Virtual California models faults in a single-layer, elastic half-space, with no viscoelastic layer. The elastic half-space assumes a homogeneous flat earth, and we do not account for rheological properties or elastic discontinuities. Furthermore, for the gravity changes shown later, we do not attenuate the modeled gravity signals according to simulated satellite observation altitude.

2.1. Fault Model

The basic components of the fault model are the fault elements and their parameters. The fault system is partitioned into the functional members of a Virtual California simulation, the fault elements. The size of the fault elements has a direct effect on the range of simulated earthquake moment magnitudes. The minimum earthquake magnitude scales with fault element size such that smaller elements allow smaller magnitudes (HEIEN and SACHS 2012).

Each fault element is given a failure stress and a constant back-slip velocity along a fixed rake vector, which always lies in the plane of the element. The model for the California fault system that we currently use is the allcal2 model, described in detail at scec.usc.edu/research/eqsims/documentation.html.

The model, shown in Fig. 1, comprises 181 fault sections corresponding to known faults in California. Each fault section is partitioned into square elements that are approximately 3 km × 3 km, for a total of 14,474 elements. In the present version of the model, we have removed the creeping section of the San Andreas fault. This section produces many events, which slows the simulation down considerably (RUNDLE et al. 2006b; HEIEN and SACHS 2012).

2.2. Fault Element Interactions

The behavior of the system is determined by interactions between elements from the Green's function and the stress release from elements during events. The code assumes long-term linear stress increase based on element–element interactions governed by Okada's implementation of elastic half-space Green's functions. More details about the physics of the simulation are described in Sect. 2.3.

Unlike physical fault systems, where fault geometry is dynamic over geologic time periods, VC makes a simplification by assuming a geometrically static fault system. This simplification reflects the goal of Virtual California simulations, to explore seismicity in today's fault systems rather than modeling their long term evolution. With a typical slip velocity of $\sim 0.5 \frac{cm}{yr}$, a simulation covering 10,000 years of simulated time would result in ~ 55 m of displacement, which is only 0.1 % the size of a single fault element.

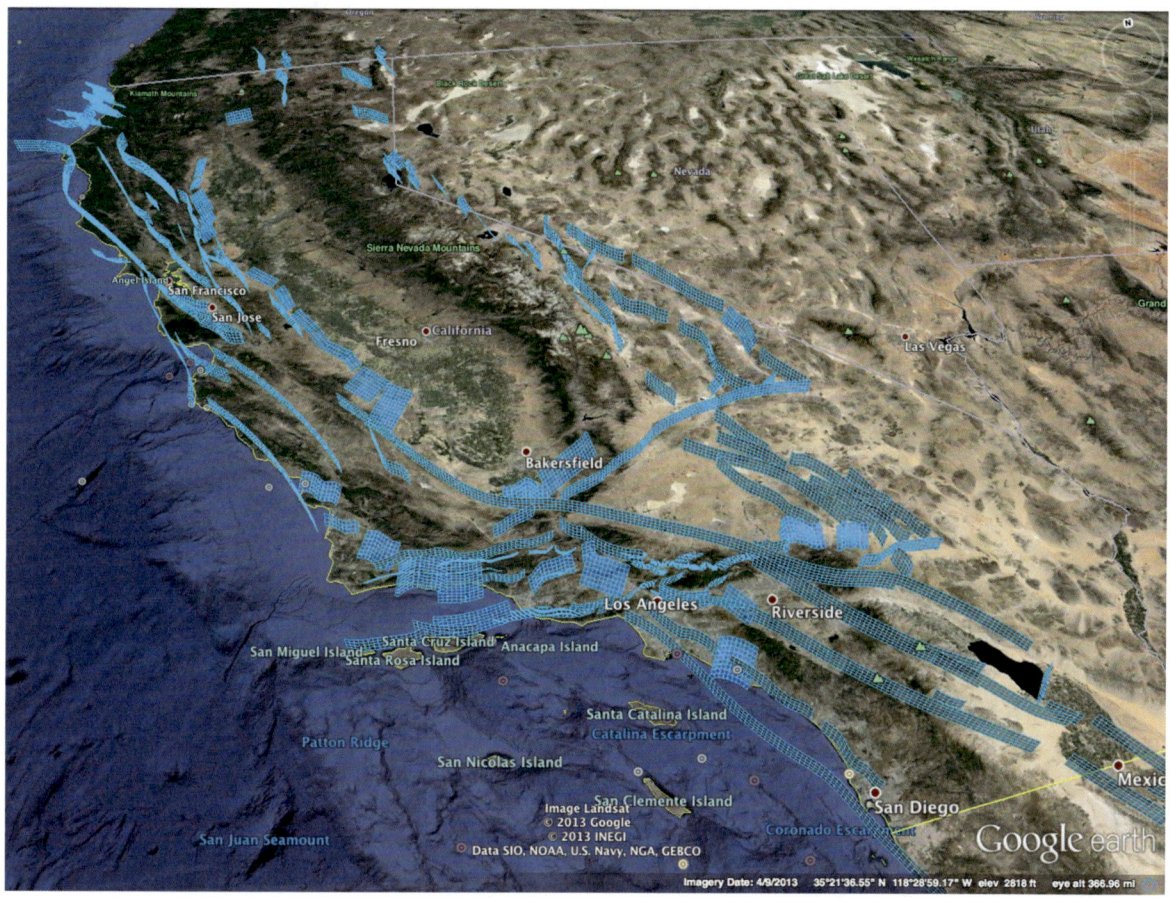

Figure 1
allcal2 California fault model based on UCERF2, meshed into 3 km × 3 km fault elements and shown above ground

A technique called "back slip" is used to model the effects of stress buildup and release between fault elements. In a back slip model, the equilibrium and initial positions of an element are the same. When building stress, a fault element is driven away from equilibrium in the opposite direction of the long range plate motion that drives the process. Once an element reaches its failure stress, it slips back toward equilibrium and releases the stress to the other elements in the fault system. In this manner, a back slip method keeps the fault system geometry static (SACHS *et al.* 2012; RUNDLE *et al.* 2006a, b; RUNDLE 1988a).

2.3. Stress Green's Functions

Interactions between fault elements depend on the relative position and orientation of each element, and are calculated using stress Green's functions (OKADA

1992) at the start of the simulation. The change in stress for an element at position x due to relative motion of all the other elements is given by RUNDLE *et al.* (2006a; b):

$$\sigma_{ij}(x,t) = \int dx'_k T^{kl}_{ij}(x-x')s^l(x',t) \qquad (1)$$

where $s^l(x',t)$ is the three-dimensional slip density of element l, $T^{kl}_{ij}(x-x')$ is the Green's function tensor, and l goes over all elements. The Einstein summation convention is used for indices k and l, and indices i, j, k, and l run over three-dimensional Cartesian coordinate axes. In VC simulations the stress field is evaluated only at element centers and slip is applied uniformly across an element's surface and along the element's rake angle as defined in the fault model. Under these assumptions, we can reduce Eq. 1 for the stress on some element (denoted as A) at time t to:

$$\sigma_{ij}^A(t) = \sum_B T_{ij}^{AB} s^B(t) \qquad (2)$$

where iteration over element B sums the contributions from all other elements, and $s^B(t)$ is the slip deficit for element B—the distance element B has slipped from its equilibrium position at time t. For our purposes we only require the shear stress along the element rake vector and normal stress perpendicular to the element. Finally, the tensor T_{ij}^{AB} reduces to T_s^A for shear stresses and T_n^A for normal stresses, which lead to the following expressions for shear and normal stresses on some element A:

$$\sigma_s^A(t) = \sum_B T_s^{AB} s^B(t) \text{ and } \sigma_n^A(t) = \sum_B T_n^{AB} s^B(t) \qquad (3)$$

The numerical values for the Green's functions are calculated via a custom implementation of Okada's half-space equations (OKADA 1992). For a more detailed discussion on the Green's function implementation, see SACHS et al. (2012).

2.4. Rupture Model

Virtual California has two main phases of simulation: the long-term stress loading phase, and the rupture propagation phase. The stress loading phase is the interseismic part of the simulation, and rupture propagation determines which elements will be involved in a simulated earthquake once the initial failing element begins to slip.

To determine when an element will begin to slip, VC uses a combined static-dynamic friction law. The law uses the Coulomb failure function (CFF) which is defined for each element A:

$$\text{CFF}^A(t) = \sigma_s^A(t) - \mu_s^A \sigma_n^A(t) \qquad (4)$$

where μ_s^A is the element's static coefficient of friction and is taken from the fault model. During long-term stress loading, an element will slip at time t_f when $\text{CFF}(t_f) = 0$ (static failure). When one or more elements slip, the simulation begins the rupture propagation phase described below.

During rupture propagation, the first element to fail slips back towards the equilibrium position. The amount of slip during this initial failure, s^A, is related

to the stress drop defined for the element in the model, $\Delta\sigma^A$, by SACHS et al. (2012):

$$s^A = \begin{cases} \frac{1}{K_L^A} \frac{N_{ef}}{S_t} (\Delta\sigma^A - \text{CFF}^A) & \text{if } N_{ef} \le S_t \\ \frac{1}{K_L^A} (\Delta\sigma^A - \text{CFF}^A) & \text{otherwise} \end{cases} \qquad (5)$$

where K_L^A is the element's stiffness or self-stress, defined as

$$K_L^A = T_s^{AA} - \mu_s^A T_n^{AA}. \qquad (6)$$

The factor $\frac{N_{ef}}{S_t}$ captures the current size of the rupture with N_{ef} representing the number of failed elements on the currently rupturing fault and S_t representing the slip-scaling threshold parameter. This factor prevents small ruptures from excessively slipping.

Once the slip is calculated for one or more elements that have statically failed (CFF = 0), a new stress state for the entire system is calculated using Eq. 3. To mimic rupture propagation in real fault systems, elements on the same fault near the failed elements are allowed to fail at a lower stress (dynamic failure) level than the value that is allowed by requiring that their CFF = 0. An element experiences dynamic failure if it is on the same fault as an already failed element and satisfies:

$$\frac{\text{CFF}_{\text{init}} - \text{CFF}_{\text{final}}}{\text{CFF}_{\text{init}}} > \eta \qquad (7)$$

where η is the dynamic triggering parameter that is either specified for the whole system or uniquely defined for each element. This parameter approximates the stress intensity factor at the tip of a propagating rupture and encourages rupture propagation.

During rupture propagation, elements that have not completely slipped back to equilibrium may fail again (potentially multiple times) and release more stress. However, elements may not slip away from equilibrium. This implies that elements cannot absorb stress released by other failed elements. This reflects our assumption that failed elements may not heal during a rupture, but also may not release all accumulated stress immediately.

It is also important to note that an element may not release all accumulated stress during a rupture, due to the slip-scaling threshold (Eq. 5) and dynamic triggering (Eq. 7), as well as the stress state of all elements prior to rupture. For example if, prior to rupture, all elements are in a high stress state then a small initial rupture can quickly propagate and become a large stress release event. However, if most elements prior to rupture are in a low stress state then a large initial rupture may quickly stop propagating, releasing relatively little stress.

3. Gravity Green's Functions

The solutions presented by OKUBO (1992) allow calculation of the vertical component of the gravity change vector for dislocations on finite faults. GRACE measures radial gravity changes along its orbit, so when it is over the fault it measures this vertical gravity component.

Since Virtual California partitions the fault system into finite fault elements embedded in an elastic half-space, the gravity Green's functions are a logical extension of the simulator's functionality. Analytic expressions for the Green's functions are given below.

3.1. Total Gravity Changes for Finite Faults

Figure 2 shows the coordinate system and variables that constitute Okubo's convention. The slip vector for a rectangular fault element of length L, down-dip width W, dip angle $0 < \delta \leq \frac{\pi}{2}$, at a depth d is given by $s = (U_1, U_2 \cos\delta - U_3 \sin\delta, -U_2 \sin\delta - U_3 \cos\delta)$.

Using this convention, Okubo derived the following Green's function for the vertical component of the total gravity changes (including changes from compression/dilatation and from uplift) on the surface ($x_3 = 0$) of a finite fault

$$\Delta g(r,s) = \{\rho G[U_1 S_g(\xi,\eta) + U_2 D_g(\xi,\eta) + U_3 T_g(\xi,\eta)] + \Delta\rho G U_3 C_g(\xi,\eta)\} \parallel -\beta \Delta h(r,s)$$
(8)

where $r = (x_1, x_2, x_3 = 0)$ is the observation point on the surface of the half-space. The gravity changes are evaluated for dislocations occurring on a finite rectangular fault, embedded in an elastic half-space (Table 1).

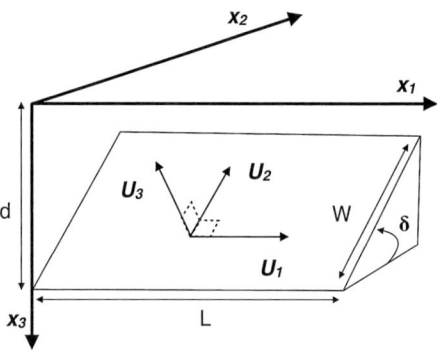

Figure 2
Coordinate and variable conventions used in Green's functions. Modified from OKUBO (1992)

We have used the double bar notation of CHINNERY (1961), where ξ and η are dummy variables and the quantity preceding the double bars is evaluated across the dislocation plane. The functions S_g, D_g, T_g, and C_g are the gravity Green's function contribution from strike-slip, dip-slip, tensile, and cavitation components, respectively. However, we only consider strike- and dip-slip faults, and their contributions are given by

$$S_g(\xi,\eta) = \ast\ast \frac{\partial \sin\delta}{\varrho} + \frac{\partial \cos\delta}{\varrho(\varrho+\eta)}$$
(9)

$$D_g(\xi,\eta) = 2I_2(\xi,\eta)\sin\delta - \frac{q\tilde{d}}{R(R+\xi)}$$
(10)

Thus, for our purposes we can simplify Eq. 8 by removing the terms associated with cavitation and tensile fault geometries, leaving us with

$$\Delta g(r,s) = \{\rho G[U_1 S_g(\xi,\eta) + U_2 D_g(\xi,\eta)]\} \parallel -\beta \Delta h(r,s).$$
(11)

The function $\Delta h(r,s)$ in Eq. 11 is Okada's Green's function for vertical displacement (OKADA 1992), and is given by

Table 1

Faulting geometries

Strike-slip	$U_1 > 0$	Left-lateral
	$U_1 < 0$	Right-lateral
Dip-slip	$U_2 > 0$	Thrusting
	$U_2 < 0$	Normal

$$\Delta h(r,s) = \frac{1}{2\pi}[U_1 S_h(\xi,\eta) + U_2 D_h(\xi,\eta) + U_3 T_h(\xi,\eta)] \,\|$$
$$(12)$$

$$S_h(\xi,\eta) = -\frac{\widetilde{d}q}{R(R+\eta)} + \frac{q\sin\delta}{R+\eta} - I_4(\xi,\eta)\sin\delta$$
$$(13)$$

$$D_h(\xi,\eta) = -\frac{\widetilde{d}q}{R(R+\xi)} - \sin\delta\tan^{-1}\left(\frac{\xi\eta}{qR}\right) \quad (14)$$
$$+ I_5(\xi,\eta)\sin\delta\cos\delta$$

$$T_h(\xi,\eta) = \frac{\widetilde{y}q}{R(R+\xi)} + \cos\delta\left[\frac{\xi q}{R(R+\eta)} - \tan^{-1}\left(\frac{\xi\eta}{qR}\right)\right]$$
$$- I_5(\xi,\eta)\sin^2\delta \qquad (15)$$

where,

$$R = \sqrt{\xi^2 + \eta^2 + q^2}$$
$$q = \sin\delta - (d - x_3)\cos\delta$$
$$\widetilde{d} = \eta\sin\delta - q\cos\delta$$

and finally,

$$I_1(\xi,\eta) = \tan^{-1}\left(\frac{-q\cos\delta + (1+\sin\delta)(R+\eta)}{\xi\cos\delta}\right)$$
$$I_2(\xi,\eta) = \tan^{-1}\left(\frac{R+\xi+\eta}{q}\right)$$
$$I_4(\xi,\eta) = (1-2\nu)[\log(R+\widetilde{d}) - \sin\delta\log(R+\eta)]\sec\delta$$
$$I_5(\xi,\eta) = 2(1-2\nu)I_1\sec\delta$$

and ν is Poisson's ratio, and is defined in terms of the Lamé parameters λ and μ as $\nu = \frac{\lambda}{2(\lambda+\mu)}$. For our fault model we have $\lambda = 3.2 \times 10^{10}$ Pa and $\mu = 3 \times 10^{10}$ Pa, which gives $\nu \sim 0.258$.

3.2. Dilatational Gravity Changes for Finite Faults

Okubo derived the following Green's function for the vertical component of the dilatational gravity changes—gravity changes arising only from density changes within the medium—on the surface $(r = (x_1, x_2, x_3 = 0))$ of a finite fault

$$\Delta g^*(r,s) = \{\rho G[U_1 S_g^*(\xi,\eta) + U_2 D_g^*(\xi,\eta) + U_3 T_g^*(\xi,\eta)] + \Delta\rho G U_3 C_g^*(\xi,\eta)\} \,\|$$
$$(16)$$

The functions S_g^*, D_g^*, T_g^*, and C_g^* are the dilatational gravity Green's function contribution from strike-slip, dip-slip, tensile, and cavitation components respectively. Again, we only consider strike- and dip-slip faults, and their contributions are given by

$$S_g^*(\xi,\eta) = I_4(\xi,\eta)\sin\delta \qquad (17)$$

$$D_g^*(\xi,\eta) = -I_5(\xi,\eta)\sin\delta\cos\delta \qquad (18)$$

Thus, for our purposes we can simplify Eq. 8 by removing the terms associated with cavitation and tensile fault geometries, leaving us with

$$\Delta g^*(r,s) = \{\rho G[U_1 S_g^*(\xi,\eta) + U_2 D_g^*(\xi,\eta)]\} \,\|. \qquad (19)$$

4. Computing Gravity Changes with VC

Following HAYES et al. (2006), we extend the capability of the newest incarnation of Virtual California to calculate the gravity Green's functions across a geometrically complex fault network. The Green's function solutions—arising from both changes in density (dilatational or compressional) and vertical displacement (free-air effect)—are computed via a custom implementation of the elastic half-space Greens' functions as described in OKUBO (1992). The sections below will present the gravity Green's functions calculated by VC for single fault elements and for an entire fault system.

4.1. Single Element Solutions

The gravity Green's function solutions for the three distinct types of fault elements used in the California model are shown in Fig. 3. The view is from the azimuth looking anti-parallel to the normal vector of the half-space. The dark lines in the figure represent the projection of each the fault element's vertical cross-section onto the surface of the half-space. The horizontal and vertical axes measure distance along and from the fault in *km*. To compute the gravity changes shown in the figure, 5*m* of slip is applied uniformly across each fault element and $\Delta g(r,s)$ is evaluated using Eq. 11.

In addition to specifying the slip, the Green's functions only require knowledge of the local fault geometry. For each fault element in Fig. 3, the parameters are $L = 10$ km, $W = 10$ km, depth to top of fault is 1 km, Poisson's ratio $v = 0.258$, and the density of the homogeneous half-space is 2,670 kg/ m^3. Strike-slip fault elements tend to have surface gravity changes an order of magnitude lower than those for normal and thrust fault elements of a given geometry and slip.

4.2. Co-seismic Solutions

Virtual California can calculate co-seismic gravity changes quickly and for arbitrarily complex fault geometries. Given the fault network geometry from the simulation input and co-seismic slip distribution from the simulation output, VC calculates gravity changes by implementing Eqs. 11 and 19 across the fault system. These patterns reveal characteristic gravity changes associated with various earthquake scenarios and, in the future, may help serve the data pipeline associated with a satellite-based fault monitoring system or other application of satellite gravitmetry.

The gravity changes constitute a vector field, but we are modeling the vertical component. GRACE measures this radial component as it flies over its observation points. The gravity signal is attenuated with distance, so the satellite would measure a smaller gravity signal. However, we do not incorporate this effect into our model at this juncture.

For the remainder of this section, we explore the possible gravity change fields associated with major earthquake scenarios in California.

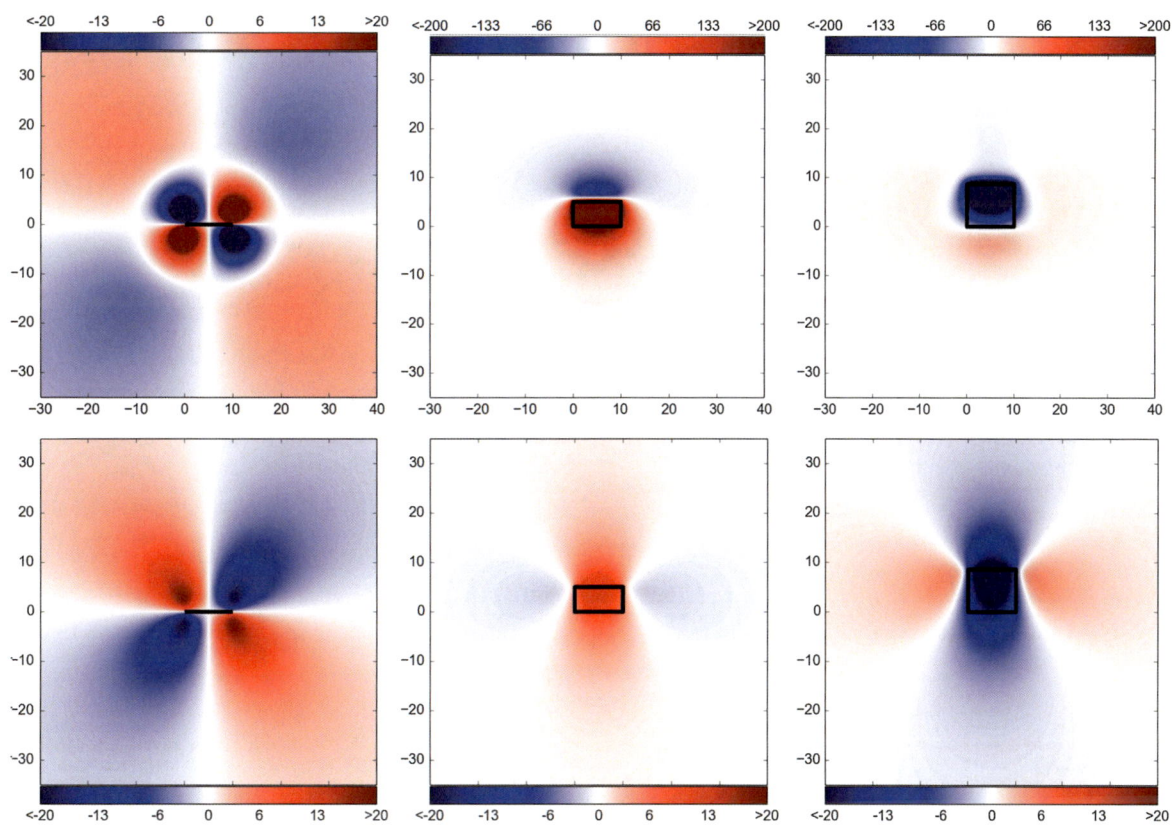

Figure 3

Gravity Green's functions solutions for the three characteristic fault element geometries. Axes denote distance along/from fault in km, the *solid black line* is the projection of the fault plane. *Top row* total gravity changes (Δg). *Bottom row* dilatational gravity changes (Δg^*). *Left* vertical right-lateral strike-slip $\delta = 90°$. *Middle* normal, $\delta = 60°$. *Right* thrust, $\delta = 30°$. The color unit is μGal, and fault parameters are specified in Sect. 4.1.

4.3. *Gravity Changes for California Earthquakes*

The first major historic earthquake we investigate here is the 1857 Fort Tejon earthquake. This southern California earthquake had an estimated magnitude of 7.73 and ruptured approximately 350 km of the southern San Andreas Fault (ZIELKE *et al.* 2012). Table 2 compares a Virtual California simulated earthquake to the Fort Tejon earthquake. Figures 4 and 5 show the solutions to the gravity Green's functions for the simulated earthquake, a

Table 2

Gravity changes at California locations for VC event 60019

	Magnitude	Avg. slip (m)	Rupture length (km)
Fort Tejon 1857	7.73	3.5	350
VC event 60019	7.81	2.62	359.9
	San Luis Obispo	Bakersfield	Los Angeles
Δg [μ Gal] (surface)	+8.541	−1.431	−9.983

Fort Tejon data from ZIELKE *et al.* (2012)

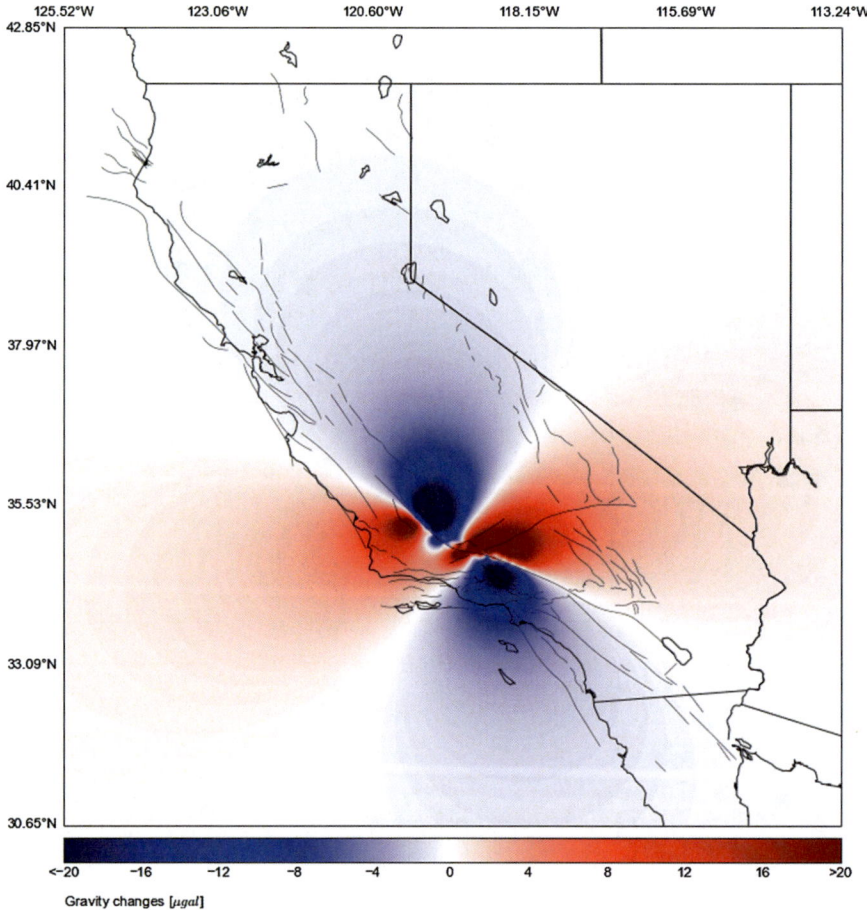

Figure 4

Green's function solutions for total surface gravity changes (Δg) for VC event 60019. Co-seismic slips and fault geometry are taken from a VC simulation with the allcal2 fault model. The system's gravity change pattern spans a few hundred kilometers and clearly reflects the right-lateral strike-slip geometry of the San Andreas Fault

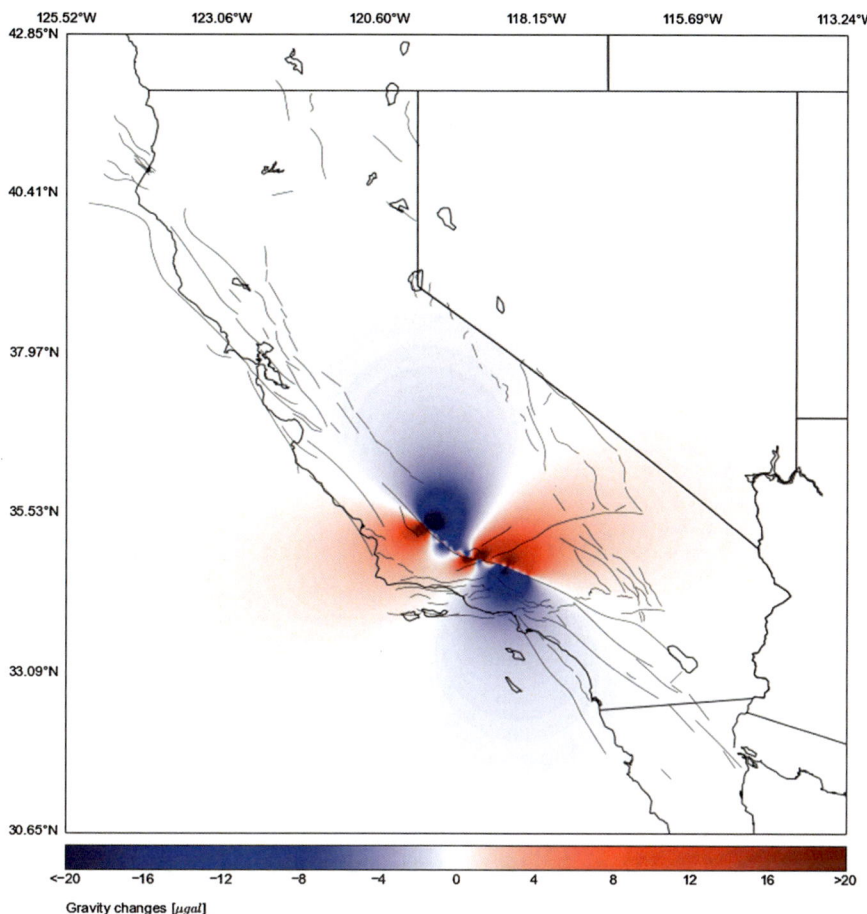

Figure 5

Dilatational gravity Green's function solutions (Δg^*) for VC event 60019

characteristic large right-lateral strike-slip earthquake that ruptures most of the southern San Andreas Fault.

The next major earthquake scenario is a large northern California earthquake that ruptures almost the entire northern San Andreas Fault (SAF). The historical example here is the 1906 San Francisco earthquake, which had an estimated magnitude of 7.9 and ruptured at least 470 km of the northern San Andreas Fault—total surface rupture length could be larger; it is not known if the offshore segment of the SAF ruptured (RUNDLE *et al.* 2005). Table 3 compares a Virtual California simulated earthquake to the great San Francisco earthquake. Figure 6 shows the solutions to the total gravity Green's function (Eq. 11) on the surface for the

Table 3

Gravity changes at California locations for VC event 231185

	Magnitude	Avg. slip (m)	Rupture length (km)
San Francisco 1906	7.9	2.0-5.0	470
VC event 231185	7.88	2.20	711.7
	Fort Bragg	San Francisco	San Jose
Δg [μ Gal] (surface)	−19.714	+19.799	+5.401

San Francisco data from RUNDLE *et al.* (2005)

simulated earthquake, a characteristic large northern California earthquake that ruptures most of the northern SAF. Figure 7 shows the solutions to the dilatational gravity Green's function (Eq. 19) as measured by a free observer, e.g. a gravity satellite.

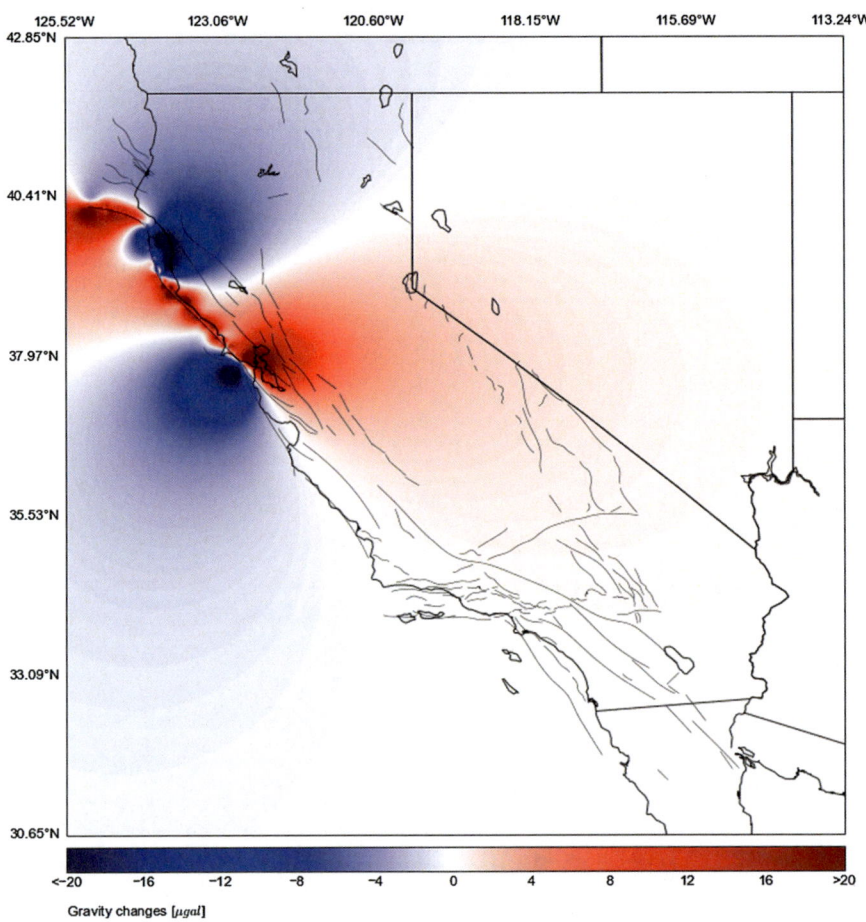

Figure 6
Surface gravity Green's function solutions (Δg) for VC event 231185

Deviations from the characteristic pattern are a direct result of anisotropic local fault geometry along the rupture length and a result of the intrinsic unpredictability of local co-seismic slips.

5. Conclusions

Co-seismic surface gravity changes—on the order of a few tens of μGal—simulated by Virtual California match the level of recorded gravity changes in California (WHITCOMB et al. 1980). Our simple model indicates that the magnitude of expected gravity signal for strike-slip faults is near the detection threshold for current satellite based gravimetry. However, for strike-slip gravity signals, the positive and negative poles are within a few hundred kilometers of each other, making detection difficult. We suspect gravimetry is better suited for monitoring dip-slip faults with large potential rupture areas like those in the great subduction zones.

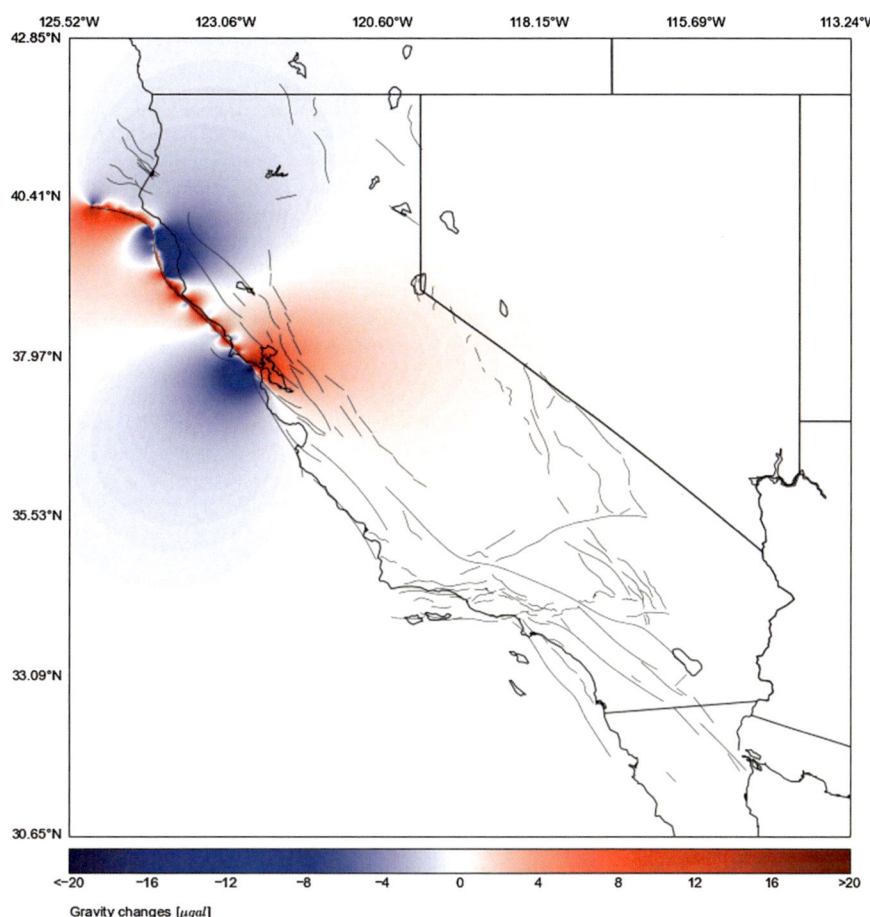

Figure 7

Dilatational gravity Green's function solutions (Δg^*) for VC event 231185

Acknowledgments

This research was supported by National Aeronautics and Space Administration (NASA) Earth and Space Science fellowship Number NNX11AL92H.

References

M. A. Chinnery. *The deformation of the ground around surface faults.* Bulletin of the Seismological Society of America, *51*(3): 355–372, 1961.

Shin-Chan Han, C. K. Shum, Michael Bevis, Chen Ji, and Chung-Yen Kuo. *Crustal dilatation observed by grace after the 2004 sumatra-andaman earthquake.* Science, *313*(5787): 658–662, 2006.

Shin-Chan Han, Jeanne Sauber, and Scott Luthcke. *Regional gravity decrease after the 2010 Maule (Chile) earthquake indicates large-scale mass redistribution.* Geophysical Research Letters, *37*(23), 2010.

T. J. Hayes, K. F. Tiampo, J. B. Rundle, and J. Fernndez. *Gravity changes from a stress evolution earthquake simulation of california.* Journal of Geophysical Research: Solid Earth, *111*(B9), 2006.

E.M. Heien and M. Sachs. *Understanding long-term earthquake behavior through simulation.* Computing in Science Engineering, *14*(5): 10–20, 2012.

Kosuke Hekiand Koji Matsuo. *Coseismic gravity changes of the 2010 earthquake in central Chile from satellite gravimetry.* Geophysical Research Letters, *37*(24), 2010.

Y. Okada. *Internal deformation due to shear and tensile faults in a half-space.* Bulletin of the Seismological Society of America, *82*(2): 1018–1040, 1992.

S. Okubo. *Gravity and potential changes due to shear and tensile faults in a half-space.* Journal of Geophysical Research: Solid Earth, *97*(B5): 7137–7144, 1992.

J. B. Rundle. *A physical model for earthquakes: 1. Fluctuations and interactions.* Journal of Geophysical Research: Solid Earth, *93*(B6): 6237–6254, 1988a.

J. B. Rundle. *A physical model for earthquakes: 2. Application to southern California.* Journal of Geophysical Research: Solid Earth, *93*(B6): 6255–6274, 1988b.

Reprinted from the journal

J. B. RUNDLE, P. B. RUNDLE, A. DONNELLAN, D. L. TURCOTTE, R. SHCHERBAKOV, P. LI, B. D. MALAMUD, L. B. GRANT, G. C. FOX, D. MCLEOD, G. YAKOVLEV, J. PARKER, W. KLEIN, and K. F. TIAMPO. *A simulation-based approach to forecasting the next great san francisco earthquake*. Proceedings of the National Academy of Sciences of the United States of America, *102*(43): 15363–15367, 2005.

JOHN B. RUNDLE. *Gravity changes and the palmdale uplift*. Geophysical Research Letters, *5*(1): 41–44, 1978.

JOHN B. RUNDLE, PAUL B. RUNDLE, ANDREA DONNELLAN, P. LI, W. KLEIN, GLEB MOREIN, D.L. TURCOTTE, and LISA GRANT. *Stress transfer in earthquakes, hazard estimation and ensemble forecasting: Inferences from numerical simulations*. Tectonophysics, *413*(12): 109–125, 2006a.

P.B. RUNDLE, J.B. RUNDLE, K.F. TIAMPO, A. DONNELLAN, and D.L. TURCOTTE. *Virtual california: Fault model, frictional parameters, applications*. In Computational Earthquake Physics: Simulations, Analysis and Infrastructure, Part I, Pageoph Topical Volumes, pages 1819–1846. Birkhäuser-Verlag, 2006b.

M. K. SACHS, E. M. HEIEN, D. L. TURCOTTE, M. B. YIKILMAZ, J. B. RUNDLE, and L.H. KELLOGG. *Virtual California Earthquake Simulator*. Seismological Research Letters, *83*(6): 973–978, 2012.

W. SUN and S. OKUBO. *Coseismic deformations detectable by satellite gravity missions: A case study of Alaska (1964, 2002) and Hokkaido (2003) earthquakes in the spectral domain*. Journal of Geophysical Research: Solid Earth, *109*(B4), 2004.

J. B. WALSH and J. R. RICE. *Local changes in gravity resulting from deformation*. Journal of Geophysical Research: Solid Earth, *84*(B1): 165–170, 1979.

J. H. WHITCOMB, W. O. FRANZEN, J. W. GIVEN, J. C. PECHMANN, and L. J. RUFF. *Time-dependent gravity in southern california, May 1974 to April 1979*. Journal of Geophysical Research: Solid Earth, *85*(B8): 4363–4373, 1980.

O. ZIELKE, J. R. ARROWSMITH, L. GRANT LUDWIG, and S. O. AKCIZ. *High resolution topography derived offsets along the 1857 fort tejon earthquake rupture trace, san andreas fault*. Bulletin of the Seismological Society of America, *102*(3): 1135–1154, 2012.

(Received May 20, 2014, revised August 14, 2014, accepted August 17, 2014, Published online September 9, 2014)

Pure Appl. Geophys. 173 (2016), 623–636
© 2015 Springer Basel
DOI 10.1007/s00024-015-1085-y

Pure and Applied Geophysics

Application of Mixture of Gaussian Clustering on Joint Facies Interpretation of Seismic and Magnetotelluric Sections

MOHAMMAD ALI SHAHRABI,[1] HOSEIN HASHEMI,[1] and MOHAMMAD KAZEM HAFIZI[1]

Abstract—Seismic and magnetotelluric (MT) methods are the most applicable geophysical methods in exploration of hydrocarbon resources. In this paper, mixture of Gaussian clustering is used to combine seismic and MT images under the scheme of Expectation/Maximization (EM) algorithm. Pre-Stack Depth Migration (PSDM) velocity, Root Mean Square (RMS) velocity and vertical gradient of RMS velocity of seismic and resistivity model of MT along 19.3 km MUN-21 profile in Munir Block that has been located in Southwest of Iran in Dezful embayment over the Seh-Qanat anticline are applied. The anticline is the most important oil trap of this area. The Expectation/Maximization (EM) method that has been applied includes: (1) creation of data vectors from the seismic and MT images using image processing techniques, (2) normalizing and mapping using Principal Component Analysis (PCA) procedure (3) unsupervised learning of dataset matrix, (4) setting the matrix in Expectation/Maximization (EM) iteration algorithm (5) remapping to physical space. The final model consists fof six classes which could be given to eight formations that belong to Eocene to Neocomian geological age. Pre-Stack Depth Migration (PSDM) velocity model obtained from seismic study on Seh-Qanat anticline only detected 2 horizons of formations, Asmari and Sarvak Formations; however, the current methodology introduces subdivision anticline into six classes by matching it to the log information of Seh-Qanat Deep-1 (SQD-1) borehole where it was excavated over the anticline with total depth of 2876 m.

Key words: Mixture of Gaussian, Expectation/Maximization clustering, seismic, magnetotelluric, joint interpretation.

1. Introduction

Since hydrocarbon resources have an important role in the economic life of every country, methods based on maximum achievements of the current acquired geophysical data have been widely promoted in the recent years (MAURER *et al.* 2010). The seismic and magnetotelluric sounding are two geophysical methods that are the most applicable approaches to identify the hydrocarbon reservoirs (ZIOLKOWSKI 2009). Each geophysical approach has its own deficiencies and limitations; and Seismic method is not an exception as well. It results in non-informative images for some geological models of substructures. On the other hand, MT method faces wave attenuation in depth in the conductive environments (BEDROSIAN *et al.* 2007).

Combining MT and seismic methods would be helpful since it provides better qualitative and quantitative results, while the concerns of scale and spatial resolution are preserved. Hence, such joint models have been developed using techniques like joint inversion of seismic travel time and DC resistivity (GALLARDO and MEJU 2007; BENNINGTON *et al.* 2010) and used for qualitative and mostly descriptive interpretation (STANLEY *et al.* 1990). In quantitative joint interpretation, some methods and disciplines were used based on a 2-D cluster analysis of the seismic and MT models (BEDROSIAN *et al.* 2004). Lithology-derived statistical approaches have also been developed to consider the correlation between the physical properties in a combined parameter space (BEDROSIAN *et al.* 2007).

Pattern recognition is a statistical approach based on mathematical discrimination and classification rules that subdivide obtained models into lithological classes (BAUER *et al.* 2012). The statistical pattern recognition has been fully discussed in text books and papers, e.g., in WEBB *et al.* (2002) and was applied for seismic and DC resistivity models by BEDROSIAN *et al.* (2007) in lithology mapping for the zones where high correlation between resistivity and velocity is available. BAUER *et al.* (2012) obtained combined models of MT resistivity and P seismic velocity using Self-Organizing Map (SOM) or Kohonen map.

———————
[1] Institute of Geophysics, University of Tehran, North Kargar Street, Post Box 14155-6466, Tehran, Iran. E-mail: md.shahrabi@ut.ac.ir

The main application of Expectation/Maximization (EM) algorithm in statistics is in the parameter estimation field. It seems that HARTLEY (1958) was the first one who introduced the concept of Expectation/Maximization (EM) and DEMPSTER et al. (1977) formalized and applied this method.

Since then, Expectation/Maximization (EM) has been applied to signal processing, pattern recognition, image restoration and reconstruction. In some cases, it has been applied for characterizing of measurement error of seismic data and provides an estimate of the mean and covariance of measurement errors (ZHAO et al. 2007). HAN et al. (2011) have proposed an application of the Expectation/Maximization (EM) algorithm to identify automatically geologic facies from seismic data with this assumption that a Gaussian distribution approximates seismic data of a geologic facies. The mean of each Gaussian model represents the average value, and the variance gives the variation of the seismic data within a facies. KUYUK et al. (2011) applied k-means, and Gaussian mixture model (GMM) analyses to classify seismic events using two-time variant parameters, the ratio of integrated powers of the velocity seismogram, and the S wave/P wave amplitude ratio. They utilized common Expectation/Maximization (EM) algorithm to derive the parameters of the mixture model distribution.

2. Methodology

Mixture of Gaussian model is an unsupervised clustering method based on probability distribution. In the mixture of Gaussian model, it is assumed that the objects in each cluster are distributed according to a Gaussian distribution. It means that each cluster is not only characterized by a mean μ_i but also by a covariance matrix Σ_i (HEIJDEN et al. 2004).

A mixture model is a distribution of the form,

$$p(x) = \sum_{j=1}^{g} \pi_j p(x; \theta_j) \qquad (1)$$

where g is the number of mixture components, $\left(\sum_{j=1}^{g} \pi_j = 1, \pi_j \geq 0 \right)$ are the mixing proportions and, $p(x; \theta_j)$, $j = 1, \ldots g$ are the component density functions which depend on a parameter vector θ_j.

There are three sets of parameters to estimate: the values of π_j, the components of θ_j, and the value of g. The component densities may be of different parametric forms and are specified using knowledge of the data generation process, if available. In the Gaussian mixture model, $p(x; \theta_j)$ is the multivariate Gaussian distribution, with

$$\theta_j = \{\mu_j, \Sigma_j\} \qquad (2)$$

Given a set of n observations $(x_1, x_2, \ldots x_n)$, the likelihood function is

$$L(\Psi) = \prod_{i=1}^{n} \sum_{j=1}^{g} \pi_j p(x_i | \theta_j) \qquad (3)$$

where Ψ denotes the set of parameters $\{\pi_1, \ldots \pi_g; \theta_1, \ldots \theta_g\}$ and we now denote the dependence of the component densities on their parameters as $p(x; \theta_j)$. In general, it is not possible to solve $\partial L / \partial \Psi = 0$ explicitly for the parameters of the model and iterative schemes must be employed. One approach for maximizing the likelihood $L(\Psi)$ is to use a general class of iterative procedures known as EM (Expectation/Maximization) algorithms, introduced in the context of missing data estimation by DEMPSTER et al. (1977).

The basic procedure is as follows. It is supposed that a set of 'incomplete' data vector $\{x\}$ is available and maximizing the likelihood $L(\Psi) = p(\{x\} | \Psi)$ is targeted. Let $\{y\}$ denote a typical 'complete' version of $\{x\}$, that is, each vector x_i is augmented by the missing values so that $y_i^T = (x_i^T, z_i^T)$. There may be many possible vectors y_i in which one can embed x_i, though there may be a natural choice for some problems. In the finite mixture case, z_i is a class indicator vector $z_i = (z_{1i}, \ldots z_{gi})^T$, where $z_{ji} = 1$ if x_i belongs to the jth component and zero otherwise.

Let the likelihood of $\{y\}$ be $g(\{y\} | \Psi)$, so that the likelihood $p(\{x\} | \Psi)$ is obtained from $g(\{y\} | \Psi)$ by integrating over all possible $\{y\}$ in which the $\{x\}$ set is embedded,

$$L(\Psi) = p(\{x\} | \Psi) = \int \prod_{i=1}^{n} g(x_i, z | \Psi) \, dz \qquad (4)$$

The EM procedure generates a sequence of estimates of form $\Psi, \{\Psi^{(m)}\}$, from an initial estimate $\Psi^{(0)}$ and consists of two steps:

E-step: evaluate

$$Q\left(\Psi, \Psi^{(m)}\right) \hat{=} E\left[\log(g(\{y\}|\Psi))|\{x\}, \Psi^{(m)}\right], \quad (5)$$

that is,

$$Q\left(\Psi, \Psi^{(m)}\right) = \int \sum \log(g(x_i, z_i|\Psi)) \\ p\left(\{z\}|\{x\}, \Psi^{(m)}\right) dz_1 \ldots dz_n \quad (6)$$

The expectation of the complete data log-likelihood, conditional on the observed data, $\{x\}$, and the current value of the parameters, $\Psi^{(m)}$

M-step: find $\Psi = \Psi^{(m+1)}$ that maximizes $Q\left(\Psi, \Psi^{(m)}\right)$. Often the solution for the M-step may be obtained in closed form.

The likelihoods of interest satisfy

$$L\left\{\Psi^{(m+1)}\right\} \geq L\left\{\Psi^{(m)}\right\} \quad (7)$$

So they are monotonically increasing.

Let us now consider the application of the EM algorithm to mixture distributions. For fully labeled data, the complete data vector y is defined to be the observation augmented by a class label; that is, $y^T = (x^T, z^T)$, where z is an indicator vector of length g with a 1 in the kth position if x is in category k and zeros elsewhere. The likelihood of y is

$$g(y|\Psi) = p(x|z, \Psi)p(z|\Psi) = p(x|\theta_k)\pi_k \quad (8)$$

which may be written as

$$g(y|\Psi) = \prod_{j=1}^{g} \left[p\left(x|\theta_j\right)\pi_j\right]^{z_j}. \quad (9)$$

Since z_j is zero expect for $j = k$, the likelihood of x is

$$p(x|\Psi) = \sum_{\text{all possible } z \text{ values}} g(y|\Psi) = \sum_{j=1}^{g} \pi_j p(x|\theta_j)$$

which is a mixture distribution. Thus, we may interpret mixture data as incomplete data where the missing values are the class labels. For n observations we have,

$$g(y_1, \ldots y_n|\Psi) = \prod_{i=1}^{n} \prod_{j=1}^{g} \left[p\left(x_i|\theta_j\right)\pi_j\right]^{z_{ji}} \quad (10)$$

with

$$\log(g(y_1, \ldots y_n|\Psi)) = \sum_{i=1}^{n} z_i^T l + \sum_{i=1}^{n} z_i^T u_i(\theta) \quad (11)$$

where the vector l has jth component $\log\left(\pi_j\right)$ has jth component $\log\left(p\left(x_i|\theta_j\right)\right)$ and z_i has components $z_{ji}. j = 1, \ldots g$, where z_{ji} are the indicator variables taking value one if pattern is in group j, and zero otherwise. The likelihood of $(x_1, \ldots x_n)$ is $L_0(\Psi)$, as follows. The steps in the basic iteration are

1. E-step: form

$$Q\left(\Psi, \Psi^{(m)}\right) = \sum_{i=1}^{n} \omega_i^T l + \sum_{i=1}^{n} \omega_i^T u(\theta) \quad (12)$$

where

$$\omega_i = E\left(z_i|x_i, \Psi^{(m)}\right) \quad (13)$$

With jth component, the probability that x_i belongs to group j given the current estimates $\Psi^{(m)}$ given by

$$\omega_{ij} = \frac{\pi_j^{(m)} p\left(x_i|\theta_j^{(m)}\right)}{\sum_k \pi_k^{(m)} p\left(x_i|\theta_k^{(m)}\right)} \quad (14)$$

2. M-step: this consists of maximizing Q with respect to Ψ. Consider the parameters π_i, θ_i in turn. Maximizing Q with respect to π_i (subject to the constraint that $\sum_{j=1}^{g} \pi_j = 1$) leads to the equation $\sum \omega_{ij} \frac{1}{\pi_j} - \lambda = 0$.

Obtained by differentiating $Q - \lambda\left(\sum_{j=1}^{g} \pi_j - 1\right)$ with respect to π_j where λ is a Lagrange multiplier. The constraint $\sum \pi_j = 1$ gives $\lambda = \sum_{j=1}^{g} \sum_{i=1}^{n} \omega_{ij} = n$ and we have the estimate of π_j as,

$$\hat{\pi}_j = \frac{1}{n} \sum_{i=1}^{n} \omega_{ij} \quad (15)$$

For normal mixtures, $\theta_i = (\mu_i, \Sigma_i)$ and we consider the mean and covariance matrix re-estimation separately. Differentiating Q with respect to μ_j and equating to zero gives

$$\sum_{i=1}^{n} \omega_{ij}\left(x_i - \mu_j\right) = 0$$

which gives the re-estimation for μ_j as

$$\hat{\mu}_j = \frac{\sum_{i=1}^n \omega_{ij} x_i}{\sum_{i=1}^n \omega_{ij}} = \frac{1}{n\hat{\pi}_j} \sum_{i=1}^n \omega_{ij} x_i \qquad (16)$$

Differentiating Q with respect to Σ_j and equating to zero gives

$$\hat{\Sigma}_j = \frac{\sum_{i=1}^n \omega_{ij} (x_i - \hat{\mu}_j)(x_i - \hat{\mu}_j)^T}{\sum_{i=1}^n \omega_{ij}}$$
$$= \frac{1}{n\hat{\pi}_j} \sum \omega_{ij} (x_i - \hat{\mu}_j)(x_i - \hat{\mu}_j)^T \qquad (17)$$

Thus, the EM algorithm for normal mixtures alternates between the E-step of estimating the ω_i (Eq. 14) and the M-step of calculating $\hat{\pi}_j$, $\hat{\mu}_j$, $\hat{\Sigma}_j$ ($j = 1, \ldots g$) given the values of ω_i [Eqs. (15), (16) and (17)]. These estimates become the estimates at stage $m + 1$ and are substituted into the right-hand side of (14) for the next stage of the iteration. The process iterates until convergence of the likelihood. (WEBB et al. 2002)

3. Petroleum Geology of the Munir Block

The Zagros mountain range is an Alpine-type Orogene created by the collision of the continental Arabian plate with segments of the Eurasian margin during Mesozoic and Cenozoic time oceanic. Arabian plate crust was subducted northward beneath Eurasia such that the continent to continent collision created locally in late Eocene, and convergence still continues today (SAHABI 2012).

The Munir Block is located southeast of the most prolific oil province of Iran, the Khuzestan, in Dezful Embayment (Fig. 1). Multiple petroleum systems are present, since at least two proven source rocks exist in the area of the Block: The Kazhdumi and Pabdeh shale sediments.

Tree main groups of reservoirs are recognized in the Khuzestan basin: the Khami group, the Bangestan Group and the Asmari Formation, that all the tree groups are recognized in Munir Block with excellent fracture permeability and locally primary porosity. The Jurassic to Early Cretaceous Khami Group consists of massive limestones and dolomites. These carbonates extend over the whole Khuzestan. In Southeast of Khuzestan the two main reservoir intervals include limestones of both Daryan and Fahlyan Formations. The seals are the Kazhdumi and Gadvanshales. Traps are Bangestan group belonged to the Mid Late Cretaceous. The main reservoir intervals are the Sarvak and the Ilam limestone. The Sarvak Formation consists of high-energy limestone deposited initially on the margins of the intra-shelf basin with high porosity and medium to high permeability.

Formation shales provide the seal. Traps are classic anticlines often asymmetric with a steep southwestern flank (REZAII 2011; MOTIEI 2010; SAHABI 2012).

Anticlines in depth vary in shape and position from the surface (REZAII 2011; MOTIEI 2010; SAHABI 2012).

The Zagros fold belt and is capped by the Gachsaran evaporates, which provide an excellent seal and consist of thick, high-energy carbonates ranging from reefal to infra-tidal limestones.

Traps are classic anticlines often asymmetric with a steep southwestern flank (REZAII 2011; MOTIEI 2010; SAHABI 2012).

4. Application of Expectation/Maximization Algorithm to Southwest Munir Block

The Expectation/Maximization (EM) algorithm is applied to seismic and MT images from a hydrocarbon exploration project in the SW Munir Block. There are four model types available: (1) Pre-Stack Depth Migration (PSDM) interval P-velocity, (2) Root Mean Square (RMS) P-velocity, (3) vertical gradient of RMS P-velocity, and (4) electrical resistivity from MT inversion. Figure 2 shows the Expectation/Maximization (EM) workflow to the MUN-21 data.

Based on litholog facies, different classes at the bottom of the anticline are considered where the Asmari Formation is located. The images were normalized in columns, and then mapping was done using principal component analysis or PCA procedure. After training of dataset matrix by multiplying of the norm matrix with the mapped matrix,

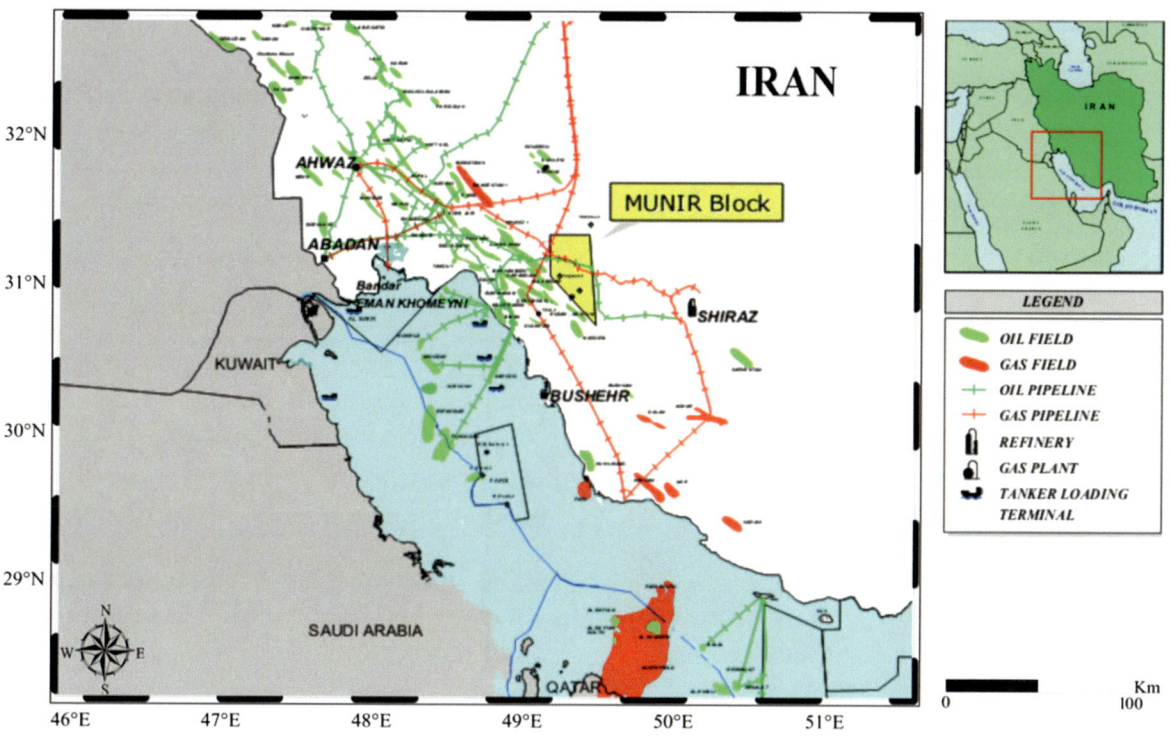

Figure 1
Location of Munir Block in Southwest of Iran

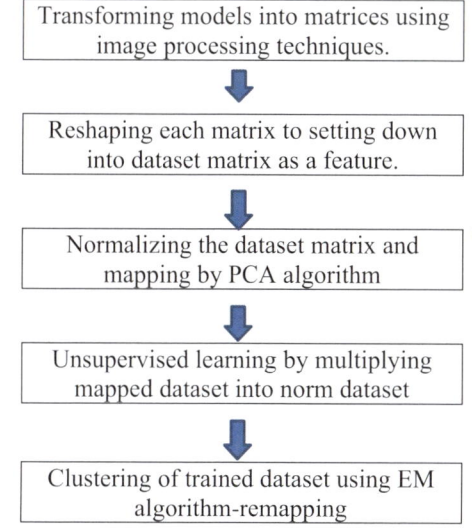

Figure 2
Schematic workflow of the Expectation/Maximization (EM) analysis

Expectation/Maximization (EM) iteration algorithm with different iteration numbers and different class numbers was run and clustering was done.

5. Geophysical Experiments in Munir Block

Seismic and MT experiments were carried out within the Munir Block by NIOC Exploration Directorate. There are several traps and hydrocarbon zones in Munir Block that the most important one is Seh-Qanat anticline, which has a symmetrical shape. The MUN-21 seismic line is located on Seh-Qanat anticline and MT stations are along it in a 19.3-km-long profile (Fig. 3). Seismic explosions with 8 kg charge size were fired from boreholes with 100 m shot spacing. The shots were recorded with a constant receiver spread along the entire profile with a sampling rate of 2 ms.

The MT experiment was carried out by deploying 28 stations with about 600 m spacing with 40 frequency values in 7 decades and period range of 0.003–2000 s. The relatively high number of frequencies applied for data gathering in each site is due to increasing of vertical resolution of obtained resistivity images. At each point, horizontal electric and magnetic field components and the vertical magnetic component were measured. Seh-Qanat Deep-1(SQD-

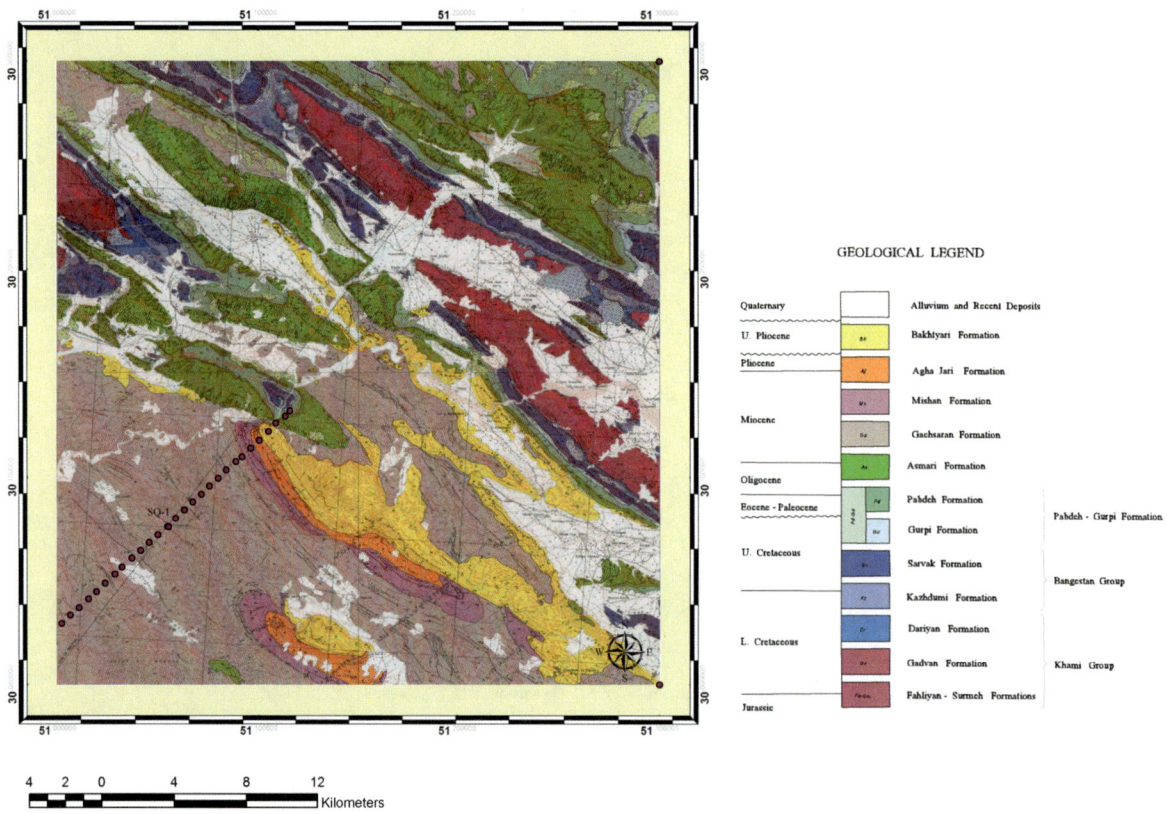

GEOLOGICAL LEGEND

Figure 3
Geological map of investigated area; location of seismic profile, MT sites and SQD-1 Borehole

1) well is located in the middle of profile that has total depth of 2876 m from the surface.

6. Seismic and MT Models

The data processing and development of PSDM velocity model of seismic are carried out and RMS velocity model and vertical gradient of it were derived (Fig. 6). MT inversion was carried out using EDI files. Correspondent minimal amplitude azimuth for all sites of MT profile at periods 0.003–20,000 s is summarized in a rose diagram shown in Fig. 4. According to this diagram, a common strike of was selected for MT profile being perpendicular to dominant directions in this diagram. Apparent resistivities and impedance phases decomposed into strike direction were used to obtain the models. Figure 5

shows the ellipticity, phase tensor and skew angle (orientation of the major axis which coincides with the strike in the 2D situation) at three different sites, which can be considered representative of the average behavior of all sites. Skew angle is smaller than 0.5 for all periods, and ellipticity is smaller than 0.1 for periods less than 1 s. Noises of all data were almost zero for short periods and a little noisy for the last decade in some sites. Both TE and TM modes with both MT polarizations were jointly inverted using the NLCG algorithm of RODI and MACKIE (2001).

The NLCG algorithm attempts to minimize an objective function that is the sum of the normalized data misfits and the smoothness of the model. The tradeoff between data misfits and model smoothness is controlled by the regularization parameter tau.

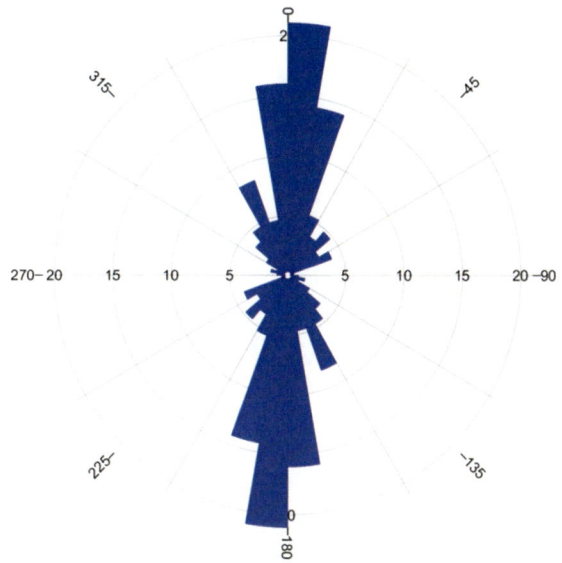

Figure 4
The rose diagram plot showing the all site strike analysis results of MT profile (MᴄNɪᴇᴄᴇ and Jᴏɴᴇs 2001) of the MT data in period range of 0.003–2000 s

Different values of tau were applied and inversion process was done by these values and best MT model was obtained by the value of tau = 4. In this paper, roughness is defined as the integral gradient of the model, assuming a standard Laplacian grid. The standard Laplacian may produce a rougher looking model, but the definition of smoothness is consistent with the model dimensions. The scale factor for roughness penalty β is set to 1.0 and α is set to 1.0 for two profiles. Selecting $\beta = 1$ means that the penalty on horizontal roughness increases with depth at the same rate as the vertical roughness. This means that the model will be smoother in both the vertical and horizontal directions as depth increases. Static shift was corrected by fitting in the inversion processing after obtaining a good fitting of the data under the assumption that static shifts are due to a zero-mean Gaussian process (Sᴀsᴀᴋɪ 2004; Oɢᴀᴡᴀ and Uᴄʜɪᴅᴀ 1996). The PSDM velocity seismic model, RMS Velocity model and vertical gradient of it has been

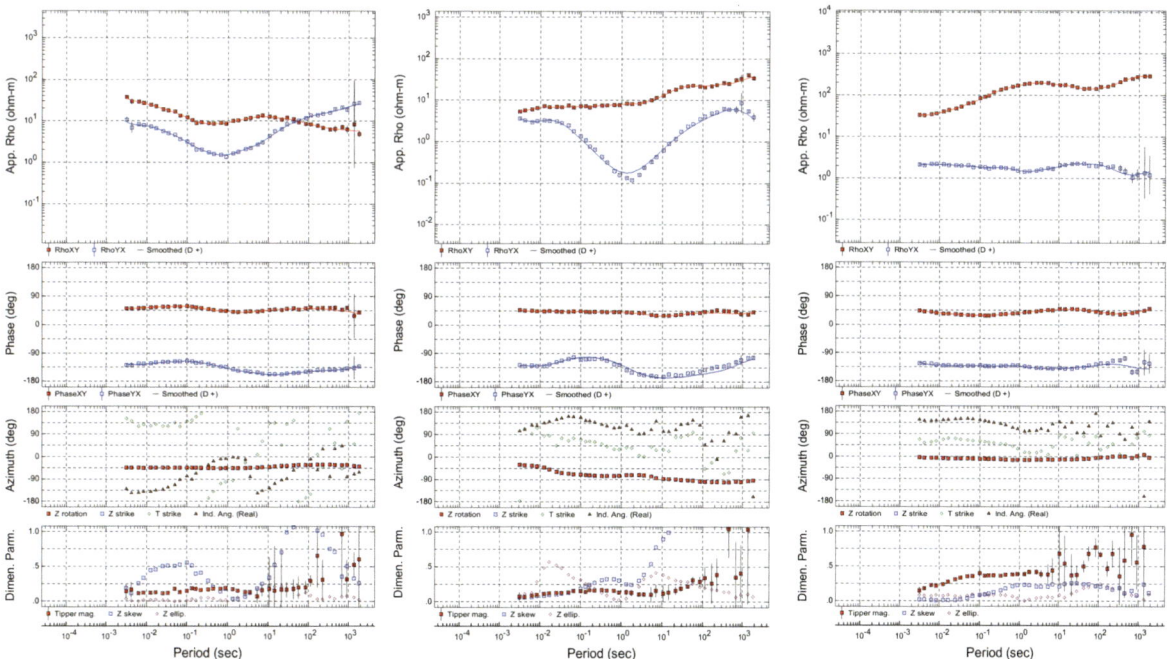

Figure 5
The fit between observed and calculated data for three representative sites selected from first (*left*), middle (*middle*), and end (*right*) of MT profile. For each site *two upper parts* are: TE for apparent resistivity and phase (*red*), TM for apparent resistivity and phase (*blue*), *solid lines* for calculated data and *squares* for observed data; and *two lower parts* are the variation of azimuth and dimension parameters for the period: Z strike, T strike and induction angle (azimuth parm) and Tipper magnitude, skew and ellipticity (dimen parm)

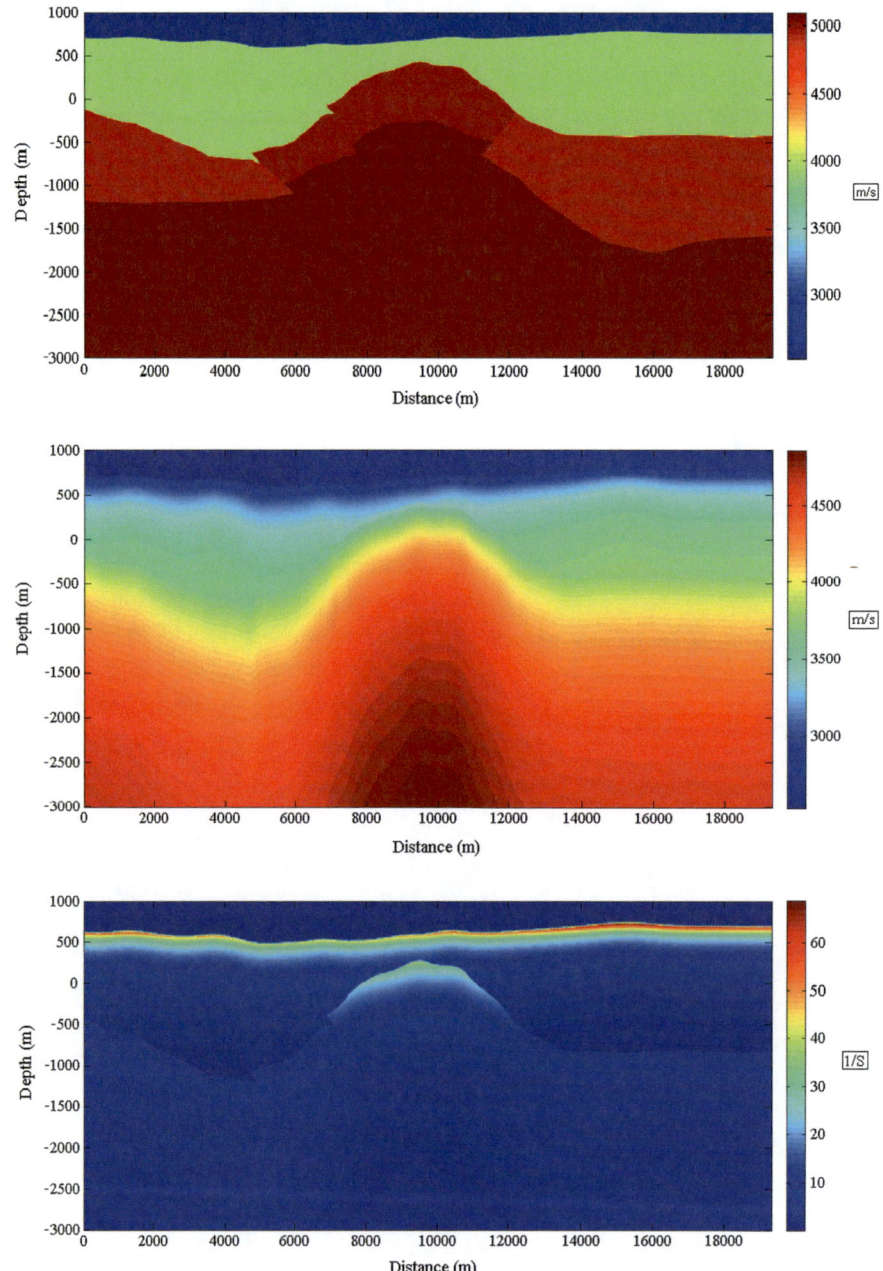

Figure 6
Images of seismic PSDM velocity (*up*), RMS velocity (*middle*) and vertical gradient of RMS velocity (*down*) applied in this study. The datum of sections is 1000 m above sea level

illustrated in Fig. 6. Furthermore, Fig. 7 shows PSDM seismic section which upper part of Fig. 6 has been achieved from it. The PSDM P–velocity seismic model can be subdivided into 4 velocity layers: Upper section is a near surface layer with lower velocity of 2.5 km/s. Second layer has velocity of 3.9 km/s. These two layers consist of Lower Miocene evaporate sediments of Gachsaran Formation, which is the cap rock of Asmari Formation. The anticline beetled with these two layers. Two deeper sections with higher

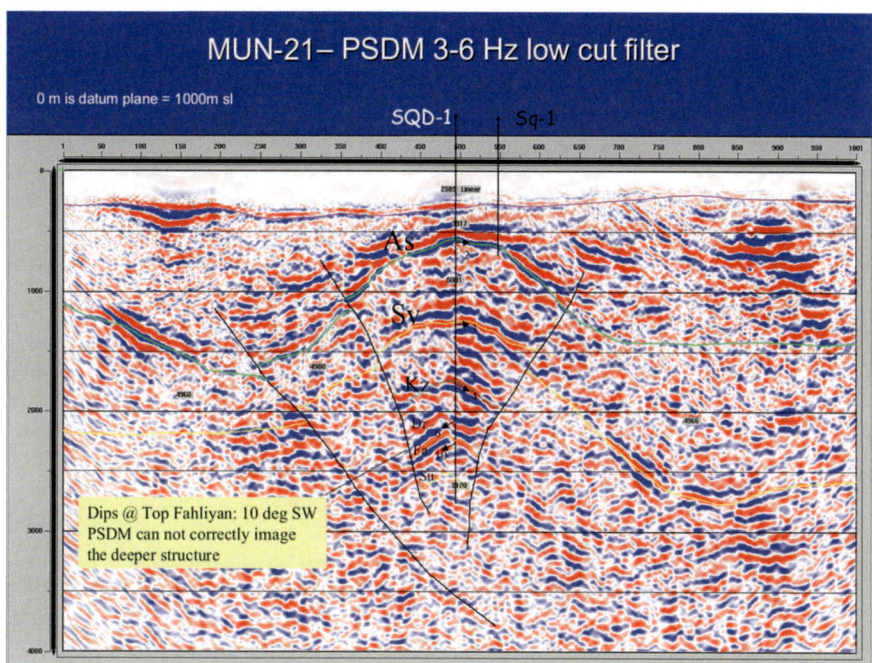

Figure 7
PSDM seismic section provided by NIOC. The PSDM velocity section was shown in upper part of Fig. 6 has been achieved from this section

velocity of 4.7–5.1 km/s form the anticline until a depth of 4 km and consist of several formations that are going to be described in interpretation section.

The resistivity model shows a high-conductivity layer at the same depth level as the small velocity layer at about 500 m depth. Zones of high conductivity are found along the profile at 3–9.5 km and 15–17 km distances and at 3–9.5 km and 15–17 km distances at surface.

Anticline has a resistivity value of 24 ohm m at top and conductivity decreases by increasing of depth underneath of the anticline. Figure 8 shows the mesh model (up) and smoothed section (down) of MT inversion result. The smoothed section model is used in Expectation/ Maximization (EM). Smoothed section datum is 1000 m above sea level contrary to mesh model that is illustrated from sea level. That is the reason for incompatibility in surfaces of two panels of this Figure.

7. Results of Expectation/Maximization Clustering Algorithm

The classification of all available data vectors and remapping in the depth section is shown in Fig. 9 for 3 different class numbers. For the purpose of better interpretation it has compared the Expectation/ Maximization (EM) clustering and remapping results with the distribution of pre-existing depth of formations obtained from Seh-Qanat Deep-1(SQD-1) borehole log data (Fig. 10). Actually, depth and thickness of each formation are known from one available log data. If obtained model from classification was verified by borehole data, it could be expanded to all the anticline. The final interpretation of the depth section is shown in Fig. 11. The remapping of the classes is superimposed to support the interpretation of the classes.

Expectation/Maximization was started with 10 iterations, and iterations gradually increased until 50. The number of classes varies between 9 and 2 in a decreasing manner to find interpretability of the final probability section. Percent of model adaptation of classes with depth and thickness of formations has been shown in Table 1. The best adaptation between joint model and pre-existing depths of formations was derived with classification of six classes with 72 % adaptation.

The anticline is interpreted in six classes as different lithotypes in Fig. 11, which can also be related to particular stratigraphic units of Fig. 9. The first 2

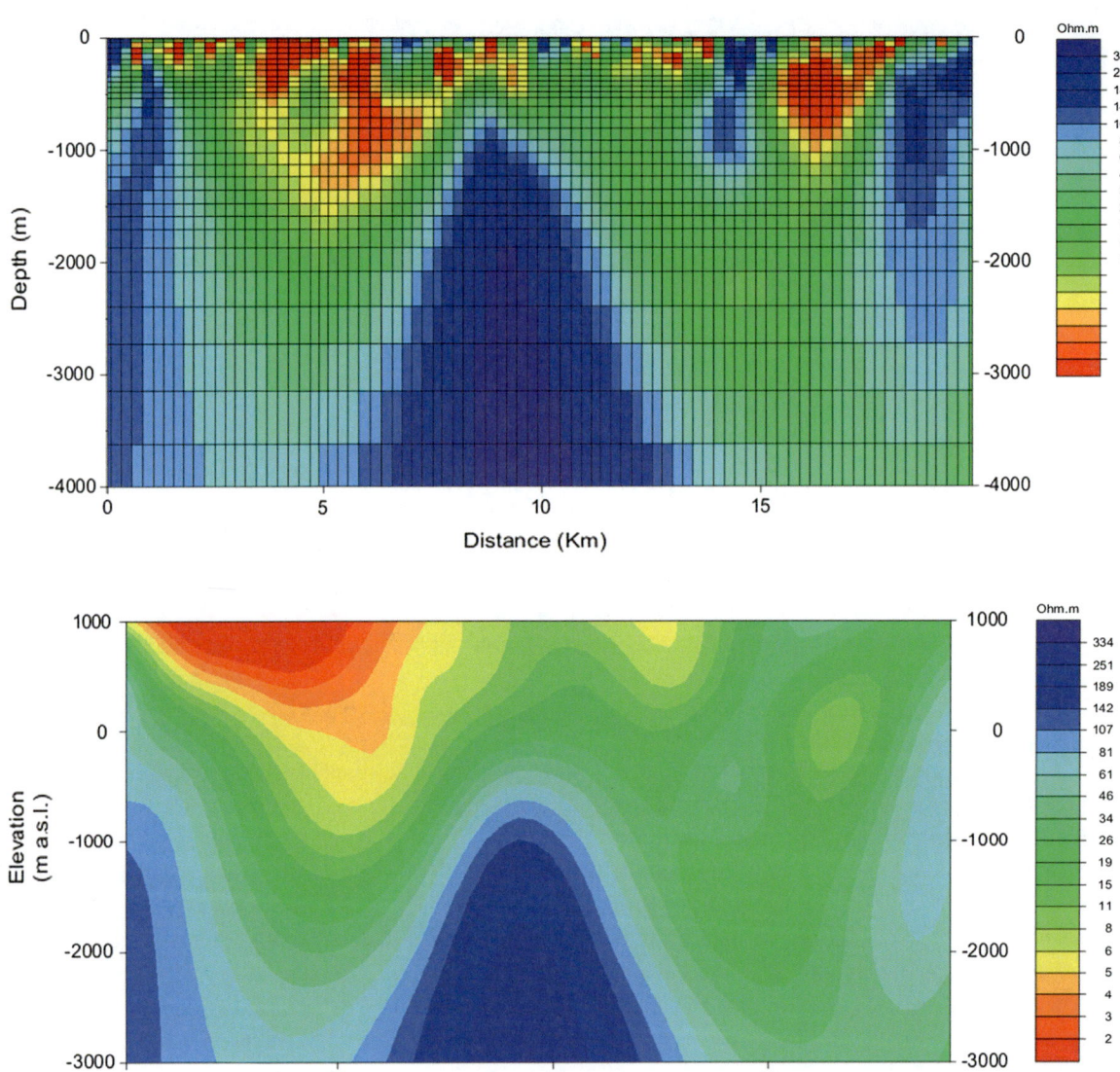

Figure 8
Resistivity models result of MT inversion; mesh model (*up*) and smoothed section model (*down*). For smoothed MT section in *lower panel* of the figure the datum is 1000 m above sea level, the mesh model datum is sea level

classes above the anticline boundary remain unchanged in all remapping sections; therefore, it has been ignored in interpretation. All the models are generally increasing with depth.

Velocity models have the velocity range between 3 and 5 km/s, which was related to carbonated materials. The electrical resistivity started with moderate values at the top of the anticline near to the surface.

It showed a tendency to lower resistivity values (depth range 0.5–1 km) and again increasing resistivity values for the deeper section. In depth range of 1–4 km, the value of resistivity changes in range of 80–330 Ohm m under the anticline.

8. Detailed Interpretation of Classes

Each class is assigned with a lithology and stratigraphics unit based on comparison with the information of formations of Seh-Qanat Deep-1

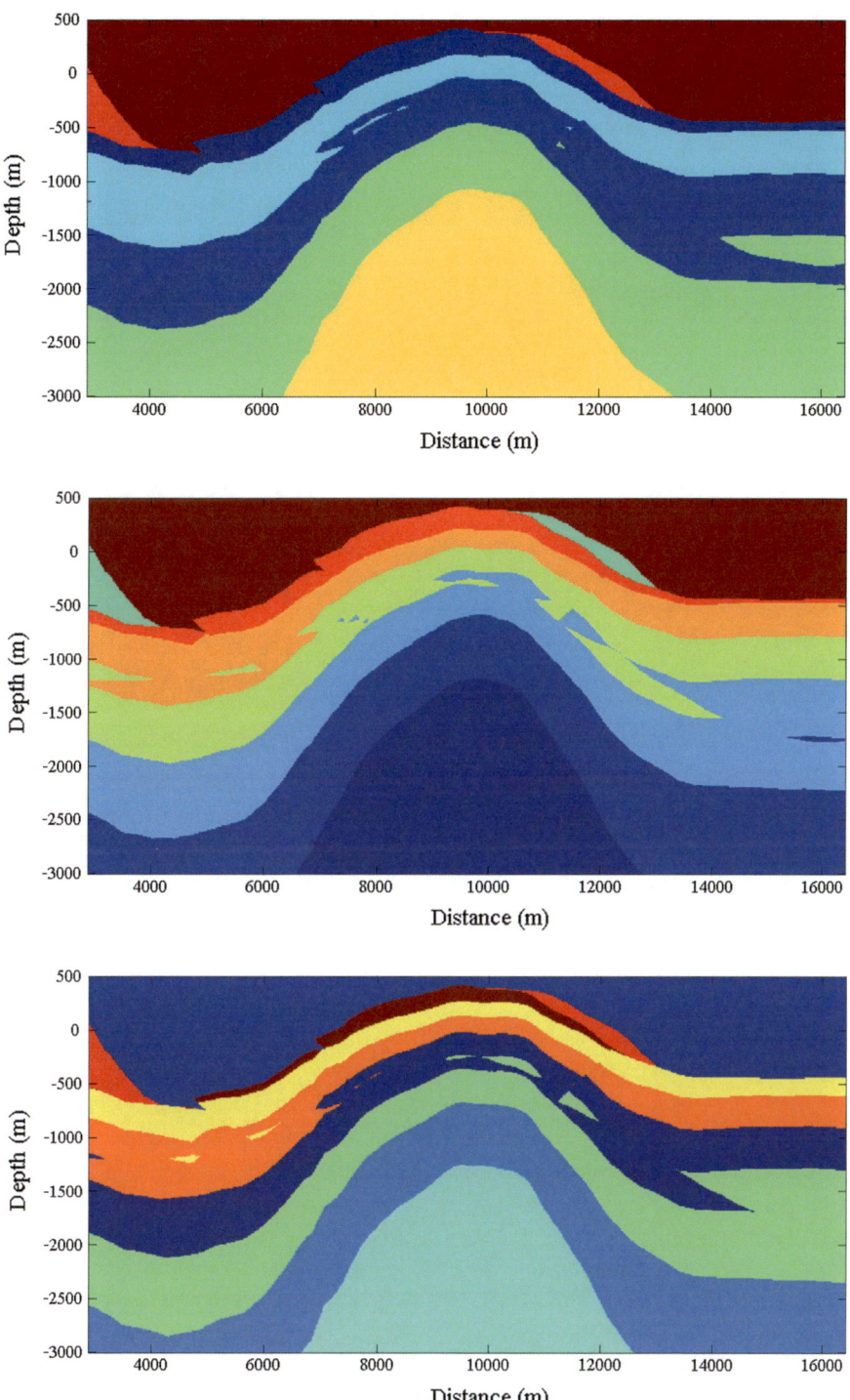

Figure 9
Remapping of classes to the depth section along the MUN-21 collocated seismic and magnetotelluric models with 3 different class numbers. Remapping of 7 classes (*up*), 8 classes (*middle*) and 9 classes (*down*). Remapping of 8 classes has the best correlation with lithology log of SQD-1 borehole (Fig. 8)

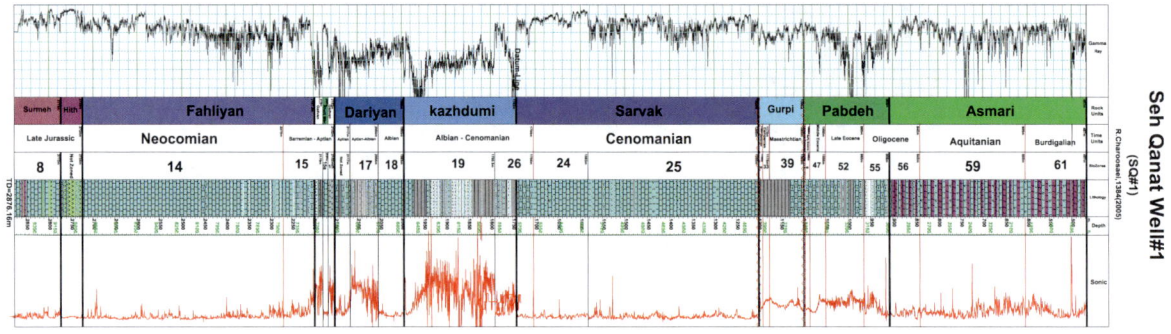

Figure 10
Lithology section of SQD-1 borehole from Top Asmari to the end of borehole. *Columns* from *left* to *right* gamma ray, rock units, time units, biozones, lithology, depth and sonic

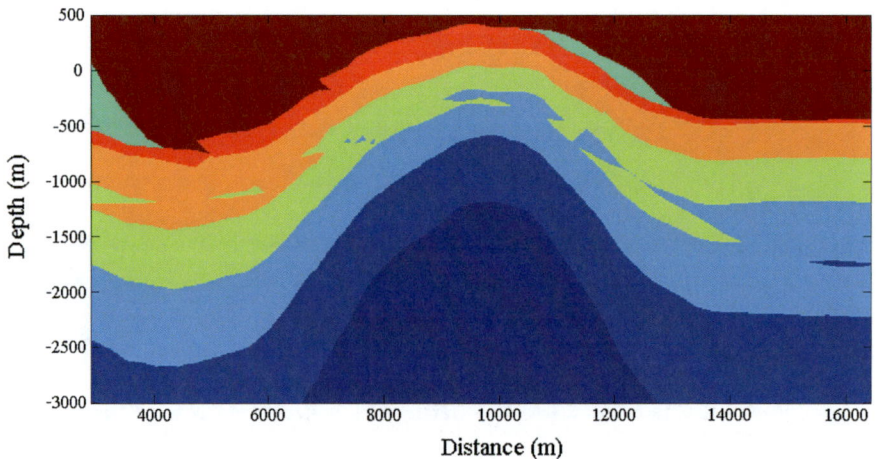

Figure 11
Final remapping of classes to the depth section along the MUN-21 collocated seismic and magnetotelluric models with 8 class numbers

borehole. This assignment is driven by the good agreement between the clustering results and the marker formations. First class is equivalent with upper Asmari regarding the depth and diameter which consists of Miocene microcrystalline (dolomudstone) to finely crystalline dolomite, in places evaporated, with rare ostracods and thin levels of anhydrite. The lithofacies suggest deposition in a moderate-energy, shallow subtidal to intertidal and supratidal settings.

Second classes are equivalent with middle and lower Asmari that belonged to early Miocene and Oligocene, which consists of dolomite cemented by anhydrite and limestone at the end of Formation. Generally, increasing of depth is corresponded with decrease of dolomite in Asmari Formation.

Pabdeh and Gurpi Formations locate in third class that consists of shale, marl and clayed limestone at the top and fine marl, shale and a little marly limestone at the bottom.

The Gurpi Formation is generally considered as a seal for the Upper Cretaceous reservoirs or as a potential source rock but in the Munir Block, the Gurpi is not buried deep enough to generate hydrocarbons. These two formations belong to Eocene and Oligocene. Fourth class is equivalent with Sarvak Formation. This Formation in upper parts is composed of bioclastic wackestone and minor packstone to Calcareous mudstone. Furthermore, in some places locally medium crystalline dolomite take places grading to mudstone sediments. The lithofacies were deposited in a moderate- to high-energy, normal marine, shallow subtidal to intertidal environment. This Formation belongs to Albian and Cenomanian geology time.

Table 1

Percent of model adaptation of classes with depth and thickness of formations

Number of iterations	Number of classes of anticline	Model adaptation (%)
50	9	42
50	8	42
50	7	57
50	6	72
50	5	28
50	4	28
50	3	42
50	2	28

Fifth class is equivalent to Kazhdumi and Daryan Formations. Kazhdumi consists of calcareous marl–shale and clay limestone. Upper Daryan Formation consists of grayish light brown to whitish, slightly argillaceous, in places chalky and locally intraclastic grainstone. The basal part comprises of bioclastic wackestone/mudstone. It was deposited in a moderate-energy, normal marine and shallow carbonate platform.

Sixth class is equivalent with Fahlyan Formation from the top that is comprised of oolite limestone, sheet to massive gray/brown limestone. Basal part locally consists of a little lime slice. The lithofacies were deposited in a low-energy, carbonate mid-ramp environment. This formation is a member of Khami group, which is isocline with Surmeh Formation at the bottom.

9. Conclusion

Pattern recognition and clustering methods are suitable tools to transform some geophysical parameter distributions into a depth section that shows a distribution of classes, which could be interpreted as different rock types. There are no limitations on the number of properties included in the analysis. It is possible to combine most of the properties available from inversion of seismic and MT data. In this example, we considered PSDM velocity, RMS velocity, vertical RMS velocity gradient from seismic and resistivity from inversion of MT data. An area of the anticline is used in models and six classes using Expectation/Maximization (EM) clustering algorithm

is derived. The agreement of the found lithological model with markers of SQD-1 borehole underlines a suitable correlation of the seismic and MT models and performance of the Expectation/Maximization (EM) clustering. PSDM velocity model only marks two horizons that correspond to Asmari and Sarvak Formations. These two formations belong to different geological times with a large interval and some formations in between with no history available.

On the other hand, PSDM velocity model could not mark formations later than Sarvak. However, we could subdivide the model by use of Expectation/Maximization (EM) clustering into six classes that have a good proportion with information obtained from SQD-1 borehole.

Acknowledgments

Particular thanks to NIOC exploration directorate that supported this study by sharing the seismic and MT information. We wish to thank Gholam Reza Peyrovian and Ahmad Reza Dehzad for their support.

REFERENCES

BAUER, K. MUNOZ, G. MOECK, I. (2012). *Pattern recognition and lithological interpretation of collocated seismic and magnetotelluric models using self-organizing maps.* Geophys. J. Int, doi:10.1111/j.1365-246X.2012.05402.x.

BEDROSIAN, P.A. UNSWORTH, M.J. EGBERT G.D. THURBER, C.H. (2004). *Geophysical images of the creeping segment of the San Andreas fault: implications for the role of crustal fluids in the earth-quake process,* Tec.phys. *385,* 137–158.

BEDROSIAN, P. A. MAERCKLIN, N. WECKMANN, U. BARTOV, Y. RYBERG, Y. RITTER, O. (2007). *Lithology-derived structure classification from the joint interpretation of magnetotelluric and seismic models,* Geophys. J. Int. *170,* 737–748.

BENNINGTON, N.L. ZHANG, H. THURBER, C.H. BEDROSIAN, P.A. (2010). Joint inversion of Vp, Vs, and resistivity at SAFOD, EOS, Trans. Am. geophys. Union, 91, Fall. Meet. Suppl., T41A–2084.

DEMPSTER, A.P. LAIRD, N.M. RUBIN, D.B. (1997). *Maximum likelihood from incomplete data via the EM algorithm,* J. R. Stat. Soc. *39,* 1–38.

HAN, M. ZHAO, Y. LI, G. REYNOLDS, A.C. (2011). *Application of EM algorithms for seismic facies classification,* Comput. Geosci., *15,* 421–429.

HARTLEY, H.O. (1958). *Maximum likelihood estimation from incomplete data,* Biometrics, *14,* 174–194.

HEIJDEN, F. DUIN, R.P.W. DE RIDDER, D. TAX, D.M.J. (2004). Classification, parameter estimation and state estimation, an engineering approach using MATLAB, Chichester: John Wiley & Sons Ltd, pp 234–241.

GALLARDO, L.A. MEJU, A. (2007). *Joint two-dimensional cross-gradient imaging of magnetotelluric and seismic travel time data for structural and lithological classification*, Geophys. J. Int. *169*, 1261–1272.

KUYUK, H. S. YILDIRIM. E. DOGAN, E. HORASAN, G. (2011). *Application of k-means and Gaussian mixture classification of seismic activities in Istanbul*, Nonlin Processes Geophys. *19*, 411–419.

MAURER, H. CURTIS, A. BOERNER, D. (2010). *Recent advances in optimized geophysical survey design.* Geophysics *75*, 75A177–75A194.

McNIECE, G.W. JONES, A.G. (2001). *Multisite, multifrequency tensor decomposition of magnetotelluric data.* Geophysics, *66*, 158–173.

MOTEIE, H. (2010). Petroleum geology of Iran, Tehran: Arian Publications.

OGAWA, Y. UCHIDA, T. (1996). *A two-dimensional magnetotelluric inversion assuming Gaussian static shift.* Geophys. J. Int. *126*, 69–76.

REZAIE, M.R. (2011). Petroleum geology of Iran, Tehran: Alavi Publications.

RODI, W., MACKIE, R.L. (2001). *Nonlinear conjugate gradients algorithm for 2-D magnetotelluric inversion.* Geophysics *66*, 174–187.

SAHABI, F. (2012). Petroleum geology, Tehran: University of Tehran.

SASAKI, Y. (2004). *Three-dimensional inversion of static-shifted magnetotelluric data.* Earth Planets Space, *56*, 239–248.

STANLEY, W.D. MOONEY, W.D. FUIS, G.S. (1990). *Deep crustal structure of the Cascade Range and surrounding regions from seismic refraction and magnetotelluric data*, J. geophys. *19*, 419–438.

WEBB, A.R. (2002). Statistical pattern recognition, Malvern: John Wiley & Sons, 41–45.

ZHAO, Y., LI, G. REYNOLDS, A.C. (2007). *Characterization of the measurement error in time-lapse seismic data and production data with an EM algorithm*, Oil Gas Sci. Technol. *62*. 181–193.

ZIOLKOWSKI, A., ENGELMARK, F. (2009). *Use of seismic and EM data for exploration, appraisal and reservoir characterization, AAPG search and discovery*, 90171 CSPG/CSEG/CWLS.

(Received March 3, 2014, revised April 10, 2015, accepted April 12, 2015, Published online May 19, 2015)

Pure Appl. Geophys. 173 (2016), 839–848
© 2014 Springer Basel
DOI 10.1007/s00024-014-0946-0

Pure and Applied Geophysics

CrossMark

Statistical Analysis of Palaeomagnetic Data from the Last Four Centuries: Evidence of Systematic Inclination Shallowing in Lava Flow Records

F. J. Pavón-Carrasco,[1] E. Tema,[2] M. L. Osete,[3,4] and R. Lanza[2]

Abstract—The main objective of this work is to compare directional (declination and inclination) volcanic and archaeomagnetic data for the last four centuries (∼1600–1990) with the historical geomagnetic predictions given by the GUFM1 model which spans from 1590 to 1990. The results show statistical agreement between archaeomagnetic data and directions given by the geomagnetic field model. However, when comparing the volcanic data with the model predictions, marked inclination shallowing is observed. This systematically lower inclination has already been observed in local palaeomagnetic studies (Italy, Mexico and Hawaii) for the 20th century, by comparing recent lava flows with the International Reference Geomagnetic Field (IGRF) model. Here, we show how this inclination shallowing is statistically present at worldwide scale for the last 400 years with mean inclination deviation around 3° lower than the historical geomagnetic field model predictions.

Key words: Inclination shallowing, Palaeomagnetism, Volcanic data, Archaeomagnetic data, Geomagnetic field models.

1. Introduction

The Earth's magnetic field undergoes changes in both space and time due to the geodynamo processes that take place in the outer core. For the last century, these variations (the so-called secular variation) have been directly recorded through geomagnetic observatories, repeat stations, satellites and airborne magnetic surveys, providing an accurate picture of the geomagnetic field behaviour (FINLAY *et al.* 2010; OLSEN *et al.* 2014). Extending knowledge of the directional geomagnetic field into the past four centuries is possible thanks to declination and inclination measurements taken on navigational routes (JONKERS *et al.* 2003). Prior to the 17th century, use of palaeomagnetic data is necessary, but the accuracy of these data is not comparable to that of instrumental data. From ∼1600 AD up to the 20th century, historical/instrumental and palaeomagnetic data together provide detailed information on geomagnetic field variations.

The historical model GUFM1 (JACKSON *et al.* 2000) provides an accurate vision of the directional geomagnetic field elements thanks to the constraint of a massive compilation of historical data from observations picked up by seamen in naval shipping (1590–1900, JONKERS *et al.* 2003) and geomagnetic observatories (last 110 years). The intensity element is only well constrained after 1840 when C.F. Gauss developed an instrument for measuring intensities.

As indicated above, there are also indirect measurements of the geomagnetic field for the same time period (last 400 years), i.e. palaeomagnetic data. We can classify palaeomagnetic data into three main types: baked clay heated archaeological artefacts, volcanic materials and lake sediment records. The first two types of data, i.e. archaeomagnetic and volcanic data, are preferred over sediment data because of the stability and origin of their remanence, commonly a thermoremanence (TRM), and because they provide spot records of the past field. On the contrary, lake sediment records are often less reliable, mainly because of their depositional remanence acquisition mechanism. They generally exhibit a

R. Lanza passed away in the summer of 2013.

[1] Istituto Nazionale di Geofisica e Vulcanologia, Via Vigna Murata 605, 00143 Rome, Italy. E-mail: javier.pavon@ingv.it

[2] Dipartimento di Scienze della Terra, Università degli Studi di Torino, Via Valperga Caluso 35, 10125 Turin, Italy. E-mail: evdokia.tema@unito.it

[3] Dpto. Física de la Tierra, Astronomía y Astrofísica I: Geofísica y Meteorología, Universidad Complutense de Madrid, Avd. Complutense s/n, 28040 Madrid, Spain. E-mail: mlosete@fis.ucm.es

[4] Instituto de Geociencias (IGEO) CSIC, UCM, Ciudad Universitaria, 28040 Madrid, Spain.

Reprinted from the journal

strongly smooth behaviour of the geomagnetic field elements and an important temporal delay in the recorded magnetic signal (TAUXE 1993). This study focusses on palaeomagnetic data with TRM magnetization.

On the one hand, archaeomagnetic data provide the most reliable record for study of the Earth's magnetic field in the past, since they are usually well dated. Palaeosecular variation curves (PSVCs) generated using archaeomagnetic data are very reliable and represent the best option to describe the variation of the geomagnetic field elements at regional scale (BUCUR et al. 1994; GALLET et al. 2002; GÓMEZ-PAC-CARD et al. 2006; KOVACHEVA et al. 2009; TEMA and KONDOPOULOU 2011).

On the other hand, volcanic rocks preserve a strong and stable magnetic signal (TRM), acquired during their cooling. This TRM is parallel to the ambient geomagnetic field at the time of eruption.

When archaeomagnetic data for a studied region/country are scarce, volcanic data have to be considered to achieve better spatial and temporal coverage (e.g. TEMA et al. 2006, 2010; HAGSTRUM and BLINMAN 2010). In this case, however, the following sources of bias in volcanic data have to be taken into account: (1) Sampling: several studies (e.g. HOLT and KIRS-CHVINK 1996) showed that significant spatial magnetic anomalies can be created by topography, and a strongly magnetized underlying terrain can affect the magnetic field direction registered in younger flows (BAAG et al. 1995; VALET and SOLER 1999; KNUDSEN et al. 2003; TANGUY and LE GOFF 2004). Collecting samples from just one site per flow does not allow one to account for this effect, which can be mitigated only if several separated sites per flow are sampled and averaged. (2) Inclination shallowing: comparison of the TRM of volcanic data with direct measurements for the last 110 years shows that often lava flows may record shallower inclination with respect to the geomagnetic field. Such error sources can potentially affect the palaeomagnetic directions despite their apparent high statistical quality (LANZA et al. 2005). (3) Dating of volcanic data: for historical lava deposits, the age assignment is based on historical texts, geochronological dating techniques (^{14}C, Ar/Ar, K/Ar or uranium-series disequilibrium methods) and/or geological constraints. The latter two

methods can be safely applied only to volcanic rocks older than Holocene (last 12 ka). In addition, use of these methods is limited by the availability of suitable materials. For instance, in an active volcanic system characterized by high eruptive frequency, palaeosols (on which radiocarbon dating is based) are rarely formed. Due to the cited limitations, dating of volcanic rocks for the last thousands of years can be particularly difficult, even though it is fundamental to define future volcanic hazards (SOLER et al. 1984; LANZA and ZANELLA 2003). One of the biases related to palaeomagnetic studies carried out using volcanic data is inclination shallowing. This shallowing bias affects inclination data, resulting in values lower than expected. This effect was first discovered in sedimentary records (e.g. KING 1955 and references therein), and during the last decades has been widely studied (e.g. KISSEL et al. 1987; BECK and SCHERMER 1994; KRIJGSMAN and TAUXE 2004; among others). In these palaeomagnetic studies, the shallowing biases are calculated by comparison with the expected inclination according to the geocentric axial dipole (GAD) hypothesis.

To study the reliability and fidelity of archaeomagnetic and volcanic data, in this paper, we compare both types of palaeomagnetic data with historical predictions given by the GUFM1 model at the same location and time as the original data. In the next section, we compile the palaeomagnetic data available for the last 400 years using the different databases and new published data. The statistical methodology followed to analyse the comparison between the palaeomagnetic data and the historical model predictions is detailed in Sect. 3, together with our results. Finally, the last section (Sect. 4) presents a discussion of the results.

2. Palaeomagnetic Data for the Last 400 years

The palaeomagnetic data used in this study were provided by the GEOMAGIA50v2 database (DONA-DINI et al. 2006; KORHONEN et al. 2008) and recent data that are not included in the cited database (DI CHIARA et al. 2012). All the archaeomagnetic and volcanic data for the last 400 years (1590–1990) with an unquestionable assigned age were included in our study. The initial dataset contained 1,220 data coming from archaeological materials and 496 coming from volcanic rocks: 175/244 declinations and 1,045/252 inclinations for archaeomagnetic/volcanic data, respectively.

We applied a filter to reject the less reliable archaeomagnetic and volcanic data. The filter was based on the statistical uncertainty of the palaeomagnetic data and the age uncertainty. The statistical parameters of the directional palaeomagnetic data are given by the angle α_{95}. This parameter is the confidence angle at 95 % probability (FISHER 1953) around the mean palaeomagnetic direction. All data with α_{95} three times larger than the mean α_{95} were rejected. The age associated with each palaeomagnetic datum also presents a confidence interval. This age uncertainty is usually given by archaeological considerations or radiocarbon dating for archaeological artefacts, and by radiocarbon dating or geological and/or historical constraints for recent volcanic data. In our case, all data with age uncertainty greater than 100 years were rejected. Data without information about the measurement and age uncertainties were also rejected. A total of 34 declination and 186 inclination archaeomagnetic data, and 19 declination

and 22 inclination volcanic data were rejected after applying the filter.

The final database presented an average α_{95} of 4.7° for the archaeomagnetic data and 2.7° for the volcanic data. The average age uncertainty for the archaeomagnetic and volcanic data was 20 and 18 years, respectively. Figure 1a and b show the spatial distribution of these two types of palaeomagnetic data. The lack of palaeomagnetic data from the Southern Hemisphere is noticeable, with no data from archaeological material and only one volcanic site in northern New Zealand. The majority of archaeomagnetic data are concentrated in Western Eurasia and North America, with a total of 141 declination and 859 inclination data for the selected time interval. The palaeomagnetic directions obtained from volcanic lava flows are located in areas with recent volcanic activity, as is the case of Hawaii, the West Coast of the USA and Mexico, Iceland, Cameroon (Africa), the Canary Islands (Spain), Italy, Japan and New Zealand. The amount of volcanic data included in this study was 225 declinations and 230 inclinations. The data are well distributed over the entire studied time interval (Fig. 1c, d). The major contribution is from archaeomagnetic inclinations.

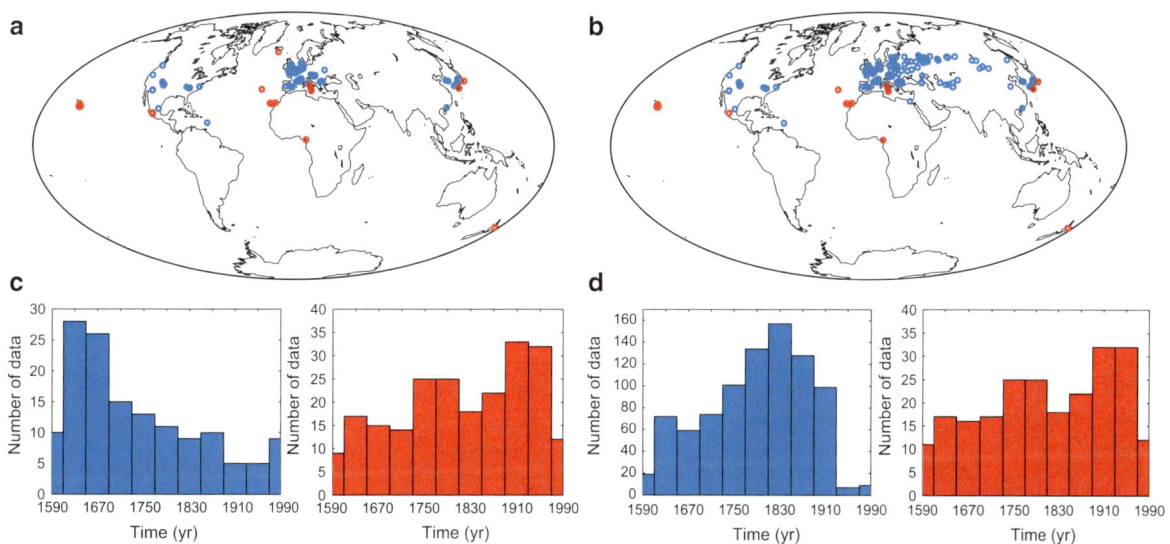

Figure 1
Spatial (**a**, declination; **b**, inclination) and temporal (**c**, declination; **d**, inclination) distribution of the archaeomagnetic (*blue*) and volcanic (*red*) data for the period 1590–1990

3. Methodology and Results

To compare the palaeomagnetic data and historical geomagnetic field predictions, we first synthesized all directional data at the same location (coordinates of the archaeological and volcanic sites) and age of both volcanic and archaeomagnetic data. Then, the residual between the synthetic and palaeomagnetic data for both declination and inclination was calculated. The synthetic data were taken from the historical GUFM1 model, which represents the most accurate approximation to the spatiotemporal evolution of the Earth's magnetic field for the last centuries (JACKSON *et al.* 2000). All the numerical values of the statistical analysis are given in Table 1.

Results obtained from the analysis of the archaeomagnetic data are shown in Fig. 2. We plot the directional archaeomagnetic data versus the synthetic data (Fig. 2a, b). A linear regression of the real and synthetic data was carried out. The statistical distribution of the residual data indicates to use the L1-norm (absolute deviation of residuals) instead of the L2-norm (classical least squares). This characteristic of the residual data was pointed out by PAVÓN-CARRASCO *et al.* (2009) in the construction of a European geomagnetic field model valid for the last 3 ka. Consequently, we followed the L1-norm to linearly fit the data. The linear regression (with error bands at 65 % confidence) of the real and synthetic data indicates statistical agreement between the archaeomagnetic and synthetic data with slope equal to 1.0 ± 0.1 for declinations and 0.98 ± 0.05 for inclinations. For comparison, we plot the contour lines showing equal values of the real and synthetic data. Both contour lines clearly lie within the 65 % confidence bands of the linear regressions.

To analyse the residual data, histograms of the residuals are also plotted in Fig. 2c, d. As expected, according to the previous results, both histograms show a symmetric shape with mean value close to 0°. We calculated the theoretical Laplace distribution with the same statistical parameters as the residual data as follows:

$$f(r/\mu, \beta) = \frac{1}{2\beta} e^{\left(-\frac{|r-\mu|}{\beta}\right)}, \qquad (1)$$

where μ is the median of the residuals (r) and β is the characteristic parameter of the Laplace distribution, being correlated with the standard deviation σ by $\sigma = \sqrt{2} \cdot \beta$. To reinforce our results, we also calculated the cumulative distribution of the residual data and the theoretical cumulative Laplace distribution as

$$F(r) = \frac{1}{2} + \frac{1}{2} \operatorname{sgn}(r - \mu) \left[1 - e^{\left(-\frac{|r-\mu|}{\beta}\right)}\right], \qquad (2)$$

where sgn is the sign function. The theoretical cumulative Laplace distribution also agrees with the cumulative residual data for both geomagnetic field elements (Fig. 2e, f). All the previous results show how the directional archaeomagnetic data for the period 1590–1990 provide, from a statistical point of view, reliable values of the geomagnetic field.

Results of a similar analysis on the volcanic data are plotted in Fig. 3. For the declinations, the linear regression (Fig. 3a) also shows agreement with the GUFM1 model predictions with a slope value of 1.00 ± 0.06. However, the linear regression for the inclination volcanic data (Fig. 3b) presents a lower value of the slope than expected: 0.81 ± 0.03. This result indicates a clear inclination shallowing effect in the volcanic inclination data. The histograms of the residuals of the volcanic data corroborate this

Table 1

Statistical parameters of the residual archaeomagnetic and volcanic data

Type of data	Geomagnetic element	n	α_{95} (°)	μ (°)	β (°)	σ (°)	m	b
A	D	141	2.8	0.9	4.5	6.3	1.00 ± 0.10	0.2 ± 1.0
A	I	859	4.7	−0.6	2.7	3.9	0.98 ± 0.05	0.0 ± 0.1
V	D	225	2.7	0.6	3.3	4.7	1.00 ± 0.06	0.1 ± 0.7
V	I	230	2.7	−2.5	2.3	3.3	0.81 ± 0.03	0.1 ± 0.1

The average α_{95} for each group of data. The statistical parameters of the residuals data: median μ, scalar parameter β and standard deviation σ. m and b are the slope and y-intercept, respectively, of the linear fit of the residuals

A archaeomagnetic data, *V* volcanic data, *D* declination, *I* inclination, *n* number of data

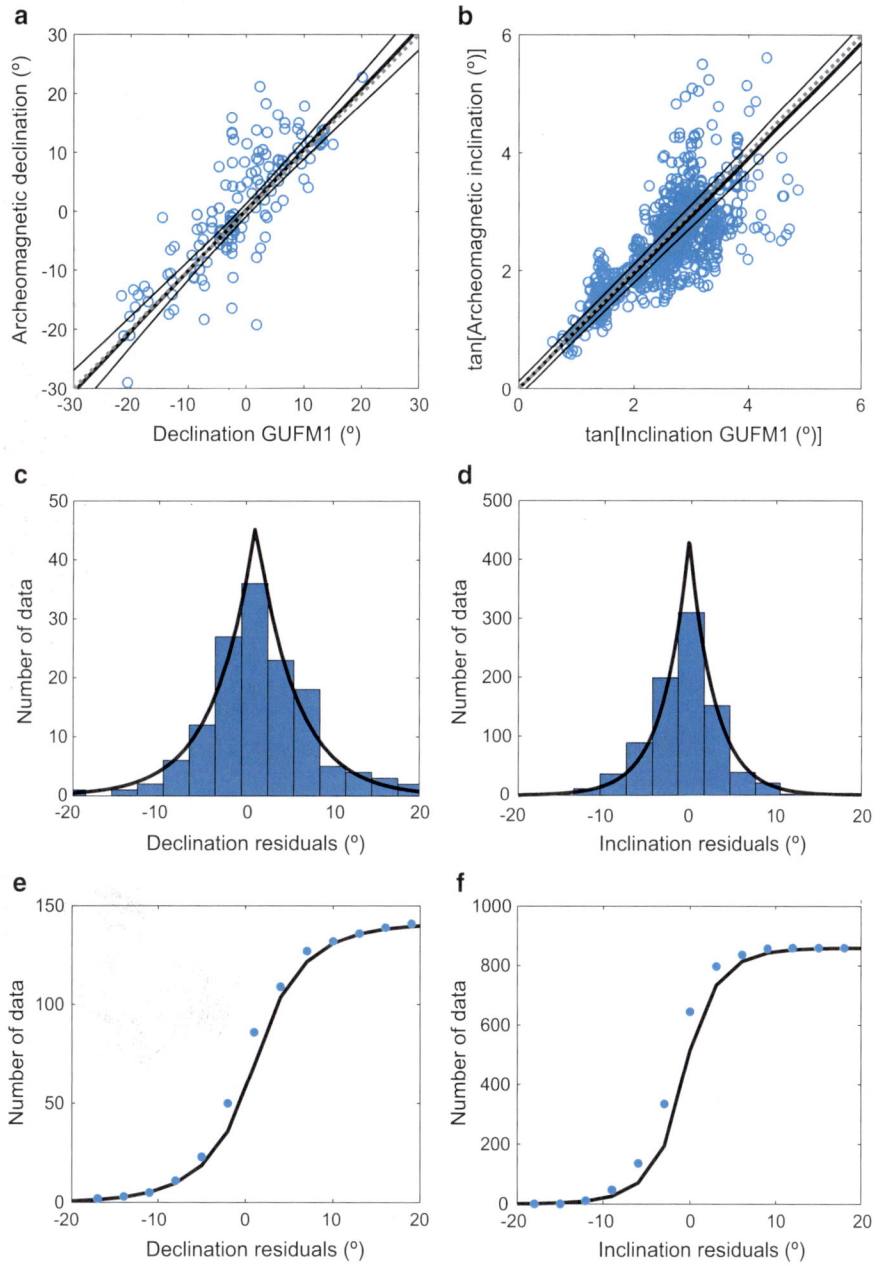

Figure 2

Statistical analysis of the archaeomagnetic data. Linear regression (*black lines* with error band at 65 %) of the archaeomagnetic and synthetic data from GUFM1 for the **a** declination and **b** inclination data. For comparison, lines of equal values (*dashed grey lines*) are also plotted. Histograms of residual data for **c** declination and **d** inclination data along with the theoretical Laplace distribution (*black line*) with the same median and β-parameter as the real data. The cumulative distribution of the residuals (*blue points* in **e** and **f**) are given together with the theoretical cumulative curve of the Laplace distribution (*black line*)

inclination shallowing, with a clear shift in the inclination residuals with an absolute median value of 2.5° (Fig. 3d), higher than the average error of the volcanic inclination data $\alpha_{65} = 1.6°$ ($\alpha_{65} = 81 \times \alpha_{95}/$ 140). This asymmetry is not observed in the histogram of the declination residuals (Fig. 3c), which present a mean value close to 0° and a symmetric shape.

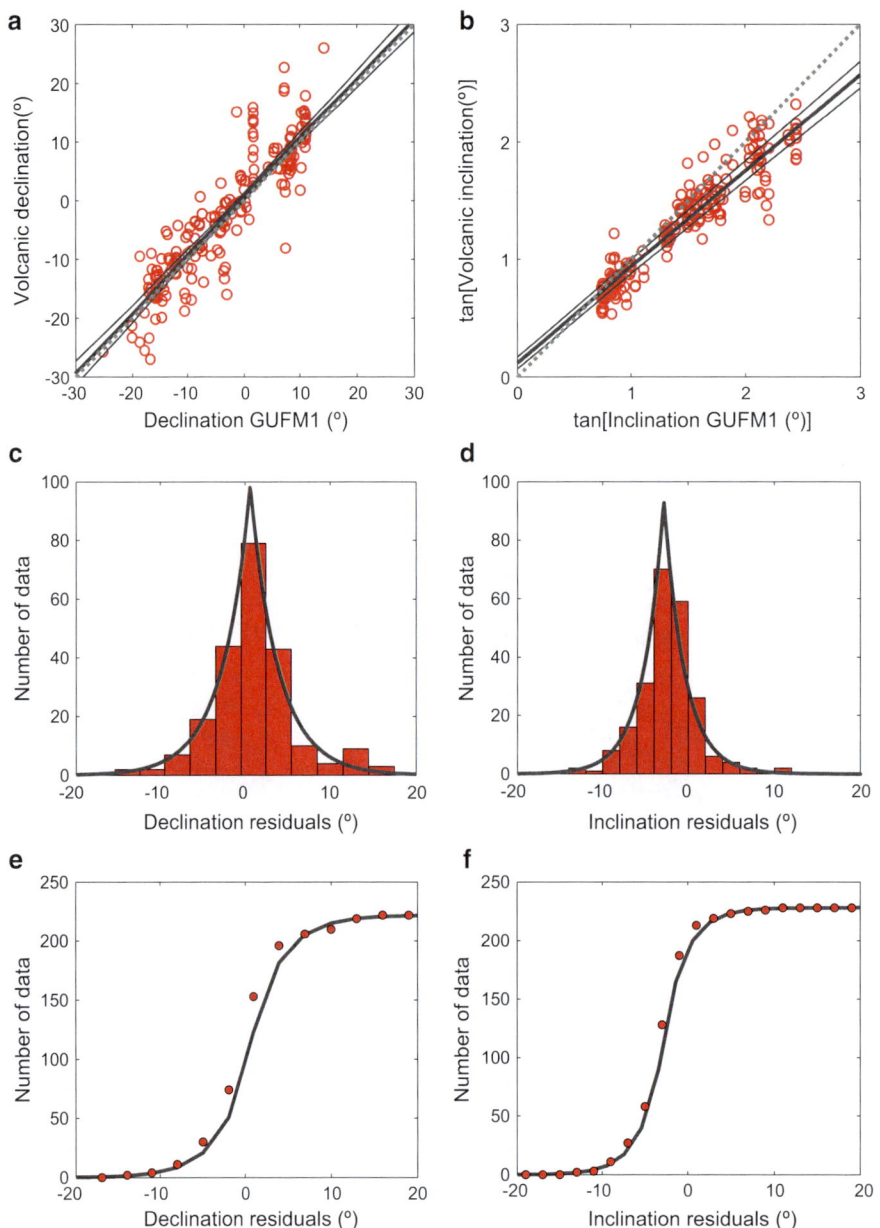

Figure 3
Statistical analysis of the volcanic data. Linear regression (*black lines* with error band at 65 %) of the archaeomagnetic and synthetic data from GUFM1 for the **a** declination and **b** inclination data. For comparison, lines of equal values (*dashed grey lines*) are also plotted. Histograms of residual data for **c** declination and **d** inclination data along with the theoretical Laplace distribution (*black line*) with the same median and β-parameter as the real data. The cumulative distribution of the residuals (*blue points* in **e** and **f**) are given together with the theoretical cumulative curve of the Laplace distribution (*black line*)

4. Discussion

The reliability of palaeomagnetic data plays an important role in accurate definition of the spatial and temporal variation of the geomagnetic field for times when direct geomagnetic measurements are not available (KORTE *et al.* 2009, 2011; LICHT *et al.* 2013; PAVÓN-CARRASCO *et al.* 2014). The last 400 years are the most appropriate temporal interval to test the reliability of palaeomagnetic records due to the

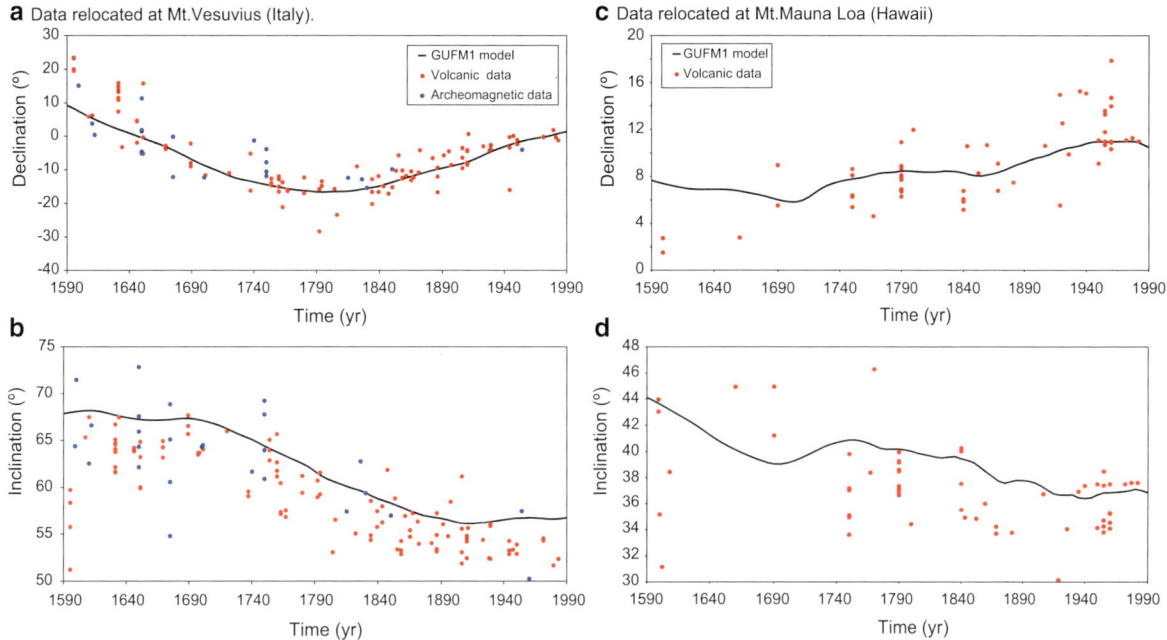

Figure 4

Palaeosecular variation curve for the declination (**a**) and inclination (**b**) at the Mt. Vesuvius (40.8°N, 14.4°E) and declination (**c**) and inclination (**d**) at Mt. Mauna Loa (19.5°N, 155.6°W) coordinates from the GUFM1 model (*black lines*). *Blue/red points* are the relocated archaeomagnetic/volcanic data located in an area of 900 km radius around Mt. Vesuvius and Mt. Mauna Loa

coexistence of both palaeomagnetic and instrumental data.

Statistically, our results show good agreement between archaeomagnetic data and GUFM1 model predictions with a clear symmetric residual distribution centred at ∼0° for both declination and inclination elements. To analyse this comparison in more detail, we focussed our study on the region with the highest density of data, i.e. the European continent (Fig. 1). In this region, we directly compared the model predictions at one reference site with all available archaeomagnetic data located in an area of 900 km radius around this reference location. To do this, we relocated the available directional archaeomagnetic data, by using the conversion via pole method (CVP, Noël and Batt 1990), from the original sites to a common reference point, which was chosen as the coordinates of Mt. Vesuvius (40.8°N, 14.4°E). The relocation error associated with the CVP does not affect our results because we are using a small region (Casas and Incoronato 2007).

A comparison of the relocated palaeomagnetic data and the GUFM1 model predictions is plotted in Fig. 4a, b. In agreement with previous results, the relocated archaeomagnetic data are distributed both over and down the synthetic palaeosecular variation curve calculated from the GUFM1 model at the Vesuvius reference site. The archaeomagnetic data present the classical dispersion characteristic of this type of data, but they show clear agreement with the synthetic curve.

We also relocated the volcanic data around 900 km from Mt. Vesuvius, which corresponds to the location of the major part of the Italian volcanic dataset (Fig. 1). The results (Fig. 4a, b) show how the temporal evolution of the declination volcanic data agrees with the model prediction, while the inclination pattern displays a clear inclination shallowing with 90 % of the values lower than the inclination of the GUFM1 predictions.

The bias of the palaeoinclination of the Italian volcanic data has also been analysed by other authors (TANGUY and LE GOFF 2004; LANZA et al. 2005) but using recent lava flows. They compared volcanic data from the last 110 years (the 20th century) with the IGRF model or with direct measurements from geomagnetic observatories. Indeed, for the last 110 years, inclination shallowing is observed for the whole Italian volcanic palaeomagnetic database (Fig. 4b). A mean value of 51.8° was obtained from 1900 to 1990, against an average value of 56.5°/56.2° given by the GUFM1/IGRF models.

A second study was carried out in the Hawaiian Archipelago, where only volcanic data are available (mostly concentrated in the last 260 years) and inclination shallowing effects in recent (∼ 1950) lava flows have been pointed out by CASTRO and BROWN (1987). The directional volcanic data were relocated at the coordinates of Mt. Mauna Loa (19.5°N, 155.6°W) on Hawaii Island. The comparison between the relocated data and the GUFM1 model predictions is plotted in Fig. 4c, d. Of the declination data, 49 % present values lower than the GUFM1 curve, showing a reasonably good fit with the declination model values. However, the percentage of lower values increases to 75 % for the inclination volcanic data, again showing a clear inclination shallowing effect for the Hawaiian volcanic data for the whole time interval.

All our previous results confirm that inclination shallowing is observed at a global scale over the last four centuries. Following TAUXE and KENT (2004), we estimated the inclination shallowing effect using the expression

$$tan(I_O) = f \cdot tan(I_R) \qquad (3)$$

where I_O is the inclination provided by the volcanic data, I_R is the inclination given by the GUFM1 model (the most reliable value of the geomagnetic field inclination), and f represents the degree of inclination shallowing, the so-called flattening factor. This value ranges from 0 (total flattening) to 1 (no flattening). In the global case, the inclination shallowing presents an f value equal to 0.81. The two regional analyses described above show a value of $f = 0.74$ for the Italian volcanic data and $f = 0.61$ for the Hawaiian volcanic data for the last 400 years.

According to different studies (BAAG et al. 1995; VALET and SOLER 1999; KNUDSEN et al. 2003; TAUXE and KENT 2004; LANZA et al. 2005), the inclination shallowing observed in recent volcanic rocks may be caused by various factors such as possible field distortion due to magnetization of the underlying terrain, the topography and/or the sampling procedure. Regional or local disturbances due to the geological structure or the topographical effect of neighbouring, magnetized rocks can create a deflection of the palaeomagnetic direction mainly affecting the inclination values. Some of these authors considered that possibly some parts of the sampled flow cooled and magnetized before the part where samples were collected. These authors indicate a protocol to mitigate this effect when calculating the mean palaeomagnetic direction by obtaining different samples of the same lava flow and averaging their palaeomagnetic directions. In addition, checks on the anisotropy of the TRM should be carried out to estimate quantitatively the inclination shallowing effect (e.g. GATTACCECA and ROCHETTE 2002).

5. Conclusions

Statistical evaluation of archaeomagnetic and volcanic data versus direct measurements of the geomagnetic field over the last 400 years showed good agreement between the archaeomagnetic data and the geomagnetic predictions based on historical/instrumental data, but a clear systematic inclination shallowing effect in the volcanic dataset. The volcanic data showed, on average, inclination values around 3° lower than the historical predictions from the most accurate geomagnetic field model (GUFM1). This bias in inclination has been pointed out by different studies focussed on lava flows of the 20th century. Our study corroborates that this effect is not only characteristic of recent volcanic data, but is also observed during the last 400 years, from 1590 to 1990, the period of validity of the GUFM1 model. The results suggest that such inclination shallowing may have also affected older time periods, where such comparison with direct/historical measurements is not possible.

Although the mean flattening deviation is low, when the accurate spatial and temporal evolution of the ancient geomagnetic field is analysed, this effect should be taken into account. In regions with additional archaeomagnetic data, such as Italy or Japan, the effect of inclination shallowing should be evaluated by comparing the inclination volcanic data with available archaeomagnetic data of similar age. In addition, when the features of the geomagnetic field are only recorded by volcanic data (with no archaeomagnetic information being available), inclination shallowing has to be taken into account, as in the case of the Hawaiian Islands. This issue could become important when palaeosecular variation curves built from volcanic data are used as a tool for dating or when volcanic data are dated by archaeomagnetism. However, the mean value of inclination shallowing reported in this study is close to the classical uncertainty of inclination data ($\sim 2.7°$), so strong modification of archaeomagnetic dating is not expected.

Acknowledgments

The authors thank Dr. Fabio Donadini for his hard work in maintaining and updating the GEOMAGIA database. They are also grateful to the Spanish research project CGL2011-24790. F.J. Pavón-Carrasco thanks ME-Fulbright contract CT-2010-0663 at the Spanish Institution FECYT. He would also like to thank Prof. Roberto Lanza for offering the opportunity to work with his team at the Università degli Studi di Torino where this study was drawn up, and for being a great professor and excellent person. This article is dedicated to his memory.

REFERENCES

BAAG, C., HELSEY, C.E., XU, S., and LIENERT, B.R. (1995), *Deflection of paleomagnetic directions due to magnetization of the underlying terrain*, J. Geophys. Res., *100*, 10013–10027.

BECK, M., and SCHERMER, E.R. (1994), *Aegean paleomagnetic inclination anomalies. Is there a tectonic explanation?* Tectonophysics, *231*, 281–292.

BUCUR, I. (1994), *The direction of the terrestrial magnetic field in France during the last 21 centuries*. Phys. Earth Planet. Inter. *87*, 95–109.

CASAS, Ll., and INCORONATO, A. (2007), *Distribution analysis of errors due to relocation of geomagnetic data using the 'Conversion via Pole' (CVP) method: implications on archaeomagnetic data*. Geophys. J. Int. *169* (2), 448–454.

CASTRO, J., and BROWN, L. (1987), *Shallow paleomagnetic directions from historical lava flows, Hawaii*, Geophys. Res. Lett., *14* (12), 1203–1206.

DI CHIARA, A., F. SPERANZA, and M. PORRECA (2012), *Paleomagnetic secular variation at the Azores during the last 3 ka*, J. Geophys. Res., *117*, B07101.

DONADINI, F., K. KORHONEN, P. RIISAGER, and L. PESONEN (2006), *Database for Holocene geomagnetic intensity information*, Eos Trans. Am. Geophys. Union, *87*(14), 137.

FINLAY, C.C., et al. (2010), *International Geomagnetic Reference Field: the eleventh generation*. Geophys. J. Int. *183*, 1216–1230.

FISHER, R.A. (1953), *Dispersion on a sphere*. Proc. R. Soc. Lond. A, *271*, 295–305.

GALLET, Y., GENEVEY, A., LE GOFF, M., (2002), *Three millennia of directional variations of the Earth's magnetic field in western Europe as revealed by archaeological artefacts*. Phys. Earth Planet. Inter., *131*, 81–89.

GATTACCECA, J., and ROCHETTE, P., (2002), *Pseudopaleosecular variation due to remanence anisotropy in a pyroclastic flow succession*. Gephys. Res. Lett., *29*, 8, 1286.

GÓMEZ-PACCARD, M., LANOS, Ph., CHAUVIN, A., MCINSTOSH, G., OSETE, M.L., CATANZARITI, G., RUIZ-MARTÍNEZ, V.C., NÚÑEZ, J.I., (2006), *The first Archaeomagnetic secular variation curve for the Iberian Peninsula. Comparison with other data from Western Europe and with global geomagnetic field models*. Geochem. Geophys. Geosyst. *7*, Q12001.

HAGSTRUM, J. T., and BLINMAN, E., (2010), *Archeomagnetic dating in western North America: An updated reference curve based on paleomagnetic and archeomagnetic data sets*, Geochem. Geophys. Geosyst., *11*, Q06009.

HOLT, J. W., and KIRSCHVINK, J.L., (1996), *Geomagnetic field inclinations for the past 400 kyr from the 1 km core of the Hawaii Scientific Drilling Project*, J. Geophys. Res. *101* (B5), 11,655–11,663.

JACKSON, A., A.R.T. JONKERS, and M.R. WALKER (2000), *Four centuries of geomagnetic secular variation from historical records*, Phil. Trans. R. Soc. Lond. A, *358*, 957–990.

Jonkers, A.R.T., Jackson, A., Murray, A. (2003), *Four centuries of geomagnetic data from historical records*, Rev. Geophys., *41*.

KING, R. F. (1955), *The remanent magnetism of artificially deposited sediments*. Mon. Not. R. Astron. Soc., Geophys. Suppl., *7*, 115–134.

KISSEL, C., LAJ, C., SENÇÖR, A.M.C., and POISSON, A. (1987), *Paleomagnetic evidence for rotation in opposite senses of adjacent blocks in northeastern Aegea and western Anatolia*. Geophys. Res. Lett., *14*, doi:10.1029/GL014i009p00907.

KNUDSEN, M.F., JACOBSEN, B.H. and ABRAHAMSEN, N. (2003), *Paleomagnetic distortion modeling and possible recovery by inversion*, Phys. Earth Planet. Inter., *135*, 55–73.

KORHONEN, K., F. DONADINI, P. RIISAGER, and L. PESONEN (2008), *GEOMAGIA50: an archeointensity database with PHP and MySQL*. Geochem. Geophys. Geosyst., *9*.

KORTE M., DONADINI, C., CONSTABLE, C.G. (2009), *The geomagnetic field for 0-3 ka, part II: a new series of time-varying global models*. Geochem. Geophys. Geosyst. *10*, Q06008.

KORTE, M., CONSTABLE, C., DONADINI, F., and R. HOLMES (2011), *Reconstructing the Holocene geomagnetic field*, Earth Planet. Sci. Lett., *312*, 497–505.

KOVACHEVA, M., Y. BOYADZIEV, M. KOSTADINOVA-AVRAMOVA, N. JORDANOVA, and F. DONADINI (2009), *Updated archaeomagnetic data set of the past 8 millennia from the Sofia laboratory, Bulgaria*, Geochem. Geophys. Geosyst., *10*, Q05002.

KRIJGSMAN, W, TAUXE L. (2004), *Shallow bias in Mediterranean paleomagnetic directions caused by inclination error*. Earth Planet. Sci. Lett. *222*, 685–695.

LANZA, R., and ZANELLA, E. (2003), *Paleomagnetic secular variation at Vulcano (Aeolian Islands) during the last 135 kyr*, Earth Planet. Sci. Lett., *213*, 321–336,

LANZA, R., MELONI, A., and TEMA, E. (2005), *Historical measurements of the Earth's magnetic field compared with remanence directions from lava flows in Italy over the last four centuries*, Phys. Earth Planet. Inter., *148*, 97–107.

LICHT, A., HULOT G., GALLET Y., THÉBAULT E. (2013), *Ensembles of low degree archeomagnetic field models for the past three millennia*. Phys. Earth Planet. Inter., *224*, 38–67.

Noël, M., and C.M. Batt (1990), *A method for correcting geographically separated remanence directions for the purpose of archaeomagnetic dating*. Geophys. J. Int., *102*, 753–756.

OLSEN, N., LÜHR, H., FINLAY, C.C., SABAKA, T.J., MICHAELIS, I., RAUBERG, J., TØNER-CLAUSEN, L. (2014), *The CHAOS-4 Geomagnetic Field Model*, Geophys. J. Int., in press.

Pavón-Carrasco, F.J., Osete, M.L., Torta, J.M., Gaya-Piqué, L.R. (2009), *A regional archaeomagnetic model for Europe for the last 3000 years, SCHA.DIF.3 K: applications to archaeomagnetic dating*. Geochem. Geophys. Geosyst., *10*, Q03013.

PAVÓN-CARRASCO, F.J., OSETE, M.L., TORTA, J.M., DE SANTIS, A. (2014), *A geomagnetic field model for the Holocene based on archaeomagnetic and lava flow data*. Earth Planet. Sci. Lett., *388*, 98–109.

SOLER, V., CARRACEDO, J.C., and HELLER, F. (1984), *Geomagnetic secular variation in historical lavas from the Canary Islands*, Geophys. J. R. Astron. Soc., *78*, 313–318.

TANGUY, J.C., and LE GOFF, M., (2004), *Distortion of the geomagnetic field in volcanic terrains: an experimental study of the Mount Etna stratovolcano*, Phys. Earth Planet. Inter., *141*, 59–70.

TAUXE, L., and KENT, D. (2004), *A Simplified Statistical Model for the Geomagnetic Field and the Detection of Shallow Bias in Paleomagnetic Inclinations: Was the Ancient Magnetic Field Dipolar?*, Geophysical Monograph Series, American Geophysical Union, *145*.

TAUXE, L. (1993), *Sedimentary records of relative paleointensity of the geomagnetic field: theory and practice*. Rev. Geophys., *31*, 319–354.

TEMA, E., and D. KONDOPOULOU (2011), *Secular variation of the Earth's magnetic field in the Balkan region during the last eight millennia based on archaeomagnetic data*, Geophys. J. Int., *186*, 603–614.

TEMA, E., I. HEDLEY, and Ph. LANOS (2006), *Archaeomagnetism in Italy: a compilation of data including new results and a preliminary Italian secular variation curve*. Geophys. J. Int., *167*, 1160–1171.

TEMA, E., GOGUITCHAICHVILI, A., and CAMPS, P. (2010), *Archaeointensity determinations from Italy: new data and the Earth's magnetic field strength variation over the past three millennia*, Geophys. J. Int., *180*, 596–608.

VALET, J.P., and SOLER, V. (1999), *Magnetic anomalies of lava fields in the Canary Islands. Possible consequences for paleomagnetic records,* Phys. Earth Planet. Inter., *115*, 109–118.

(Received March 29, 2014, revised September 17, 2014, accepted September 24, 2014, Published online October 18, 2014)

Pure Appl. Geophys. 173 (2016), 849–860
© 2014 Springer Basel
DOI 10.1007/s00024-014-0981-x

▌Pure and Applied Geophysics

Ocean Surface Geostrophic Circulation Climatology and Annual Variations Inferred from Satellite Altimetry and GOCE Gravity Data

J. M. Sánchez-Reales,[1] M. I. Vigo,[1] and M. Trottini[2]

Abstract—We have studied, for the first time, variations in absolute surface geostrophic currents (SGC) using satellite data only. The proposed approach combines 18 years' altimetry data, which provide reliable measurements of absolute sea level (ASL), with a gravity field and steady-state ocean circulation explorer geoid model to obtain dynamic topography, and achieves unprecedented precision and accuracy. Our proposal overcomes the main limitations of existing approaches based solely on altimetry data (which suffer from lack of an independent reference for derivation of ASL maps), and approximations based on in-situ data (which are characterized by a sparse and inhomogeneous coverage in time and space). Features of annual variations of SGC are also addressed. As a result of our study we provide new absolute SGC climatology in the form of a 52-week data set of surface current fields, gridded at quarter degree longitude and latitude resolution and resolving spatial scales as short as 140 km. For presentation, this data set is averaged monthly and the results, presented as monthly climatology, are compared with climatology based on in-situ observations from drifter data.

Key words: Surface geostrophic circulation, dynamic topography, gravity field and steady-state ocean circulation explorer (GOCE), altimetry.

1. Introduction

Knowledge of variations of surface geostrophic currents (SGC) is crucial for understanding ocean dynamics, their driving mechanisms, and ocean mass and heat transport. In the recent past, ocean surface currents have mainly been estimated by use of drifter data from satellite-tracked drifters placed in all the ocean basins, within the global drifter program (Niiler *et al.* 2003; Lumpkin and Pazos 2007), which replaced historical ship drift (Richardson 1989).

More recently, several studies have demonstrated the added value of altimetry data in study of the flow fields of the oceans. For example, Scharffenberg and Stammer (2010) showed the potential of the tandem mission Jason-TOPEX/Poseidon to enable understanding of the spatial structure of annual flow changes in a global context. Despite their unquestionable value, these contributions based on altimetry data are not suitable for study of absolute SGC, because they lack an independent estimate of the geoid for use as a reference for the altimetry-derived absolute sea level (ASL) maps.

The success of the gravity field and steady-state ocean circulation explorer (GOCE) mission has enabled this limitation to be overcome, because it provides an independent geoid representation with unprecedented spatial resolution. Sánchez-Reales *et al.* (2011, 2012), Bingham *et al.* (2011) and Knudsen *et al.* (2011) have recently proved the ability of GOCE to improve previous geodetic mean dynamic topography, and the corresponding benefits for ocean circulation studies.

In this paper we exploit these recent advances to estimate annual cycle and climatology of the major SGC in the oceans by combining 18 years altimetry data, which enable reliable measurement of ASL, with a GOCE geoid model to obtain a mean dynamic topography (DT) with unprecedented precision and accuracy. Our interest in annual variations is because these are the most relevant variations of SCG (at least for areas where the geostrophic flow is substantial).

The paper is organized as follows. Data and methodology are described in the next section. Annual climatology and cycle of SGC derived from altimetry and GOCE are discussed in the Sect. 7,

---
[1] Department of Applied Mathematics, University of Alicante, Alicante, Spain. E-mail: vigo@ua.es
[2] Department of Statistics and Operational Research, University of Alicante, Alicante, Spain.

where a comparison with the corresponding clima-tology and cycle obtained from in-situ measurements from drifter buoys is also presented. Concluding remarks are outlined in the Sect. 11.

2. Data and Methodology

2.1. Geoid

We use the third generation of GOCE data to determine the geoid; in particular, we use the earth gravity model (EGM) solution produced by the time-wise approach. The time-wise approach offers a GOCE-only model in a rigorous sense. That is, no external gravity field information is included, and thus the solution is solely based on GOCE data (PAIL et al. 2010, 2011). The GOCE gravity model is given after application of the usual corrections (EGG-C 2009; KURTENBACH et al. 2009). To match the grid of altimetry data we evaluate the GOCE geoid also in a quarter-degree grid.

Earth's gravity model solution is given as a set of normalized Stokes coefficients completed up to a maximum degree and order 232. This corresponds to a potential resolving power of approximately 86 km for the geoid. SÁNCHEZ-REALES et al. (2012) showed that, because of the nature of the GOCE, this resolution is sufficient to resolve the mean geo-strophic flow for middle-to-high latitudes, but not for the equator, which remains a noisy band for which some filtering is needed (described in the Sect. 5).

2.2. Sea Surface Height

Weekly and monthly maps of absolute sea level are estimated by restoring the sea level anomalies (SLA) to the mean sea surface (MSS) of reference.

Sea level maps were provided by AVISO (http://www.aviso.oceanobs.com, downloaded 3 May 2011) as a weekly merged solution from several altimetry satellites (ERS-1/2, Topex/Poseidon (T/P), ENVISAT and Jason-1/2) for the time span 1992/10/14 to 2010/12/01. These maps are given as anomalies in relative to CLS01-MSS (HERNANDEZ

and SCHAEFFER 2001), which is provided with a two-degree resolution mesh. To obtain ASL from these anomalies, the CLS01-MSS was redistributed to a quarter-degree mesh via a 2D cubic splines inter-polation, and added back to each SLA map to make feasible computation of the DT and their derived SGC. The usual corrections had already been applied to all data sets (SSALTO/DUACS USER HANDBOOK 2011).

For weekly climatology, data are redistributed (via weighted averages) to weeks corresponding to days 1–7, 8–14, … and so on. Averaging the weekly maps, by week of the year, for the 18-year period from December 1992 to November 2010 yields the weekly climatology.

For monthly climatology, monthly maps were obtained as weighted averages of weekly maps from AVISO. This produces 18 maps for each month. The average of these monthly maps, by month, yields the monthly climatology.

2.3. Time-Variable Surface Geostrophic Currents

The absolute DT (ADT) map for each time (either weekly or monthly) is obtained by subtracting the geoid, N, from the corresponding absolute sea level map, i.e.:

$$\mathrm{ADT}(t) = \mathrm{ASL}(t) - N \qquad (1)$$

where t represents time (in weeks or months). The ADT(t) in Eq. (1) can be readily expressed in the spectral domain (BINGHAM et al. 2008, HUGHES and BINGHAM 2006). Although SÁNCHEZ-REALES et al. (2012), proved that GOCE data can resolve space scales as short as 86 km for middle-to-high latitudes, solving for circulation at low latitudes requires a high level of filtering. Therefore, Gaussian smoothing of 140 km of half-wave length was applied to both the ASL(t) and N to ensure homogenization of the length scales and remove omission errors in the geoid (KNUDSEN et al. 2011).

The SGC speed $U_s = u_s + iv_s$ in terms of the zonal or eastward component, u_s, and the meridional or northward component, v_s, along the east (x), and north (y) directions, follows immediately via the geostrophic equation for the balance of the pressure gradient force and the Coriolis frequency:

$$u_s(t) = -\frac{g}{f}\frac{\partial ADT(t)}{\partial y},$$
$$v_s(t) = \frac{g}{f}\frac{\partial ADT(t)}{\partial x}, \qquad (2)$$

where g is gravity, and $f = 2\Omega \sin \varphi$ is the Coriolis frequency, which depends on the latitude φ (Ω denotes the rate of rotation of the earth). The Coriolis frequency vanishes at the equator; therefore numerical computation of Eq. (2) becomes unstable when $\varphi \to 0$. We treat this problem by estimating the SGC for the equatorial band [5°S, 5°N] by following the methodology proposed in (LAGERLOEF et al. 1999). Let U_g denote the geostrophic component of the circulation:

$$\text{if } U_g = -gZ \qquad (3)$$

where $U_g = u_g + iv_g$ and $Z = \frac{\delta ADT}{\delta y} + i \cdot \frac{\delta ADT}{\delta x}$. Denoting by U_β the geostrophic velocities computed with a β-plane approximation ($f = \beta \cdot y$), using the derivative of Eq. (3):

$$\beta U_\beta + \beta y \frac{\partial U_\beta}{\partial y} = ig\frac{\partial Z}{\partial y} \qquad (4)$$

where $U_\beta = u_\beta + i\, v_\beta$ in the equatorial band. We approximate the solution of Eq. (4) by use of a polynomial expansion:

$$Z_\beta = Z_0 + Z_1\varphi + Z_2\varphi^2 + Z_3\varphi^3$$
$$U_\beta = \frac{ig}{\beta L}\left(Z_1 + Z_2\varphi + Z_3\varphi^2\right) \qquad (5)$$

where $L \approx 111$ km represents 1° at the equator. The SGC are then obtained by weighting U_s and U_β by:

$$\left(u_g, v_g\right) = \omega_\beta\left(u_\beta, v_\beta\right) + \omega_s(u_s, v_s) \qquad (6)$$

where $\omega_\beta = 1$ and $\omega_s = 0$ at the equator, $\omega_\beta \to 0$ and $\omega_s \to 1$ at 5°N and 5°S. We defined the weighting functions as:

$$\omega_\beta = \exp\left[-(\varphi/\varphi_c)^2\right]$$
$$\omega_s = 1 - \omega_\beta \qquad (7)$$

where the latitudinal length scale φ_c (here $\varphi_c = 2.2°$) is selected to distribute the weights in Eq. (6).

2.4. Drifter Buoy Measurements

A monthly climatology for the ocean surface currents based on in-situ drifter buoy measurements is provided by the Global Drifter Program on a one degree grid covering latitudes between 73°S and 85°N (http://www.aoml.noaa.gov/phod/dac/drifter_climatology.html; LUMPKIN and GARRAFFO 2005; NIILER 2001; SYBRANDY and NIILER 1991). Drifter

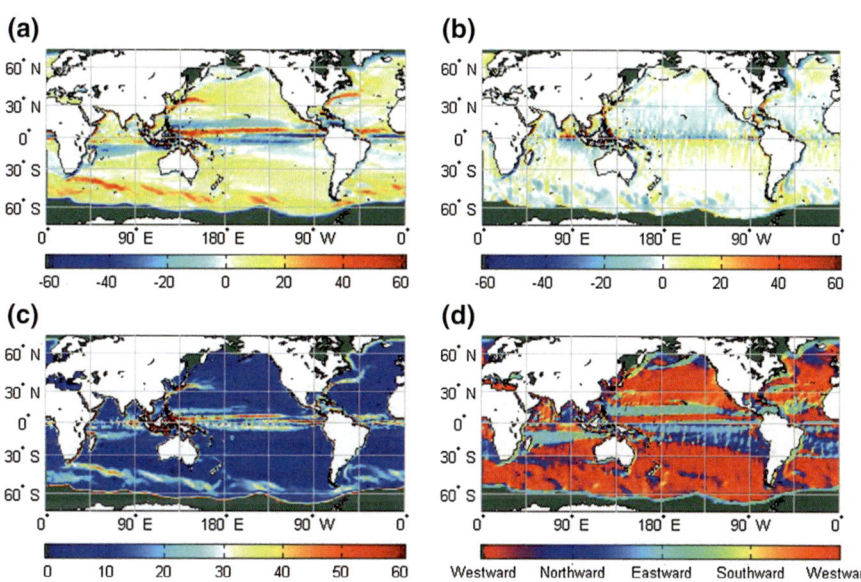

Figure 1
Annual mean SGC climatology estimated from altimetry and GOCE data: **a** zonal component; **b** meridional component; **c** velocity; **d** direction

observations include geostrophic currents and several other signals (tide currents, Ekman currents, inertial currents, and high-frequency ageostrophic currents), hence the drifter data must be corrected to achieve consistency for comparison with the geostrophic velocities derived from the geodetic DT. This solution is provided when daily winds from NCEP/NCAR re-analysis are interpolated on to the drifter positions and used to estimate and remove the slip associated with direct wind forcing (NIILER and PADUAN 1995; PAZAN and NIILER 2000). Wind stress and the local Coriolis frequency were used to estimate the Ekman component (RALPH and NIILER 1999), which is subtracted to generate a separate Ekman-free surface geostrophic velocity field. Finally, a 5-day low pass filter was used to remove inertial and tidal currents and residual high-frequency ageostrophic currents.

Note that, in contrast with GOCE-derived currents, which are determined from a well-defined time span, the drifter observations are obtained from approximately a thousand drifters non-uniformly distributed in space and time. Therefore some artefactual discrepancies may be present when comparing the SGC estimates, particularly for areas with strong inter-annual to decadal variations (QIU and CHEN 2005).

3. Results

3.1. Climatology of Surface Geostrophic Currents

The annual mean climatology, which is an approximation to the mean SGC for the 18-year period, is shown in Fig. 1. Figure 1a, b show the

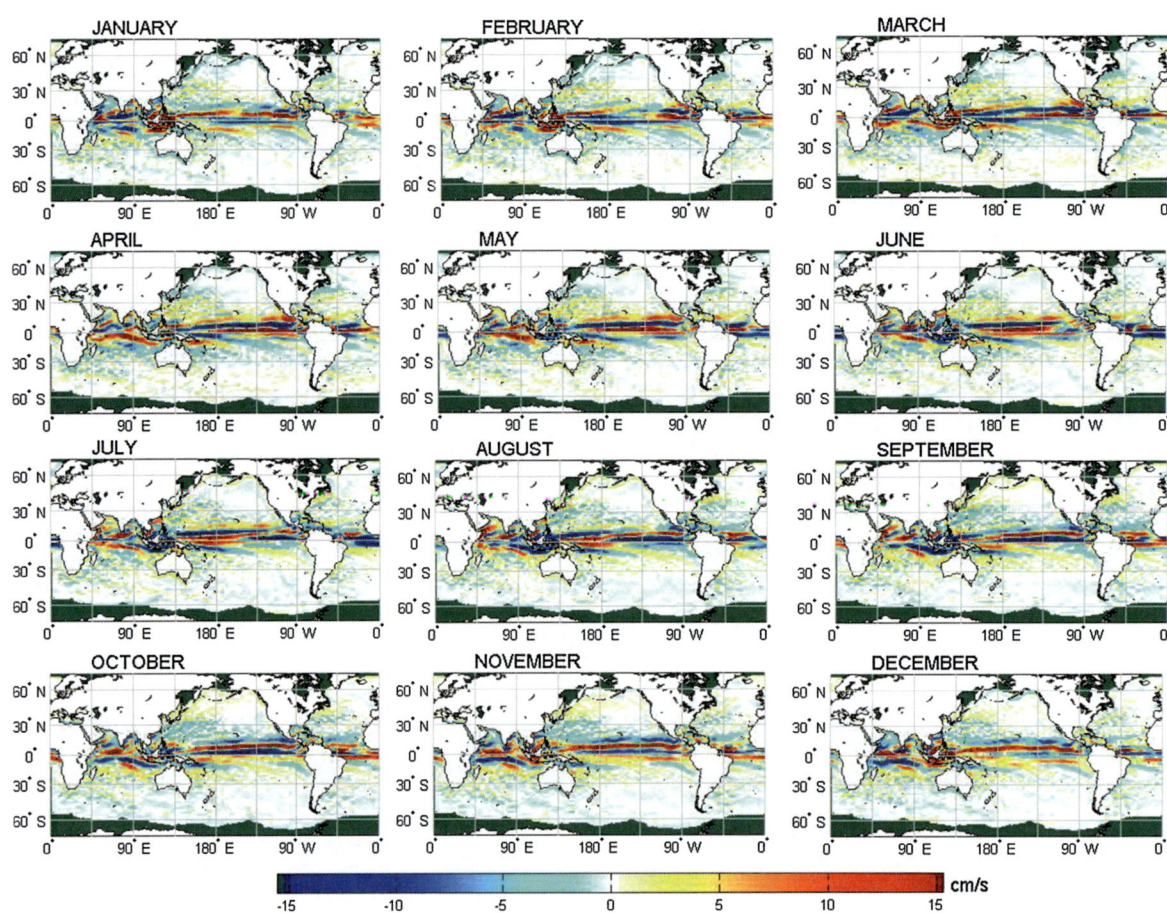

Figure 2
Monthly SGC climatology anomalies for the zonal component estimated from altimetry and GOCE data

zonal and meridional components of the SGC annual mean climatology in cm/s. The velocity map (cm/s) and the direction map for the annual mean climatology are presented in Fig. 1c, d. The results are in accordance with previous studies of the mean SGC (SÁNCHEZ-REALES *et al.* 2012; KNUDSEN *et al.* 2011). The estimated velocities, Fig. 1c, reach values higher than 40–50 cm/s in the western boundary currents, that is, the Kuroshio and the Gulf stream located at the North Pacific and North Atlantic Oceans, respectively. Large velocities higher than 30–40 cm/s are also observed at the Southern Ocean Circulation System, where the Antarctica Circumpolar Current and the Falkland current can be clearly identified. Equatorial currents are also well defined in the Pacific and Atlantic basins. To better illustrate monthly variations of the SGC the annual mean climatology,

shown in Fig. 1, is removed from the data. Figures 2 and 3 show monthly climatology anomalies for the zonal and the meridional components of the SGC, respectively. The strongest variations are observed in the zonal component along the low-latitude currents, for which velocities peak at more than 15 cm/s in absolute value. The three branches at the equatorial pacific formed by the north equatorial current, the equatorial counter current, and the south equatorial current are modulated in the seasonal cycle. These three currents can be identified clearly in October in the same direction as the mean flow. This is the month in which anomalies (with respect to annual mean climatology) are higher in the same direction than the mean flow, and, therefore, anomalies intensify the velocity of the mean current. From November, the anomaly flow remains in the direction

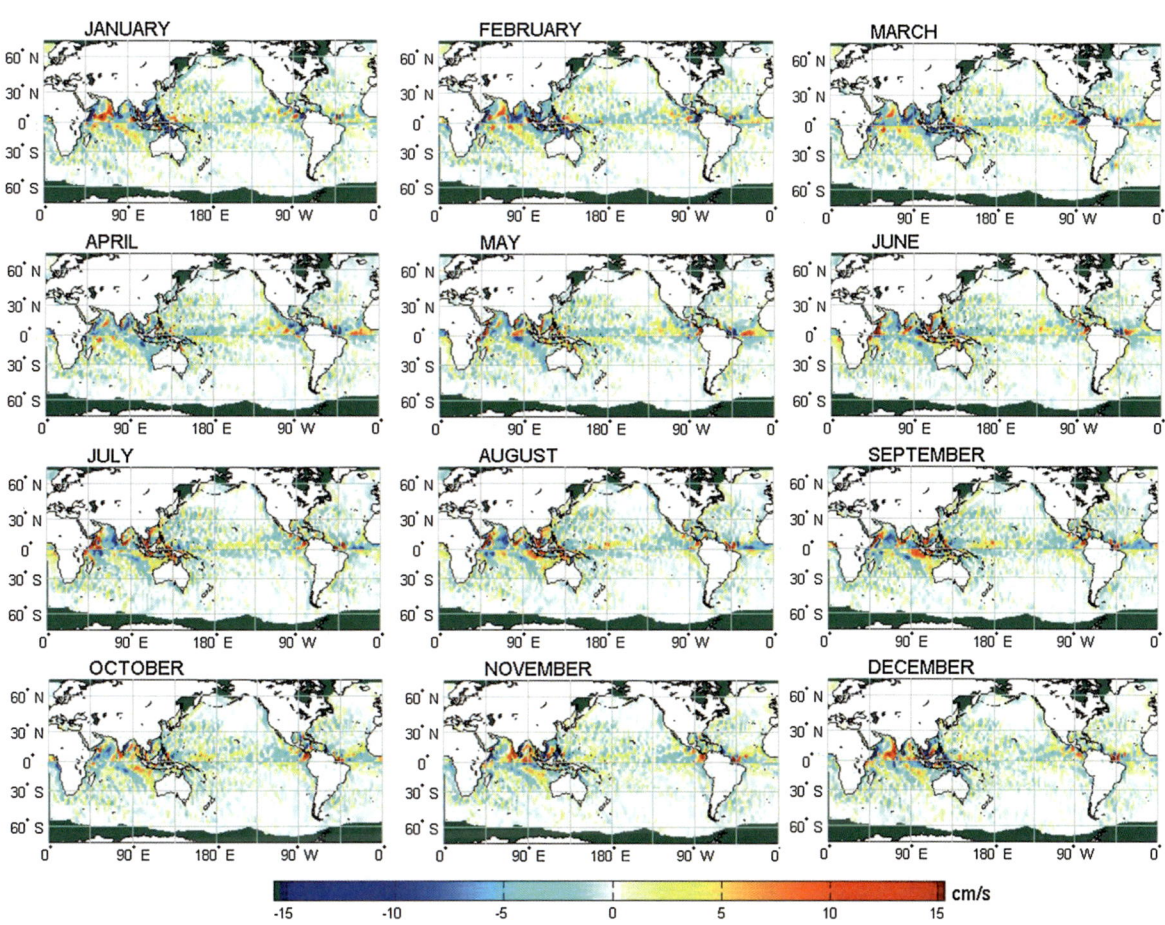

Figure 3
Monthly SGC climatology anomalies for the meridional component estimated from altimetry and GOCE data

of the mean flow but intensities decrease, becoming weaker and weaker throughout December–February until, in March, the anomalies flow is reversed and flows in the direction opposite to the mean flow for the three branches. This results in a decrease in the absolute velocity of these three currents (the mean flow is decelerated by the anomaly flow) through May to June, when anomalies reach the maximum velocity along the direction opposite to the mean flow. From June, velocities start to accelerate (with the anomalies turning into the mean flow direction) until they reach their maxima in September–October. For the meridional component, intensities are much smaller, except for the Kuroshio and Agulhas current. Most of activity accumulates in the main currents areas, the equator band, and the Indian Ocean. In the latter, seasonality is more prominent in the northern part than in the southern part.

Seasonal variability (standard deviation) of the geostrophic flow of the monthly SGC for both the zonal and the meridional components is shown in Fig. 4. Notice that the color scale is designed to resolve regional features rather than extreme values and saturates at 15 cm/s. As expected from the strong annual cycle observed in Fig. 2, the equator band is the area with the highest variability for the zonal component (Fig. 4a) with peaks above 20 cm/s (saturated dark red in Fig. 4). However, significant variability can also be observed in the major current areas:

- the Kuroshio current running north-eastward being the boundary current in the North Pacific Ocean;
- the gulf stream flowing north-eastward along the North American coast being the boundary current in the North Atlantic Ocean; and
- the Antarctica Circumpolar Current along the Antarctica coast from south Africa to Australia

All reach values of approximately 10 cm/s. This high variability is related to the "chaotic" nature of these areas responding to shorter time-scale variations rather than to a clear seasonal cycle (Fu et al. 2010; Fu 2009, 2007). In opposition to the zonal map, the meridional component map

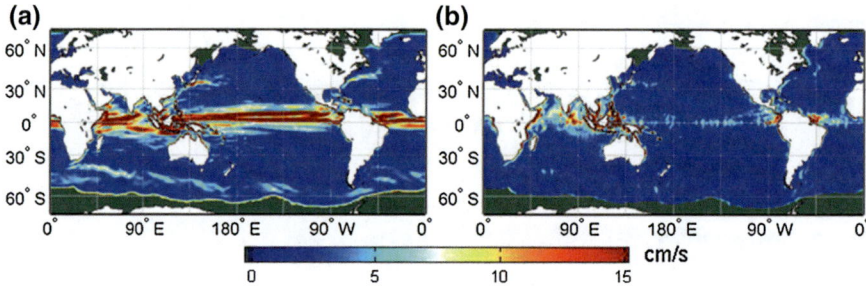

Figure 4
Standard deviation of the monthly SGC climatology estimated from altimetry and GOCE data: **a** zonal component; **b** meridional component

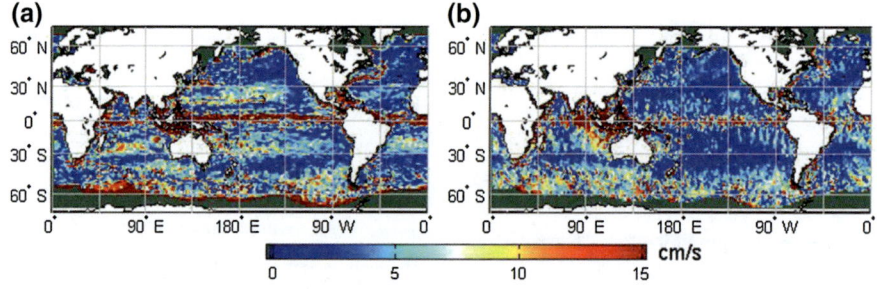

Figure 5
Absolute differences between annual mean SGC climatology estimated from altimetry and GOCE data and the corresponding climatology obtained from drifter data: **a** zonal component; **b** meridional component. The *color scale* is designed to resolve regional features rather than extreme values and saturates at 15 cm/s

(a)
(b)
60° N
30° N
0°
30° S
60° S
0°
90° E
180° E
90° W
cm/s
0
5
10
15

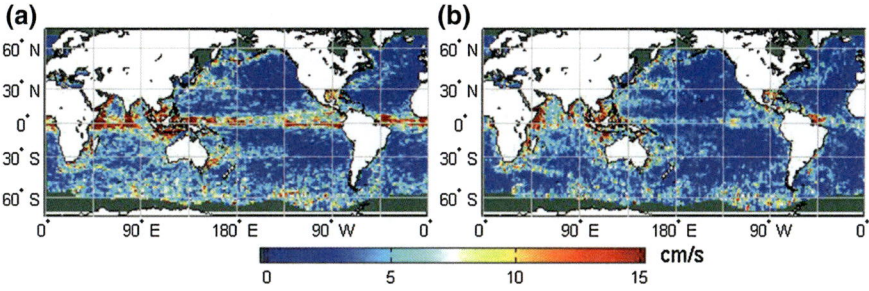

Figure 6
RMS differences between annual mean SGC from altimetry and GOCE data for the 18 years of this study, and the corresponding annual mean values obtained from drifter data: **a** zonal component; **b** meridional component

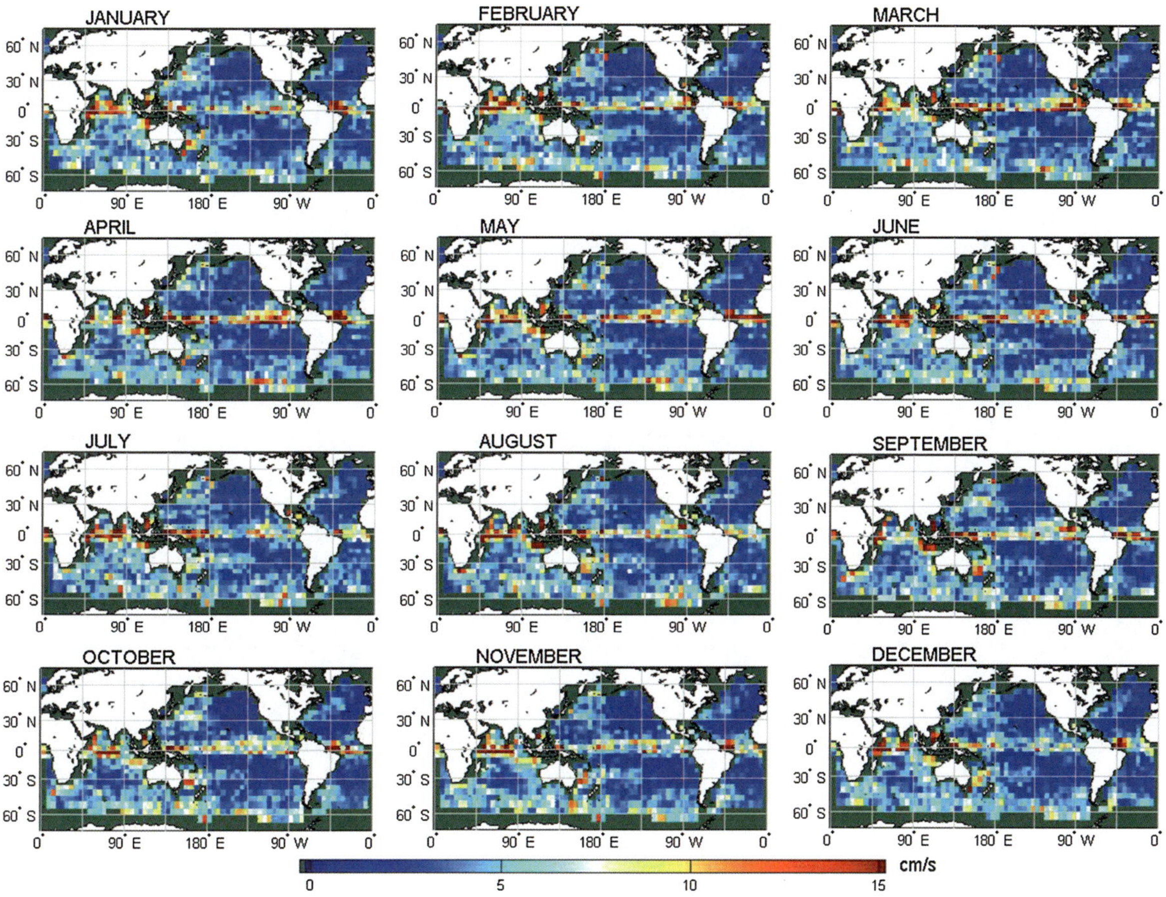

JANUARY
FEBRUARY
MARCH
APRIL
MAY
JUNE
JULY
AUGUST
SEPTEMBER
OCTOBER
NOVEMBER
DECEMBER
60° N
30° N
0°
30° S
60° S
90° E
180° E
90° W
cm/s
0
5
10
15

Figure 7
RMS monthly differences (for each 5° bin) between altimetry and GOCE and drifter-based SGC climatology for the zonal component

(Fig. 4b) shows much weaker variability reaching relatively high values in the Indian Ocean and the Equatorial Atlantic. The gulf stream and the Kuroshio currents can be discerned in the map but the Antarctic Circumpolar Current can barely be identified.

3.2. Comparison with Drifters

Climatology from altimetry and GOCE, shown in Figs. 1, 2 and 3, was compared with the resulting climatology obtained from drifter measurements. Figure 5 shows maps of absolute differences between annual mean climatology from altimetry and GOCE (Fig. 1a, b) and the corresponding climatology obtained from drifters for both the zonal (Fig. 5a) and meridional (Fig. 5b) components of the flow. Notice that the color scale is designed to resolve regional features rather than extreme values and saturates at 15 cm/s. Differences larger than 15 cm/s (saturated to dark red in Fig. 5) are mainly observed in:

– a narrow equatorial band, partly because of the geodetic approximation to the geostrophic flow in this area (β-plane);
– the gulf stream; and
– the Southern circulation system where the observed differences could be because satellite altimeters have a weaker coverage of the sea

surface at higher latitudes, and drifter measurements are contaminated by strong winds and ocean waves (NIILER et al. 2003).

Areas with absolute differences of 8–10 cm/s can be identified in the Southern Atlantic and North and South Pacific basins. These differences are located in areas where velocities are close to zero, and are mainly because of differences between zonal components estimated from satellite and drifter data. As shown by SÁNCHEZ-REALES et al. (2012), the estimated mean velocities obtained from both GOCE and GRACE data in combination with altimetry are higher than observed velocities from drifter data in these regions. This could be because small-scale circulation details are removed in the filtering process, but further studies are needed to confirm this.

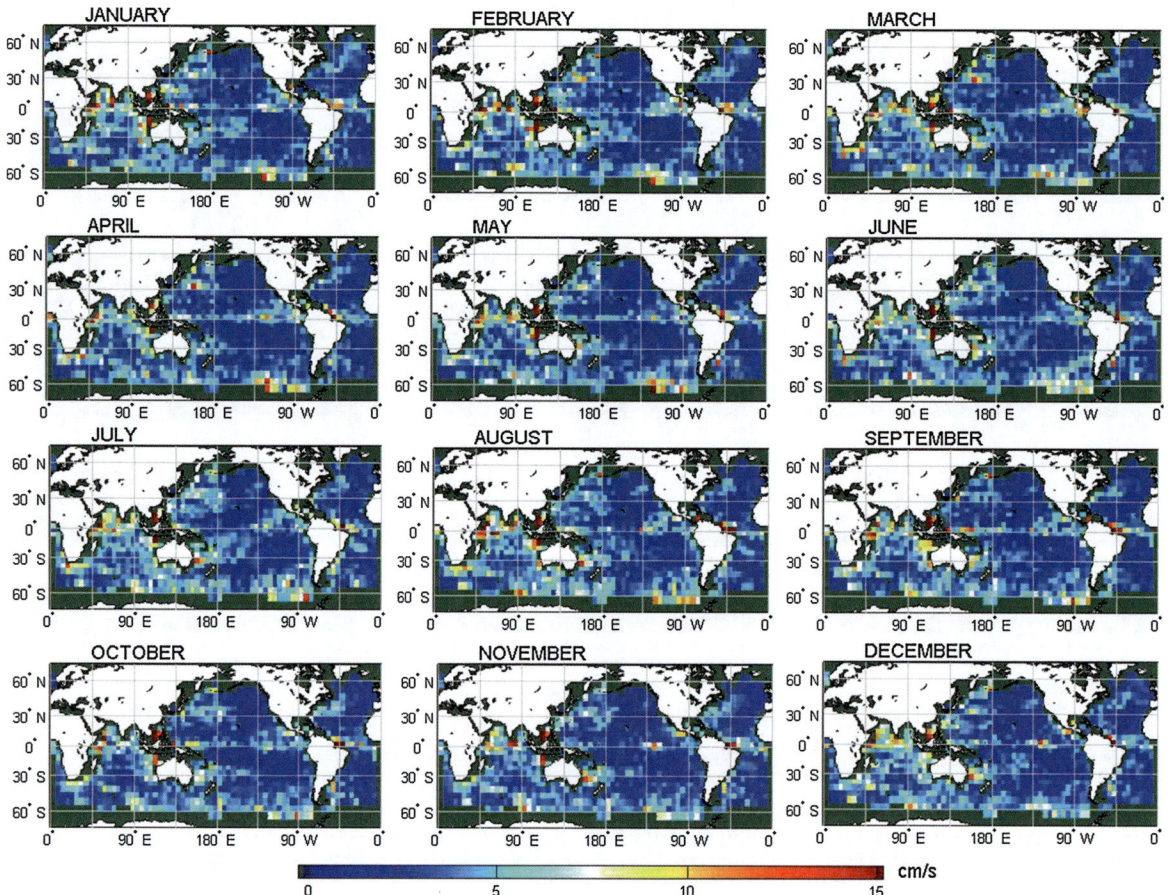

Figure 8
RMS monthly differences (for each 5° bin) between altimetry and GOCE and drifter-based SGC climatology for the meridional component

Figure 6a shows the root mean square (RMS) differences between annual mean values for the zonal component of the SGC from altimetry and GOCE data for the 18 years of this study and the corresponding annual mean values obtained from drifter data. The same information for the meridional component is shown in Fig. 6b. In general, RMS differences are smoother than absolute differences in Fig. 5, and remarkable differences (approx. 15 cm/s for the zonal component) are mainly located along the equatorial band and some coastal areas in the Indian Ocean. Climatology estimated from both datasets are in good agreement.

RMS (for each 5-degree bin) monthly differences between altimetry and GOCE and drifter-based climatology is shown in Figs. 7 and 8 for the zonal and meridional components, respectively. In general, higher RMS values coincide with larger anomalies (Figs. 2, 3) in both space and time, e.g. the equatorial Pacific Ocean in April–May, as could be expected.

The high RMS values observed in the North Indian Ocean can be explained by the geostrophic monsoon currents that do not form, or decay, across the basin all at once. Instead, patches of the currents appear or decay at different times. Higher RMS differences occur during the summer and winter monsoon currents whereas lower RMS values are observed for March–April and September, which are the transition months, in agreement with regional studies (SHANKAR et al. 2002). In contrast, the Antarctic Circumpolar Current is weakly identified, by RMS differences of approximately 5 cm/s (differences here could be because of the same factors as for the mean SGC).

3.3. Annual Variations

The amplitudes of the annual cycle, estimated as half the range of the climatology, in the zonal (u) and meridional (v) components of the flow are shown in Fig. 9a, d respectively. As expected, spatial

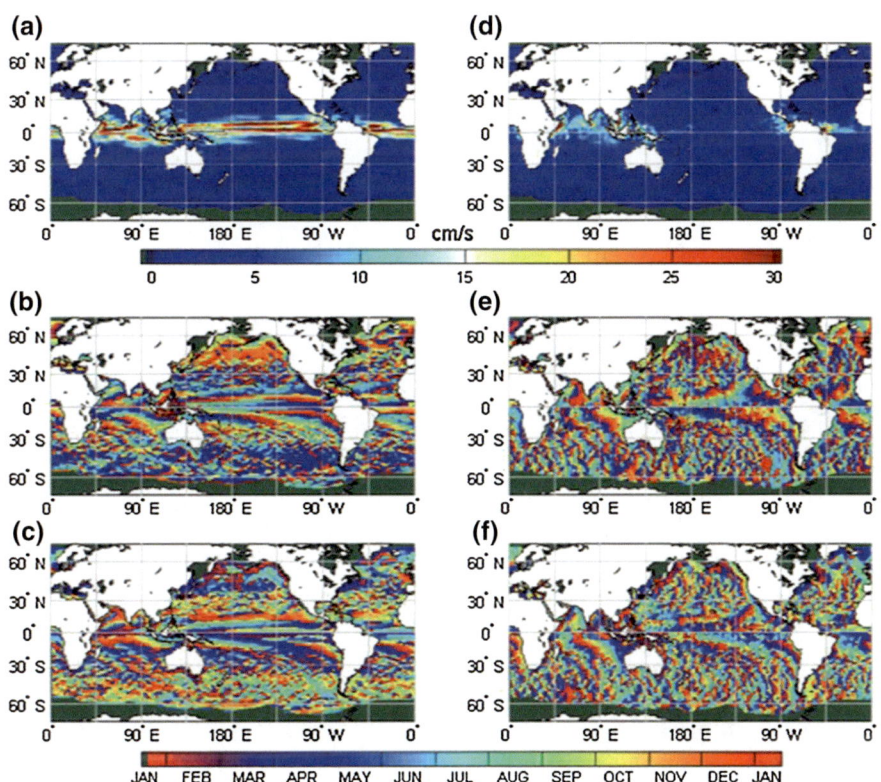

Figure 9

Annual variation of SGC derived from altimetry and GOCE. Annual amplitude, half range: **a** zonal component; **d** meridional component. Month at which the maximum is achieved: **b** zonal component; **e** meridional component. Month at which the minimum is achieved: **c** zonal component; **f** meridional component

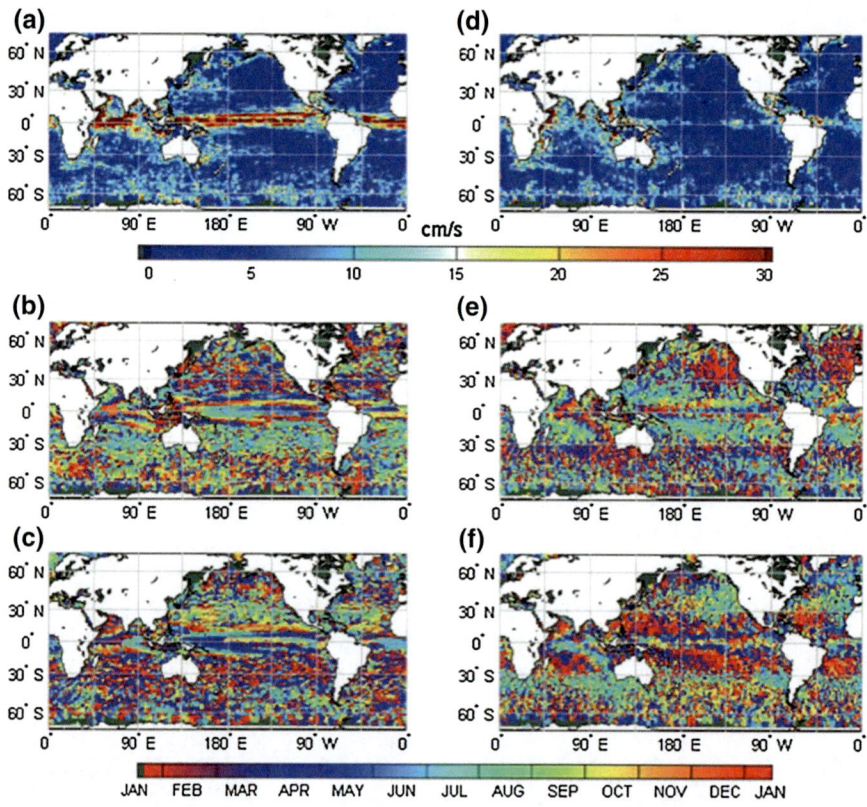

Figure 10
Annual variation of SGC as measured by use of drifters. Annual amplitude, half range: **a** zonal component; **d** meridional component. Month at which the maximum is achieved: **b** zonal component; **e** meridional component. Month at which the minimum is achieved: **c** zonal component; **f** meridional component

distribution of the annual amplitude reflects the corresponding distribution of the standard deviation in Figs. 4a, b. Except for the equator band, the major current areas are characterized by very low seasonal amplitudes, because intra-seasonal variability in these areas is a result of higher time-frequencies rather than a strong annual harmonic.

For the zonal component, most of major equatorial currents—the equatorial counter current (ECC), the north equatorial current (NEC), and the south equatorial current (SEC)—seasonal amplitudes are higher than 20 cm/s, with values up to 40 cm/s at some regions mainly for the ECC. Notice that the color scale is designed to resolve regional features rather than extreme values and saturates at 30 cm/s. In contrast, the meridional component seems to be characterized by a much weaker annual cycle, with the highest amplitudes (>10 cm/s) in the Western Equatorial Atlantic, the ECC in the Indian Ocean, and Agulhas currents.

Figures 9b, e show the months in which the maximum value in the seasonal cycle is reached at each location for the zonal and the meridional components of the geostrophic flow. Similarly, Figs. 9c, f show the months in which the minimum value in the seasonal cycle is reached at each location for the zonal and meridional components of the geostrophic flow. These maps might be interpreted as the phase of the seasonal cycle. Areas of interest are those in which the seasonal cycle is significant. It is apparent that areas characterized by the strongest amplitudes in the Indian ocean reach their maximum values in opposite phases—approximately December in the northern hemisphere and approximately May in the southern hemisphere. The sense of the circulation is clearly captured for the Eastern Equatorial Pacific where some kind of propagation can be appreciated at the NEC-ECC system.

Seasonal cycle results based on altimetry and GOCE are in good agreement with similar results based on drifter data, shown in Fig. 10. Amplitudes and pattern distribution (Fig. 10a, d) agree well for low latitudes. Nevertheless, drifter data furnish larger seasonal cycles for the Kuroshio, Gulf Stream, and Antarctic Circumpolar Current than the altimetry and GOCE. This is because solving the SGC on the global scale requires a high level of filtering to solve the equator band. In this instance we used a Gaussian filter of 140 km half-wave length, which removes small-scale circulation details in these areas.

4. Conclusions

In this work, for the first time, we have obtained weekly and monthly SGC climatology using only satellite data from altimetry and GOCE missions. The proposed approach combines 18 years of altimetry data with an independent estimate of the geoid based on the third release of GOCE data. Our approach overcomes the main limitations of existing approaches based solely on altimetry data (which suffer from lack of an independent estimate of the geoid which can be used as a reference for the altimetry-derived ASL maps), and approximations based solely on in-situ data (which are characterized by sparse and inhomogeneous coverage in time and space).

The annual variations of the SGC that we report are in agreement with several regional studies, and enable them to be placed in a global context. High, significant variability is observed for the SGC for the equator band and for the major current areas (the Kuroshio, Gulf Stream, ACC, and the Malvinas currents) mainly because of the zonal component. Most of this variability is related to inter-seasonal variations except for some areas in the equator band, as is also observed for the Eastern Pacific and the Atlantic Ocean, where the seasonal cycle explains most of the variance of the currents.

Our study shows the annual cycle is very strong for the equatorial band of the zonal component. The NEC, ECC, and NECC are clearly identified by seasonal amplitudes >20–30 cm/s. The estimated climatology is in general agreement with that obtained from in-situ drifter buoy measurements.

Study of variability on a smaller scale would require finer resolution of the geoid and use of more advanced filtering methods. Further research on anisotropic filters could provide a useful tool to address these problems.

REFERENCES

BINGHAM R., HAINES K., and HUGHES C. (2008) *Calculating the ocean's mean dynamic topography from a mean sea surface and a geoid*, J Atmos Ocean Tech, 25(10),1808–1822. doi:10.1175/2008JTECHO568.1.

BINGHAM RJ., KNUDSEN P., ANDERSEN O., and PAIL R. (2011) *An initial estimate of the North Atlantic steady-state geostrophic circulation from GOCE*. Geophys Res Lett, 38:L01606. doi:10.1029/2010GL045633.

EGG-C (EUROPEAN GOCE GRAVITY CONSORTIUM) (2009). GOCE high level processing facility: GOCE level 2 product data handbook. GO-MA-HPF-GS-0110.

FU, L.-L., CHELTON, D.B. LE TRAON P.-Y., and MORROW R. (2010). *Eddy dynamics from satellite altimetry*. Oceanography 23(4):14–25.

FU, L.-L. (2009), *Pattern and velocity of propagation of the global ocean eddy variability*, J. Geophys. Res., 114, C11017, doi:10.1029/2009JC005349.

FU, L.-L. (2007), *Intraseasonal Variability of the Equatorial Indian Ocean Observed from Sea Surface Height, Wind, and Temperature Data*. J. Phys. Oceanogr., 37, 188–202.

HERNANDEZ F. and SCHAEFFER P. (2001). The CLS01 MSS: A validation with the GSFC00.1 surface. Tech. rep., CLS, Ramonville St Agne, 14 pp.

HUGHES, C. and BINGHAM, R. (2006). *An oceanographers guide to GOCE and the geoid*. Ocean Sci. Discuss., 3, 1543–1568. doi:10.5194/osd-3-1543-2006.

KNUDSEN, P., ANDERSEN O., BINGHAM R. and RIO M.-H. (2011), *A global mean dynamic topography and ocean circulation estimation using a preliminary GOCE gravity model*. J. Geodesy, doi:10.1007/s00190-011-0485-8.

KURTENBACH, E., MAYER-GÜRR T., and EICKER, A. (2009). *Deriving daily snapshots of the Earth's gravity field from GRACE L1B data using Kalman filtering*, Geophys. Res. Lett., 36, L17102, doi:10.1029/2009GL039564.

LAGERLOEF, G. S. E., G. T. MITCHUM, R. B. LUKAS, and P. P. NIILER (1999), *Tropical Pacific near-surface currents estimated from altimeter, wind, and drifter data*, J. Geophys. Res., 104(C10), 23, 313–23, 326, doi:10.1029/1999JC900197.

LUMPKIN R., and GARRAFFO Z., (2005) *Evaluating the Decomposition of Tropical Atlantic Drifter Observations*. J Atmos Ocean Techn I 22, 1403–1415. doi:10.1175/JTECH1793.1.

LUMPKIN, R., and M. PAZOS, (2007), Measuring surface currents with Surface Velocity Program drifters: the instrument, its data, and some recent results, Chapter 2 of "Lagrangian Analysis and Prediction of Coastal and Ocean Dynamics", ed. A. Griffa, A. D. Kirwan, A. Mariano, T. Özgökmen, and T. Rossby, Cambridge University Press.

NIILER, P. P. and J. D. PADUAN (1995), *Wind-driven motions in the northeast Pacific as measured by Lagrangian drifters*. J. Phys. Oceanogr. 25, 2819–2830.

NIILER, P. P., (2001), The world ocean surface circulation. In Ocean Circulation and Climate, G. Siedler, J. Church and J. Gould, eds., Academic Press, Volume 77 of International Geophysics Series, 193–204.

NIILER, P. P., N. A. MAXIMENKO and J. C. MCWILLIAMS, (2003), *Dynamically balanced absolute sea level of the global ocean derived from near-surface velocities observations.* Geophys. Res. Lett., *30*(22) 2164, doi:10.1029/2003GRL018628.

PAIL R., GOIGINGER H., MAYRHOFER R., SCHUH W.D., BROCKMANN J.M., KRASBUTTER I., HÖCK E., and FECHER T. (2010) GOCE grafity field model derived from orbit and gradiometry data applying the time-wise method. Proc. ESA Living Planet Symposium. Bergen, Norway, 28 June-2 July 2010 (ESA SP-686, December 2010).

PAIL R., BRUINSMA S., MIGLIACCIO F., FOERSTE C., GOIGINGER H., SCHUHAND W.D., HOECK E., REGUZZONI M., BROCKMANN J.M., ABRIKOSOV O., VEICHERTS M., FECHER T., MAYRHOFER R., KRASBUTTER I., SANSO F., and TSCHERNING C. (2011) *First GOCE gravity field models derived by three different approaches.* J Geodesy, 85:819–843. doi:10.1007/s00190-011-0467-x.

PAZAN, S. E. and P. P. NIILER, (2000), *Recovery of near-surface velocity from undrogued drifters.* J. Atmos. Oceanic Technol. *18*, 476–489.

QIU B. and S. CHEN, (2005), *Variability of the Kuroshio extension jet, recirculation gyre, and mesoscale eddies on decadal time scales.* J. Phys. Oceanogr., *35*, 2090–2103.

RALPH, E. A. and P. P. NIILER, (1999), *Wind-driven currents in the Tropical Pacific.* J. Phys. Oceanogr. *29*, 2121–2129.

RICHARDSON, P. L. 1989. *Worldwide ship drift distributions identify missing data.* J. Geophys. Res. *94*(C5):6169–6176, doi:10.1029/JC094iC05p06169.

SÁNCHEZ-REALES, J. M., VIGO, M. I., JIN, S. G., and CHAO, B. F. (2011), Global Surface Geostrophic Currents from Satellite Altimetry and GOCE. Proc. 4th International GOCE User Workshop. Munich, Germany, 31 March–1 April 2011 (ESA SP-696, July 2011).

SÁNCHEZ-REALES, J.M., VIGO, M.I., JIN, S.G., and CHAO, B.F. (2012), *Global Surface Geostrophic Ocean Currents Derived from Satellite Altimetry and GOCE* Geoid. Marine Geodesy. 35:sup1, 175–189.

SCHARFFENBERG, M. G., and D. STAMMER, (2010), *Seasonal variations of the large-scale geostrophic flow field and eddy kinetic energy inferred from the TOPEX/Poseidon and Jason-1 tandem mission data,* J. Geophys. Res., *115*, C02008, doi:10.1029/2008JC005242.1.

SHANKAR D., P.N. VINAYACHANDRAN, and A.S. UNNIKRISHNAN, (2002), *The monsoon currents in the north Indian Ocean.* Progress in Oceanography *52*, 63–120.

SSALTO/DUACS USER HANDBOOK (2011) M(SLA) and M(ADT) Near-Real Time and Delayed-Time. SALP-MU-P-EA-21065-CLS.

SYBRANDY, A. L. and P. P. NIILER, (1991), WOCE/TOGA Lagrangian drifter construction manual. WOCE Rep. 63, SOI Ref. 91/6, 58 pp, Scripps Inst. of Oceanogr., La Jolla, Calif.

(Received March 28, 2014, revised October 10, 2014, accepted November 4, 2014, Published online November 29, 2014)

Pure Appl. Geophys. 173 (2016), 861–870
© 2015 Springer Basel
DOI 10.1007/s00024-015-1154-2

Consistency Problems in the Improvement of the IAU Precession–Nutation Theories: Effects of the Dynamical Ellipticity Differences

ALBERTO ESCAPA,[1] JOSÉ M. FERRÁNDIZ,[1] TOMÁS BAENAS,[1] JUAN GETINO,[2] JUAN F. NAVARRO,[1] and SANTIAGO BELDA–PALAZÓN[1]

Abstract—The complexity of the modeling of the rotational motion of the Earth in space has produced that no single theory has been adopted to describe it in full. Hence, it is customary using at least a theory for precession and another one for nutation. The classic approach proceeds by deriving some of the fundamental parameters from the precession theory, like, e.g., the dynamical ellipticity H_d, and then using those values in the nutation theory. The former IAU 1976 precession and IAU 1980 nutation theories followed that scheme. Along with the improvement of the accuracy of the determination of Earth orientation parameters, IAU 1980 was superseded by IAU2000, based on the application of the MHB2000 transfer function to the previous rigid Earth analytical theory REN2000. The latter was derived while the precession model IAU 1976 was still in force, therefore it used the corresponding values for some of the fundamental parameters, as the precession rate, associated to the dynamical ellipticity. The new precession model P03 was adopted as IAU 2006. That change introduced some inconsistency since P03 used different values for some of the fundamental parameters that MHB2000 inherited from REN2000. Besides, the derivation of the basic Earth parameters of MHB2000 itself comprised a fitted variation of the dynamical ellipticity adopted in the background rigid theory. Due to the strict requirements of accuracy of the present and coming times, the magnitude of the inconsistencies originated by this twofold approach is no longer negligible as earlier, hence the need of discussing the effects of considering slightly different values for H_d in precession and nutation theories.

Key words: Earth rotation, precession and nutation, reference systems, celestial mechanics.

1. Introduction

Earth rotation is presently a very active field of research, in which mathematical and physical methods concur to satisfy the very stringent demands of accuracy resulting from a broad set of applications to various fields. A comprehensive description of the geodetic problems and the expected societal benefits which led to the establishment of the requirements of such high accuracy can be seen, e.g., in PLAG and PEARLMAN's book (2009) on GGOS, the global geodetic observing system promoted by the International Association of Geodesy (IAG).

To respond to the scientific challenges associated with the rapidly increasing requirements associated to geodetic observations (PLAG *et al.* 2009), GGOS demands improved consistency to all IAG products at the Horizon 2020, as well as and accuracy of the order of 1 mm to the frames of reference, besides a stability in time of 0.1 mm/year. The former accuracy in position, measured on the Earth surface, corresponds roughly to an angle of 30 μas from the Earth's center. The realizations of the terrestrial reference frames (TRF) depends on complex processes and are highly correlated with the determination of the Earth orientation parameters (EOP), which provide the time dependent rotation relating the terrestrial to the celestial reference frames (CRF)—see, e.g., HEINKELMANN *et al.* (2015a, b).

From the observational side, the accuracy and performance of the major Space Geodesy techniques contributing to the EOP determination is noticeably increasing. A good example is provided by the deployment of a new generation of very long baseline interferometry (VLBI) stations compliant with the

---
[1] Department of Applied Mathematics, University of Alicante, P.O. Box 99, 03080 Alicante, Spain. E-mail: alberto.escapa@ua.es; jm.ferrandizleal@ua.es; tomas.baenas@ua.es; jf.navarro@ua.es; santiago.belda@ua.es
[2] Department of Applied Mathematics, University of Valladolid, 47011 Valladolid, Spain. E-mail: getino@maf.uva.es

GGOS/IAG 2010 specifications, the so-called VGOS. It can be expected that more accurate EOP time series will be produced in a few years, even at a sub-daily rate, following the experience of continuous VLBI campaigns (Nilsson et al. 2010). That would be useful to overcome deficiencies in the models used to describe diurnal and sub-diurnal variations of EOPs (Böhm et al. 2012).

In this context, the International Astronomical Union (IAU) and the IAG set up a new Joint Working Group on Theory of Earth Rotation (JWG_ThER). It was aimed at promoting the development of theories of Earth rotation that are fully consistent and that agree with observations and provide predictions of EOP with the accuracy required to meet the needs of the near future; its operation started in 2013. Consistency at the targeted level of accuracy is one of the main topics of interest since the beginning of the activities of that JWG.

Within this framework, we examine the effect stemmed by the simultaneous assignment of more than one value to a same parameter in different components of the precession and nutation theories. Since the modeling of the Earth rotation is extremely complex, its motion is usually described considering one theory for its long-term behavior, precession, and other for the short-term one, nutation. This twofold approach has the consequence that the parameters entering in the description are not given exactly the same values in precession and nutation theories, as it should be obviously. This lack of consistency might originate numerical differences that would be incompatible with nowadays accuracy requirements. The situation is even more complicated, because the current nutation model, IAU2000, is, as well, made up of two different theories, namely, the rigid Earth theory REN2000 (Souchay et al. 1999) and the non-rigid transfer function of MHB2000 (Mathews et al. 2002) that is applied to REN2000 terms.

In particular, we focus on the slightly differences considered in precession and nutation models for the numerical values of the dynamical ellipticity[1] H_d,

with $H_d = (C - \bar{A})/C$, being C the polar and $\bar{A}$ the mean equatorial moments of inertia of the Earth, since it is a key parameter in the determination and understanding of the rotation of the Earth.

Basically, the main differences arise because the dynamical ellipticity considered in the current precession model, IAU 2006 (Hilton et al. 2006), introduced a time-dependent part, due to the J_2 time rate, not considered in IAU2000 nutation model. In turn, second-order effects coming from the nutation–nutation coupling, in the sense of perturbation theories, induce a net contribution to the value of precession, hence producing a change of H_d not taken into account in IAU 2006 precession model. Next, we discuss the scope of those inconsistencies and estimate their numerical influence in the precession and nutation of the Earth.

The paper is structured as follows. In Sect. 2 we sketch the main features of the official standards of precession and nutation adopted by IAU and in force nowadays. We show the potential sources of inconsistencies motivated by the co-existence of three different dynamical models that serve as the basis of the nutation and precessional ones. The adjustments of the nutational part to the precessional one, to achieve higher levels of consistency, are examined in Sect. 3; whereas, those induced by improvements in the mathematical methods of solution or some selected, previously unaccounted, geophysical effects on the precession and nutations are presented in Sect. 4. Finally, we draw some general conclusions about their implications on the current requirements of accuracy of precession–nutation theories.

2. Background: Current Precession and Nutation Models

The General Assemblies (GA) of the International Astronomical Union (IAU) held in 2000 and 2006 approved transcendental Resolutions relative to the precession and nutation models that are widely used

---

[1] Some authors prefer the denomination of dynamical flattening (e.g., Huang et al. 2014). Here, we follow the terminology employed in Kinoshita (1977), Moritz and Mueller (1987) or Williams (1994), among others. Different names such as

Footnote 1 continued
mechanical ellipticity can also be found (e.g., Heisanken and Moritz 1967).

in Astronomy and Space or Earth Observation Sciences.

Specifically, the XXIVth IAU GA (2000) Resolution B1.6 endorsed the IAU2000A precession–nutation model. IAU2000A was intended for users needing accuracy at the 0.2 mas (mili-arcsecond) level. Its nutational component consists of the convolution of the MHB2000 transfer function with a pre-existing rigid Earth model, REN2000. Its precessional component is basically that of IAU 1976 (LIESKE et al. 1977), updated with corrections to the precession rates and offsets. The development of new expressions for precession consistent with IAU2000A was then encouraged.

Six years later, the XXVIth IAU GA (2006) Resolution B1 adopted the IAU 2006 precession model. The precession component of the IAU2000A precession–nutation model was replaced by the P03 precession theory (CAPITAINE et al. 2003), although the model developed by FUKUSHIMA (2003) also provided a very complete description of the precessional motion, including a new parameterization of the angles that is one of the alternatives recommended in the IERS Conventions 2010. P03 model gave new values to the precession rate in longitude and the obliquity of the equator, more accurate than those of LIESKE et al. (1977), including also a conventional J_2 time rate and polynomial expansions of higher degree for the fundamental angles, e.g.,

$$\dot{J}_2 = -3 \times 10^{-9} \mathrm{cy}^{-1},$$
$$\psi_A = 5038''.481507t - 1''.0790069t^2 - 0''.00114045t^3$$
$$+ 0''.000132851t^4 - 0''.000000951t^5,$$
$$\epsilon_A = 84381''.406000 - 46''.836769t - 0''.0001831t^2$$
$$+ 0''.00200340t^3 - 0''.000000576t^4 - 0''.0000000434t^5.$$
$$(1)$$

Here, ψ_A and ϵ_A are the angles providing the precession of the equator in longitude and obliquity (CAPITAINE et al. 2003), respectively, and t is measured in Julian centuries since JD2000.0.

At the highest levels of precision, the replacement of the precessional part of IAU2000 by P03 is not trivial: the three main components of the IAU2000 and IAU 2006 precession–nutation theories (REN2000, MHB2000, P03) assign slightly different values to several astronomical and geophysical

parameters with identical definition, due in part to their updating. That source of inconsistency was already pointed in a first analysis by CAPITAINE et al. (2005).

At this point, it is convenient to recall some basic features of the implied theories. The rigid Earth theory REN2000 is mainly analytical and provides expressions for both precession and nutation, unlike the non-rigid theories based on the transfer function approach. It uses the Hamiltonian formalism, following the method first developed by KINOSHITA (1977), both regarding the used canonical variables and the derivation of the solution by means of perturbations methods (in particular the Lie–Hori canonical one, HORI 1966, KINOSHITA 1977); the tide generating potential (first-order gravitational disturbing potential) is expanded in terms of the canonical Andoyer's variables and the five classic Delaunay's angles appearing in the analytical solutions to the lunar and planetary orbital motions, which implies the dependence of most of the long-period nutation terms on those fundamental arguments, denoted as

$$\Theta_i = m_{i1}l + m_{i2}l' + m_{i3}F + m_{i4}D + m_{i5}\Omega. \quad (2)$$

The subindex i represent a 5-tuple of integers m_{ij}; l, g and h are the usual Delaunay variables of the Moon (M); l', g' and h' are those of the Sun (S); $F = l + g$, $D = l + g + h - l' - g' - h'$, and $\Omega = h - \lambda$. Here, λ is the longitude of the node of the plane orthogonal to the angular momentum axis with respect to the ecliptic, reckoned in direct sense.[2]

At the first-order of perturbation, the nutations in longitude, $\Delta\psi$, and obliquity, $\Delta\epsilon$, are given as sum of the so-called Poisson and Oppolzer terms, which provide, respectively, the perturbations due to action of the disturbing bodies, mainly Moon and Sun, on the angular momentum axis and the differential perturbations of the figure axis with respect to the former.

In the following expressions, originally developed in this way by KINOSHITA (1977), the Poisson terms are those depending on functions $B_{i;p}$ of the auxiliary angle I_0, while the Oppolzer terms contains functions

---

[2] Notice that we use the notation of GETINO and FERRÁNDIZ (1991) for the canonical variables, so that λ corresponds to h in Kinoshita's notation.

$C_{i;p}$. Recall that $I_0 = -\epsilon_A(t)$ according to the perturbation method, although the strictly quasi-periodic nutation terms result from substituting $I_0 = -\epsilon_0$, the constant term of ϵ_A in Eq. (1). Namely, it turns out that (KINOSHITA 1977)

$$\Delta\psi = -\frac{1}{\sin I_0} \sum_{p=M,S} k_p \sum_{i\neq 0} \frac{1}{n_i} \frac{dB_{i;p}}{dI}(I_0) \sin \Theta_i$$
$$+ \frac{1}{\sin I_0} \sum_{p=M,S} k_p \sum_i \sum_{\tau=\pm 1} \frac{\tau C_{i;p}(I_0,\tau)}{n_\mu - \tau n_i} \sin \Theta_i,$$
$$\Delta\epsilon = -\frac{1}{\sin I_0} \sum_{p=M,S} k_p \sum_{i\neq 0} \frac{1}{n_i} m_{i5} B_{i;p}(I_0) \cos \Theta_i$$
$$+ \sum_{p=M,S} k_p \sum_i \sum_{\tau=\pm 1} \frac{C_{i;p}(I_0,\tau)}{n_\mu - \tau n_i} \cos \Theta_i. \qquad (3)$$

Here $n_i = \dot{\Theta}_i$ are the time derivatives of the nutation arguments Θ_i, n_μ is the mean motion (or frequency) of the Andoyer variable μ, close to the conventional mean value of the angular velocity of the Earth, ω_E, and k_M, k_S are the parameters introduced by KINOSHITA (1977), defined by

$$k_p = \frac{3Gm_p}{\omega_E a_p^3} H_d, \quad p = M, S \text{(Moon, Sun)}, \qquad (4)$$

which are proportional to the Earth's dynamical ellipticity H_d. The gravitational constant is denoted as G, m_p stands for the mass of body p, and a_p is the semi-major axis of its orbit.

The functions $B_{i;p}$ and $C_{i;p}$ depend on the orbital coefficients $A_i^{(j)}$ computed first by KINOSHITA (1977), then by KINOSHITA and SOUCHAY (1990) and afterwards by NAVARRO (2001), from the analytical solutions ELP2000 and VSOP87 (CHAPRONT-TOUZÉ and CHAPRONT 1983; BRETAGNON 1988)

$$B_{i;p}(I) = -\frac{1}{6}\left(3\cos^2 I - 1\right)A_{i;p}^{(0)} - \frac{1}{2}\sin 2I A_{i;p}^{(1)}$$
$$-\frac{1}{4}\sin^2 I A_{i;p}^{(2)},$$
$$C_{i;p}(I,\tau) = -\frac{1}{4}\sin 2I A_{i;p}^{(0)} + \frac{1}{2}(1 + \tau\cos I)$$
$$\times (-1 + 2\tau\cos I)A_{i;p}^{(1)}$$
$$+\frac{1}{4}\tau\sin I(1 + \tau\cos I)A_{i;p}^{(2)}. \qquad (5)$$

With respect to the precessional motion, the main component, p_A', of the general precession in longitude rate, p_A, is given by

$$p_A' = p_M' + p_S' = 3H_d\left[\left(\frac{m_M}{m_M + m_E}\right)\left(\frac{n_M^2}{\omega_E}\right)M_0 \right.$$
$$\left. + \left(\frac{m_S}{m_S + m_E + m_M}\right)\left(\frac{n_S^2}{\omega_E}\right)S_0\right]\cos\epsilon_A. \qquad (6)$$

It is attributable to the first-order secular solution (KINOSHITA and SOUCHAY 1990), and the original notation there is kept: constants M_0 and S_0 are the orbital coefficients $A_{(0,0,0,0,0);M}^{(0)}$ and $A_{(0,0,0,0,0);S}^{(0)}$, respectively, m_B represent the mass of each body ($B = E, M, S$) and n_B its relative mean motion. The observed value of the precession rate adopted in REN2000 was the one recommended by IAU at that time (the Lieske's constant, not the value updated in 2006). It was used to estimate p_A' and then to derive the value of H_d, and the resultant Kinoshita's constants k_M and k_S, depending also of the system of astronomical constants in force those years (LIESKE et al. 1977).

SOUCHAY and KINOSHITA (1996) already provided convenient expressions to update the value of the dynamical ellipticity as other parameters would vary, i.e.,

$$\delta H_d = 6.4947 \times 10^{-7}\delta p_A + 6.8812 \times 10^{-9}\delta\epsilon_A$$
$$- 4.4969 \times 10^{-3}\delta M_0 - 2.0708 \times 10^{-3}\delta S_0$$
$$- 0.17973\delta\mu + 1.03583 \times 10^{-3}\delta\rho, \qquad (7)$$

where here $\delta\mu$ and $\delta\rho$ represent the variations associated to the mass ratios between the Moon, the Earth, and the Sun.

The estimated value of p_A' (and thus H_d) changes not only along with variations of the observed precession rate, p_A, but also with the changes of the precession of the ecliptic and various higher order contributions to precession, here denoted globally as p_S (either by the updating of the values of some accounted effects or by the addition of new contributions), since $p_A' + p_S = p_A$.

Non-rigid nutation theory MHB2000 is basically derived by convolving a certain nutation vector $\bar{\eta}_R(\sigma; e_R)$, a complex linear combination of the nutations in longitude and obliquity, computed from the previous REN2000 solution after decomposing each periodic nutation in prograde and retrograde parts. After introducing a generalized transfer function $T(\sigma; e|e_R)$, the amplitude of each nutation term of the non-rigid Earth is given by $\bar{\eta}(\sigma;e) = T(\sigma;e|e_R)\bar{\eta}_R(\sigma;e_R)$. In those expressions $e = (C - \bar{A})/\bar{A}$ is a parameter close to the dynamical ellipticity H_d and σ is a forcing frequency relative to the rotating Earth, related to the time derivative of the nutation arguments.

The basic geophysical Earth model in MHB2000 has an elastic mantle, a fluid outer core (FOC), and a solid inner core (SIC), and its main features are gathered in a set of basic Earth parameters (BEP), some of them being fitted to observations. Effects of anelasticity and oceans were treated as corrections to some BEP, although, e.g., they could have been alternatively incorporated into the integration of the dynamical equations of non-rigid Earth nutation by introducing complex elastic Lame parameters (HUANG et al. 2001) and non-free stress boundary condition on the outer surface (HUANG 2001). The model also considers electromagnetic and viscous couplings, or torques exerted by the fluid outer core on the mantle and on the inner core at the core–mantle boundary (CMB) and inner-core boundary (ICB), respectively.

The MHB2000 transfer function is given by a resonance formula depending on the normal mode frequencies σ_α of the basic Earth model and the relevant forcing frequencies σ

$$T(\sigma;e|e_R) = \frac{e_R - \sigma}{e_R + 1} N_0 \left[1 + (1 + \sigma) \left(\sum_{\alpha=1}^{4} \frac{N_\alpha}{\sigma - \sigma_\alpha} \right) \right],$$

$$N_0 = \frac{H_d}{H_{dR}}. \tag{8}$$

Let us remark that in the previous formula H_{dR} stands for the dynamical ellipticity of REN2000, which is allowed to be given a new value H_d in the computation of the MHB2000 transfer function for the sake of a better fitting of the theoretical to the VLBI observed amplitudes of a selected set of main nutation terms, which gave raise to the introduction of the additional parameter N_0. However, MHB2000 final nutation series were derived with a different transfer function, owing to a similar functional dependence but in which some of the geophysical parameters were changed from constants (N_α and σ_α) to linear functions of time (Q_α and s_α) in the diurnal band, also in order to improve the fitting to observations.

It may be not out of place to point out that nowadays MHB2000 model presents some short-comings. For instance, it was early recognized (GETINO and FERRÁNDIZ 2000) that the transfer function treatment of second-order effects, i.e., nutation–nutation couplings proportional to H_d^2, is not correct, since this method is essentially linear—a fact also pointed out by MATHEWS et al. (2002). Yet another problem recently discussed is referred to the electromagnetic coupling at the CMB estimated in MHB2000 (2002). That mechanism was proposed to fill the dominant gap between the values of the period of the free core nutation (FCN) normal mode and the amplitude of retrograde annual nutation term either derived from theory or from observations, and stressed as one of the main advantages of that model over all other non-rigid Earth nutation models.

However, that coupling has been studied further and suspected broken by other works, e.g., HUANG et al. (2011). In that investigation it is used the same model and values of the electromagnetic parameters crossing CMB as in BUFFETT et al. (2002). It was found that the contribution of this coupling to the retrograde annual nutation, and the change of the FCN period as well, is approximately one order of magnitude smaller than what is required, and also expected in BUFFETT et al. (2002), to fill the difference between the observations and the said theoretical models. Hence, there are other mechanisms that must play a more important role than previously considered in MHB2000 like, for example, viscous coupling, topographic coupling, etc. (see HUANG et al. 2011 and references therein).

As a consequence of the foregoing description of the current IAU Earth rotation theory, the adoption of the precession model IAU2006 compels the adjustment of the IAU2000 nutation series to keep consistency within the set of successive IAU Resolutions (B1.6 adopted in 2000, B1 and B2 in 2006). In its turn, a right treatment of different non-rigid couplings of the nutation model, i.e., nutation–nutation cross effects, entails the appearance of new contributions to the precession rates. Both issues lead to differences in the employed values of the dynamical ellipticity H_d and are revised in the next sections.

3. Changes of IAU2000 Nutation Theory from IAU2006 Precession Theory

As explained in the last section, the current IAU precession and nutation official models stem from separate dynamical theories. Needless to say that this is a valid approach, but current accuracy requirements demand a detailed analysis about the way in which these two parts are matched, in order to ensure the highest levels of consistency.

Next, we discuss the changes induced by the IAU2006 precession theory on the IAU2000 nutation theory. A comprehensive study of all the involved effects can be found in ESCAPA et al. (2014). ESCAPA et al. (2015) provide the theoretical background and technical details necessary to formulate those effects within the Hamiltonian framework.

The adjustments considered in this work are due to the new features considered in the P03 precession model arising from the inclusion of the J_2 time rate, $\dot{J}_2$, in the modeling of the J_2 coefficient. Since the effects to be discussed here are at the micro-arcsecond (μas) level in the amplitude of each argument Θ_i of the nutation series, we consider a first-order theory that accounts for the nutations of the angular momentum axis (Poisson terms), incorporating into the modeling the J_2 time rate. As it is well-known those nutations are common for one, two, and three-layer Earth models (MORITZ and MUELLER 1987).

The time evolution of the angular momentum axis is constructed by means of the Hamiltonian formalism. In the expression of the Hamiltonian, the Kinoshita's parameter k_p, which is proportional to the Earth's dynamical ellipticity, is replaced by a time-dependent function taking into account the H_d time rate induced by the J_2 time rate. At first order, it has the form

$$k_p = k_{p,0}\left(1 + t\frac{\dot{H}_d}{H_d}\right) \simeq k_{p,0}\left(1 + t\frac{\dot{J}_2}{J_2}\right), \quad (9)$$

where $k_{p,0}$ denotes now the time-independent Kinoshita's parameter for Moon and Sun introduced in Eq. (3).

The computation of this effect (ESCAPA et al. 2014, 2015), following the perturbation equations, leads to the appearance of mixed secular terms (proportional to t, previously reported in CAPITAINE

et al. 2005), and to in- and out-of-phase nutations. Namely, the following expressions to be summed to Eq. (3) are obtained

$$d_{j_2}\Delta\psi = -\sum_{p=M,S}\left[t\frac{\dot{J}_2}{J_2}\frac{k_{p,0}}{\sin I_0}\sum_{i\neq 0}\frac{1}{n_i}\frac{dB_{i;p}}{dI}(I_0)\sin\Theta_i\right.$$
$$\left. +\frac{\dot{J}_2}{J_2}\frac{k_{p,0}}{\sin I_0}\sum_{i\neq 0}\frac{1}{n_i^2}\frac{dB_{i;p}}{dI}(I_0)\cos\Theta_i\right],$$

$$d_{j_2}\Delta\epsilon = -\sum_{p=M,S}\left[t\frac{\dot{J}_2}{J_2}k_{p,0}\sum_{i\neq 0}\frac{m_{i5}}{n_i}\frac{B_{i;p}(I_0)}{\sin I_0}\cos\Theta_i\right.$$
$$\left. -\frac{\dot{J}_2}{J_2}k_{p,0}\sum_{i\neq 0}\frac{m_{i5}}{n_i^2}\frac{B_{i;p}(I_0)}{\sin I_0}\sin\Theta_i\right]. \quad (10)$$

To be consistent in the development of the theory, the inclusion of the J_2 time rate forces the consideration of the time rate of the orbital coefficients $A_i^{(0,1,2)}$ (KINOSHITA 1977), due to the secular variation of Sun eccentricity (ESCAPA et al. 2014, 2015). Analytically its treatment is similar to that of J_2 rate, providing also out-of-phase nutations and mixed secular terms with formulas similar to Eq. (10).

Let us note that the mixed secular terms in Eq. (10) are proportional to $\dot{J}_2/J_2$, both in longitude and obliquity, in agreement with the result derived in CAPITAINE et al. (2005). Strictly speaking those expressions are only valid for the first-order terms of the nutations, proportional to H_d, since some of the second-order terms are proportional to k_p^2, i.e., to H_d^2, and the rescaling factor to be considered would be different from $\dot{J}_2/J_2$.

In contrast, the out-of-phase terms have not been considered previously and provide *new* relevant contributions at the μas level given by

$$d_{j_2}\Delta\psi = -1.4\cos\Omega, \quad d_{j_2}\Delta\epsilon = -0.8\sin\Omega. \quad (11)$$

In these expressions, and subsequent ones unless otherwise specified, the amplitudes are given in micro-arcseconds.

Table 1 includes all the main related corrections arising from the J_2 time rate, comparing the mixed secular terms with the ones computed previously and currently included in IERS Conventions 2010.

The P03 precession model also introduced other changes in the numerical values of some precession parameters, which were different from LIESKE et al.

Table 1

Corrections due to the inclusion of J_2 time rate

Θ_i					Period	Secular mixed terms (μas/cJ)				Out-of-phase (μas)	
l	l'	F	D	Ω	Days	$d\psi(t \times \sin)$		$d\epsilon(t \times \cos)$		$d\psi(\cos)$	$d\epsilon(\sin)$
0	0	0	0	1	−6798.36	47.8[a]	48.0[b]	−25.6[a]	−25.6[b]	−1.4[b]	−0.8[b]
0	0	0	0	2	−3399.18	−0.6	−0.6	–	–	–	–
0	0	2	−2	2	182.62	3.7	3.5	−1.6	−1.5	–	–
0	0	2	0	2	13.66	0.6	0.6	–	–	–	–

[a]Capitaine *et al.* (2005)

[b]Escapa *et al.* (2014)

(1977). Specifically, the updating of the value of the obliquity $I_0 = -\epsilon_A$ originates significative contributions, above the micro-arcsecond level, that should be envisaged as corrections to the IAU2000A nutation series in order to ensure its consistency with the IAU2006 precession. Corrections of this kind were first derived by Capitaine *et al.* (2005) who provide the terms

$$d_{\epsilon_A}\Delta\psi = -8.1\sin\Omega - 0.6\sin(2F - 2D + 2\Omega) \tag{12}$$

gathered in Sect. 5.6.3 of the IERS Conventions 2010. A more detailed study falls out of the scope of this paper, which focuses on effects associated to the variations of the dynamical ellipticity. However, for the sake of completeness, let us say that the former corrections were completed by Escapa *et al.* (2014, 2015). According to the premises stated in those references, the total effect is

$$d_{\epsilon_A}\Delta\psi = (-15.6 - 8.1t)\sin\Omega$$
$$- 0.6\sin(2F - 2D + 2\Omega), \quad d_{\epsilon_A}\Delta\varepsilon = 0.8\cos\Omega. \tag{13}$$

4. Unaccounted Contributions to the Earth's Precession and Effects on Nutations

With the precedent notations, the precession of the equator is made up of the main linear term $p'_A t$ and various second order, smaller contributions. Let us recall that the expression "second order terms" is used in the literature to designate contributions of very different kinds. First, it means terms of various physical origins that produce linear perturbations of

small magnitude. There are also second-order terms in the sense of the perturbation methods, which are quadratic in the main perturbation parameter, which we can assimilate to H_d, so arising from the crossing of first-order perturbations—asymptotically speaking. There were included in REN2000 for the rigid model, but a linear convolution with a transfer function is meaningless or doubtful since it does not fit well with the quadratic essence of those terms, as was pointed out in Sect. 2.

Second-order terms accounted in the precession theory have stayed unchanged since several years. A classic source is the paper by Williams (1994); Kinoshita and Souchay (1990) computed the full (quadratic) second-order effects for the rigid Earth, clarifying some differences of terminology and casting of terms with respect to Williams in their 1997 paper (Souchay and Kinoshita 1997). Those former contributions are the same shown by Capitaine *et al.* (2003)—apart from some slight adaptations.

However, a few, significant, new contributions have been found in the last years. The most remarkable are due to: (1) non-rigid quadratic second-order mathematical solution (Ferrándiz *et al.* 2004, 2007); (2) effects of the mantle anelasticity (Lambert and Mathews 2006 coupled to oceans, as well as Ferrándiz *et al.* 2012; Baenas 2014, uncoupled). Following the analytical procedure described above, the Hamiltonian approach provided a systematic procedure to compute both kinds of terms, which guarantees the consistency of the results; the method proceeds by adding the new relevant terms to the basic Hamiltonian and then applying the same perturbation method to the extended Hamiltonian.

Table 2

Corrections due to unaccounted effects on precession

Effect (diff. WRT rigid model)		Δ Longitude	Δ Ellipticiy	Δ Obliquity
Earth model	Perturb. order	dp_ψ (mas/cy)	δH_d (ppm)	dp_ϵ (mas/cy)
Poincaré	2	−23.1578	4.8	–
Fluid core + elastic mantle	2	−12.5673	2.5	–
Mantle anelasticity	1	6.0059	−1.2	0.7748

Table 3

Corrections due to indirect effects on nutation

Θ_i					Period	Poincaré model (μas)		Fluid core + elasticity (μas)	
l	l'	F	D	Ω	(days)	$\Delta\psi$ (sin)	$\Delta\epsilon$ (cos)	$\Delta\psi$ (sin)	$\Delta\epsilon$ (cos)
0	0	0	0	1	−6798.36	−82.59	44.19	−44.64	23.93
0	0	0	0	2	−3399.18	1.00	−0.43	–	–
0	0	2	−2	2	182.62	−6.32	2.75	−3.42	1.49
0	0	2	0	2	13.66	−1.09	0.47	–	–

4.1. *Unaccounted Effects on Non-Rigid Earth Precession*

After P03 precession model was adopted, several research works have identified non-negligible effects on non-rigid Earth precession. Their addition to the existing models is equivalent to a suitable change of the determined (or observed) value of precession, producing thus an equivalent change of the value of the main coefficient p'_A. Therefore, the derived parameter H_d varies accordingly and the whole set of nutations suffers and indirect effect along with that change. Notice that, the quadratic rigid second-order effects on precession being part of REN2000, they must be subtracted from those computed from a more realistic non-rigid Earth model to derive the effective variations of them. Table 2 shows examples of such kind of new contributions, extracted from Ferrándiz et al. (2007) and Baenas (2014), considering different elastic responses of the Earth's mantle.

4.2. *Indirect Effects On Nutation*

Indirect effects on nutation are those resulting from the changes in the value of the main precession parameter H_d, as discussed above. Their derivation is simple, since it suffices to compute the linear approximation by multiplying the nutation

amplitudes by the factor $(H_d + \delta H_d)/H_d$, either in the fully non-rigid Hamiltonian approach or the rigid Hamiltonian + non-rigid transfer function approaches. The main terms arising from $(\delta H_d)/H_d$ are displayed in Table 3. Note that some of them reach tens of μas, well above the accuracy objective pursued nowadays.

5. *Conclusions*

Current research on Earth rotation modeling, encouraged by the IAU/IAG Joint Working Group on Theory of Earth Rotation, has to face the problem of obtaining very demanding accuracy goals, set as 30 μas and 3 μas/year in orientation and velocity, respectively, as shown, e.g., in Ferrándiz and Gross (2014a, b). This is the same very stringent level of accuracy first established by the GGOS of the IAG, see, e.g., Plag and Pearlman (2009) for a comprehensive description of the geodetical problems and the expected societal benefits which led to such high-accuracy requirements, including the geodetical monitoring of the global change. The consistency of the current precession and nutation theories (derived successively in time as a set of three main pieces, based on mathematically different approaches) is an obvious requirement, although it is not fully achieved at such level of accuracy, but additional corrections must be added to the existing ones. Although only few nutation terms require not negligible corrections due to the differences of the dynamical ellipticity values used in the IAU2000 and IAU2006 models, as shown in this paper, those corrections cannot be ignored any more, since they are needed to ensure the consistency of the current precession–nutation models.

Acknowledgments

The authors thank the editor M. Charco, C.L. Huang and a second anonymous referee for their valuable advice, which helped to improve the manuscript. This work has been partially supported by the Spanish government under Grant AYA2010-22039-C02-02 from Ministerio de Economía y Competitividad (MINECO) and Generalitat Valenciana project GV/2014/072.

REFERENCES

BAENAS, T. (2014), Contribuciones al estudio analítico del movimiento de rotación de una Tierra deformable (in Spanish), PhD thesis, University of Alicante.

BÖHM, S., BRZEZINSKI, A., SCHUH, H. (2012), *Complex demodulation in VLBI estimation of high frequency Earth rotation components*, J. Geodyn., *62*, 56–58.

BRETAGNON, P. (1988), *Planetary theories in rectangular and spherical variables. VSOP87 solution*, Astron. Astrophys. *202*, 304–315.

BUFFET, B.A., MATHEWS, P.M., HERRING, T.A. (2002), *Modeling of nutation and precession: Effects of electromagnetic coupling*, J. Geophys. Research *107*, No. B4, 2070.

CAPITAINE, N., WALLACE, P.T., CHAPRONT, J. (2003), *Expressions for IAU2000 precession quantities*, Astron. Astrophys. *412*, 567–586.

CAPITAINE, N., WALLACE, P.T., CHAPRONT, J. (2005), *Improvement of the IAU2000 precession model*, Astron. Astrophys. *432*, 355–367.

CHAPRONT-TOUZÉ, M., CHAPRONT, J. (1983), *The lunar ephemeris ELP–2000*, Astron. Astrophys. *124*, 50–62.

ESCAPA, A., GETINO, J., FERRÁNDIZ, J., BAENAS, T. (2014), On the changes of the IAU2000 nutation theory stemming from IAU 2006 precession theory, In: Capitaine, N. (ed) Proceedings of the Journées 2013 Systèmes de Référence Spatio-Temporels, Observatoire de Paris, 148–151.

ESCAPA, A., GETINO, J., FERRÁNDIZ, J. M., BAENAS, T. (2015), Dynamical matching between IAU2000 nutation theory and IAU 2006 precession theory (in preparation)

FERRÁNDIZ, J. M. and GROSS, R.S. (2014), The goal of the IAU/IAG Joint Working Group on the Theory of Earth rotation, In: Capitaine, N. (ed) Proceedings of the Journées 2013 Systèmes de Référence Spatio-Temporels, Observatoire de Paris, 139–143.

FERRÁNDIZ, J. M. and GROSS, R.S. (2014), The New IAU/IAG Joint Working Group on Theory of Earth Rotation, IAG Symp 143 (to appear).

FERRÁNDIZ, J.M., NAVARRO, J.F., ESCAPA, A., GETINO, J. (2004), *Precession of the nonrigid Earth: effect of the fluid outer core*, Astron. J. *128*, 1407–1411.

FERRÁNDIZ, J.M., NAVARRO, J.F., ESCAPA, A., GETINO, J., BAENAS, T. (2007), Influence of the mantle elasticity on the precessional motion of a two-layer Earth model, In: Lemaître, A. (ed) The rotation of celestial bodies, Press. Universitaires de Namur, 9–14.

FERRÁNDIZ, J.M., BAENAS, T., ESCAPA, A. (2012), Effect of the potential due to lunisolar deformations on the Earth precession. Geophysical Research Abstracts, 14, EGU2012-6175.

FUKUSHIMA, T. (2003), *A new precession formula*, Astron. J. *126*, 494–534.

GETINO, J., FERRÁNDIZ, J. M. (1991), *A Hamiltonian Theory for an Elastic Earth: Elastic Energy of Deformation*, Celes. Mech. *51*, 17–34.

GETINO, J., FERRÁNDIZ, J. M. (2000), Towards models and constants for sub-microarcsecond astrometry, In: Johnston, K. J., McCarthy, D. D., Luzum, B. J., Kaplan, G. H. (eds) Proceedings of IAU Colloquium 180, U. S. Naval Observatory, Washington, DC, USA, 236–241.

HEINKELMANN, R., BELDA-PALAZÓN, S., FERRÁNDIZ, J.M., SCHUH, H. (2015) How consistent are the current conventional celestial and terrestrial reference frames and the conventional Earth orientation parameters?, IAG Symp (Proc. of Reference Frames for Applications in Geosciences, REFAG2014, accepted 2015)

HEINKELMANN, R., BELDA-PALAZÓN, S., FERRÁNDIZ, J.M., SCHUH, H. (2015) The consistency of the current conventional celestial and terrestrial reference frames and the conventional EOP series, In: Capitaine, N., Malkin, Z. (eds) Proceedings of the Journées 2014 Systèmes de Référence Spatio-Temporels, 224–225

HEISKANEN, W. A., MORITZ, H., Physical Geodesy (W. H. Freeman & Co. Ltd. 1967).

HILTON, J.L., CAPITAINE, N., CHAPRONT, J., FERRÁNDIZ, J.M., FIENGA, A., FUKUSHIMA, T., GETINO, J., MATHEWS, P., SIMON, J.L., SOFFEL, M., VONDRAK, J., WALLACE, P., WILLIAMS, J. (2006), *Report of the International Astronomical Union Division I Working Group on Precession and the Ecliptic*, Celest. Mech. Dyn. Astron. *94*, 351–367. doi:10.1007/s10569-006-0001-2.

HORI, G. I. (1966), *Theory of general perturbations with unspecified canonical variables*, Publ. Astron. Soc. Jpn. *18*, 287–296.

HUANG, C. L. (2001), *The scalar boundary conditions for the motion of the elastic Earth to second order in ellipticity*, Earth, Moon, and Planets, *84*, 125–141.

HUANG, C. L., JIN, W. J., LIAO, X. H. (2001), *A new nutation model of a non-rigid Earth with ocean and atmosphere*, Geophys. J. Int., *146*, 126–133.

HUANG, C. L., DEHANT, V., LIAO, X. H., VAN HOOLST, T., ROCHESTER, M.G. (2011), *On the coupling between magnetic field and nutation in a numerical integration approach*, J. Geophys. Res., *116*, No. B03403.

HUANG, C. L., LIU, C. J., LIU, Y. (2014), A generalized theory of the figure of the Earth: application to the moment of inertia and global dynamical flattening, In: Capitaine, N. (ed) Proceedings of the Journées 2013 Systèmes de Référence Spatio-Temporels, Observatoire de Paris, 156–159.

KINOSHITA, H. (1977), *Theory of the rotation of the rigid Earth*, Celest. Mech. *15*, 277–326.

KINOSHITA, H., SOUCHAY, J. (1990), *The theory of the nutation for the rigid-Earth model at the second order*, Celest. Mech. *48*, 187–265.

LAMBERT, S.B., MATHEWS, P.M. (2006), *Second-order torque on the tidal redistribution and the Earth's rotation*, Astron. Astrophys. *453*, 363–369.

LIESKE, J., LEDERLE, T., FRICKE, W., MORANDO, B. (1977), *Expression for the precession quantities based upon the IAU (1976) system of astronomical constants*, Astron. Astrophys. *58*, 1–16.

MATHEWS, P.M., HERRING, T.A., BUFFET, B.A. (2002), *Modeling of nutation and precession: New nutation series for nonrigid Earth and insights into the Earth's interior*, J. Geophys. Research *107*, No. B4, 2068.

MORITZ, H., MUELLER, I., Earth Rotation (Frederic Ungar 1987).

NAVARRO, J.F. (2001), Teoría analítica de la rotaci ón de la Tierra rígida mediante manipulación simbólica espec ífica (in Spanish), PhD thesis, University of Alicante.

NILSSON, T., BÖHM, J., SCHUH, H.(2010), *Sub-diurnal earth rotation variations observed by VLBI*, Artificial Satellites, *45*, No. 2. doi:10.2478/v10018-010-0005-8.

PETIT, G. and LUZUM, B.,(eds.) IERS Conventions 2010, IERS Technical Note 36, Verlag des Bundesamtes für Kartographie und Geodäsie, Frankfurt am Main.

PLAG, H.-P., GROSS, R., ROTHACHER, M. (2009), *Global Geodetic Observing System for Geohazards and Global Change*. Geosciences, BRGM's journal for a sustainable Earth, *9*, 96–103.

PLAG, H.-P. and PEARLMAN, M. (eds) (2009), Global Geodetic Observing System: Meeting the Requirements of a Global Society on a Changing Planet in 2020, Springer–Verlag, Berlin–Heidelberg.

SOUCHAY, J. and KINOSHITA, H. (1996), *Corrections and new developments in rigid earth nutation theory. I. Lunisolar influence including indirect planetary effects*, Astron. Astrophys. *312*, 1017–1030.

SOUCHAY, J. and KINOSHITA, H. (1997), *Corrections and new developments in rigid-Earth nutation theory. II. Influence of second-order geopotential and direct planetary effect*, Astron. Astrophys. *318*, 639–652.

SOUCHAY, J., B. LOSLEY, H. KINOSHITA, FOLGUEIRA, M. (1999), *Corrections and new developments in rigid earth nutation theory. III. Final tables "REN-2000" including crossed-nutation and spin-orbit coupling effects*. Astron. Astrophys. Suppl. Ser., *135*, 111–131.

WILLIAMS, J.G. (1994), *Contributions to the Earth's obliquity rate, precession, and nutation*. Astron. J. *108*, 2, 711–724.

(Received December 30, 2014, revised July 9, 2015, accepted July 17, 2015, Published online August 18, 2015)

Pure Appl. Geophys. 173 (2016), 871–884
© 2015 Springer Basel
DOI 10.1007/s00024-015-1050-9

∎ **Pure and Applied Geophysics**

Improving Surface Geostrophic Current from a GOCE-Derived Mean Dynamic Topography Using Edge-Enhancing Diffusion Filtering

J. M. Sánchez-Reales,[1] O. B. Andersen,[2] and M. I. Vigo[1]

Abstract—With increased geoid resolution provided by the gravity and steady-state ocean circulation explorer (GOCE) mission, the ocean's mean dynamic topography (MDT) can be now estimated with an accuracy not available prior to using geodetic methods. However, an altimetric-derived MDT still needs filtering in order to remove short wavelength noise unless integrated methods are used in which the three quantities are determined simultaneously using appropriate covariance functions. We studied nonlinear anisotropic diffusive filtering applied to the oceańs MDT and a new approach based on edge-enhancing diffusion (EED) filtering is presented. EED filters enable controlling the direction and magnitude of the filtering, with subsequent enhancement of computations of the associated surface geostrophic currents (SGCs). Applying this method to a smooth MDT and to a noisy MDT, both for a region in the Northwestern Pacific Ocean, we found that EED filtering provides similar estimation of the current velocities in both cases, whereas a non-linear isotropic filter (the Perona and Malik filter) returns results influenced by local residual noise when a difficult case is tested. We found that EED filtering preserves all the advantages that the Perona and Malik filter have over the standard linear isotropic Gaussian filters. Moreover, EED is shown to be more stable and less influenced by outliers. This suggests that the EED filtering strategy would be preferred given its capabilities in controlling/preserving the SGCs.

Key words: Edge enhancing diffusion, GOCE, mean dynamic topography filtering, geostrophic currents.

1. Introduction

Surface geostrophic currents (SGCs) have a strong directional behaviour along the gradients of the mean dynamic topography (MDT). Here, we address geodetic MDTs, which are the surfaces obtained by subtracting a geoid height (N) from an altimetric mean sea surface (MSS). Filtering of the MDT with standard linear isotropic filters, such as the Gaussian filter (Wahr *et al.* 1998), leads to signal loss in areas where sharp gradients are smoothed (Bingham 2010). This results in considerable attenuation of the geostrophic flow. In this sense, Bingham (2010) showed that non-linear isotropic filtering processes are preferred as smoothing methods in order to detect and keep the high values of the gradients that we could call "edges" of the surface. In the approach presented here, we take the development one step further and apply an anisotropic filter that allows us to control both the magnitude and the direction, instead of just the magnitude of the diffusion flux, which is highly desirable for the filtering of MDT surfaces.

The mean SGCs are derived from the ocean MDT by

$$u = -\frac{g}{f}I_y; \quad v = \frac{g}{f}I_x \qquad (1)$$

where (u, v) are the velocities of the flow in the eastward and northward (x, y) directions, respectively. In (1), g represents (normal) acceleration due to gravity, f is the Coriolis parameter and $(I_x, I_y) = \nabla I$ is the gradient of the MDT in the (x, y) directions. Given that the Coriolis parameter vanishes close to the equator, the solution in the equator band must be estimated following other approaches.

A satellite-derived MDT requires an altimetric mean sea surface (MSS) and a geodetically determined geoid (N); subsequently, the MDT is determined as MDT = MSS − N. Fundamental limitations of satellite-derived MDT are the spectral inconsistency between the MSS and the geoid (Slobbe *et al.* 2012; Albertella and Rummel 2009) and the high noise for shorter wavelengths in the

[1] Department of Applied Mathematics, University of Alicante, Alicante, Spain. E-mail: jms.reales@ua.es
[2] Danish National Space Center, DTU Space, Kongens Lyngbyv, Denmark.

satellite-derived gravitational models (PAIL et al. 2010, 2011; TAPLEY et al. 2005). We will focus on the short wavelength noise in this paper and illustrate how an anisotropic filter that takes into account directional information can address the problem of noise reduction.

From (1) a simple relation follows between the MDT and the SGC: level lines in the MDT are streamlines of the geostrophic flow, or, similarly, the ocean currents flow perpendicular to the gradients of the MDT. The latter has a clear consequence. Because the ocean currents are directional, anisotropic rather than isotropic filtering will be preferred.

Geostrophic flow has anisotropic behaviour (emphasized when the current increases). It is composed of two directions: the one that the current follows, and the direction across the flow. Because the Gaussian filter is isotropic, it weighs all directions equally, depreciating any kind of information in this sense. This implies that applying a Gaussian filter potentially attenuates the current signal in areas of strong currents, consequently diminishing velocities. As an alternative, it would be preferable in this case to use non-linear filtering (along rather than across the gradient).

BINGHAM (2010) and BINGHAM et al. (2011) showed how a non-linear isotropic diffusion filter adapted from PERONA and MALIK (1990) was useful in preserving strong SGCs in the Gulf Stream region of the western North Atlantic. They found that the estimated current speeds using non-linear filtering can have twice the amplitude when compared with usual Gaussian filtering. However, the Perona and Malik filter (hereafter PMF) is still limited, since only the magnitude of the diffusion flux can be controlled and not its direction. Moreover, strong gradients due to strong noise (outliers) can be detected as an edge and preserved by the filter, as will be seen in this paper.

The PMF can be generalized to the edge-enhancing diffusion (EED) filtering. A regularization parameter is used in this case to control how strongly the MDT is pre-filtered in order to smooth the noise when determining the magnitude and direction of the flux. The use of a diffusion tensor matrix will rotate the filtering's direction along the current's direction (rather than along the gradient) and will fix the problems related to local errors that can arise when using the PMF. In this work, we illustrate these two main advantages of EED filtering over PMF filtering when applied to a satellite-derived ocean MDT for the determination of its associated SGC.

In the next section, the anisotropic diffusion concept is introduced. We briefly describe diffusive filtering and then introduce EED filtering. SGC velocities derived from the filtered MDTs are used to compare the various approaches to filtering. For this, we centre our attention to the Northwestern Pacific Ocean. Data and methodology is introduced in Sect. 3, and additional detail regarding the MDTs as used in this study are provided. Finally, the main results are discussed in Sect. 4 along with a comparison with an independent data field. Conclusions are summarized in Sect. 5.

2. Edge Enhancing Diffusion

Equation 1 described how the mean SGC can be derived from the slope of the MDT. Amplitude of the SGC is determined from (1) by $\|\mathrm{SGC}\| = \frac{g}{f}\sqrt{I_x^2 + I_y^2}$, while $\theta_{\mathrm{SGC}} = \arctan(-I_x/I_y)$ represents the direction along the current. This is equivalent to saying that the vector $[I_x\ I_y]$ is directed across the flow (or the edge of the MDT) and $[-I_y\ I_x]$ is directed along the flow. Following the discussion above, strong currents are derived from strong gradients in the MDT. Therefore, we are interested in preserving strong gradients while the noise is attenuated. For this, we should reduce the influence that locations across the stream flow (MDT edges) have in the filtering process.

An anisotropic diffusion filter (WEICKERT 1998) allows us to set the behaviour of a diffusive process by defining characteristics over a diffusion tensor. Diffusion is expressed by the continuity equation

$$\frac{dI}{dt} = \mathrm{div}(D \cdot \nabla I) \qquad (2)$$

where t is a pseudo-time variable controlling the degree of filtering of a grid I; dI/dt denotes the derivative of the grid I through an iterative process t times; $\nabla I = (I_x, I_y)$ is the gradient along the x and y directions; div is the divergence $(\partial/\partial x + \partial/\partial y)$; and D is the diffusion tensor (a positive definite symmetric matrix). The product $D \cdot \nabla I$ can be physically interpreted as a flux that attempts to equilibrate concentration differences ∇I.

For the case in which D is a diagonal matrix with equal and constant elements d, both directions of the gradient are equally weighted in (2) and, therefore, the diffusive filter would be equivalent to applying a standard Gaussian filter with $\sqrt{2td}$ of half wavelength (CRANK 1975, pp. 11).

PERONA and MALIK (1990) proposed the use of a function $g(\|\nabla I\|^2)$ instead of constant values in the diagonal of D. This function $g(.)$, when chosen properly, will preserve and enhance the edges of I. For that, $g(.)$ needs to be monotonically decreasing, making the flux $g(\|\nabla I\|^2) \cdot \nabla I$ increase monotonically up to a certain threshold k, and decrease monotonically from there. They proposed several functions that satisfy such a requirement and showed how they provided similar results. Based on their results, we use

$$g(s^2) = \frac{1}{1 + (s/k)^2}. \tag{3}$$

The constant k is an important parameter defined by the user in order to mark the sensitivity of the filter to the local gradient. Details on a practical application of the PMF are provided in Appendix 1.2.

The PMF used in BINGHAM (2010) does not control the diffusion's direction but only the magnitude of the diffusion's flux. In the study presented here, the PMF is extended to a more advanced version of the diffusion filtering by considering the rotation of the diffusion's direction. Since the PMF is directed along the gradient of the MDT, in the case of a noisy MDT, the gradient will not represent the direction of the flow. As we are interested in filtering the MDT with the objective of determining the SGC, it would be desirable here to rotate the diffusion's direction toward the SGC direction. Moreover, the presence of strong noise makes the PMF strongly dependent on the k value as well as highly influenced by local errors, since accidental noise-related "jumps" in the gradient can be detected as an edge and, therefore, maintained through the iterations.

The direction of the mean SGC can be approximated from a smoothed version of the MDT that will also reduce non-white noise in the original MDT. Then, by introducing a regularization parameter σ we

will be able to use in (2) a suitable diffusion tensor D that will address both problems in the PMF: setting the direction of the diffusion flux to coincide with the direction of the flow and to reduce the influence that non-white noise has in determining the magnitude of diffusion flux. This parameter permits the determination of the magnitude and direction of the flux from a smoothed gradient that removes local noise influence and rotates the diffusion direction to the SGC direction.

EED (WEICKERT 1996, 1998) is a filtering method that sets the diffusion tensor D in (2) to reduce the diffusivity across the edges. This is made by defining the structure tensor

$$J_\sigma = K_\sigma * J_0, \tag{4}$$

where J_0 is the Hessian of I, K_σ is a Gaussian smoothing with σ half wavelength, and $*$ denotes the convolution of the two matrices. This smoothed structure tensor is used to minimize the influence of local errors in the data for the determination of flow directions. It can be decomposed as

$$J_\sigma = (\begin{matrix} v_1 & v_2 \end{matrix}) \begin{pmatrix} \lambda_1 & 0 \\ 0 & \lambda_2 \end{pmatrix} \begin{pmatrix} v_1 \\ v_2 \end{pmatrix} \tag{5}$$

where λ_1, λ_2 are the eigenvalues of J_σ associated with the eigenvectors v_1 and v_2, respectively. Each eigenvalue can be interpreted as the portion of flow variance in the direction of its associated eigenvector. The first eigenvector (corresponding to the largest eigenvalue) is a local average over the directions of maximum variance, which is along the flow. Thus, if $\lambda_1 \geq \lambda_2$ then v_1's direction is along the flow and v_2's direction is across the flow.

This information is used to establish the diffusion tensor matrix D as

$$D = (\begin{matrix} v_1 & v_2 \end{matrix}) \begin{pmatrix} \mu_1 & 0 \\ 0 & \mu_2 \end{pmatrix} \begin{pmatrix} v_1 \\ v_2 \end{pmatrix} \tag{6}$$

where μ_1 is obtained from λ_1 as $\mu_1 = g(\lambda_1)$, where $g(.)$ is given in (3), and μ_2 is set to be $\mu_2 = 1$. Note that the PMF is a particular EED case when σ is set to 0 since the unfiltered structure tensor J_0 has eigenvalues $\|\nabla I\|^2$ and 0 associated to $[I_x \; I_y]$ and $[-I_y \; I_x]$, respectively.

In (4), σ acts as a regularization parameter. Since ∇I is smoothed for length scales smaller than σ, detecting the magnitude and direction of the flow is insensitive to noise in such scales. Given that spatial scales smaller than σ are ignored, residual noise will also be smoothed out, reducing its influence into the final result. When the short scale noise is reduced, the SGC direction is better delimitated, which will help to enhance edge-like structures in the MDT. This makes the filter less affected by the choice of the k parameter in (3) when σ increases. A detailed description of EED for a practical implementation is provided in Appendix 1.3.

3. Data

Here we use the first release of ESA's GOCE geoid data computed by the 'direct method' (PAIL et al. 2011) as available through the GOCE Virtual Online archive (http://eo-virtual-archive1.esa.int/). This data is given in spherical harmonics complete up to degree and order 240. It allows a determination of the Earth's geoid with a theoretical spatial resolution of approximately an 83-km half wavelength. However, the noise-to-signal content increases dramatically for the higher degrees. All scales shorter than this spatial resolution are not contained in the final surface, implying that sharp edges in the real surface will appear as smoothed edges. Moreover, a ringing effect due to truncation is reflected in the surface hiding the true signal.

In order to derive the MDT, an MSS with the same spatial resolution as the geoid is required. Here, we used an MSS obtained from the CLS01 (HER-NANDEZ and SCHAEFFER 2001) extended to an 18-year period using sea level anomalies provided by AVISO (Aviso, http://www.aviso.oceanobs.com, version 29 Mar 2011). Both the MSS and the geoid are provided with the usual corrections applied, referring to the same tidal model and projected on the same reference ellipsoid. A complete description of the GOCE products and applied corrections are given in detail in EGG-C (2009).

The MDT is derived as the difference between the MSS and the geoid. For this, either the geoid must be brought into the space domain as a grid of geoid height points or the MSS must be brought into the spectral domain as a set of spherical harmonics (see

BINGHAM et al. 2008). Following (BINGHAM 2010), these two different approaches are from now on referred to as the point-wise approach and the spectral-wise approach, respectively. It is important to note, that the altimetric MSS has a much higher resolution (roughly 10–15 km) compared with the GOCE-derived geoid (roughly 80 km). This difference in resolution leads to a non-random noise when N is subtracted from the MSS to obtain the MDT.

One way of making the MSS and N contain the same spectral content would be to truncate the MSS spherical expansion to the same maximum harmonic degree as the geoid. Two limitations arise here: firstly, the MSS is only defined over the ocean areas, making the spectral expansion of the MSS ill-conditioned; secondly, the truncation of the spherical harmonics at a certain degree leads to Gibbs phenomena. These limitations can be solved by combining the MSS and N as suggested in the approach by BINGHAM et al. (2008) or HUGHES et al. (2006). In this approach, the missing MSS data (land, polar gaps, small islands, etc.) is filled with geoid data in order to obtain the spectral representation of the MSS over the entire sphere. By completing the MSS in uncovered areas with the same geoid we used to determine the MDT, the Gibbs phenomena is strongly reduced.

We then have an unfiltered MDT resolving space scales limited by the geoid resolution. Two main error sources call for low pass filtering before the SGC can be derived from such an MDT. Because the spectral consistency between the MSS and geoid is matched only in theory, it might be expected that short wavelength gravity signals that are not captured by the geoid model are still being reflected in the altimetric MSS. Moreover, the signal-to-noise ratio dramatically decreases with increasing degrees, particularly for the geoid. Here, we investigate two filtering methods to smooth the noise from the MDT when the objective is to determine the SGC.

The first method, referred to as the spectral-wise method, is a convenient way to obtain an MDT, since spectral inconsistency has been theoretically solved; i.e., all signals corresponding to harmonic degrees higher than that of the GOCE geoid are omitted when the difference is computed. Therefore, we test and compare the filter for a spectral-wise MDT. When we hereinafter refer to the spectral-wise MDT we assume spectral consistency has been theoretically achieved using this approach. Hence, this surface will not be particularly noisy and we refer to it as the "easy case" for filtering.

The second method is referred to as the point-wise MDT. Here, we assume that the geoid and MSS have been evaluated point-wise in the space domain on a 0.5° grid and the MDT has been computed as the point-wise difference. Hence, all signals to the resolution of the MSS are retained by the method that makes the resulting MDT far noisier. This leads to residual noise due to the spectral inconsistency between the MSS and the geoid, which is not necessarily isotropic. In this sense, a point-wise MDT allows us to evaluate the anisotropic filter behavior in the presence of non-isotropic noise and represents a harder test than a spectral-wise MDT. A point-wise MDT would never be used in a practical calculation of the MDT given the reasons above. However, to explore the filter's behavior under harder conditions, we decided to use a point-wise MDT because results can be expected and easily interpreted, instead of a theoretical example that would eventually lack practical/physical meaning.

For the investigation, the MDT is practically evaluated on a half degree grid in which the land areas are set to NaN. This also means that scales down to 55 km are maintained at the equator. The SGC are computed from the MDT according to (1). Note that the Coriolis parameter vanishes close to the equator. For this reason, the solution in the band [5°S 5°N] was estimated as proposed by LAGERLOEF et al. (1999).

The PMF is frequently strongly forced by local errors. These make setting the parameter of the filter difficult, and sometimes it leads to results being not as good as would be desirable. One of the advantages of EED is the low influence that these local errors have on the filtering process, given that the gradient is implicitly smoothed by a Gaussian filter controlled by σ in (4). To illustrate this, in the following section we determine the MDT using both approaches, considering the spectral-wise MDT as an easy case for the filtering strategy and the point-wise MDT as a hard case.

The filtering was tested in the Northwestern Pacific area [1°N 40°N] × [120°E 190°E], shown in Fig. 1. In this region, three SGC systems are found: the strong Kuroshio Current, running north-eastward along the intersection between the Eurasian and the Philippine plate and being directed out to the Pacific Ocean from the Eastern Japanese coast; the westward North Equatorial Current (NEC), entering the area from the east at latitudes of around 10°N; and the Equatorial Counter Current (ECC), running eastward close to the Equator.

Figure 2 illustrates the MDTs derived using the spectral-wise and point-wise methods as well as the derived geostrophic currents. It is clearly seen how the point-wise method creates a much noisier MDT surface than the spectral-wise method due to dominant high frequency signals. For the point-wise-evaluated surface, bathymetry is reflected in the MDT as it appears in the MSS (ANDERSEN and KNUDSEN 2009), but not in the geoid due to the lower resolution of the GOCE geoid. This kind of noisy signal represents a hard test for the filter since this kind of non-random, non-isotropic signal must be smoothed out. Furthermore, bathymetric signals are frequently directional.

One way of validating the filtering is to compare estimated currents with currents from independent observations. We use a 1° × 1° long-term climatology record (LUMPKIN and GARRAFFO 2005) coming from in situ measurements for the near-surface circulation provided by satellite-tracked surface drifting buoys (SYBRANDY and NIILER 1991; NIILER 2001). The surface geostrophic velocity field (after the Ekman component is removed) is provided by the Global Drifter Program (GDP) assembly center (http://www.aoml.noaa.gov/phod/dac/drifter_climatology.html for a full description of the data field).

4. Analysis of the Filters

The difference between the three filters (Gaussian, PMF, and EED) and the ability of each filter to approximate the observed SGC are evaluated. We apply the following different approaches.

In order to perform an objective comparison of the filters, we should try to ensure that all three filtered surfaces are reflecting the same signals corresponding to the same degree of filtering. To

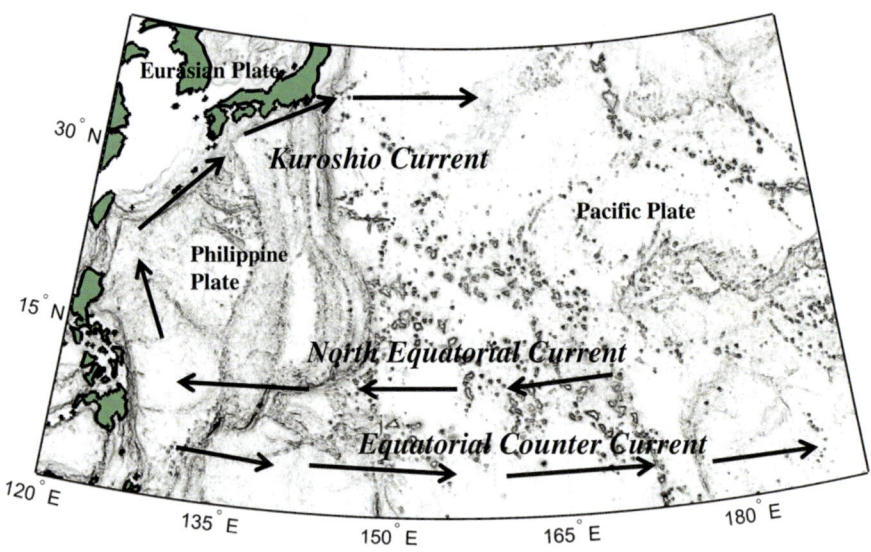

Figure 1

The Northwestern Pacific Ocean area of study. The three major currents systems (Kuroshio, North Equatorial and Equatorial Counter Currents) are displayed by *arrows*. Bathymetric features are *shaded* in *black* and major tectonic features are also indicated

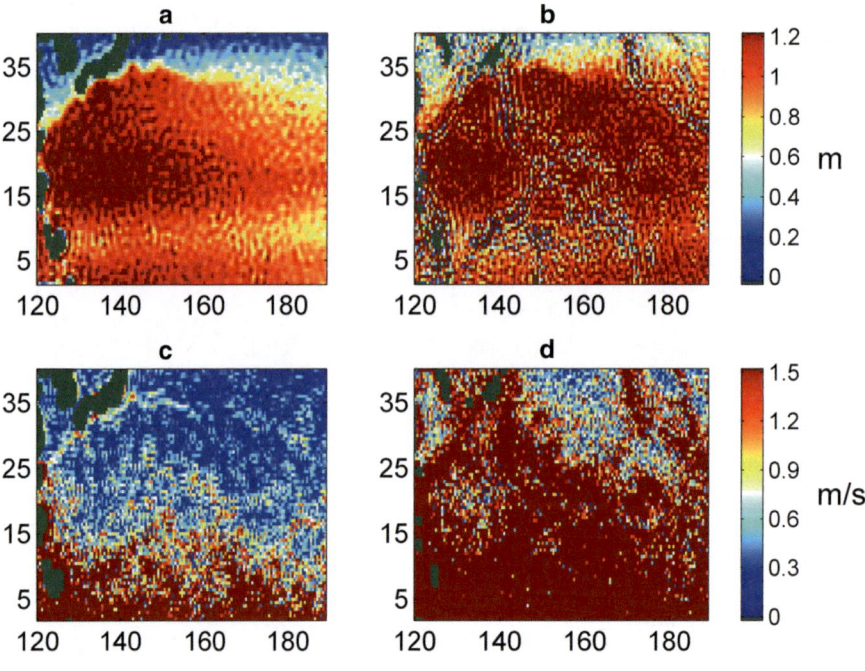

Figure 2

Unfiltered MDT as obtained by **a** the spectral-wise and **b** the point-wise methods. SGC derived from **c** the spectral-wise MDT and **d** the point-wise MDT

determine the degree of filtering, we can use the Gaussian filtered surface as reference. For the iterative filters, the degree of filtering increases with the number of iterations called t. In areas with a smooth gradient, where no dominating directions are found, non-isotropic filters should have a similar behaviour as the Gaussian filter and, therefore, (at some point of the iterative process) the EED or PMF should provide similar results to the Gaussian filter. This point determines the maximum number of iterations, T, for the EED and the PMF.

We selected the area [20°N 30°N] × [170°E 190°E] for this determination, as this area is, in theory, a zone free of currents, implying that the gradient should be smooth as the observed signal can be considered as noise. Then, the T is chosen as the value of t minimizing the root mean square (RMS) of the difference between the magnitude of the SGC from a non-isotropic-filtered MDT and the magnitude of the SGC from a Gaussian-filtered MDT. We will refer this criterion to stop the filter as the "common degree" criterion.

Removing the noise from the MDT also requires different degrees of filtering depending on the geographical region in which we are working, since higher latitudes provide denser satellite data and, therefore, more accurate results. KNUDSEN et al. (2011) reported a half-wave length of 140 km for the Gaussian filter when applied to a global MDT estimate that is similar to the findings in this study. At higher latitudes, SANCHEZ-REALES et al. (2012) showed how all major currents (located at middle latitudes) are accurately depicted using a half wavelength of just 83 km. For our investigation in the Western Pacific close to the Equator and for filtering the Kuroshio and the NEC, we found a half wavelength of 111 km (1° of resolution in the equatorial band) reasonable (Fig. 3b). This picture clearly depicts flow directions over the entire area and, therefore, we also set $\sigma = 111$ km to smooth the structure tensor in (4).

In order to investigate the capabilities of each filter, we compare these with in situ observations. Because of different inherent characteristics, the requirements to find the optimal filtering could also be different. Consequently, we follow a different strategy here. Instead of fixing a certain degree of filtering, we define the T for which the SGC best matches the in situ measurements, considering these as "true" values. Here, we select a zone with strong flow, more specifically, the area [30°N 40°N] × [125°E 155°E] where the Kuroshio flows along the ridge produced by the intersection of the Eurasian and Philippine plates and the Japanese coast (Fig. 1). Then, the value of T is set as the t minimizing the RMS of the difference between the magnitude of the SGC derived from a filtered MDT and the magnitude of the in situ velocities. We hereinafter refer to this criterion as the "optimum SGC approximation" criterion.

We compare the PMF and EED results with the Gaussian filter in Fig. 3. Here, three different half wavelength values, 83.3 km (best available), 111.1, and 140 km (globally required) are shown. The Gaussian filter is performed in the spectral domain (see Appendix 1.1).

For the anisotropic filters, the k parameter must be fixed. As discussed above and since the structure tensor is regularized in (4) by $\sigma = 111$-km half wavelength, the EED provides similar results for a wide range of k. Contrary to that, the PMF (or EED with $\sigma = 0$) is strongly dependent on the choice of k and in some cases it is difficult to find a value. In this case, all results from the EED process are shown with a k relative to the PMF requirements initially chosen by trial and error: Once the filter is run and T is chosen as described above, results have to be checked for t in a neighborhood of T while looking for extremely sharp edges (indicating a higher value of k would be more adequate) or excessive smoothing

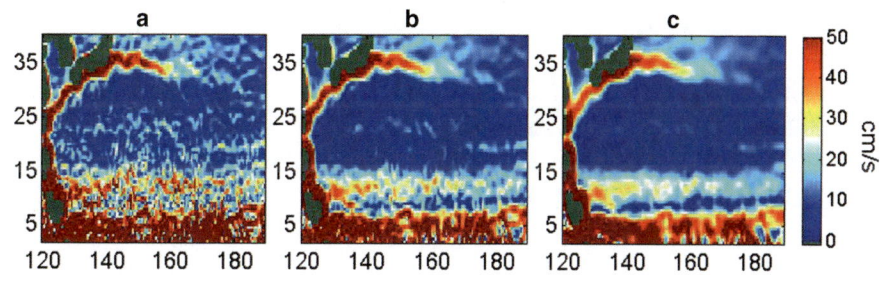

Figure 3

MDT as filtered by a Gaussian filter with a half wavelength of **a** ∼83 km; **b** ∼111 km; and **c** 140 km

of the stronger signals (indicating that a smaller value of k would be more adequate).

In the following, results from MDTs obtained by the spectral-wise and the point-wise methods are provided to represent the easy and the hard case for the filtering process.

4.1. Filter Capabilities

4.1.1 Spectral-Wise MDT (Easy Case)

The MDT determined by the spectral-wise method is not a very noisy surface (Fig. 2a) and particularly not for latitudes higher than $\sim 25°N$. We found by trial and error that a value of $k = 0.3$ is a good value to obtain representative results with the PMF.

To compare the filters, the same degree of filtering is defined for the three filters, the same degree as the one determined by a Gaussian filter with an ~ 111-km half wavelength. Figure 4a, b show the results of the EED and PMF where a "common degree" of filtering is reached for $T = 67$ and $T = 17$, respectively. Their differences are displayed in Fig. 4c. Some differences arise precisely on the stream flow

along the Kuroshio. The negative value implies that the PMF provides higher values than EED for the estimated velocity amplitudes just on the crest of the current. Moreover, the stream flow becomes wider by EED than by the PMF as inferred from the positive yellow value all along both sides of the crest.

Most of these differences disappear if we design the filters to stop by the optimum SGC approximation criterion rather than the same degree of filtering. Figures 4d, e show the magnitude of the velocities using EED, $T = 49$, and PMF, $T = 16$, respectively. Again, their differences are plotted in Fig. 4f. Now, the Kuroshio Current is hardly identified in the difference maps and only minor differences arise at the eastern coast of Japan due to enhancement of the central ridge by the PMF. In general, it can be seen how most of the differences are positive, corresponding to higher velocities determined by EED compared with the PMF. This is because EED maintains shorter wavelengths than the PMF does, as can be observed when comparing Fig. 4d with Fig. 4e.

The results for the less noisy MDT do not allow us to favour EED over the PMF or vice versa. With this easy case, we have tried to highlight the differences

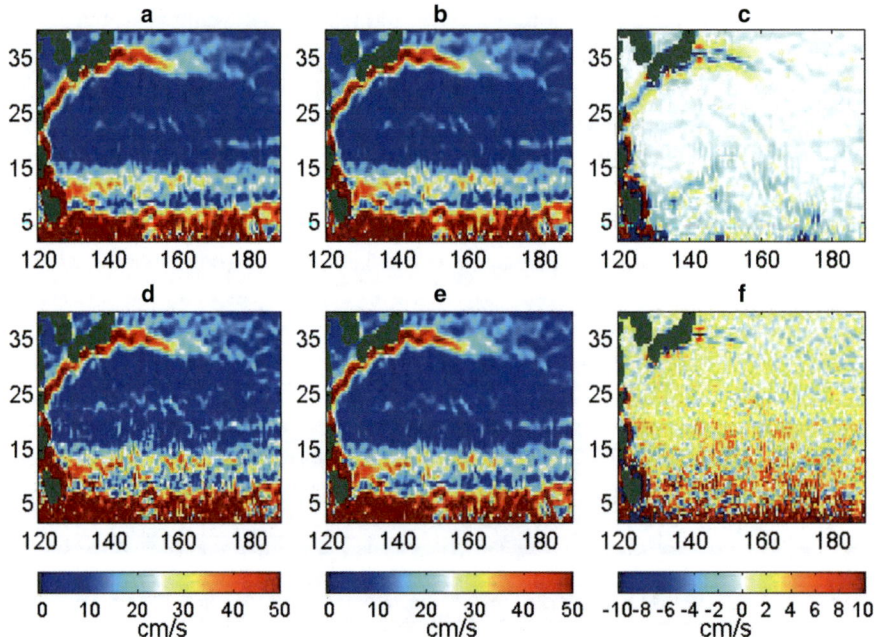

Figure 4

SGC velocities derived from the spectral-wise MDT (Fig. 2a) as filtered by the "common degree" criterion for the **a** EED and **b** PMF approaches; and **c** their differences. **d** EED- and **e** PMF-induced results by the "optimum approximation" criterion, and **f** their differences

between the filters. EED seems to retain weaker signals that are removed by the PMF; and the PMF tends to enhance the ridges of a stream line of flux. Nevertheless, a hard case (noisy surface) will show the real advantages of the EED method over the PMF in the following.

4.1.2 Point-Wise MDT (Hard Case)

MDT obtained by the point-wise method is very noisy as it retains some short-scale signals from the MSS that are not present in the much smoother geoid, and, consequently, these signals act here as non-random, non-isotropic noise. Since parts of the Kuroshio current flow along strong gradients of the submarine topography, and these are reflected in the short wavelengths of the MSS, the point-wise difference MSS-N can affect the estimated SGC. Figure 5 shows the SGC velocities from EED- and PMF-filtered MDTs and their differences, respectively. In this case, we found that a $k = 0.8$ is a good value for the PMF. Here, we show the results corresponding to the optimum SGC approximation criterion found for $T = 42$ for EED and $T = 21$ for the PMF. A simple visual inspection of the results shows a clearly more physical behaviour of EED filtering. The PMF-filtered surface introduces several unrealistic signals mainly located along the areas where the differences in resolution between the MSS and the geoid are strong due to a strong MSS/geoid signal (along the intersection between the Pacific and Philippine plates, see Fig. 1).

The EED approach is shown to be much closer to its corresponding easy case in Fig. 4d than the PMF-filtered results. The PMF, on the contrary, shows several artificial discrepancies between this hard case, Fig. 5b, and its corresponding easy case, Fig. 4e. This fact illustrates the main advantage of the EED approach compared with the PMF approach: the relatively low influence of the local errors. As discussed above, the PMF does not control the diffusion's direction but only the magnitude of the diffusion flux from a noisy MDT. On the contrary, the EED controls both magnitude and direction of diffusion from a regularized MDT, implying that EED filters in the direction of the flow, diminishing the influence of the individual errors.

4.2. Comparison with Drifter Data

To provide additional validation of the results, comparisons will be made with in situ measurements provided by drifting buoys. BINGHAM (2010) reported improvement of the PMF diffusive filtering approach over conventional Gaussian smoothing at regions of strong currents. We firstly center our attention on the EED and PMF filtering techniques and their ability to determine the SGC by comparing the results induced from filtered MDTs, as considered in previous sections, with independent SGCs computed using only drifter measurements. Figure 6 shows the RMS differences, by latitude, between results provided by each diffusive filter and drifter data. Figures 6a, b, correspond to results provided by the common degree

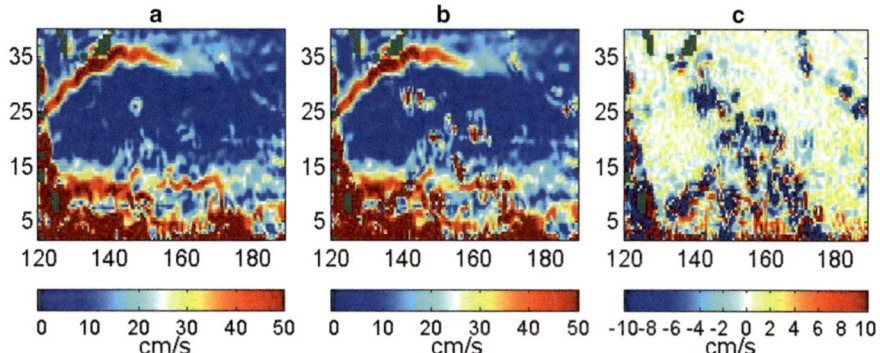

Figure 5
SGC velocity amplitudes inferred from the point-wise MDT (Fig. 2b) as filtered by **a** EED and **b** PMF approaches when N is selected to obtain the best approximation of velocity magnitudes for the Kuroshio flow provided by in situ measurements. **c** Differences between the EED and PMF approaches

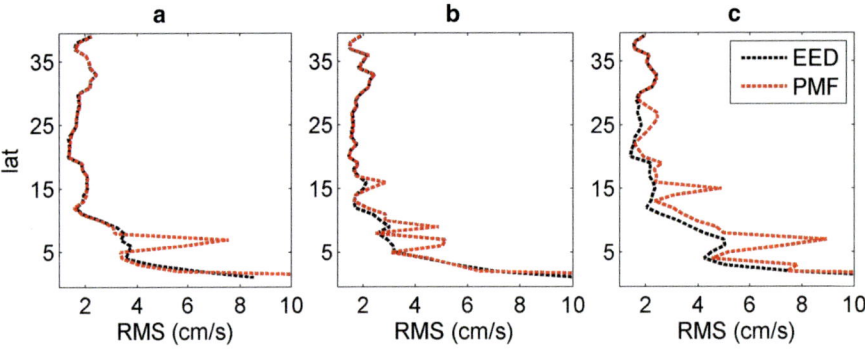

Figure 6

RMS of differences between in situ and SGC velocity amplitudes inferred from the spectral-wise MDT for **a** "common degree" (Fig. **4a–c**), and **b** "optimum approximation" (Fig. **4d–f**) criteria; and from the point-wise MDT for a **c** "optimum approximation" (Fig. **5a–c**) criterion for both the PMF (in *red*) and EED (in *black*) filters

criterion, spectral-wise case (from Fig. 4a–c) and point-wise case (from Fig. 4d–f), respectively.

From Fig. 6a, b, it can be observed that at higher latitudes, PMF and EED results are very similar, and the RMS is smoothly increasing the lower the latitude for EED, while for the PMF, some RMS peaks were found around 15°N and below. Those peaks are due to short wavelengths or spatially uncorrelated noise that influences the PMF's performance. This kind of error is more evident in the point-wise case, but still appeared in the spectral-wise case (the single peak around 6°N can be clearly identified in Fig. 4c around 6°N 175°E). This advantage of the EED filter over the PMF approach is emphasized in Fig. 6c, which shows the residuals when the optimum filtering criterion is applied to both filters. It is immediately obvious that EED and the PMF provide similar results at the reference region used to set the filters, with latitude limits of 30–40°N. Out of this reference region, latitudes 30°N and lower, the RMS is always reduced by the EED filter in comparison to the PMF. The highest improvements occur where short wavelength noise is present, as evidenced by the smoothness of the EED filter RMS curve in relation to the PMF RMS curve at lower latitudes (Fig. 6c).

In order to compare the Gaussian, PMF and EED filters from a more general point of view, we also estimated the RMS differences between the results obtained by the three filters and results by drifter data for the full area (with limits 1°–40°N, 120°–190°E), and in all the scenarios presented in previous sections. We summarize these results in Fig. 7 as

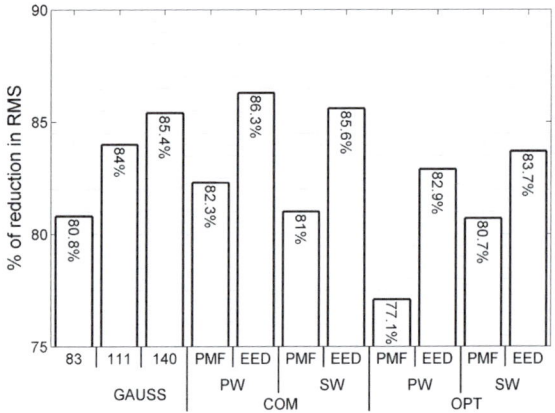

Figure 7

Percentage of reduction in RMS differences with buoy measurements

the percentage of RMS reduction for each of the cases. It can be observed how in each case (except the PMF for the hardest test, PW-OPT) results are within a range of approximately 5 % and, in all cases, EED enhances the PMF results. For the common degree criterion, the EED and PMF filters were set to have a degree of filtering similar to a Gaussian filter with a 111-km half wavelength (with an RMS reduction of 84 %), and in this case we found that the EED filter improves the Gaussian results while the PMF does not. On the other hand, for the optimum criterion filtering approach, we found that both diffusion filters could do no better in terms of global RMS reduction than Gaussian smoothing, due to short wavelength noise that is maintained out of the reference region. Despite this, EED shows similar percentages of RMS reduction as a Gaussian filter with a 111-km half wavelength. In general, EED results are shown to be better than those from a Gaussian filter defined with a half wavelength shorter than about 111 km and poorer than those from a Gaussian smoothing with higher cut-off frequency. At this point, it is important to take into account the blurring effect a high half wavelength value for a Gaussian filter will produce on those signals that eventually are the target of study (edges).

Quantifying differences, EED improves noise reduction about 3–5 % (in terms of RMS differences) when compared to the PMF; and around 1.5–2 % when compared to a similar degree of filtering as that of the Gaussian filter (111 km). These differences in percentage of RMS reduction are amplified when shorter wavelengths are used in a Gaussian filter, as would be required to preserve narrower stream currents. These differences increase up to 5–6 % in favor of the EED when compared to an 83-km half wavelength Gaussian filter used by SANCHEZ-REALES *et al.* (2012) for estimating SGC at higher latitudes far from the equator. Cases showing the best EED performance provided a noise reduction similar to a Gaussian filter with a 140-km half wavelength.

5. Conclusions

Non-linear isotropic filtering approaches, such as the PMF, have been previously proven to be more accurate than isotropic Gaussian filters for MDT filtering in order to preserve the signal at strong gradient areas (BINGHAM 2010). In this study, we present an alternative approach based on EED filtering as a step further in nonlinear anisotropic filtering applied to the ocean MDT, and its associated currents. The EED is a generalization of the PMF by introducing both a regularization parameter (to smooth the noise when determining the direction and magnitude of the flow) and a structure tensor matrix (that will rotate the direction of the filtering to be directed along the flow). Therefore, the proposed EED filter keeps all the PMF advantages over the standard Gaussian smoothing, preserving the signal at the strongest gradients. In addition, EED provides some advantages over the PMF.

EED filtering is less influenced by outliers, short wavelengths or spatially uncorrelated errors. This is particularly evident when there are strong signals in the MDT. Given the PMF only uses the nearest neighbours to filter a single location, errors at such neighbours are retained through the iterations leading to errors at the final surface. On the contrary, the EED filters in the direction of the flow, attenuating the individual errors' influence due to the fact that the magnitude and direction of the diffusion is determined from a smoothed surface.

Both the EED and PMF depend on a user-defined k parameter. This parameter can be interpreted as a kind of threshold from where the gradient will be considered as an edge. In addition, the EED needs a regularization parameter σ to be defined. This σ permits smoothing of the MDT for a better approximation to the true value of the direction and magnitude of the SGC. These values are then used by the EED to filter the MDT to preserve the strong current flow as well the edge-like structures. Contrary to that, the direction and magnitude of diffusion for the PMF is determined from the MDT gradient. Therefore, the PMF is strongly conditioned to the strength and structure of the noise in the MDT. This makes the PMF strongly sensitive to variations of the k parameter, providing quite different results for small changes in k. Contrary to that, since the noise is smoothed for all scales shorter than σ, the EED provides similar results for a wider range of k values, making the filtering design more robust.

For the studied region in the Northwestern Pacific Ocean, we found small differences between the EED, PMF and Gaussian filters in terms of noise reduction (RMS differences). In all cases, EED improves the PMF results. In the hard case, the EED performs as well as a Gaussian filter with a half wavelength of approximately 111 km, and in the standard case, the EED provided similar results as a Gaussian filter with a half wavelength of approximately 140 km.

The advantages of EDD over the PMF filtering strategy suggest that EED should be preferred, particularly in the presence of a noisy signal, when applied to an ocean MDT with the objective of determining the SGC. According to the results in this study, the EED provide a higher global RMS reduction in relation to PMF when validating the results with in situ measurements. Similarly, RMS by latitudes are shown to be comparable for both diffusive filtering at the reference region, and increasing with lower latitudes, with a smooth behaviour for the EDD, while the PMF peaks at noisier areas (e.g., 6°N and 15°N). Results provided by the PMF for the spectral-wise MDT showed some strong discrepancies when compared with the point-wise MDT-induced velocities.

The main disadvantage of EED over the PMF is connected to the number of mathematical operations involved in the process due to the increased complexity of the method. In addition, smoothing is much slower for EED (for most of the test, we performed the required EED of around 2–4 times the number of iterations of the PMF to converge to some predefined criterion). This leads to a drastic increase in the computational time: while the PMF can perform 300 iterations in seconds on any conventional computer, EED requires around 2 h.

Here, we have investigated the capabilities of EED in preserving strong gradients of a geodetic MDT in order to determine the SGC. This study was carried out by focusing on three strong currents in the Northwestern Pacific Ocean, finding promising results. Nevertheless, further work is required in order to study the behaviour of the EED filter for shorter space scales and to determine weaker currents.

Acknowledgments

This work is supported by projects AYA2009-07981 and CGL2010-12153-E from the Spanish Department of Science and Innovation (MICINN). ESA is acknowledged for the use of the GOCE data; NOAA for the use of drifter data; and AVISO for the use of the altimetric data.

Appendix: Practical Implementation of the Filters

1.1 Gaussian Filter

The Gaussian filter was applied in the spectral domain as proposed by JEKELI (1981). This is by filtering the spherical harmonic coefficients W_n by,

$$W_{n+1} = -\frac{2l-1}{b}W_n + W_{n-1} \qquad (7)$$

where n denotes the harmonic degree; $W_0 = 1/2\pi$, $W_1 = W_0(1 + \exp(-2b))/(1 - \exp(-2b)) - W_0/b$, and $b = \log 2/(1 - \cos(r/a))$, r being the half wavelength and a the mean equatorial radius of the Earth.

1.2 Perona and Malik Filter

Following the same notation in Sect. 2, the PMF sets (2) as

$$I_t = \mathrm{div}(g(|\nabla I|^2) \cdot \nabla I) \qquad (8)$$

where g(.) was defined in (3). A simple discretization of (8) is provided by PERONA and MALIK (1990):

$$I_{i,j}^{t+1} = I_{i,j}^t + dt[c_N \Delta_N I_{i,j} + c_S \Delta_S I_{i,j} + c_E \Delta_E I_{i,j} + c_W \Delta_W I_{i,j}]^t \qquad (9)$$

which defines the iterative process to filter the original grid $I_0 = $ MDT for any location (i, j). In (9), Δ_m indicates the nearest neighbor differences for $m = $ N,S,E,W (north, south, east, and west; e.g. $\Delta_N = I_{i+1,j} - I_{i,j}$). The conduction coefficients cm are estimated from (3), being $c_m = g(\Delta_m)$. Here the time-step is required to be $0 < dt \le 1/4$ for the numerical scheme to be stable. It was set to $dt = 0.1$ as for the EED.

1.3 Edge Enhancing Diffusion

To perform the EED filter we first determine the structure tensor J_0 and convolve it with a Gaussian filter K_σ to obtain the regularized structure tensor J_σ. In this case, the pre-filtering implied by K_σ is carried out in the geographical domain and applied to MDTs determined as described in data section. Weighting the eigenvalues from J_σ by (3) the diffusion tensor D is determined as described in Sect. 2. Then, if

$$D = \begin{pmatrix} a & b \\ b & c \end{pmatrix}, \qquad (10)$$

the diffusion Eq. (2) can be rewritten as

$$I_{i,j}^{t+1} = I_{i,j}^t + dt[\partial_x(a\partial_x I_{i,j}) + \partial_x(b\partial_y I_{i,j}) + \partial_y(b\partial_x I_{i,j}) + \partial_y(c\partial_y I_{i,j})]^t \qquad (11)$$

which defines the iterative process to filter the original grid $I_0 = $ MDT for any location (i, j). Partial differential equations in (11) were discretized as following central differences. That is,

$$\partial_x(a\partial_x I_{i,j}) = \frac{1}{2}\left[a_{i+1,j}\frac{I_{i+1,j}-I_{i,j}}{2}+a_{i-1,j}\frac{I_{i-1,j}-I_{i,j}}{2}\right] \tag{12}$$

$$\partial_x(b\partial_y I_{i,j}) = \frac{1}{4}\left[b_{i+1,j}\frac{I_{i+1,j+1}-I_{i+1,j-1}}{4}\right. \\ \left.+b_{i-1,j}\frac{I_{i-1,j+1}-I_{i-1,j-1}}{4}\right] \tag{13}$$

$$\partial_y(b\partial_x I) = \frac{1}{4}\left[b_{i,j+1}\frac{I_{i+1,j+1}-I_{i-1,j+1}}{4}\right. \\ \left.+b_{i,j-1}\frac{I_{i+1,j-1}-I_{i-1,j-1}}{4}\right] \tag{14}$$

$$\partial_y(c\partial_y I) = \frac{1}{2}\left[c_{i,j+1}\frac{I_{i,j+1}-I_{i,j}}{2}+c_{i,j-1}\frac{I_{i,j-1}-I_{i,j}}{2}\right] \tag{15}$$

The time-step dt in (11) is required to be $dt < \left(\sum_l 1/d_l^2\right)^{-1}$ where d_l denotes distance from $I_{i,j}$ to any location I_l involved in its filtering process (WEICKER and BENHAMOUDA 1997). As we are applying EED to filter $I_{i,j}$ from its eight nearest neighbors, $dt < 1/6$. We set $dt = 0.1$ for the PMF.

REFERENCES

ALBERTELLA A., and RUMMEL R. (2009) *On the spectral consistency of the altimetric ocean and geoid surface: a one-dimensional example*. J Geodesy, *83*:805–815. doi:10.1007/s00190-008-0299-5.

ANDERSEN O.B., and KNUDSEN P. (2009) *DNSC08 mean sea surface and mean dynamic topography*. J Geophys Res, *114*, C11. doi:10.1029/2008JC005179.

BINGHAM R., HAINES K., and HUGHES C. (2008) *Calculating the ocean's mean dynamic topography from a mean sea surface and a geoid*, J Atmos Ocean Tech, *25* (10),1808–1822. doi:10.1175/2008JTECHO568.1.

BINGHAM R.J. (2010) *Nonlinear anisotropic diffusive filtering applied to the ocean's mean dynamic topography*. Remote Sens Lett, *1*:4, 205–212. doi:10.1080/01431161003743165.

BINGHAM RJ., KNUDSEN P., ANDERSEN O., and PAIL R. (2011) *An initial estimate of the North Atlantic steady-state geostrophic circulation from GOCE*. Geophys Res Lett, *38*:L01606. doi:10.1029/2010GL045633.

CRANK, J. (1975) The mathematics of diffusion. (Oxford Science Publications, 1975).

EGG-C (European GOCE Gravity Consortium) (2009) GOCE high level processing facility: GOCE level 2 product data handbook (http://www.esa.int/esaLP/GTCVCKSC_LPgoce_0.html).

HERNANDEZ F., and SCHAEFFER P. (2001) The CLS01 MSS: A validation with the GSFC00.1 surface. Tech rep., CLS, Ramonville St Agne, 14 pp.

HUGHES C.W., and BINGHAM R.J. (2006) An oceanographers guide to GOCE and the geoid. Ocean Sci Discuss, *3*, 1543–1568. doi:10.5194/osd-3-1543-2006.

JEKELI C. (1981) Alternative methods to smooth the Earth's gravity field, Tech. Rep. 327, Dep. of Geod. Sci. and Surv., Ohio State Univ., Columbus.

KNUDSEN P., ANDERSEN O.B., BINGHAM R.J., and RIO M.H. (2011), *A global mean dynamic topography and ocean circulation estimation using a preliminary GOCE gravity model*. J Geodesy. doi:10.1007/s00190-011-0485-8.

LAGERLOEF G.S.E., MITCHUM G.T., LUKAS R.B., and NIILER P.P. (1999) *Tropical Pacific near-surface currents estimated from altimeter, wind, and drifter data*, J Geophy Res, *104*(C10), 23,313–23,326. doi:10.1029/1999JC900197.

LUMPKIN R., and GARRAFFO Z., (2005) *Evaluating the Decomposition of Tropical Atlantic Drifter Observations*. J Atmos Ocean Techn I *22*, 1403–1415. doi:http://dx.doi.org/10.1175/JTECH1793.1.

NIILER P.P. (2001) The world ocean surface circulation. In Ocean Circulation and Climate, G. Siedler, J. Church and J. Gould, eds., Academic Press, Volume 77 of International Geophysics Series, 193–204. doi:10.1016/S0074-6142(01)80119-4.

PAIL R., GOIGINGER H., MAYRHOFER R., SCHUH W.D., BROCKMANN J.M., KRASBUTTER I., HÖCK E., and FECHER T. (2010) *GOCE grafity field model derived from orbit and gradiometry data applying the time-wise method*. Proc. ESA Living Planet Symposium. Bergen,Norway, 28 June–2 July 2010 (ESA SP-686, December 2010).

PAIL R., BRUINSMA S., MIGLIACCIO F., FOERSTE C., GOIGINGER H., SCHUHAND W.D., HOECK E., REGUZZONI M., BROCKMANN J.M., ABRIKOSOV O., VEICHERTS M., FECHER T., MAYRHOFER R., KRASBUTTER I., SANSO F., and TSCHERNING C. (2011) First GOCE gravity field models derived by three different approaches. J Geodesy, *85*:819–843. doi:10.1007/s00190-011-0467-x.

PERONA P., and MALIK J. (1990) *Scale-Space and Edge Detection Using Anisotropic Diffusion*. IEEE transactions on Pattern Analysis and Machine Intelligence, *12*, pp. 629–639.

SANCHEZ-REALES J.M., VIGO M.I., JIN S.G., and CHAO B.F. (2012) *Global Surface Geostrophic Ocean Currents Derived from Satellite Altimetry and GOCE Geoid*. Marine Geodesy, *35*:sup1, 175–189.

SLOBBE D.C., SIMONS F.J., and KLEES R. (2012) *The spherical Slepian basis as a means to obtain spectral consistency between mean sea level and the geoid*. J Geodesy. doi:10.1007/s00190-012-0543-x.

SYBRANDY A.L., and NIILER P.P. (1991) WOCE/TOGA Lagrangian drifter construction manual. WOCE Rep. 63, SOI Ref. 91/6, 58 pp, Scripps Inst. of Oceanogr., La Jolla, Calif.

TAPLEY B., RIES J., BETTADPUR S., CHAMBERS D., CHENG M., CONDI F., GUNTER B., KANG Z., NAGEL P., PASTOR R., PEKKER T., POOLE S., and WANG F. (2005) *GGM02-An improved Earth gravity field model from GRACE*. J. Geodesy, doi:10.1007/s00190-005-0480-z.

WAHR, J., MOLENAAR, M., and BRYAN, F., 1998. *Time-Variability of the Earth's Gravity Field: Hydrological and Oceanic Effects and*

Their Possible Detection Using GRACE, J. Geophys. Res., *103*, 30205–30230.

WEICKERT J. (1996) *Theoretical foundations of anisotroic diffusion in image processing*. Computing, Suppl.11, pp. 221–236.

WEICKERT J., and BENHAMOUDA B. (1997) Why the Perona-Malik filter works. Københavns Universitet, Datalogisk Institut.

WEICKERT J. (1998) Anisotropic Diffusion in Image Processing. ECMI. B.G. Teubner, Stuttgart.

(Received March 28, 2014, revised February 4, 2015, accepted February 5, 2015, Published online February 14, 2015)

Pure Appl. Geophys. 173 (2016), 885–907
© 2015 Springer Basel
DOI 10.1007/s00024-015-1185-8

❙Pure and Applied Geophysics

A Lagrange-Galerkin *hp*-Finite Element Method for a 3D Nonhydrostatic Ocean Model

PEDRO GALÁN DEL SASTRE[1] and RODOLFO BERMEJO[2]

Abstract—We introduce in this paper a Lagrange-Galerkin *hp*-finite element method to calculate the numerical solution of a nonhydrostatic ocean model. The Lagrange-Galerkin method yields a Stokes-like problem the solution of which is computed by a second-order rotational splitting scheme that separates the calculation of the velocity and pressure, the latter is decomposed into hydrostatic and nonhydrostatic components. We have tested the method in flows where the nonhydrostatic effects are important. The results are very encouraging.

Key words: Navier-Stokes equations, nonhydrostatic ocean models, Lagrange-Galerkin, *hp*-finite element method, second-order backward difference formula, rotational based splitting.

1. Introduction

Since the early stages of computational oceanography in the mid 60s of the past century, numerical ocean models are based on the hydrostatic primitive equations (HPE). It is well accepted that features of the ocean circulation whose spatial scales range from 10 km to 10,000 km, see MARSHALL *et al.* (1997), are accurately described by the HPE models. This range of scales comprises important phenomena such as the wind-driven ocean gyres and the geostrophic eddies and rings associated with its hydrodynamic stability. However, the hypotheses of the HPE models break down for spatial scales extending from 10 to 1 km where phenomena that are important in regional, coastal, and climate models, such as wind- and buoyancy-driven turbulence in the upper mixed layer, deep water convection, and baroclinic instability, take place. In oceanography, the nonhydrostatic models consist of the fully 3D Navier-Stokes equations formulated in a rotating framework for a fluid of variable density (under the assumptions that the ocean is a slightly compressible Newtonian fluid and the validity of the Boussinesq approximation). These equations are completed with advection–diffusion equations for temperature and salinity plus the equation of state for the density. So that, the nonhydrostatic models are more general than HPE models and can be applied to study both small- and large-scale phenomena. To be aware of the generality of the nonhydrostatic models, it is worth recalling that the HPE models start with the 3D Navier-Stokes equation and make use of the Boussinesq and incompressibility assumptions, plus the thin layer approximation and the assumption of hydrostatic balance in the third component of the momentum equations. The consequences of the HPE assumptions [see MARSHALL *et al.* (1997) for a thorough explanation] are (1) assuming the rigid lid condition holds at the upper surface of the ocean, that is, the vertical component of the velocity $w = 0$ at the upper surface, implies that the constraint, div $\mathbf{u} = 0$, where $\mathbf{u} = (u, v, w)$ is the velocity vector, becomes a non-local constraint of the form

$$\int_{-H(x,y)}^{0} \left(\frac{\partial u}{\partial x}(x, y, z) + \frac{\partial v}{\partial y}(x, y, z) \right) \mathrm{d}z = 0,$$

where H is a continuous function that describes the bottom topography; (2) the vertical component of the velocity of the flow w is a diagnostic variable that has to be calculated through the vertical integration of the divergence of the horizontal components (u, v), that is,

[1] Departamento de Matemática Aplicada, E.T.S. Arquitectura, Universidad Politécnica de Madrid, Avda. Juan de Herrera 4, 28040 Madrid, Spain. E-mail: pedro.galan@upm.es
[2] Departamento de Matemáticas del Área Industrial, E.T.S.I. Industriales, Universidad Politécnica de Madrid, c/ José Gutiérrez Abascal 2, 28006 Madrid, Spain.

$$w(x, y, z) = \int_z^0 \left(\frac{\partial u}{\partial x}(x, y, s) + \frac{\partial v}{\partial y}(x, y, s) \right) ds,$$

consequently, the numerical calculation of w is prone to large numerical errors (because it is well known that the error in the calculation of the numerical derivative of approximated variables might be large, particularly in the case that the horizontal velocity is not smooth enough); (3) the pressure at the surface level is calculated by solving a two-dimensional Poisson equation. From a computational point of view, and taking into consideration the computer technology of 50–60 years ago, an era in which the computers had a limited storage memory and a low speed to perform arithmetic calculations, HPE models were perhaps the only affordable choice to numerically simulate the large scale ocean dynamics for the following reasons: (1) they model large scale phenomena, so that it is not necessary to use very fine grids and, consequently, does not require a large storage memory capacity; (2) the compromise between reduced capacity for storage and low speed for arithmetic calculations leads to the use of explicit schemes, in particular, the so-called leap-frog scheme (a second-order scheme), as time integrators, and the splitting of the equations for the (u, v) components of the velocity in barotropic and baroclinic modes. The barotropic mode is essentially described by a set of 2D equations whose integration requires, for stability reasons, a shorter time step than the baroclinic mode which is described by a set of 3D equations.

The steady and rapid developments in both computer technology and numerical methods have opened the possibility to employ nonhydrostatic models for ocean studies, see to this respect CASULLI and ZANOLLI (2002), KANARSKA et al. (2007), MARSHALL et al. (1997). All these models use finite volume for space discretization and explicit schemes for the advective, Coriolis and diffusive terms, except CASULLI and ZANOLLI (2002) that uses a semi-implicit scheme for the vertical viscous terms. This fact imposes, for stability reasons, a restriction on the size of the time step Δt that sometimes may be quite severe. The pressure is expressed as sum of the surface pressure and the hydrostatic, baroclinic, and nonhydrostatic components; the calculations of the surface pressure and the nonhydrostatic component are carried out via

a fractional step procedure that reduces the overall order of the method to first order in time.

To integrate the equations of the nonhydrostatic model, we propose a second-order projection method, which was introduced and analyzed in GUERMOND and SHEN (2003), that we extend for non-isotropic fluids, in combination with a Lagrange-Galerkin (LG) method for time discretization along the trajectories of the fluid particles and higher order hp-finite element method for the discretization in space of the differential operators. The method was introduced by the authors in GALÁN DEL SASTRE and BERMEJO (2011) to integrate the Navier-Stokes equations for isotropic fluids with constant density. The discretization of the total derivative operator backward in time along the fluid trajectories yields a scheme that has a large stability region, so that the size of Δt can be chosen for accuracy rather than for stability reasons. On the other hand, the use of projection (or splitting as they are also known) methods to integrate the incompressible Navier-Stokes equations aims to overcome the difficulty posed by the fact that the velocity is coupled with the pressure through the divergence equation. Projection methods decouple at each time step the pressure and velocity into a series of elliptic problems that can be solved efficiently, in particular, when projection methods are combined with LG methods because in this case the velocity is the solution of a well conditioned symmetric elliptic problem. Then, regarding the method for the nonhydrostatic model, the projection method yields a Poisson equation for the nonhydrostatic component of the pressure, and a well-conditioned linear symmetric elliptic problem for the velocity. The first projection methods were introduced by CHORIN (1968) and TEMAM (1969) and they can be viewed as fractional step methods of order $O(\Delta t)$. Since then several authors have proposed different projection methods with the main idea of improving the order while keeping the efficiency, see GUERMOND et al. (2006) for a fine overview of the literature on projection methods for incompressible flows.

The type of hp-finite element used in our method is the one generated by modal expansions. We have chosen modal expansions in contrast to nodal expansion as many authors do, see for instance ISKANDARANI et al. (2003), where the authors develop a 3D *hp*-finite element model of the HPE, for the following reasons: (1) the basis is generated by a hierarchical set of polynomials and this yields an algorithm for the calculations associated to the LG method more efficient than the one produced by nodal basis, see GALÁN DEL SASTRE and BERMEJO (2011); (2) the stiffness matrix of the Laplace operator is quasi diagonal when the basis of the *hp*-finite element is modal; this property is very relevant regarding the efficiency to solve the equations for the pressure and the velocity, particularly, in 3D models.

In this paper, we show the first results of a new model under development so that we focus on the behavior of our method on density driven flows because they can be hard tests and, which is more important, are relevant for many ocean phenomena such as internal wave formation, deep water convection, slope bottom currents, water interchange through the Strait of Gibraltar, and so on.

The organization of the paper is as follows. In Sect. 2, we introduce the nonhydrostatic ocean model equations. A description of the *hp*-finite element method and the Lagrange-Galerkin formulation is done in Sect. 3, where both methods are combined to solve numerically the model. In Sect. 4, some tests are presented to show the efficiency of the proposed numerical approach.

2. Governing Equations

Let $D \subset \mathbb{R}^3$ be a bounded ocean domain with sufficiently smooth boundary ∂D and let $[0, \Theta]$ be a time interval. As we mentioned above, we shall consider in this paper that the upper surface of the ocean is rigid, which means that the vertical component of the velocity w is zero there. Let $\omega \subset \mathbb{R}^2$ be an open bounded domain with smooth boundary $\partial \omega$ and let $H : \omega \to \mathbb{R}^+$ be a piecewise C^1 function such that there exist positive constants H_1 and H_2 that satisfy $H_2 \geq H(x, y) \geq H_1 > 0$. The domain D can be defined as

$$D = \left\{ (x, y, z) \in \mathbb{R}^3 : (x, y) \in \omega \text{ and } -H(x, y) < z < 0 \right\}.$$

The boundary $\partial D = \Gamma_s \cup \Gamma_b \cup \Gamma_l$ where $\Gamma_s = \omega \times \{0\}$ represents the upper surface of the ocean, $\Gamma_b = \left\{ (x, y, z) \in \mathbb{R}^3 : (x, y) \in \omega \text{ and } z = -H(x, y) \right\}$ denotes

the bottom boundary, and $\Gamma_l = \partial D - (\Gamma_s \cup \Gamma_b)$ represents the lateral boundary of the ocean. In geographical terms, x and y denote the east-west and south-north directions, respectively, and z is positive upward. Let $\mathbf{u} = (u, v, w), p, T, S$ and ρ denote the velocity, pressure, temperature, salinity, and density, respectively; the nonhydrostatic equations that describe the variation of these variables in $D \times (0, \Theta)$ are:

Momentum equation:

$$\begin{cases} \dfrac{D\mathbf{u}}{Dt} - \text{div}\,(A\nabla\mathbf{u}) + \mathbf{f} \times \mathbf{u} = -\dfrac{1}{\rho_0}\nabla p + \dfrac{\rho}{\rho_0}\mathbf{g}, & \text{in } D \times (0, \Theta) \\[2mm] \text{div } \mathbf{u} = 0, & \text{in } D \times (0, \Theta) \\[2mm] \mathbf{u}_{|\Gamma_b \cup \Gamma_l} = 0,\ A_v\dfrac{\partial u}{\partial \mathbf{n}}_{|\Gamma_s} = \tau_x,\ A_v\dfrac{\partial v}{\partial \mathbf{n}}_{|\Gamma_s} = \tau_y,\ w_{|\Gamma_s} = 0, \\[2mm] \mathbf{u}(0) = \mathbf{u_0}. \end{cases}$$

$$(1)$$

Conservation equation for temperature

$$\begin{cases} \dfrac{DT}{Dt} - \text{div }(K\nabla T) = 0, & \text{in } D \times (0, \Theta) \\[3mm] (K\nabla T)\mathbf{n}_{|\Gamma_b \cup \Gamma_l} = 0,\ k_v\dfrac{\partial T}{\partial \mathbf{n}}_{|\Gamma_s} = Q_T, \\[3mm] T(0) = T_0. \end{cases}$$

$$(2)$$

Conservation equation for salinity

$$\begin{cases} \dfrac{DS}{Dt} - \text{div }(K\nabla S) = 0, & \text{in } D \times (0, \Theta) \\[3mm] (K\nabla S)\mathbf{n}_{|\Gamma_b \cup \Gamma_l} = 0,\ k_v\dfrac{\partial S}{\partial \mathbf{n}}_{|\Gamma_s} = Q_S, \\[3mm] S(0) = S_0. \end{cases}$$

$$(3)$$

Equation of state

$$\rho = \rho(T, S, z), \qquad (4)$$

where

$$A = \begin{pmatrix} A_h & 0 & 0 \\ 0 & A_h & 0 \\ 0 & 0 & A_v \end{pmatrix} \text{ and } K = \begin{pmatrix} k_h & 0 & 0 \\ 0 & k_h & 0 \\ 0 & 0 & k_v \end{pmatrix},$$

$\tau = (\tau_x, \tau_y)$ is the prescribed wind stress, $\mathbf{g} = (0, 0, -g)$ is the gravity vector, $\mathbf{f} = (0, 0, f)$ and f is the Coriolis parameter, ρ_0 is a constant reference density, $\mathbf{u_0}, T_0$ and S_0 are the prescribed initial condition for velocity, temperature, and salinity, Q_T and Q_S denote the prescribed temperature and salinity fluxes at the surface and $\frac{D}{Dt}$, is the material derivative, i.e., for any sufficiently smooth function Ψ,

$$\frac{\mathrm{D}\Psi}{\mathrm{D}t} = \frac{\partial \Psi}{\partial t} + \mathbf{u} \cdot \nabla \Psi.$$

Note also that this set of equations is closed by the equation of state (4) that computes the density of the sea water depending on the temperature and salinity (UNESCO et al. 1981).

In what follows, we use the notation:

$$(\varphi, \phi) = \int_D \varphi(x)\phi(x)\mathrm{d}x$$

for any $\varphi, \phi \in L^2(D)$ (functional space of square integrable functions in D), and

$$(\nabla\varphi, \nabla\phi) = \int_D \nabla\varphi(x)\nabla\phi(x)\mathrm{d}x$$

for any function $\varphi, \phi \in H^1(D)$ [subspace of $L^2(D)$ of functions with first derivatives in $L^2(D)$]. Furthermore, we introduce the following spaces that we need below:

$$\mathbf{V} = \big\{\mathbf{v} = (v_1, v_2, v_3) \in H^1(D)^3 :$$
$$v_{1|\Gamma_l \cup \Gamma_b} = v_{2|\Gamma_l \cup \Gamma_b} = 0 \text{ and } v_{3|\partial D} = 0\big\}$$

and

$$H\left(D, \frac{\partial}{\partial z}\right) = \left\{v \in L^2(D) : \frac{\partial v}{\partial z} \in L^2(D)\right\},$$

$$H_0\left(D, \frac{\partial}{\partial z}\right) = \left\{v \in L^2(D) : \frac{\partial v}{\partial z} \in L^2(D), v_{|\Gamma_s} = 0\right\}.$$
$$(5)$$

Some important properties as well as the definition of the trace of the functions of $H\left(D, \frac{\partial}{\partial z}\right)$ are explained in "Appendix 1".

3. Numerical Approximation

In this work, we propose to solve numerically Eqs. (1)–(4) using the *hp*-finite element method for space discretization and the Lagrange-Galerkin method for time discretization.

3.1. The hp-Finite Element Method; Generalities

Let $D_h = \{R_j\}_{j=1}^{N_e} \subset \mathbb{R}^3$ be a partition of D, where N_e is the number of elements in the partition and R_j is a hexahedral element for any $1 \leq j \leq N_e$. For any j, we

define $h_j = \mathrm{diam}\, R_j$, $\alpha_j = \sup\{\mathrm{diam}\, B : B \subset R_j,$ B being a ball} and $h = \min_{1 \leq j \leq N_e} h_j$. Then, we assume that any partition D_h satisfies: (1) $\bar{D} = \bigcup_{j=1}^{Ne} R_j$; (2) any face of R_j is either a face of any other element R_i, $i \neq j$, or a subset of ∂D; (3) there exists a real constant $\sigma > 0$ (that does not depend on the partition D_h) such that $\frac{h_j}{\alpha_j} < \sigma$ for all $j = 1, 2, \ldots, N_e$.

Let $\hat{R} = [-1, 1]^3$ be the reference element and let $F_j : \hat{R} \to R_j$ be a continuous bijective transformation. We define the set of functions $P_m(R_j)$ by

$$P_m(R_j) = \{p \in C(R_j) : p = \hat{p} \circ F_j^{-1},$$
$$\hat{p} \in P_m([-1, 1]) \otimes P_m([-1, 1]) \otimes P_m([-1, 1])\},$$

where $P_m([-1, 1])$ is the set of polynomials of degree $\leq m$ defined on the interval $[-1, 1]$. Then, the finite element space V_h is defined by

$$V_h = \big\{v_h \in C(\bar{D}) : v_{h|R_j} \in P_m(R_j)\big\}.$$

Moreover, we also introduce the following subspaces of V_h:

$$V_{h0} = \big\{v_h \in V_h : v_{h|\partial D} = 0\big\},$$
$$V_{hl+b,0} = \big\{v_h \in V_h : v_{h|\Gamma_b \cup \Gamma_l} = 0\big\},$$
$$V_{hs,0} = \big\{v_h \in V_h : v_{h|\Gamma_s} = 0\big\}.$$

Since the finite element space V_h is finite dimensional, there exists a basis $\{\phi_k\}_{k=1}^{N_n} \subset V_h$, where $N_n = \dim V_h$. Note that this basis can be chosen depending only on a basis for $P_m([-1, 1])$ that is usually set up with either nodal or modal expansion. When using the nodal expansion, the basis for $P_m([-1, 1])$ consists of $m + 1$ Lagrange polynomials associated to $m + 1$ nodal points. On the other hand, the modal expansion is associated with a hierarchical set of polynomials in the sense that if B_m is the basis of $P_m([-1, 1])$, $B_m \subset B_{m+1}$ for all m, see Fig. 1. This expansion is very useful when implementing $p-$adaptivity. For a general description of this method, see KARNIADAKIS and SHERWIN (1999) or ŠOLÍN et al. (2004).

From the computational point of view, and depending on the algorithm to be implemented, both expansions have their own advantages. In this work, we propose the Lagrange-Galerkin method for the discretization of the material derivative, and for this case, the modal expansion is better because the

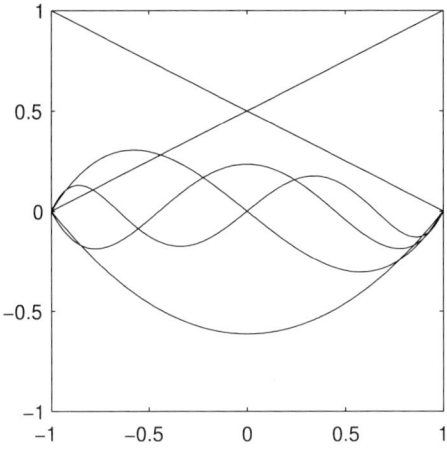

Figure 1
Modal basis for polynomials of degree $m = 5$

implementation results in a faster algorithm as was shown in GALÁN DEL SASTRE and BERMEJO (2011).

3.2. The Lagrange-Galerkin Method; Generalities

The Lagrange-Galerkin method discretizes backward in time the material derivative of any function Ψ using the so-called characteristics curves. To fix ideas, let $\Delta t > 0$ be the time step of the numerical scheme, and let $t_n = n\Delta t$. Then for any $x \in D$ and $n \in \mathbb{N} \cup \{0\}$, we define $X(x, t_{n+1}; \cdot) : (0, \Theta) \to D$ such that

$$\begin{cases} \dfrac{dX}{dt}(x, t_{n+1}; t) = \mathbf{u}(X(x, t_{n+1}; t), t), \\ X(x, t_{n+1}; t_{n+1}) = x. \end{cases}$$

Using the definition of $X(x, t_{n+1}; t)$, the material derivative of the function Ψ can be written as follows:

$$\frac{D\Psi}{Dt}(X(x, t_{n+1}; t), t) = \frac{d}{dt}(\Psi(X(x, t_{n+1}; t), t)),$$

then, one can use any classical formula to discretize the material derivative. As we explain below, in this work we propose to use a second-order BDF formula, so that

$$\frac{D\Psi}{Dt}(x, t_{n+1}) \simeq \frac{\frac{3}{2}\Psi^{n+1}(x) - 2\Psi^{n*}(x) + \frac{1}{2}\Psi^{(n-1)**}(x)}{\Delta t},$$

where the following notation is used:

$$\Psi^{n+1}(x) = \Psi(X(x, t_{n+1}; t_{n+1}), t_{n+1}) = \Psi(x, t_{n+1}),$$
$$\Psi^{n*}(x) = \Psi(X(x, t_{n+1}; t_n), t_n),$$
$$\Psi^{(n-1)**}(x) = \Psi(X(x, t_{n+1}; t_{n-1}), t_{n-1}).$$

When combining the Lagrange-Galerkin method and the finite element method, usually the velocity function $\mathbf{u}$ is not known analytically, but the finite element approximation is known at some time instants t_j for $0 \leq j \leq n$. If $\mathbf{u}_h^n$ denotes the finite element approximation of $\mathbf{u}(x, t_n)$, we propose to compute the approximation of $X(x, t_{n+1}; t_{n-j})$, for $j = 0, 1$, using the following fourth-order Runge-Kutta method: at time t_{n+1}, $n \geq 1$, given $\mathbf{u}_h^n$ and $\mathbf{u}_h^{n-1}$, calculate

$$K_1 = \Delta t \mathbf{u}_h^{n+1}(x),$$
$$K_2 = \Delta t \mathbf{u}_h^{n+\frac{1}{2}}\left(x - \frac{1}{2}K_1\right),$$
$$K_3 = \Delta t \mathbf{u}_h^{n+\frac{1}{2}}\left(x - \frac{1}{2}K_2\right),$$
$$K_4 = \Delta t \mathbf{u}_h^n(x - K_3),$$
$$X_h(x, t_{n+1}; t_n) = x - \frac{1}{6}(K_1 + 2K_2 + 2K_3 + K_4),$$

where $\mathbf{u}_h^{n+1}(\cdot)$ and $\mathbf{u}_h^{n+\frac{1}{2}}(\cdot)$ are extrapolated by formulas of order two, see GALÁN DEL SASTRE and BERMEJO (2011) for details. We denote this approximation by $X_h^{n+1, n-j}$. Furthermore, in what follows, we use the notation:

$$\Psi_h^{n*}(x) = \Psi_h^n\left(X_h^{n+1, n}(x)\right),$$
$$\Psi_h^{(n-1)**}(x) = \Psi_h^{n-1}\left(X_h^{n+1, n-1}(x)\right).$$

Note that when the Lagrange-Galerkin method is used in combination with the finite element method, the computation of the integrals (Ψ_h^{n*}, v_h) and $(\Psi_h^{(n-1)*}, v_h)$ is required for any function $v_h \in V_h$, and more specifically when $v_h = \phi_k$ for any $k = 1, 2, \ldots, N_n$. These integrals are calculated by a Gauss quadrature rule of high order, so that,

$$\left(\Psi_h^{n*}, \phi_k\right) = \sum_{j=1}^{N_e} \int_{R_j} \Psi_h^{n*}(x)\phi_k(x)dx$$
$$= \sum_{j=1}^{N_e} \int_{\hat{R}} \hat{\Psi}_h^{n*}(\hat{x})\hat{\phi}_k(\hat{x})|J_j(\hat{x})|d\hat{x}$$

where $\hat{\Psi}_h^{n*} = \Psi_h^{n*} \circ F_j$, $\hat{\phi}_k = \phi_k \circ F_j$ and J_j being the Jacobian determinant of the transformation F_j. Finally, we approximate

$$\int_{\hat{R}} \hat{\Psi}_h^{n*}(\hat{x}) \hat{\phi}_k(\hat{x}) |J_j(\hat{x})| d\hat{x} \simeq \sum_{i=1}^{NPG} \omega_i \hat{\Psi}_h^{n*}(\hat{x}_i) \hat{\phi}_k(\hat{x}_i) |J_j(\hat{x}_i)|$$

where ω_i and $\hat{x}_i \in \hat{R}$ are the weights and points, respectively, of the numerical quadrature in the reference element and NPG is the number of points used in the quadrature formula. In this paper, we use the Gauss-Legendre quadrature rule with $(m + 2) \times (m + 2)$ points, m being the degree of polynomials in V_h, so that the Gaussian quadrature integration calculates exactly for polynomials of degree $2m + 3$ in each coordinate direction. Following MORTON *et al.* (1988), the Lagrange-Galerkin method should be implemented with higher order quadrature rules to maintain stability and convergence properties, which is ensured with the number of Gaussian points that we consider. Note that the computation of $\hat{\phi}_k(\hat{x}_i) = \phi_k(F_j\hat{x}_i)$ can be done once for all, whereas the computation of $\hat{\Psi}_h^{n*}(\hat{x}_i) = \Psi_h^{n*}(F_j\hat{x}_i)$ has to be done every time step using the Runge-Kutta method explained above.

Summing up, let $\{x_i^j\}_{i=1,...,NPG}^{j=1,...,N_e}$, where $x_i^j = F_j\hat{x}_i$, be the Gaussian quadrature points in D_h, then

$$\left(\Psi_h^{n*}, \phi_k \right) \simeq \sum_{j=1}^{N_e} \sum_{i=1}^{NPG} \omega_i \Psi_h^{n*}(x_i^j) \hat{\phi}_k(\hat{x}_i) |J_j(\hat{x}_i)|$$

$$= \sum_{i=1}^{NPG} \omega_i \hat{\phi}_k(\hat{x}_i) \sum_{j=1}^{N_e} \Psi_h^{n*}(x_i^j) |J_j(\hat{x}_i)|. \tag{6}$$

3.3. *Numerical Approximation of the Nonhydrostatic Navier-Stokes Equations*

For the numerical approximation of Eqs. (1)–(4), we propose a numerical scheme based on the implicit second-order Backward Difference Formula (BDF). Although some other implicit second-order schemes could be considered, such as Crank-Nicholson method, BDF methods offer good stability properties in stiff problems and avoid the computation at time t_n of the Jacobian matrix of the transformation $x \to X_h(x, t_{n+1}; t_n)$ that arise in Lagrange-Galerkin or semi-Lagrangian methods when evaluating the explicit terms.

Then, the discretization of the conservation of temperature equation is

$$\begin{cases} \dfrac{\frac{3}{2}T^{n+1} - 2T^{n*} + \frac{1}{2}T^{(n-1)**}}{\Delta t} - \text{div}\,(K\nabla T^{n+1}) = 0, & \text{in } D \\ (K\nabla T^{n+1})\mathbf{n}_{|\Gamma_b \cup \Gamma_l} = 0, k_v \dfrac{\partial T^{n+1}}{\partial \mathbf{n}}\bigg|_{\Gamma_s} = Q_T^{n+1} \end{cases}, \tag{7}$$

and analogously for S. Note that this discretization requires the computation of T^{n*} and $T^{(n-1)**}$ (analogously for S), or more specifically the computation of $\left(T_h^{n*}, \phi_k\right)$ and $\left(T_h^{(n-1)**}, \phi_k\right)$. This computation is carried out using the numerical quadrature (6).

The discretization for $\mathbf{u}$ is more involved, but the idea is similar, so we intend to use the following discretization:

$$\begin{cases} \dfrac{\frac{3}{2}\mathbf{u}^{n+1} - 2\mathbf{u}^{n*} + \frac{1}{2}\mathbf{u}^{(n-1)**}}{\Delta t} - \text{div}\,(A\nabla\mathbf{u}^{n+1}) + \mathbf{f} \times \mathbf{u}^{n+1} = -\dfrac{1}{\rho_0}\nabla p^{n+1} + \dfrac{\rho^{n+1}}{\rho_0}\mathbf{g}, & \text{in } D \\ \text{div}\,\mathbf{u}^{n+1} = 0, & \text{in } D \\ \mathbf{u}^{n+1}_{|\Gamma_b \cup \Gamma_l} = 0, A_v \dfrac{\partial u^{n+1}}{\partial \mathbf{n}}\bigg|_{\Gamma_s} = \tau_x^{n+1}, A_v \dfrac{\partial v^{n+1}}{\partial \mathbf{n}}\bigg|_{\Gamma_s} = \tau_y^{n+1}, w^{n+1}_{|\Gamma_s} = 0. \end{cases} \tag{8}$$

In doing so, we have several difficulties: (1) since the Coriolis term is implicit, the matrix associated with the resulting algebraic system is not symmetric; (2) there is a large variation in the scale for the values of p; (3) $\mathbf{u}^{n+1}$ must satisfy the divergence-free condition; and (4) we have a coupled system for $\mathbf{u}^{n+1}$ and p^{n+1}.

To handle (1), we propose an explicit treatment of the Coriolis term (see "Appendix 2"). Regarding (2), we first note that $p \sim p_s + \rho g z$; here p_s is the pressure at the surface Γ_s, ρ the density of the ocean, g the gravity constant, and z the depth. So, taking into account that $p_s \sim 10^4 - 10^5$ Pa , $\rho \sim 10^3$ kg m^{-3} and $g \sim 10$ m s^{-2}, one can see that $p \sim p_s \sim 10^4 - 10^5$ Pa at $z = 0$ m, whereas $p \sim 10^7$ Pa at $z \sim 1000$ m ; this can lead to large numerical errors when solving p^{n+1}. Then, following CASULLI (1999), CASULLI and ZANOLLI (2002), MARSHALL *et al.* (1997) or MARSHALL *et al.* (1997), we write the pressure

$$p^{n+1} = q^{n+1} + r^{n+1},$$

where r^{n+1} is defined by the following problem:

$$\begin{cases} \dfrac{\partial r^{n+1}}{\partial z} = -g\rho^{n+1}, & \text{in } D, \\ r^{n+1}_{|\Gamma_s} = 0. \end{cases} \tag{9}$$

The function r^{n+1} is the so-called hydrostatic pressure, whereas $q^{n+1} = p^{n+1} - r^{n+1}$ is known as the nonhydrostatic pressure. Note that the hydrostatic pressure can be calculated separately from q^{n+1}, p^{n+1} and $\mathbf{u}^{n+1}$ since it only requires ρ^{n+1}. Moreover r^{n+1} retains the difference of scales of the pressure due to the density and the depth, whereas in q^{n+1} appears the nonhydrostatic forces of the fluid.

Furthermore, to avoid numerical errors when computing r^{n+1} we write the density function as

$$\rho^{n+1} = \rho_0\left(1 + (\rho')^{n+1}\right), \tag{10}$$

$$\begin{cases} \dfrac{\partial r^{n+1}_1}{\partial z} = -g\rho_0, & \text{in } D, \\ r^{n+1}_{1|\Gamma_s} = 0, \end{cases} \qquad \begin{cases} \dfrac{\partial r^{n+1}_2}{\partial z} = -g\rho_0(\rho')^{n+1}, & \text{in } D, \\ r^{n+1}_{2|\Gamma_s} = 0, \end{cases}$$

$$\begin{cases} \dfrac{\partial \bar\rho^{n+1}}{\partial z} = -(\rho')^{n+1}, & \text{in } D, \\ \bar\rho^{n+1}_{|\Gamma_s} = 0. \end{cases} \tag{11}$$

It is straightforward to prove that

$$r^{n+1} = r^{n+1}_1 + r^{n+1}_2 = r^{n+1}_1 + g\rho_0\bar\rho^{n+1}$$

and

$$\frac{\partial r^{n+1}}{\partial x} = g\rho_0 \frac{\partial \bar\rho^{n+1}}{\partial x},$$
$$\frac{\partial r^{n+1}}{\partial y} = g\rho_0 \frac{\partial \bar\rho^{n+1}}{\partial y}.$$

Remark 1 The functions r^{n+1} and $\bar\rho^{n+1}$ can be defined either using (9) and (11) or, as it is usually done in the literature, using the following:

$$r^{n+1}(x, y, z) = g \int_z^0 \rho^{n+1}(x, y, s)\,ds$$

and

$$\bar\rho^{n+1}(x, y, z) = \int_z^0 (\rho')^{n+1}(x, y, s)\,ds.$$

Using an explicit treatment of the Coriolis term and the definition of the hydrostatic and nonhydrostatic pressure, Eq. (8) becomes

$$\begin{cases} \dfrac{\frac{3}{2}\mathbf{u}^{n+1} - 2\mathbf{u}^{n*} + \frac{1}{2}\mathbf{u}^{(n-1)**}}{\Delta t} - \text{div}\,(A\nabla\mathbf{u}^{n+1}) + \mathbf{f}\times\mathbf{u}^{n+1}_e = -\dfrac{1}{\rho_0}\nabla q^{n+1} - \dfrac{1}{\rho_0}\nabla r^{n+1} + \dfrac{\rho^{n+1}}{\rho_0}\mathbf{g}, & \text{in } D, \\ \text{div}\,\mathbf{u}^{n+1} = 0, & \text{in } D, \\ \mathbf{u}^{n+1}_{|\Gamma_b\cup\Gamma_l} = 0, A_v\dfrac{\partial u^{n+1}}{\partial\mathbf{n}}\bigg|_{\Gamma_s} = \tau_x, A_v\dfrac{\partial v^{n+1}}{\partial\mathbf{n}}\bigg|_{\Gamma_s} = \tau_y, w^{n+1}_{|\Gamma_s} = 0, \end{cases} \tag{12}$$

where ρ_0 is a reference density value and ρ' is the deviation of the density respect to the reference value ρ_0. Thus, we further introduce the following functions:

where $\mathbf{u}^{n+1}_e$ is an extrapolation of the velocity at t_{n+1}.

We still have a system with $\mathbf{u}^{n+1}$ coupled with q^{n+1} via the divergence-free condition that $\mathbf{u}^{n+1}$ must satisfy. To break this coupling, we propose a splitting scheme based on GUERMOND and SHEN (2003), GUERMOND and SHEN (2003), or KARNIADAKIS et al. (1991), also referred to as the rotational form of the velocity-correction scheme, see GUERMOND et al. (2006) for a thorough description of many different kind of splittings for the Navier-Stokes equations. The main idea of the velocity-correction scheme is that for each time step, the projection substep is computed at first and velocity can be calculated using the pressure of the projection substep. Both velocity-correction and pressure-correction schemes are usually described for homogeneous isotropic fluids, i.e., $\rho = \rho_0$ and $A = \nu I$, where I is the identity matrix. However, the ocean is not a homogeneous fluid and is not considered isotropic since the horizontal length scale is much larger than the vertical one. Although these splittings can also solve heterogeneous fluid, see GUERMOND and QUARTAPELLE (2000), the stiffness matrix must be recalculated at each time step in the projection substep to compute the pressure; this can be computationally prohibitive, specially when using higher order polynomials.

Regarding the heterogeneity of the ocean, and noting that this heterogeneity only appears in the buoyancy term in Eq. (1) or (12) thanks to the Boussinesq approximation, the stiffness matrix can be computed once and for all since it remains unchanged.

On the other hand, when a fluid is isotropic, the main idea of the rotational form of the velocity-correction scheme is to replace the viscous term $-\Delta \mathbf{u}^{n+1}$ by $\nabla \times \nabla \times \mathbf{u}^{n+1} = -\Delta \mathbf{u}^{n+1} + \nabla \operatorname{div} \mathbf{u}^{n+1}$ in the projection substep. This procedure avoids the non-physical boundary condition for the pressure $\frac{\partial p^{n+1}}{\partial \mathbf{n}}|_{\partial D} = \cdots = \frac{\partial p^0}{\partial \mathbf{n}}|_{\partial D}$ and improves the convergence estimates of the method GUERMOND and SHEN

(2003). Since ocean is not considered an isotropic fluid, we further extend this idea and replace the viscous term $-\operatorname{div}(A\nabla \mathbf{u}^{n+1})$ by $\nabla \times (A\nabla) \times \mathbf{u}^{n+1} = -\operatorname{div}(A\nabla \mathbf{u}^{n+1}) + A\nabla \operatorname{div} \mathbf{u}^{n+1}$ (note that in both cases, isotropic and non-isotropic fluid, the idea is motivated by the identity of the rotational term and that we are searching for divergence-free solutions).

Thus, the splitting scheme is as follows:

1. Set $\mathbf{u}_e^{n+1} = 2\mathbf{u}^n - \mathbf{u}^{n-1}$ and compute $q^{n+1} \in H^1(D)$ that satisfies

$$\Delta t (\nabla q^{n+1}, \nabla v) = \rho_0 \left(2\mathbf{u}^{n*} - \frac{1}{2}\mathbf{u}^{(n-1)**}, \nabla v \right)$$
$$+ \rho_0 \left(-\Delta t \nabla \times (A\nabla) \times \mathbf{u}_e^{n+1} - \Delta t \mathbf{f} \times \mathbf{u}_e^{n+1}, \nabla v \right)$$
$$- \Delta t (\nabla r^{n+1}, \nabla v) + \Delta t \mathbf{g}(\rho^{n+1}, \nabla v)$$

for all $v \in H^1(D)$.
2. Compute $\mathbf{u}^{n+1} \in \mathbf{V}$ such that

$$\frac{3}{2}(\mathbf{u}^{n+1}, \mathbf{v}) + \Delta t (A\nabla \mathbf{u}^{n+1}, \nabla \mathbf{v})$$
$$= \left(2\mathbf{u}^{n*} - \frac{1}{2}\mathbf{u}^{(n-1)**} - \Delta t \mathbf{f} \times \mathbf{u}_e^{n+1}, \mathbf{v} \right)$$
$$- \Delta t \frac{1}{\rho_0}(\nabla q^{n+1}, \mathbf{v}) - \Delta t \frac{1}{\rho_0}(\nabla r^{n+1}, \mathbf{v})$$
$$+ \Delta t \frac{1}{\rho_0}\mathbf{g}(\rho^{n+1}, \mathbf{v})$$

for all $\mathbf{v} \in \mathbf{V}$.

Remark 2 We have described the splitting of the numerical scheme with the variational formulation, but instead, one could write both substeps in the following equivalent formulation:

1. Set $\mathbf{u}_e^{n+1} = 2\mathbf{u}^n - \mathbf{u}^{n-1}$ and solve $\hat{\mathbf{u}}^{n+1}$ and q^{n+1} such that:

$$\begin{cases} \dfrac{\frac{3}{2}\hat{\mathbf{u}}^{n+1} - 2\mathbf{u}^{n*} + \frac{1}{2}\mathbf{u}^{(n-1)**}}{\Delta t} + \nabla \times (A\nabla) \times \mathbf{u}_e^{n+1} + \mathbf{f} \times \mathbf{u}_e^{n+1} = -\dfrac{1}{\rho_0}\nabla q^{n+1} - \dfrac{1}{\rho_0}\nabla r^{n+1} + \dfrac{\rho^{n+1}}{\rho_0}\mathbf{g}, & \text{in } D, \\ \operatorname{div} \hat{\mathbf{u}}^{n+1} = 0, \text{ in } D \\ \hat{\mathbf{u}}^{n+1} \cdot \mathbf{n}_{|\partial D} = 0 \end{cases}$$

2. Find $\mathbf{u}^{n+1} \in \mathbf{V}$ such that:

$$\begin{cases} \dfrac{\frac{3}{2}\mathbf{u}^{n+1} - \frac{3}{2}\hat{\mathbf{u}}^{n+1}}{\Delta t} - \text{div}\left(A\nabla\mathbf{u}^{n+1}\right) - \nabla \times (A\nabla) \times \mathbf{u}_e^{n+1} = 0, \text{ in } D \\ \mathbf{u}^{n+1}_{|\Gamma_b \cup \Gamma_l} = 0, A_v\dfrac{\partial u^{n+1}}{\partial \mathbf{n}}_{|\Gamma_s} = \tau_x, A_v\dfrac{\partial v^{n+1}}{\partial \mathbf{n}}_{|\Gamma_s} = \tau_y, w^{n+1}_{|\Gamma_s} = 0, \end{cases}$$

Furthermore, one can observe from the second substep that

$$\left(\text{div}\left(A\nabla\mathbf{u}^{n+1}\right) + \nabla \times (A\nabla) \times \mathbf{u}_e^{n+1}\right) \cdot \mathbf{n}_{|\partial D} = 0,$$

and thus, using substep one and the fact that $p^{n+1} = q^{n+1} + r^{n+1}$, one can get

$$\frac{\partial p^{n+1}}{\partial \mathbf{n}}_{|\partial D} = \left(g\rho^{n+1} + \rho_0 \, \text{div}\left(A\nabla\mathbf{u}^{n+1}\right)\right) \cdot \mathbf{n}_{|\partial D}$$

which is a consistent boundary condition for the pressure.

Note that the computation of $\mathbf{u}^{n+1}$ and q^{n+1} using this splitting requires r^{n+1} or $\bar{\rho}^{n+1}$, the calculation of this variable is not trivial in unstructured meshes. However, using the definition (11), it follows that for any function $v \in H\left(D, \frac{\partial}{\partial z}\right)$,

$$\int_D \frac{\partial \bar{\rho}^{n+1}}{\partial z}\frac{\partial v}{\partial z} = -\int_D (\rho')^{n+1}\frac{\partial v}{\partial z}.$$

Hence, knowing that $\bar{\rho}^{n+1}_{|\Gamma_s} = 0$, we calculate $\bar{\rho}^{n+1}$ as the solution of the following variational problem: find $\bar{\rho}^{n+1} \in H_0\left(D, \frac{\partial}{\partial z}\right)$ such that

$$\int_D \frac{\partial \bar{\rho}^{n+1}}{\partial z}\frac{\partial v}{\partial z} = -\int_D (\rho')^{n+1}\frac{\partial v}{\partial z} \quad \text{for all } v \in H_0\left(D, \frac{\partial}{\partial z}\right). \tag{13}$$

This approach is also used in LABEUR and PIETRZAK (2005).

Existence and uniqueness of problem (13) is not trivial, see "Appendix 1".

Finally, the scheme is as follows:

1. Compute $X_h^{n*}\left(x_i^j\right)$ and $X_h^{(n-1)**}\left(x_i^j\right)$ for all $j = 1, 2, \ldots N_e$ and $i = 1, 2, \ldots NPG$.
2. Compute $\mathbf{u}_h^{n*}, \mathbf{u}_h^{(n-1)**}, T_h^{n*}, T_h^{(n-1)**}, S_h^{n*}, S_h^{(n-1)**}$ at the Gaussian quadrature points $\{x_i^j\}_{i=1,\ldots,NPG}^{j=1,\ldots,N_e}$.
3. Using (7), compute $T_h^{n+1} \in V_h$ such that

$$\frac{3}{2}\left(T_h^{n+1}, v_h\right) + \Delta t\left(K\nabla T_h^{n+1}, \nabla v_h\right)$$
$$= \left(2T_h^{n*} - \frac{1}{2}T_h^{(n-1)**}, v_h\right) + \Delta t \int_{\Gamma_s} Q_T v_h$$

for all $v_h \in V_h$ and $S_h^{n+1} \in V_h$ such that

$$\frac{3}{2}\left(S_h^{n+1}, v_h\right) + \Delta t\left(K\nabla S_h^{n+1}, \nabla v_h\right)$$
$$= \left(2S_h^{n*} - \frac{1}{2}S_h^{(n-1)**}, v_h\right) + \Delta t \int_{\Gamma_s} Q_S v_h$$

for all $v_h \in V_h$.
4. Compute T_h^{n+1} and S_h^{n+1} at the Gaussian quadrature points $\{x_i^j\}_{i=1,\ldots,NPG}^{j=1,\ldots,N_e}$ and, using the equation of state (4) and the definition of ρ' (10), compute $\left(\rho_h'\right)^{n+1}$ at the Gaussian quadrature points.
5. Compute $\bar{\rho}_h^{n+1} \in V_{hs,0}$ such that

$$\left(\frac{\partial \bar{\rho}_h^{n+1}}{\partial z}, \frac{\partial v_h}{\partial z}\right) = \left(-(\rho_h')^{n+1}, \frac{\partial v_h}{\partial z}\right)$$

for all $v_h \in V_{hs,0}$.
6. Define $\mathbf{u}_{he}^{n+1} = 2\mathbf{u}_h^n - \mathbf{u}_h^{n-1}$.
7. Compute $\omega_{he}^{n+1} = (A\nabla) \times \mathbf{u}_{he}^{n+1}$.
8. Compute

$$\mathbf{F}_h^{n+1} = -\nabla \times \omega_{he}^{n+1} - \mathbf{f} \times \mathbf{u}_{he}^{n+1}$$
$$-\frac{1}{\rho_0}\nabla r_h^{n+1} + \frac{\rho_h^{n+1}}{\rho_0}\mathbf{g}$$

at the Gaussian quadrature points $\{x_i^j\}_{i=1,\ldots,NPG}^{j=1,\ldots,N_e}$. Note that

$$F_{h1}^{n+1} = -\left(\frac{\partial \omega_{he,3}^{n+1}}{\partial y} - \frac{\partial \omega_{he,2}^{n+1}}{\partial z}\right) + fv_{he}^{n+1} - g\frac{\partial \bar{\rho}_h^{n+1}}{\partial x},$$

$$F_{h2}^{n+1} = -\left(\frac{\partial \omega_{he,1}^{n+1}}{\partial z} - \frac{\partial \omega_{he,3}^{n+1}}{\partial x}\right) - fu_{he}^{n+1} - g\frac{\partial \bar{\rho}_h^{n+1}}{\partial y},$$

$$F_{h3}^{n+1} = -\left(\frac{\partial \omega_{he,2}^{n+1}}{\partial x} - \frac{\partial \omega_{he,1}^{n+1}}{\partial y}\right).$$

9. Compute $q_h^{n+1} \in V_h$ such that

$$\Delta t \int_D \nabla q_h^{n+1}\nabla v_h = \rho_0 \int_D \left(2\mathbf{u}_h^{n*} - \frac{1}{2}\mathbf{u}_h^{(n-1)**} + \Delta t\mathbf{F}_h^{n+1}\right)\nabla v_h$$

for all $v_h \in V_h$.
10. Compute

$$\mathbf{G}_h^{n+1} = -\mathbf{f} \times \mathbf{u}_e^{n+1} - \frac{1}{\rho_0}\nabla p_h^{n+1} + \frac{\rho_h^{n+1}}{\rho_0}\mathbf{g}$$

at the Gaussian quadrature points $\{x_i^j\}_{i=1,\ldots,NPG}^{j=1,\ldots,N_e}$. Note that

$$G_{h1}^{n+1} = fv_{he}^{n+1} - \frac{1}{\rho_0} \frac{\partial}{\partial x} \left(q_h^{n+1} + \rho_0 g \bar{\rho}_h^{n+1} \right),$$

$$G_{h2}^{n+1} = -fu_{he}^{n+1} - \frac{1}{\rho_0} \frac{\partial}{\partial y} \left(q_h^{n+1} + \rho_0 g \bar{\rho}_h^{n+1} \right),$$

$$G_{h3}^{n+1} = -\frac{1}{\rho_0} \frac{\partial q_h^{n+1}}{\partial z}.$$

11. Compute $\mathbf{u}_h^{n+1} \in V_{hl+b,0} \times V_{hl+b,0} \times V_{h0}$ such that

$$\frac{3}{2}\left(\mathbf{u}_h^{n+1}, \mathbf{v}_h \right) + \Delta t \left(A \nabla \mathbf{u}_h^{n+1}, \nabla \mathbf{v}_h \right)$$

$$= \left(2\mathbf{u}_h^{n*} - \frac{1}{2}\mathbf{u}_h^{(n-1)**} + \Delta t \mathbf{G}_h^{n+1}, \mathbf{v}_h \right)$$

$$+ \Delta t \int_{\Gamma_s} \tau \cdot \mathbf{v}_h$$

for all $\mathbf{v}_h \in V_{hl+b,0} \times V_{hl+b,0} \times V_{h0}$.

Remark 3

(a) All the integrals defined on the whole domain D that appear on the right-hand side of the equations are calculated by (6).

(b) The integrals defined on the surface Γ_s are calculated by Gaussian quadrature rules over the elements of Γ_s.

(c) As in GUERMOND and SHEN (2003), it is also possible to avoid the calculation of the term $\nabla \times (A\nabla) \times \mathbf{u}_e^{n+1}$ using a modification of the splitting proposed [see GUERMOND *et al.* (2006) for details], but the resulting scheme also requires $\mathbf{u}^{(n-1)*}, \mathbf{u}^{(n-2)**}, \mathbf{u}^{(n-2)*}$ and $\mathbf{u}^{(n-3)**}$ at time level $n+1$.

4. Numerical Results

We now carry out some numerical tests to show how our numerical model can handle ocean flows. In this first step of our model, we focus on density driven flows that can result in hard problems. To this end, we perform some test problems, two proposed in HAIDVOGEL and BECKMANN (1999) to compare four numerical models, and another one as a modelization of the physical experiment carried out in HORN *et al.* (2001). Finally, we conduct a test problem as an idealization of the dynamics involved in the Strait of Gibraltar in which density driven flows occur.

4.1. Gravitational Adjustment of a Density Front

The first test we consider is the gravitational adjustment of a two-density layer system proposed in HAIDVOGEL and BECKMANN (1999). The problem domain is 64-km long and 20-m deep. The initial conditions for temperature is $20°$ C in the whole domain, whereas for salinity is 30 psu for $x > 32$ km and 36.5605 psu otherwise. This choice of temperature and salinity generates two water masses with a density contrast of 5 kg m^{-3} at time $t = 0$ (see Fig. 2). At this time, the vertical wall separating the two fluids is lifted up and the fluids start moving mainly by the action of gravity.

We carried out three simulations considering a discretization similar to that used in HAIDVOGEL and BECKMANN (1999), where 0.5 km is fixed as horizontal resolution and 1 m as vertical resolution. In our simulations, we consider uniform meshes with element sizes of 1 km×2 m, 2 km×4 m and 4 km×$\frac{20}{3}$ m, and polynomial degrees $m = 2, 4$ and 8, respectively.

The coefficients used in all the simulations are $A_h = 500$ m^2 s^{-1}, $A_v = 10^{-4}$ m^2 s^{-1}, $k_h = 10$ m^2 s^{-1} and $k_v = 10^{-5}$ m^2 s^{-1}, and the time step is $\Delta t = 300$ s. The density distributions after 10 h of simulation are shown in Fig. 3 for the three simulations. Note that the low values of the viscosity parameters used in the simulations together with the low resolution meshes produce some small

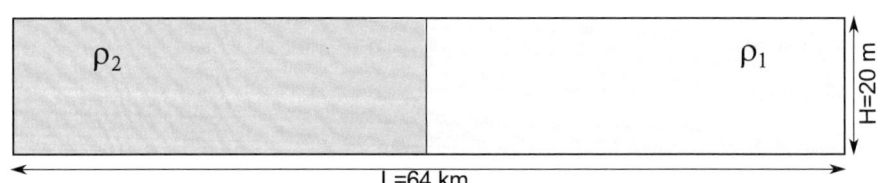

Figure 2
Initial configuration for the gravitational adjustment problem ($\rho_2 > \rho_1$)

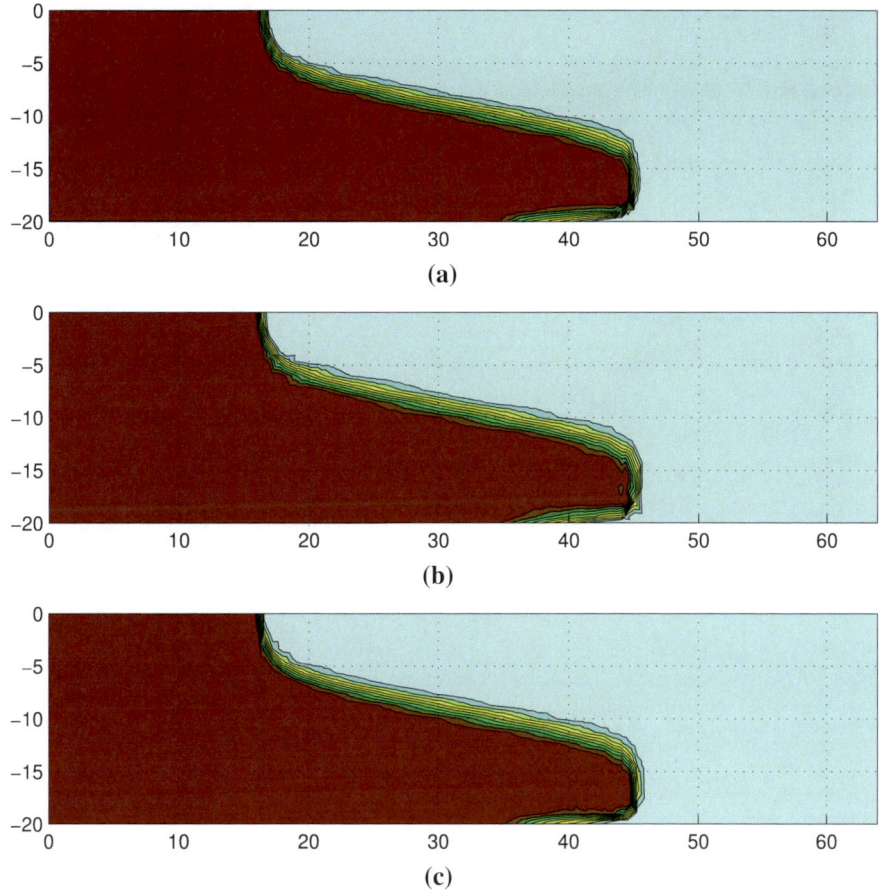

Figure 3
Solution after 10 h of numerical simulation with three different meshes: **a** 1 km × 2 m, $m = 2$; **b** 2 km× 4 m, $m = 4$; **c** 4 km × $\frac{20}{3}$ m, $m = 8$. The contour lines are separated every 0.5 kg m^{-3}

oscillations particularly for higher order finite elements, i.e., $m > 2$.

In Fig. 4a, we show the solution of the same problem but using the mesh with resolution 2 km × 4 m and $m = 10$ instead of $m = 4$. We can see how the results do not exhibit oscillations because the mesh can handle the solution of the problem with this initial discontinuity.

In HAIDVOGEL and BECKMANN (1999), this test is solved using four numerical models (MOM, MICOM, SCRUM, and SEOM) with different parameters and advection schemes, and with free slip boundary condition at the bottom. In the solution obtained in the majority of the experiments, the front advances about 20 km in 10 hours of simulations. In all our simulations, the front moves slightly less than 20 km. We presume that this discrepancy comes from the

different bottom boundary conditions used in our model, $\mathbf{u}_{|\Gamma_b} = 0$, i.e., no slip boundary conditions. However, we have also run an experiment with free slip boundary condition at the bottom and the result is shown in Fig. 4b. We see that the front advances about 20 km in this simulation. Note that the no slip boundary condition at the bottom makes the problem harder because the bottom boundary layer must be resolved properly, as is done in our simulations.

We also perform several simulations to further compare the results of our numerical model with those reported in ISKANDARANI *et al.* (2005). For this purpose, we use the three meshes with different values of A_h and k_h and compute the maximum and minimum values of the density after 10 hours of simulation. These simulations are carried out with slip boundary conditions that are the condition used

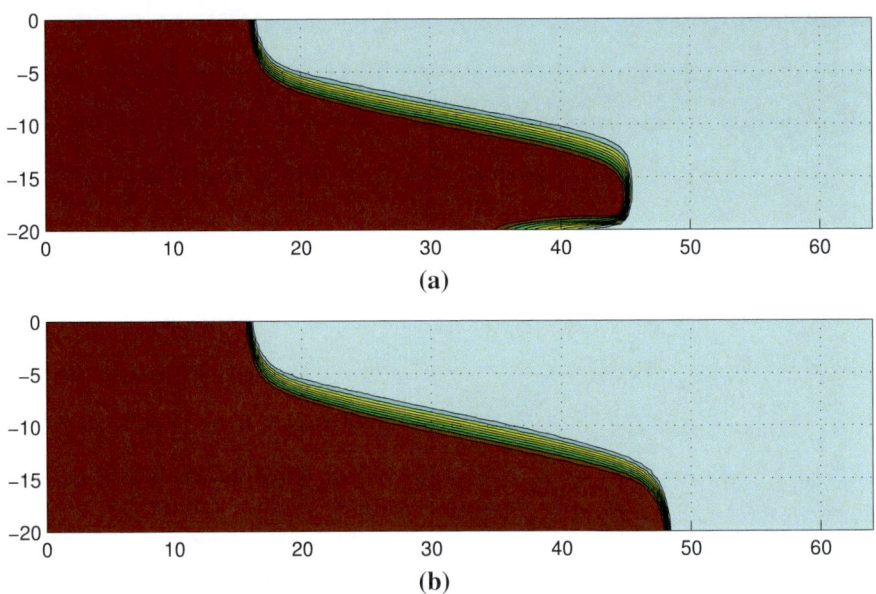

(a)

(b)

Figure 4

Solution after 10 h of numerical simulation with the mesh 2 km × 4 m, $m = 10$ using no slip boundary conditions (**a**), and slip boundary conditions (**b**). The contour lines are separated every 0.5 kg m^{-3}

Table 1

Difference between the maximum and minimum value of the density at $t = 10$ h and the minimum value of density at initial time

k_h(m^2 s^{-1})	$A_h = 500$ m^2 s^{-1}		$A_h = 100$ m^2 s^{-1}		$A_h = 50$ m^2 s^{-1}	
1 km × 2 m ($m = 2$)						
10	−0.2617	5.2635	−0.8413	5.8396	−1.3411	6.3388
50	−0.0384	5.0386	−0.0654	5.0658	−0.0703	5.0707
100	8.4×10^{-8}	5.00004	-7.8×10^{-11}	5.00004	3.3×10^{-11}	5.00004
2 km × 4 m ($m = 4$)						
10	−0.5124	5.5151	-1.5857	6.5882	−1.4840	6.4899
50	−0.0100	5.0101	−0.0729	5.0731	−0.0648	5.0668
100	-2.5×10^{-5}	5.00006	−0.0039	5.0041	−0.0964	5.0097
4 km × $\frac{20}{3}$ m ($m = 8$)						
10	−0.1537	5.1560	−0.8397	5.8503	−1.3955	6.4157
50	−0.0152	5.0154	−0.0270	5.0265	−0.0512	5.0523
100	−0.0003	5.0003	−0.0041	5.0042	−0.0056	5.0063

in ISKANDARANI *et al.* (2005). In Table 1 we show at time $t = 10$ h the values of ρ_{hmax} and ρ_{hmin}, where

$$\rho_{hmax} = \max_{x \in D} \rho_h(x, 10) - \min_{x \in D} \rho_h(x, 0),$$

$$\rho_{hmin} = \min_{x \in D} \rho_h(x, 10) - \min_{x \in D} \rho_h(x, 0).$$

Theoretically, the values of ρ_{hmax} and ρ_{hmin} should always remain between 0 and 5 kg m^{-3}, but the numerical solution may exhibit spurious oscillations. Following FERZIGER and PERIC (1999, p. 142–143), one can argue that these oscillations may be due to

the use of the BDF2 schemes, particularly when Δt is large; however, we think that the oscillations are due to the fact that the Lagrange-Galerkin method does not satisfy the maximum principle in advection–diffusion equations. We can see in the table that the simulations computed with large values of A_h and k_h exhibit very small oscillations; nevertheless when $k_h = 50$ m^2 s^{-1} and mainly when $k_h = 10$ m^2 s^{-1}, the solutions exhibit larger spurious oscillations. This is a well-known behavior when hp-finite element method is used and either the mesh

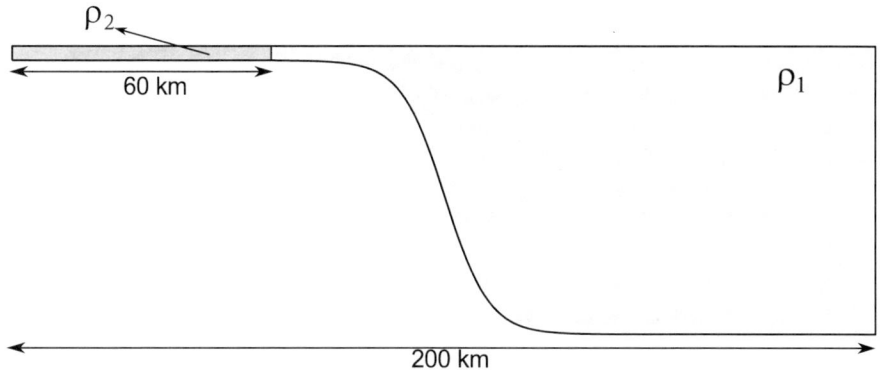

Figure 5
Initial configuration for the gravitational adjustment over a slope problem ($\rho_2 > \rho_1$)

is not sufficiently fine or the degree of polynomial is not large enough [see for instance ŠOLÍN *et al.* (2004) and KARNIADAKIS and SHERWIN (1999)]. Our results are better than those reported in ISKANDARANI *et al.* (2005) for both, the Discontinuous and Continuous Galerkin Method, that these authors used in their calculations; we presume that this is due to the Lagrange-Galerkin method.

4.2. Gravitational Adjustment Over a Slope

The second test, also proposed in HAIDVOGEL and BECKMANN (1999), is a gravitational adjustment of a front over a slope. This test is very similar to the first one, but now the domain simulates a steep bathymetry (Fig. 5) that makes the dense water forms a narrow ribbon moving down the slope and at the same time a plume of dense water is formed. For this test, the bathymetry is simulated by the function

$$H(x) = h_{\min} + \frac{1}{2}(h_{\max} - h_{\min})\left(1 + \tanh\left(\frac{x - x_0}{L_s}\right)\right),$$

where $h_{\min} = 200$ m, $h_{\max} = 4000$ m, $x_0 = 100$ km and $L_s = 10$ km. The bathymetry descends 3800 m in about 20 km. The initial conditions for this test are the same as in the previous one, the only difference is the position where the front is located, for this problem is $x = 60$ km.

The coefficients used in the simulations are $A_h = 1000$ m^2 s^{-1}, $A_v = 10^{-3}$ m^2 s^{-1}, $k_h = 100$ m^2 s^{-1} and $k_v = 10^{-3}$ m^2 s^{-1}, whereas the time step is $\Delta t = 150$ s. We carry out four simulations with different meshes and degrees of polynomials.

Three of these meshes are regularly structured meshes in which the vertical discretization is following-terrain, see Fig. 6a–c, while the fourth mesh is unstructured in areas where the bottom topography has a strong gradient, whereas it is structured in areas where the bottom topography is flat, see Fig. 6d.

The mesh in Fig. 6a consists of 200×40 elements and is used with polynomial degree $m = 2$, and this is equivalent to a horizontal resolution of 0.5 km and an average vertical resolution of 50 m when the depth is 4000 m. The mesh used with polynomial degree $m = 4$ is displayed in Fig. 6b and has 100×10 elements, which gives the same horizontal resolution as the mesh described before, and an averaged vertical resolution of 100 m in the deeper part of the domain. Regarding the mesh for polynomial degree $m = 6$, see Fig. 6c, it has 32×5 elements, which is equivalent to a horizontal resolution of about 1 km and an average vertical resolution of about 130 m when the depth is close to 4000 m. Finally, the unstructured mesh is built in such a way that we join the structured part of the mesh when $x < 64$ km (just before the slope) with only 5 elements in the vertical, and the other structured part of the mesh when $x > 136$ km (just after the slope) with 10 elements in the vertical. The total number of elements in this mesh is 2910. The purpose of working with the unstructured mesh is to see the behavior of the solution when the number of vertical elements in the whole domain is variable; for example, this mesh has 5 elements in the vertical in the shallow region and 10 elements in the vertical in the deep region.

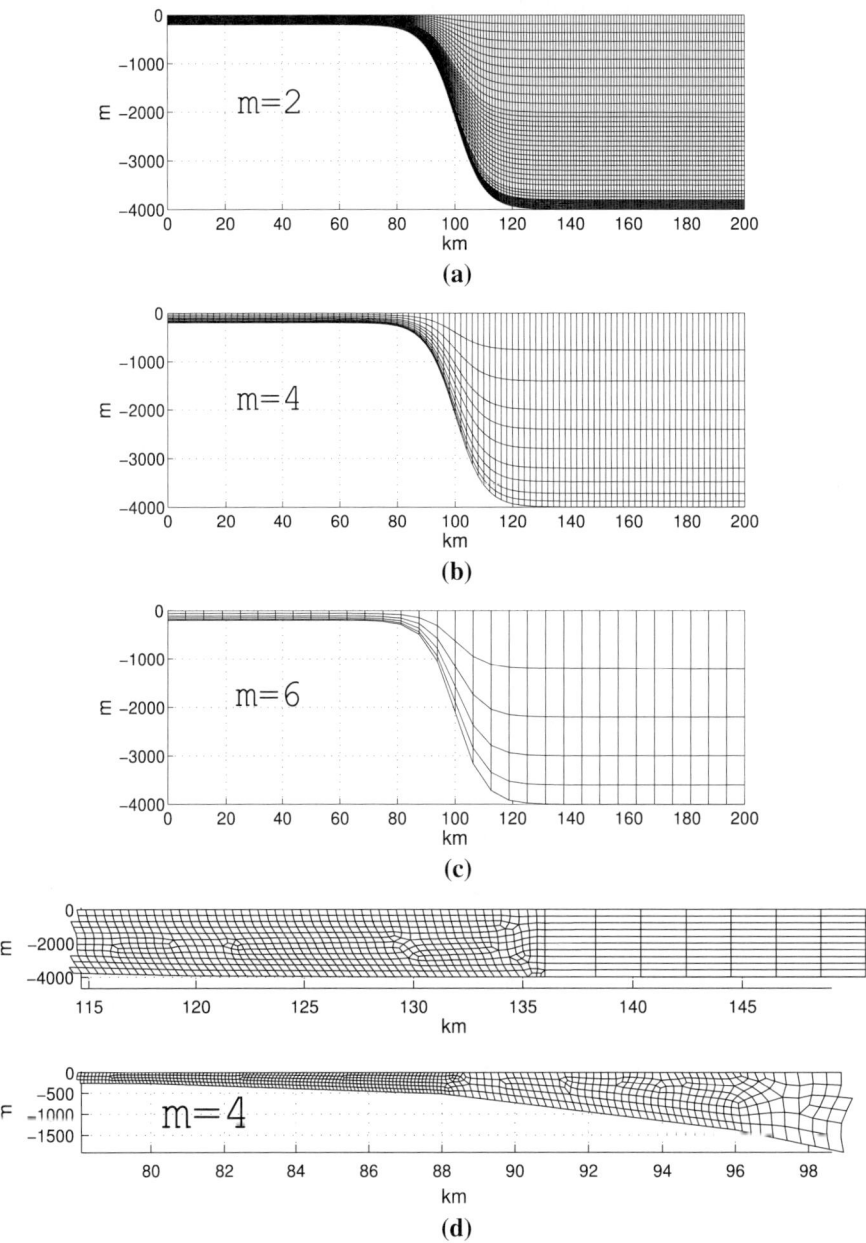

Figure 6

The finite element meshes used in the gravitational adjustment over a slope problem with $m = 2$ (**a**), $m = 4$ (**b**), $m = 6$ (**c**) and local detail of the unstructured mesh with $m = 4$ (**d**)

In Fig. 7, we show the results obtained when using the mesh in Fig. 6a. We observe how the model can properly reproduce the plume formation moving down the slope that reaches the bottom of the domain after 12.5 h approximately. Note that the plume descends slightly slower than in the results reported in HAIDVOGEL and BECKMANN (1999) obtained with hydrostatic models, and the results of HEGGELUND et al. (2004) where a comparison of a hydrostatic and a nonhydrostatic model is performed. As we pointed out in the first test, we presume that the slower motion is due to the boundary layer at the bottom for the non-slip boundary condition at Γ_b.

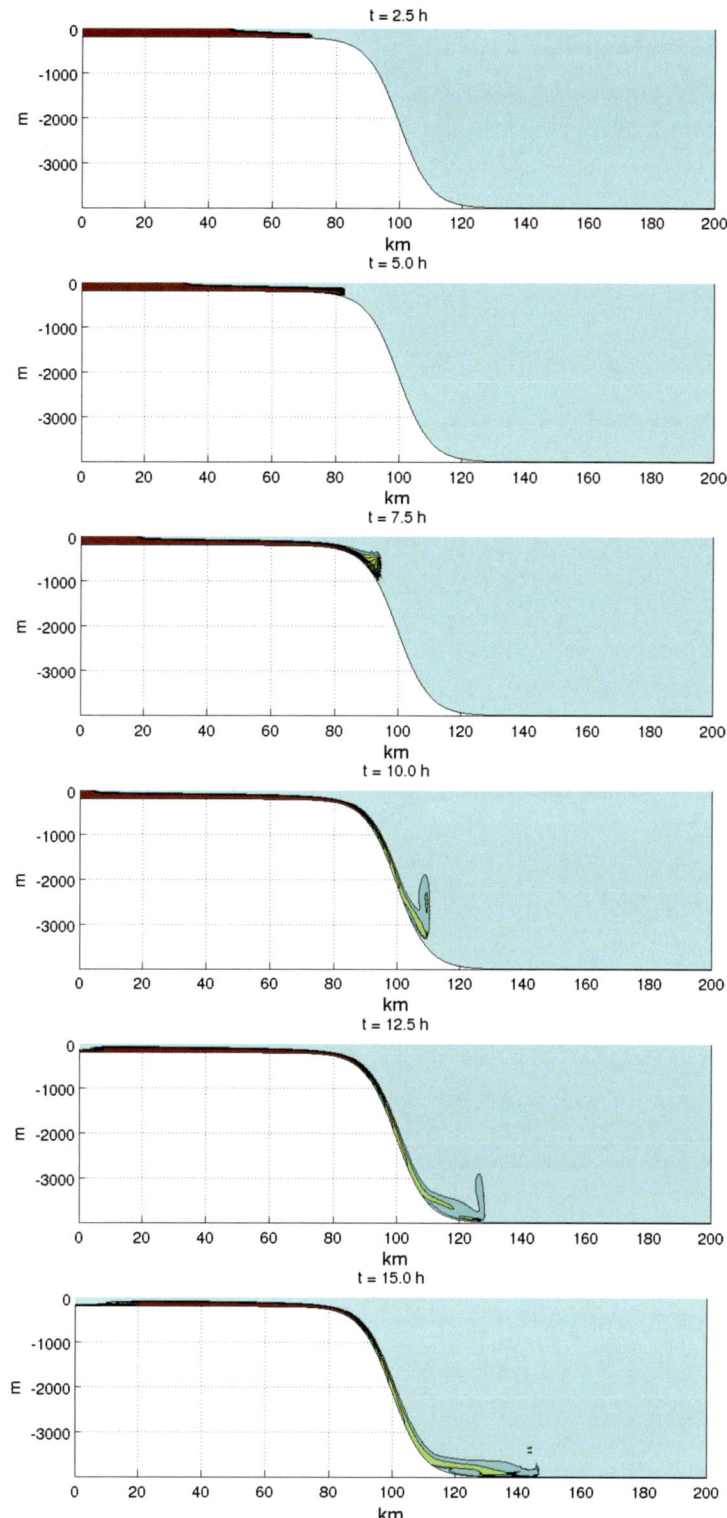

Figure 7
Salinity distribution at different time instants for the simulation with the mesh in Fig. 6a. The contour lines are separated every 0.65 psu

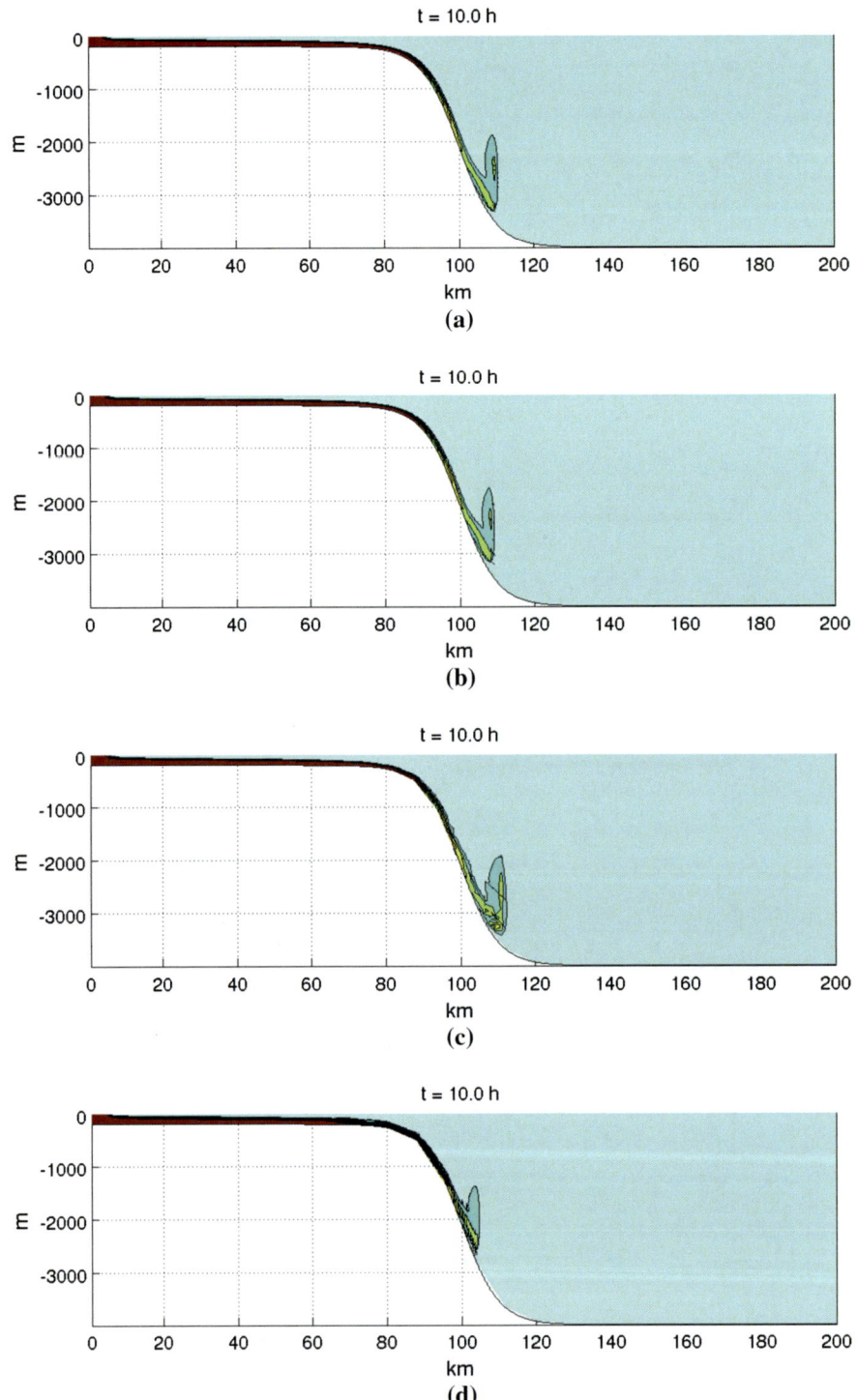

Figure 8
Salinity distribution at $t = 10$ h for the simulation with the meshes in Fig. 6. The contour lines are separated every 0.65 psu

In Fig. 8, we show the results of the four simulations at $t = 10$ h. We see that the plume formed when the dense water moves down is qualitatively the same in all the simulations, with slightly differences in Fig. 8c (corresponding to the mesh of Fig. 6c) and Fig. 8d (corresponding to the unstructured mesh, Fig. 6d). We presume that these discrepancies are due to the different resolution of the meshes. The solution in Fig. 8c is obtained using a horizontal resolution of 1 km instead of 0.5 km as in the meshes in Fig. 6a, b. Regarding the solution in Fig. 8d, we can note that the solution suffers from numerical diffusion probably caused by the low regularity of its elements compared to the regularity of the elements in the structured meshes.

4.3. Internal Seiche Waves

Nonhydrostatic models are essential tools for studying wave fronts [see KANARSKA *et al.* (2007) for a comparison of hydrostatic and nonhydrostatic results]. In this test, we perform a simulation based on the experimental work presented in HORN *et al.* (2001). In this work, the authors study the internal seiche waves that can be observed in many lakes and inner seas by a series of laboratory experiments; the idea is to compare the different regimes that the seiche waves can present. Our aim in this test is to compare the numerical results produced by the model with the experimental ones.

The experiments consist of a closed tank 6-m long and 0.29-m deep that could rotate about a horizontal axis through its center. At initial time, the tank is filled of two water masses with different density and rotated an angle α, see Fig. 9a. After that, the tank is set horizontally and internal waves are generated due to the gravity. In HORN *et al.* (2001), the authors

claim that depending on the tilt angle α and the depth of the lower layer h, there exist different regimes in the evolution of the internal waves.

We perform numerical simulations with $h = 0.203$ and $\alpha = 1.5°, 1.0°, 0.75°$ and $0.5°$. Following HORN *et al.* (2001), the difference in the density for the two water masses is $\Delta\rho = 20$ kg m^{-3}, so that at initial time, see Fig. 9b, the initial temperature is $20°$ C homogeneously and the initial salinity

$$S_0(x,z) = 12 + 13.2\left(1 + \tanh\left(-\left(\frac{z + h - (x - 3)\tan\alpha}{\delta_\rho}\right)\right)\right),$$

with $\delta_\rho = 0.005$ as the interface thickness. The mesh consists on 100×10 regular rectangular elements with polynomial degree $p = 8$. The viscosity parameters for these simulations are $A_h = 10^{-5}$ m^2 s^{-1}, $A_v = 5 \times 10^{-6}$ m^2 s^{-1}, $k_h = 10^{-5}$ m^2 s^{-1} and $k_v = 5 \times 10^{-7}$ m^2 s^{-1} and the time step is $\Delta t = 0.1$ s. In Fig. 10, the evolution of interface displacement at the center of the domain is shown for the simulations that were carried out in the present work and the experiments in HORN *et al.* (2001) for comparison. Note that for the numerical simulations, the interface is calculated as the vertical distance y such that $S(3.0, -0.29 + h + y) = 25.0$ psu. The numerical results obtained reproduced properly the internal waves observed in each experiment, although we can observe very small discrepancies in the amplitude and period of these waves. Nevertheless we must remark that these discrepancies are a consequence of the small differences between the numerical simulation and the experiments; for example, the initial density contrast is 20 kg m^{-3} in the numerical simulations, whereas in the experiments is 20 ± 2 kg m^{-3}.

One of the main advantages of nonhydrostatic models is that they can reproduce some important

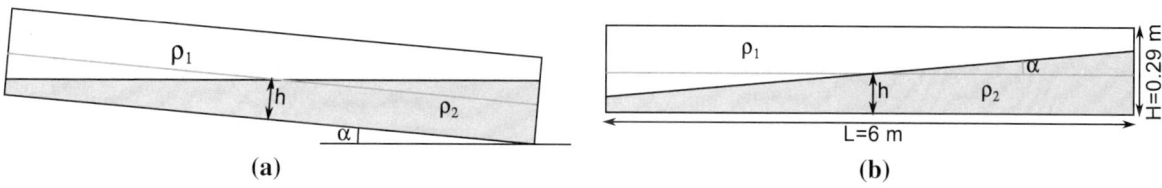

(a)   **(b)**

Figure 9

Initial configuration for the internal seiche problem ($\rho_2 > \rho_1$): **a** experimental work in HORN *et al.* (2001) and **b** numerical simulation in present work

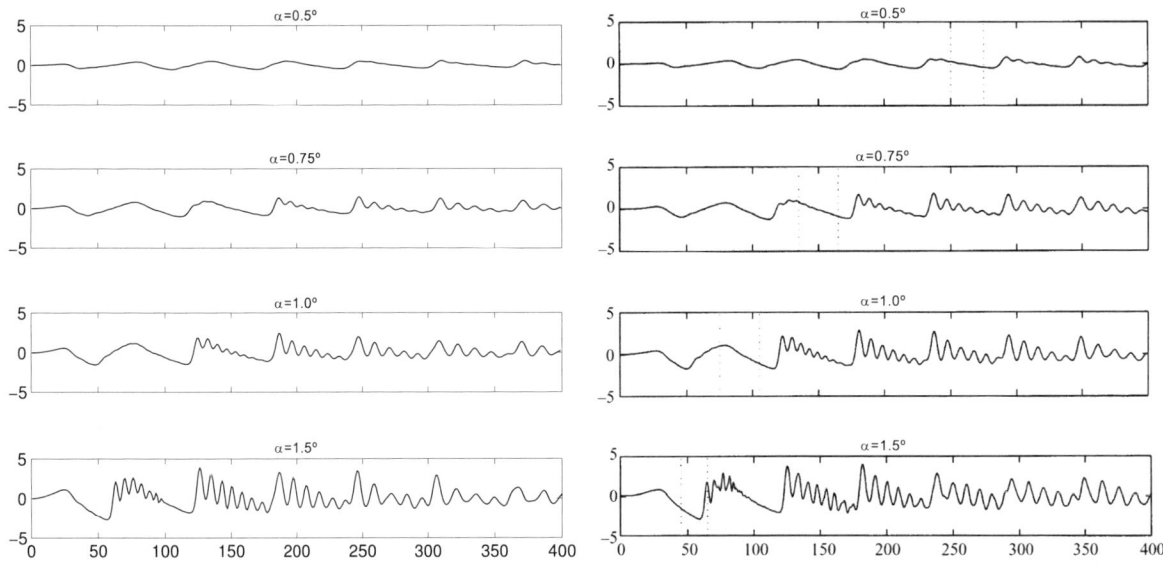

Figure 10
Interface displacement at the center of the domain for the simulations of the present work (*left*) and the experiments in HORN *et al.* (2001)
(*right*)

Figure 11
Solution for $h = 0.145$ and $\alpha = 85°$ at $t = 5, 10, 21$ s showing Kelvin-Helmholtz instabilities

features such as Kelvin-Helmholtz instabilities. We further perform a new simulation with an initial condition given by $h = 0.145$, $\alpha = 85°$ and $\delta_\rho = 0.1$. With this initial configuration, the fluid enters in a different regime and develops Kelvin-Helmholtz instabilities. In Fig. 11, we show the solution of this numerical simulation at different time instants.

4.4. Water Exchange Through the Strait of Gibraltar

This test is devoted to study the water exchange that takes place in the Strait of Gibraltar between the two different water masses: the Mediterranean water characterized by its high density and the Atlantic water, with a salinity much lower than the Mediterranean water, which makes the Atlantic water be less dense. The density difference

between the Atlantic and Mediterranean waters drives the fresh Atlantic water eastward in the upper layer, whereas in the lower layer, the denser Mediterranean water flows westward toward the Atlantic Ocean.

In this test, we propose a three-dimensional numerical experiment to simulate the exchange of these two water masses through the Strait of Gibraltar. Similarly to the two-dimensional tests described above, we set up initial conditions for temperature and salinity, T_0 and S_0, simulating the two water masses that we suppose initially separated at 5.85°W. The temperature and salinity are supposed to be horizontally homogeneous but vertically stratified. The profiles for both variables describe an idealized profile in the Atlantic Ocean and in the Mediterranean Sea, see Fig. 12.

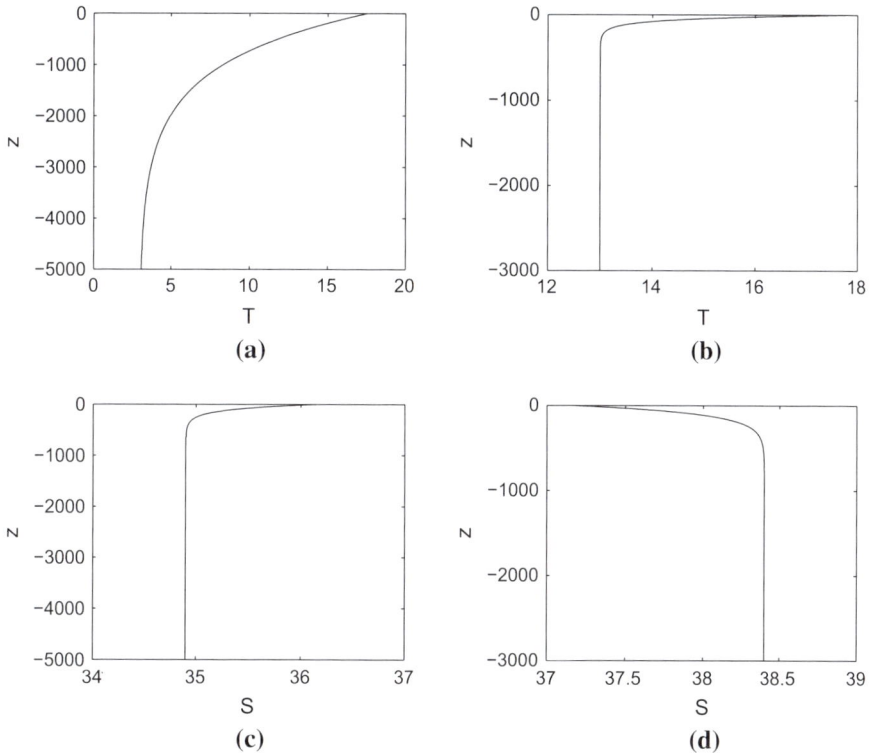

Figure 12
Vertical profiles of the initial conditions for temperature and salinity in the Atlantic Ocean (**a, c**), and the Mediterranean Sea (**b**) and (**d**)

Although the Coriolis force plays an important role in geophysical flows, since the area of the simulation is relatively small and narrow, we have not considered Coriolis force in this simulation. The other parameters of the simulation are $A_h = 1000$ m^2 s^{-1}, $A_v = 10^{-3}$ m^2 s^{-1}, $k_h = 500$ m^2 s^{-1}, $k_v = 10^{-5}$ m^2 s^{-1} and a time step of $\Delta t = 600$ s.

The generation of the hexahedral mesh starts with the quadrilateral mesh of the horizontal ocean surface in a way that each surface quadrilateral is the upper face of a hexahedral column that extends down to the bottom; the lateral faces of such a hexahedral column being perpendicular to the ocean surface plane. Next, the hexahedral column is divided into elementary

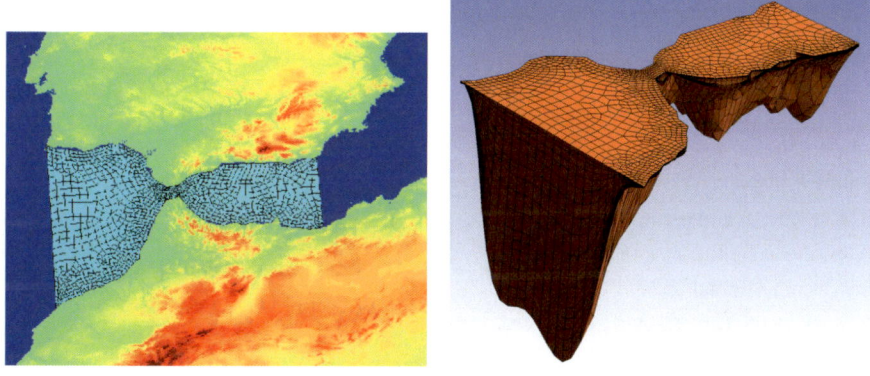

Figure 13
The finite element mesh used in the Gibraltar Strait problem ($m = 2$)

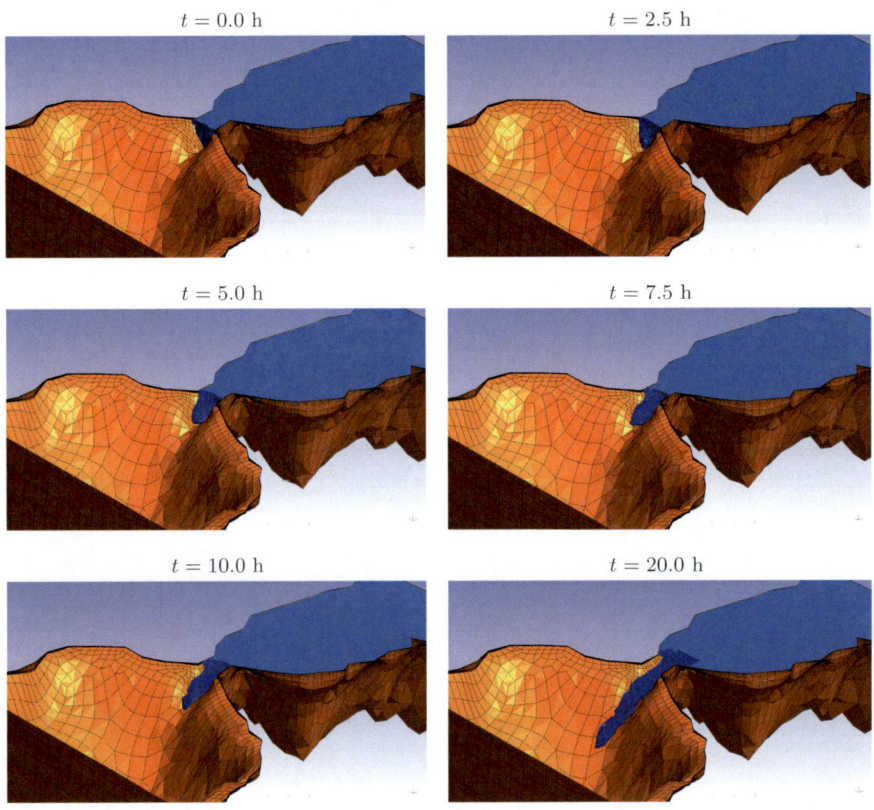

Figure 14
Isosurfaces of salinity at different time instants. The isosurface values are 37 and 38.5 psu

hexahedra by using a terrain following transformation. Notice that the number of elementary hexahedra is the same in each hexahedral column (see Fig. 13). The horizontal resolution of the elements ranges from 5 km (in the Strait of Gibraltar) to 20 km, whereas in the vertical, the mesh consists of 18 levels. In the simulation, we use $m = 2$, so this mesh gives an equivalent resolution that ranges from 2.5 km to 10 km in the horizontal and 37 vertical levels. Summing up, the mesh has 1183 quadrilateral elements in the surface and 4979 horizontal nodes, whereas it has 21,294 hexahedral elements and 184,223 nodes.

In Fig. 14, we show salinity isosurfaces for 37 psu and 38.5 psu at different time instants. As in the gravitational adjustment over a slope test, we observe how the denser water of the Mediterranean Sea moves down the slope towards the Atlantic basin. The velocity in the east-west direction is about 1 m s^{-1} (not shown) and agrees well with the results obtained in SANNINO et al. (2002) where a similar test is performed.

Acknowledgments

The first author research has been partially funded by the Spanish Economy and Competitivity Ministry and the European Regional Development Fund, through grant CGL2013-47261-R. Both authors would like to acknowledge the comments of one of the referees, who has helped in improving the paper.

Appendix 1: Existence and Uniqueness of the Density Problem

This appendix is related to the existence and uniqueness of the variational problem (13) that is used in the numerical scheme to compute the function $\bar{\rho}$ defined in (11). In a general framework, this can be written as follows. Given $f \in L^2(D)$, find $u \in H_0(D, \frac{\partial}{\partial z})$ such that

$$\int_D \frac{\partial u}{\partial z}\frac{\partial v}{\partial z} = \int_D f \frac{\partial v}{\partial z} \qquad \text{for all } v \in H_0\left(D, \frac{\partial}{\partial z}\right). \tag{14}$$

The space $H\left(D, \frac{\partial}{\partial z}\right)$ defined in (5) is a Hilbert space with norm

$$\|v\|_{H(D,\frac{\partial}{\partial z})} = \sqrt{\|v\|^2 + \left\|\frac{\partial v}{\partial z}\right\|^2}.$$

The subspace $H_0(D, \frac{\partial}{\partial z})$ is well defined because following TEMAM (1977), we can define a trace operator $\gamma : H(D, \frac{\partial}{\partial z}) \to H^{-\frac{1}{2}}(\Gamma_s)$ the kernel of which is the space $H_0(D, \frac{\partial}{\partial z})$. It is easy to see that $\left\|\frac{\partial u}{\partial z}\right\|$ is a norm in $H_0(D, \frac{\partial}{\partial z})$. Now applying the Lax-Milgram Lemma, it follows that problem (14) has a unique solution.

Appendix 2: Impact of Extrapolated Coriolis Term on the Overall Stability

Roughly speaking, Eq. (8) is the semi-discrete version of

$$\frac{Du}{Dt} - \text{div } (A\nabla u) + \mathbf{f} \times \mathbf{u} = \mathbf{r},$$

and the finite element formulation of this equation yields

$$M\frac{D\mathbf{u}_h}{Dt} + S\mathbf{u}_h + M(\mathbf{f} \times \mathbf{u}_h) = M\mathbf{r}_h, \tag{15}$$

where M and S are the mass and stiffness matrices, respectively, the coefficients of which are given by

$$m_{ij} = \int_D \phi_i\phi_j,$$

$$s_{ij} = \int_D A\nabla\phi_i\nabla\phi_j,$$

and where $\mathbf{u}_h$ and $\mathbf{r}_h$ are the finite element approximation of $\mathbf{u}$ and $\mathbf{r}$, respectively. Since M is a non-singular matrix, we can write (15) as

$$\frac{D\mathbf{u}_h}{Dt} + M^{-1}S\mathbf{u}_h + \mathbf{f} \times \mathbf{u}_h = \mathbf{r}_h$$

The stability of the latter equation can be studied as the stability of the simplified ODE

$$y' = \lambda y$$

where $y = u + iv$ and $\lambda = a + bi$ are complex function and number, respectively (note that, for this specific case, a corresponds to the eigenvalues of the matrix $M^{-1}S$ and b to the Coriolis parameter f). This ODE can be discretized using the second-order BDF formula, as in (8):

$$\frac{\frac{3}{2}y^{n+1} - 2y^n + \frac{1}{2}y^{n-1}}{\Delta t} = \lambda\left(2y^n - y^{n-1}\right).$$

The stability diagram is shown in Fig. 15a. Since the smallest eigenvalue of $M^{-1}S$ is much smaller than f, λ is very close to the imaginary axis, so the extrapolation formula could be unstable.

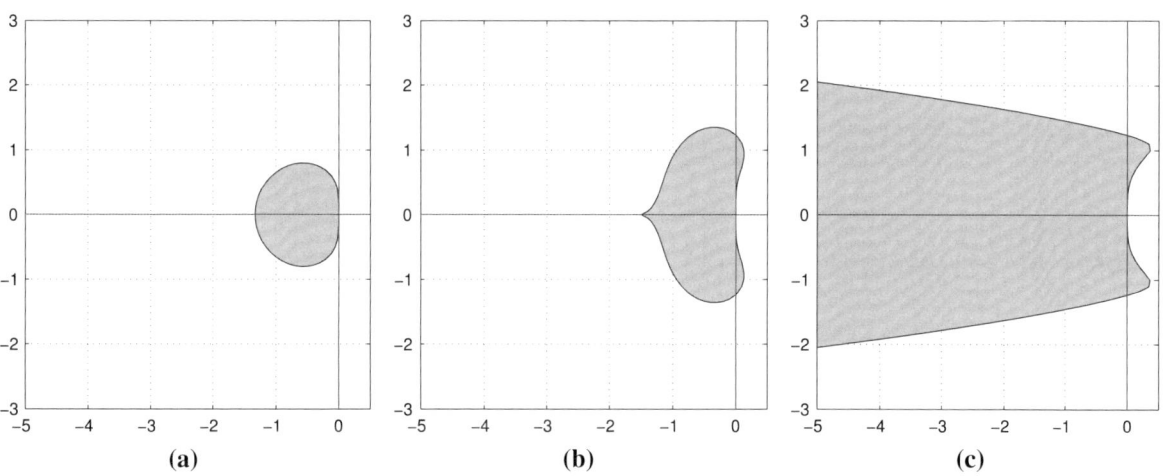

Figure 15
Stability region of **a** explicit second-order BDF method, **b** predictor–corrector second-order BDF method and **c** predictor–corrector second-order BDF method with implicit treatment of viscous term (see "Appendix 2")

To extend the stability region, one can use a prediction-correction treatment of this term, i.e.,

$$\frac{\frac{3}{2}\hat{y}^{n+1} - 2y^n + \frac{1}{2}y^{n-1}}{\Delta t} = \lambda\left(2y^n - y^{n-1}\right),$$

$$\frac{\frac{3}{2}y^{n+1} - 2y^n + \frac{1}{2}y^{n-1}}{\Delta t} = \lambda\hat{y}^{n+1}.$$

The stability diagram for the predictor–corrector scheme is shown in Fig. 15b.

A more stable scheme corresponding to the implicit treatment of the viscous term is the following:

$$\frac{\frac{3}{2}\hat{y}^{n+1} - 2y^n + \frac{1}{2}y^{n-1}}{\Delta t} = a\hat{y}^{n+1} + bi\left(2y^n - y^{n-1}\right),$$

$$\frac{\frac{3}{2}y^{n+1} - 2y^n + \frac{1}{2}y^{n-1}}{\Delta t} = ay^{n+1} + bi\hat{y}^{n+1}.$$

The stability diagram of which is represented in Fig. 15c.

This predictor–corrector applied to the proposed splitting becomes: for $j = 1, 2$,

1. Set $\mathbf{u}_{e,j}^{n+1} = 2\mathbf{u}^n - \mathbf{u}^{n-1}$ for $j = 1$ or $\mathbf{u}_{e,j}^{n+1} = \mathbf{u}_{e,j-1}^{n+1}$ for $j = 2$, and compute $q_j^{n+1} \in H^1(D)$ that satisfies:

$$\Delta t\left(\nabla q_j^{n+1}, \nabla v\right) = \rho_0\left(2\mathbf{u}^{n*} - \frac{1}{2}\mathbf{u}^{(n-1)**}, \nabla v\right)$$
$$+ \rho_0\left(-\Delta t\nabla \times (A\nabla) \times \mathbf{u}_{e,j}^{n+1} - \Delta t\mathbf{f} \times \mathbf{u}_{e,j}^{n+1}, \nabla v\right)$$
$$- \Delta t(\nabla r^{n+1}, \nabla v) + \Delta t\mathbf{g}\left(\rho^{n+1}, \nabla v\right)$$

for all $v \in H^1(D)$.

2. Compute $\mathbf{u}_j^{n+1} \in V$ such that

$$\frac{3}{2}\left(\mathbf{u}_j^{n+1}, \mathbf{v}\right) + \Delta t\left(A\nabla\mathbf{u}_j^{n+1}, \nabla\mathbf{v}\right)$$
$$= \left(2\mathbf{u}^{n*} - \frac{1}{2}\mathbf{u}^{(n-1)**} - \Delta t\mathbf{f} \times \mathbf{u}_{e,j}^{n+1}, \mathbf{v}\right)$$
$$- \Delta t\frac{1}{\rho_0}\left(\nabla q_j^{n+1}, \mathbf{v}\right) - \Delta t\frac{1}{\rho_0}\left(\nabla r^{n+1}, \mathbf{v}\right)$$
$$+ \Delta t\frac{1}{\rho_0}\mathbf{g}\left(\rho^{n+1}, \mathbf{v}\right)$$

for all $\mathbf{v} \in \mathbf{V}$.

Finally, set $\mathbf{u}^{n+1} = \mathbf{u}_2^{n+1}$.

REFERENCES

V. CASULLI. *A semi-implicit finite difference method for non-hydrostatic, free-surface flows.* Int. J. Numer. Meth. Fluids, 30(4):425–440, 1999.

V. CASULLI and P. ZANOLLI. *Semi-implicit numerical modeling of nonhydrostatic free-surface flows for environmental problems.* Math. Comput. Modelling, 36(9-10):1131–1149, 2002.

A.J. CHORIN. *Numerical solution of the Navier-Stokes equations.* Math. Comp., 22:745–762, 1968.

J.H. FERZIGER and M. PERIC Computational method for fluid dynamics. 2nd Edition, Springer, 1999.

P. GALÁN DEL SASTRE and R. BERMEJO. *A comparison of semi-Lagrangian and Lagrange-Galerkin hp-FEM methods in convection-diffusion problems.* Commun. Comput. Phys., 9(4):1020–1039, 2011.

J. L. GUERMOND, P. MINEV, and J. SHEN. *An overview of projection methods for incompressible flows.* Comput. Methods Appl. Mech. Engrg., 195(44-47):6011–6045, 2006.

J.-L. GUERMOND and L. QUARTAPELLE. *A projection FEM for variable density incompressible flows.* J. Comput. Phys., 165(1):167–188, 2000.

J. L. GUERMOND and J. SHEN. *A new class of truly consistent splitting schemes for incompressible flows.* J. Comput. Phys., 192(1):262–276, 2003.

J. L. GUERMOND and J. SHEN. *Velocity-correction projection methods for incompressible flows.* SIAM J. Numer. Anal., 41(1):112–134 (electronic), 2003.

D.B. HAIDVOGEL and A. BECKMANN. Numerical ocean circulation modeling. Imperial College Press, 1999.

Y. HEGGELUND, F. VIKEBO, J. BERNTSEN, and G. FURNES. *Hydrostatic and non-hydrostatic studies of gravitational adjustment over a slope.* Cont. Shelf Res., 24(18):2133–2148, DEC 2004.

D.A. HORN, J. IMBERGER, and G.N. IVEY. *The degeneration of large-scale interfacial gravity waves in lakes.* J. Fluid Mech., 434:181–207, MAY 10 2001.

M. ISKANDARANI, D. B. HAIDVOGEL, and J. C. LEVIN. *A three-dimensional spectral element model for the solution of the hydrostatic primitive equations.* J. Comput. Phys., 186(2):397–425, 2003.

M. ISKANDARANI, J.C. LEVIN, B.J. CHOI, and D.B. HAIDVOGEL. *Comparison of advection schemes for high-order h-p finite element and finite volume methods.* Ocean Model., 10(1-2):233–252, 2005.

Y. KANARSKA, A. SHCHEPETKIN, and J. C. McWILLIAMS. *Algorithm for non-hydrostatic dynamics in the regional oceanic modeling system.* Ocean Model., 18(3-4):143–174, 2007.

G. KARNIADAKIS, M. ISRAELI, and S.A. ORSZAG. *High-order splitting methods for the incompressible Navier-Stokes equations.* J. Comput. Phys., 97(2):414–443, 1991.

G. KARNIADAKIS and S.J. SHERWIN. Spectral/hp element methods for CFD. Numerical Mathematics and Scientific Computation. Oxford University Press, New York, 1999.

R.J. LABEUR and J.D. PIETRZAK. *A fully three dimensional unstructured grid non-hydrostatic finite element coastal model.* Ocean Model., 10:51–67, 2005.

J. MARSHALL, A. ADCROFT, C. HILL, L. PERELMAN, and C. HEISEY. *A finite-volume, incompressible Navier Stokes model for studies of the ocean on parallel computers.* J. Geophys. Res., *102*(C3):5753–5766, MAR 15 1997.

J. MARSHALL, C. HILL, L. PERELMAN, and A. ADCROFT. *Hydrostatic, quasi-hydrostatic, and nonhydrostatic ocean modeling.* J. Geophys. Res., *102*(C3):5733–5752, MAR 15 1997.

K.W. MORTON, A. PRIESTLEY, and E. SÜLI. *Stability of the Lagrange-Galerkin method with nonexact integration. RAIRO Modél.* Math. Anal. Numér., *22*(4):625–653, 1988.

G. SANNINO, A. BARGAGLI, and V. ARTALE. *Numerical modeling of the mean exchange through the Strait of Gibraltar.* J. Geophys. Res., *107*(C8):9–1–9–24, 2002.

R. TEMAM. *Sur l'approximation de la solution des équations de Navier-Stokes par la méthode des pas fractionnaires. II.* Arch. Rational Mech. Anal., 33:377–385, 1969.

R. TEMAM. Navier-Stokes equations. Theory and numerical analysis. North-Holland Publishing Co., Amsterdam, 1977. Studies in Mathematics and its Applications, Vol. 2.

UNESCO. *Tenth report of the joint panel on oceanographic tables and standards.* UNESCO Technical Papers in Marine Sci., *36*, 1981.

P. ŠOLÍN, K. SEGETH, and I. DOLEŽEL. Higher-order finite element methods. Studies in Advanced Mathematics. Chapman & Hall/CRC, Boca Raton, FL, 2004. With 1 CD-ROM (Windows, Macintosh, UNIX and LINUX).

(Received April 1, 2014, revised June 25, 2015, accepted September 29, 2015, Published online October 31, 2015)

Pure Appl. Geophys. 173 (2016), 909–922
© 2015 Springer Basel
DOI 10.1007/s00024-015-1196-5

Pure and Applied Geophysics

CrossMark

Two-Dimensional Coupled Distributed Hydrologic–Hydraulic Model Simulation on Watershed

MIGUEL CEA[1,2] and MARTIN RODRIGUEZ[1]

Abstract—The objective of this work is to develop a coupled distributed model that enables to analyze water movement in watershed as well as analyze the rainfall-runoff. More specifically, it allows to estimate the various hydrologic water cycle variables at each point of the watershed. In this paper, we have carried out a coupled model of a distributed hydrological and two-dimensional hydraulic models. We have incorporated a hydrological rainfall-runoff model calculated by cell based on the Soil Conservation Service (SCS) method to the hydraulic model, leaving it for the hydraulic model (GUAD2D) to conduct the transmission to downstream cells. The goal of the work is demonstrate the improved predictive capability of the coupled Hydrological-Hydraulic models in a watershed.

Key words: Saint-Venant equations, Finite volume method, Coupled model, Hydrologic model, Distributed hydraulic model.

1. Introduction

The development of coupled hydrologic-hydraulic models is justified, as they can be a useful tool to achieve a watershed management from the water resources point of view, flood prevention, etc. A good balance between prediction capacity, the computational cost, data requirements and sensitivity to the model parameters should be a key objective.

The main objective of this models is to write a drainage works construction project which improves flood problems in the watershed. The partial objectives are drawing hydrologic study, hydraulic study, sediment transport study, risk analysis, environmental study and proposal of measures to improve floods in a watershed. The specific objectives are to determinate flows of rivers which cross some city to identify flood problems, quantify damage and suggest measures to reduce this damage (CAMPOS *et al.* 2004a, b).

The goal of the work is demonstrate the improved predictive capability of the coupled Hydrological-Hydraulic models in a watershed, because rainfall in situ is principally responsible for damage in many cases.

The Saint-Venant system is commonly used to model flows in rivers; this system describes the flow as a conservation laws with and additional source term, which was introduced in DE SAINT-VENANT (1871). The hydraulic model simulates the water transfer introduced upstream of the modeled area. This model resolves Saint Venant equations on an unstructured mesh by finite volume in each time step and applying progressive methods and Godunov method to bring stability to the system (CUNGE *et al.* 1980; BERMÚDEZ and VÁZQUEZ 1994; BERMÚDEZ *et al.* 1998; BRUFAU *et al.* 2002; RAVIART and GODLEWSKY 1996; GUDONOV 1959; LEVEQUE 1998).

The distributed model enables better results, because it allows to consider the spatial distribution of rainfall on the one hand, and to consider issues affecting spatial distribution of runoff production an propagation on the other hand [(SIMPA (RUIZ 1999), MIKE SHE (REFSGAARD and STORM 1995), TOPMODEL (BEVEN and KIRKBY 1979).

This model enables to include the spatial and temporal variability of rainfall, which may consider with some reasonable accuracy the location of the storm and the area of the watershed that it covers, which implies that these elements are not part of uncertainty in the results. Thus, a distributed model allows obtaining the temporal flooding evolution on the watershed elements during the same period of analysis.

[1] INCLAM, S.A. Ingeniería del Agua, C/ Samaria 4, 28009 Madrid, Spain. E-mail: miguel.cea@inclam.com; martin.rodriguez@inclam.com

[2] Departamento de Matemáticas, Universidad Autónoma de Madrid, Cantoblanco, 28049 Madrid, Spain.

Reprinted from the journal

Within these models we find distributed parameter models, which are characterized in that watershed, which is divided into homogeneous cells where a global model is applied. Thus the overall response of watershed is calculated from the contributions of the cells [(DHMS (Palacios-Vélez and Cuevas-Renaud 1989), CATFLOW (Maurer 1997)].

Therefore, the overall scale effects are due to the sum of the individual effects of each cell scale, mainly due to

1. The pattern of spatial distribution of input variables and parameters.
2. The spatial interpolation method used for aggregation of data at higher spatial scales.
3. The mathematical structure of the model that describes the process.
4. The spatio-temporal structure of the data used for parametrization.

In a traditional hydrologic model, it is considered that much of the storm water runoff is produced by excess precipitation. This is called direct or surface runoff, and the portion of the precipitation volume that occurs is called net or effective precipitation. Using a transfer function, the volume becomes direct runoff and subsequently the cost basis is added, thus the total runoff hydrograph is obtained. When we calculate the runoff cell level, we take into account four processes: precipitation, loss, transformation of excess rainfall in direct runoff and cost base. About the cell, we believe that each process occurs evenly, and therefore that the process responds to such leaves as an aggregate model [MIKE SHE (Refsgaard and Storm 1995), IHDM (Morris 1980; Rogers et al. 1985)].

2. Hydraulic Model

The water movement is governed by the fundamental mass and momentum conservation principles. This approach has combined several assumptions that define the shallow water model. This section will give a brief description of the physical model associated with the hydrodynamic simulation in 2D.

The essential fact is that the thickness of the water layer is small. The basic assumptions of this model are:

1. The vertical pressure distribution is hydrostatic or vertical acceleration is small.
2. The friction losses are similar in transient flow and in steady flow.
3. The average slope of the channel bed is so small that the tangent can be estimated by the angle and the measurements in the bed are equivalent to the measurements in the horizontal plane.

This is a non-linear hyperbolic system of conservation laws, which is expressed by the following partial differential equations system (Bermúdez et al. 1998):

$$\begin{cases} \dfrac{\partial h}{\partial t} + \dfrac{\partial hu}{\partial x} + \dfrac{\partial hv}{\partial y} = 0 \\[2mm] \dfrac{\partial hu}{\partial t} + \dfrac{\partial}{\partial x}\left(hu^2 + g\dfrac{h^2}{2}\right) + \dfrac{\partial huv}{\partial y} = gh(S_{0x} - S_{fx}) \\[2mm] \dfrac{\partial hv}{\partial t} + \dfrac{\partial huv}{\partial x} + \dfrac{\partial}{\partial y}\left(hv^2 + g\dfrac{h^2}{2}\right) = gh(S_{0y} - S_{fy}) \end{cases}$$

$$(1)$$

where h is water depth, hu and hv are the flow rates along directions spatial coordinates, S_{0x}, S_{0y} are the bed slopes

$$S_{0x} = -\frac{\partial z}{\partial x}, S_{0y} = -\frac{\partial z}{\partial y} \qquad (2)$$

S_{fx}, S_{fy} are the friction losses along the two coordinate directions

$$S_{fx} = \frac{n^2 u \sqrt{u^2 + v^2}}{h^{4/3}}, S_{fy} = \frac{n^2 v \sqrt{u^2 + v^2}}{h^{4/3}} \qquad (3)$$

and n is the Manning's roughness coefficient. This coefficient is determined from experimental measurements or estimated from the value tables (Chow et al. 1994; Manning 1891).

Conservation laws described above in (1) can be written in vector form as (Stoker 2011):

$$\frac{\partial U}{\partial t} + \frac{\partial F(U)}{\partial x} + \frac{\partial G(U)}{\partial y} = S(U), \qquad (4)$$

where

$$U = (h, hu, hv)^T$$

$$F = \left(hu, hu^2 + \frac{gh^2}{2}, huv \right)^T \quad G = \left(hv, huv, hv^2 + \frac{gh^2}{2} \right)^T$$

$$S = \left(0, gh(S_{0x} - S_{fx}), gh(S_{0y} - S_{fy}) \right)^T$$

$$(5)$$

or conservative formulation as

$$\frac{\partial U}{\partial t} + \vec{\nabla} E(U) = \vec{\nabla} T(U) \tag{6}$$

where $E = E(F, G)$ y $T = T(U)$.

We used the finite volume method for solving the problem (4) because this method is better adapted to the hyperbolic equations than finite element method or finite differences (CUNGE et al. 1980; BERMÚDEZ et al. 1998; VÁZQUEZ-CENDÓN 1999; MURILLO et al. 2005, 2006).

In order to do this, we now introduce a discretization of this problem as follows:

For any $h > 0$, we consider a triangulation $\mathcal{T}_h = \left\{ (\tau_i^h)_{i \in I_h} \right\}$ of $\mathcal{C}$ made of finite elements τ_i^h so that

$$\mathcal{C} = \overset{\circ}{\bigcup_{i \in I_h} \tau_i^h},$$

where $\overset{\circ}{A}$ denotes the interior of $A \subset \mathbf{R}^2$, To this end, we suppose that the triangulations are uniformly regular, that is

$$\exists \sigma > 0 \text{ s.t. } \forall h > 0, \quad \tau_i^h \in \mathcal{T}_h, \quad 0 < \frac{h}{\rho_i} \leq \sigma,,$$

where the grid size h is defined as the maximum diameter of the elements τ_i^h and ρ_i is the radius of the largest ball contained in τ_i^h, (usual mesh of GUAD2D in Fig. 1).

We integrate in Ω in each cell of Eq. (4) to compute the variables value in the centers of the cells, applying the Gauss Theorem to the second and third term and approximating the contour integrals $\partial \Omega$ over the sum of the edges k, in each cell Ω_i:Alginet model without rain

$$\frac{\partial}{\partial t} \int_{\Omega_i} U(x, y) d\Omega + \sum_{k=1}^{NE} \int_{e_k}^{e_{k+1}} \delta E(x, y)_k \mathbf{n}_k dl$$

$$= \sum_{k=1}^{NE} \int_{e_k}^{e_{k+1}} \delta T(x, y)_k \mathbf{n}_k dl, \tag{7}$$

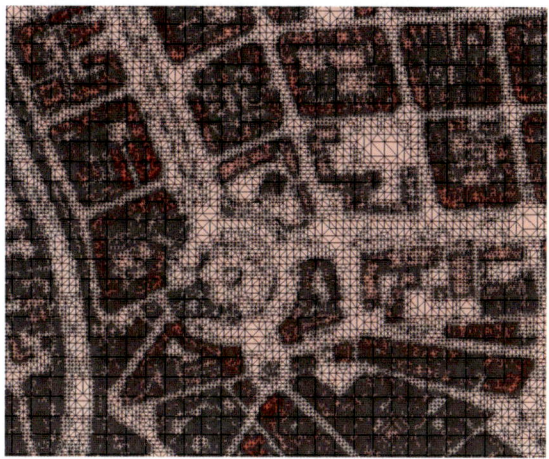

Figure 1
Usual mesh of GUAD2D

where $\delta E = E_j - E_i$, $\delta T = T_j - T_i$, j is the function value on the neighbor cells Ω_j.

The mathematical properties of the hyperbolic system include the existence of a Jacobian matrix, $\mathbf{J}_n$, (ROE 1981) defined as

$$\mathbf{J}_n = \frac{\partial (E \cdot n)}{\partial U} = \frac{\partial F}{\partial U} \mathbf{n}_x + \frac{\partial G}{\partial U} \mathbf{n}_y. \tag{8}$$

The problem can be reduced to a one-dimensional Riemann problem projected onto the normal direction at each cell edge (RAVIART and GODLEWSKY 1996), thus:

$$\sum_{k=1} \delta E_k \mathbf{n}_k l_k = \sum_{k=1} \tilde{\mathbf{J}}_{n,k} \delta U_k l_k \tag{9}$$

that their properties:

$$\tilde{\mathbf{J}}_{n,k} \tilde{e}_k^m = (\tilde{\lambda} \tilde{e})_k^m \tag{10}$$

Then, we use the flux difference procedure, the difference in vector U across the edge cell is projected onto the matrix eigenvectors basis:

$$\delta U_k = U_{j,o} - U_{i,o} = \sum_{m=1}^{3} (\alpha \tilde{e})_k^m \tag{11}$$

which allow us to write the flux us follows:

$$\sum_{k=1}^{NE} \tilde{\mathbf{J}}_{n,k} \delta U_k l_k = \sum_{k=1}^{NE} \sum_{m=1}^{3} (\tilde{\lambda}^m \alpha^m \tilde{e}^m)_k l_k. \tag{12}$$

As all variables defined in each cell are uniform, the contour integral over the source term is approximated by:

$$\sum_{k=1}^{NE} \int_{e_k}^{e_{k+1}} \delta T_k \mathbf{n}_k dl = \sum_{k=1}^{NE} \delta T_k \mathbf{n}_k l_k. \qquad (13)$$

Finally, the numeric scheme can be written as follows:

$$U_i^{n+1} = U_i^n - \sum_{k=1}^{NE} \sum_{m=1}^{3} ((\tilde{\lambda}^- \alpha - \beta^-)\tilde{e})_k^m l_k \frac{\Delta t}{A_i}, \qquad (14)$$

the β^m are defined as

$$\beta^1 = -\frac{\tilde{c}}{2}(\delta z + d_n S_f), \quad \beta^1 = -\beta^3, \quad \beta^2 = 0 \quad (15)$$

being the eigenvalues

$$\tilde{\lambda}^1 = \tilde{u} \cdot \mathbf{n} + \tilde{c}, \quad \tilde{\lambda}^2 = \tilde{u} \cdot \mathbf{n}, \quad \tilde{\lambda}^3 = \tilde{u} \cdot \mathbf{n} - \tilde{c} \quad (16)$$

with

$$\tilde{u}(u_j, u_i) = \frac{u_j \sqrt{h_j} + u_i \sqrt{h_i}}{\sqrt{h_j} + \sqrt{h_i}}, \quad \tilde{v}(v_j, v_i) = \frac{v_j \sqrt{h_j} + v_i \sqrt{h_i}}{\sqrt{h_j} + \sqrt{h_i}},$$

$$\tilde{c}(h_j, h_i) = \sqrt{g \frac{(h_j + h_i)}{2}}$$
$$(17)$$

and the pertinent eigenvectors:

$$\tilde{e}^1 = \begin{pmatrix} 1 \\ \tilde{u} + \tilde{c}\mathbf{n}_x \\ \tilde{v} + \tilde{c}\mathbf{n}_y \end{pmatrix}, \quad \tilde{e}^2 = \begin{pmatrix} 0 \\ -\tilde{c}\mathbf{n}_y \\ \tilde{c}\mathbf{n}_x \end{pmatrix}, \quad \tilde{e}^3 = \begin{pmatrix} 1 \\ \tilde{u} - \tilde{c}\mathbf{n}_x \\ \tilde{v} - \tilde{c}\mathbf{n}_y \end{pmatrix}$$
$$(18)$$

resulting in the following expression for the values of α^m:

$$\alpha^{1,3} = \frac{\delta h}{2} \pm \frac{1}{2\tilde{c}}(\delta q - \tilde{u}\delta h)\mathbf{n}, \quad \alpha^2 = \frac{1}{\tilde{c}}(\delta q - \tilde{u}\delta h)\mathbf{n}_t, \qquad (19)$$

where $q = (hu, hv)$ and $\mathbf{n}_t = (-\mathbf{n}_y, \mathbf{n}_x)$.

In order to compare the GUAD 2D with respect to other software packages (see Table 1), for example, SOBEK (ABAZI 2005) and Mike 21 (VANDERKIMPEN et al. 2008), the computational times were obtained with and Intel Xeon 1.7GH Processor and 1GB RAM memory (GARROTE et al. 2008).

3. Hydrological Model

The model we have developed uses as hydrological method the calculation runoff per cell based on

the SCS method (CRONSHEY 1986), leaving it for the GUAD-2D model to take the transmission to downstream cells for strictly hydraulic methods. This system has the advantage of incorporating the geomorphological variables of watershed automatically without relying on formulations of the watershed models or other distributed models (RUIZ 1999; REFSGAARD and STORM 1995; BEVEN and KIRKBY 1979; MORRIS 1980, etc.).

Therefore, the model calculates the cell surface runoff by applying the SCS method, and it is necessary to characterize the precipitation and the lost function area. The first is from the areal rainfall of each of the return periods and the second is defined by the curve number (CN) and initial abstraction. From net runoff in each cell the accumulated runoff is computed.

The input data of the model are:

1. The hyetograph of the modeled area.
2. The curve number is incorporated into the model through a GRID of curve numbers, so that each cell has a curve number.
3. The initial losses expressed as a percentage k of S^1 and common for the entire modeled area.

The SCS method is widely used for its ability to estimate the parameters from soil and vegetation data (CRONSHEY 1986). It is based on the water balance equation and two fundamental hypotheses. The first hypothesis states that the ratio between the volume of direct runoff and the maximum potential runoff is equal to the ratio between the real infiltration and maximum potential infiltration. The second hypothesis states that the initial infiltration is a fraction of the potential retention. Also, assume the existence of a threshold runoff, $I_a{}^2$, below which the rainfall do not cause runoff. The value of I_a states from S by the formula:

$$I_a = k \times S \qquad (20)$$

In turn, the value of S is related to the curve number through the formula:

---

[1] S is the maximum possible retention of the soil expressed in depth units (mm or in).
[2] Initial surface moisture storage capacity, initial abstraction depth units (mm or in).

Table 1

Computational times by different methods

	SOBEK	Mike 21	GUAD 2D
Numerical scheme	Finite differences	Finite differences	Finite volumes
Hydraulic jump	111 sg	262 sg	21 sg
Channel contraction	182 sg	160 sg	44 sg
Dam break	9876 sg	Inconsistent result	235 sg
Flooding event (2200 ha)	119324 sg	Unfeasible computational time	54264 sg

$$S = \frac{25400}{CN} - 254 \text{ metric units}$$

$$\text{or} \qquad (21)$$

$$S = \frac{1000}{CN} - 10 \text{ imperial units}$$

The hypothesis of the SCS method is that the relationship between real and potential quantities are equal, or what is the same:

$$\frac{F_a}{S} = \frac{P_e}{P - I_a}, \qquad (22)$$

where P_e is the direct runoff, which is always less than or equal to the accumulated rainfall up to said, P (mm or in). F_a is the additional depth of water retained in the watershed that is less than or equal to the maximum potential retention, S. By the continuity principle we know that

$$P = P_e + I_a + F_a. \qquad (23)$$

Therefore, with Eqs. (22) and (23) we obtain the basic equation for calculation direct runoff from a storm to a given time by the formula

$$P_e = \frac{(P - I_a)^2}{P - I_a + S}. \qquad (24)$$

The net runoff to incorporate in that increment time can be obtained by difference with the runoff of the above time increment. This value is added directly to the resulting depth of the hydraulic calculation, proceeding to pass the hydraulic calculation of the next interval.

Therefore, the most important part in the coupling of the two methods is produced in the following form. We incorporate this to the hydraulic model explained in the previous section. Taking the equation system (14) we have

$$h_i^{n+1} = h_i^n - \Delta t \sum_{k=1}^{NE} \Psi_i^k, \qquad (25)$$

and we introduce compute of net runoff (24) and we obtain

$$h_i^{n+1} = h_i^n - \Delta t \sum_{k=1}^{NE} \Psi_i^k - \frac{1}{10^3} \left[(P_e)_i^n - (P_e)_i^{n-1} \right], \qquad (26)$$

with

$$\Psi_i^k = \sum_{m=1}^{3} \left((\tilde{\lambda}^- \alpha - \beta^-) \tilde{e} \right)_k^m \frac{l_k}{A_i}.$$

Equation (25) can be generalized as follows:

$$h_i^{n+1} = h_i^n - \Delta t \sum_{k=1}^{NE} \Psi_i^k - I(i, \Delta t), \qquad (27)$$

where the function $I(i, \Delta t)$ is any runoff model for predicting infiltration of rainfall losses in hydrologic modeling, for example:

- Horton model (HORTON 1940)

$$I = I_{min} + (I_0 - I_{min}) \exp^{-kt},$$

where I_0 is the initial infiltration rate, I_{min} is the minimum infiltration rate and k is decay coefficient.
- Philip model (PHILIP 1957)

$$I = S t^{1/2} + At,$$

where S is the sorptivity and A is a parameter with dimension of saturated hydraulic conductivity (A is equivalent to I_{min} in Horton model).
- Green-Ampt model (CHOW *et al.* 1994)

$$I = k_s \left(1 + \frac{(h + \Psi)\Delta\theta}{L_0 \cdot \Delta\theta + F} \right)$$

$$F = \int_0^t f dt \qquad L = L_0 + \frac{F}{\Delta\theta} \qquad \delta\theta = \phi - \theta_i,$$

where h is the depth, k_s is the saturated soil permeability, Ψ is the cushion in the non-saturated soil region, $\Delta\theta$ the humidity content change of the soil as the saturated front, θ_i the soil's initial humidity content, ϕ the soil's total porosity and L is the depth of the saturated region.

As seen from the three phenomena normally simulated in watersheds hydrologic models (calculation of surface runoff, transport to the end of the watershed and transport by network channels) the hydrological model GUAD-2D uses only the first phenomenon, which is the calculation of surface runoff, leaving the other two processes in the hydraulic part of the model.

This methodology has the advantage of reflecting the morphology of watershed much more precise, obviating the use of formulas, such as Concentration Time or Muskingum factors (Ruiz 1999; Refsgaard and Storm 1995), that represent a very simplistic way of how the water flows through watershed.

4. Numerical Results

With the model proposed in the previous chapter, we try to see the functioning of the study area in future situation. For this case there is no real solution with which to compare the solution, but both models are highly tested separately, both the hydraulic model (for example, the models of Table 1) and other real case (Olivera et al. 2008; Ferrer-Juliá et al. 2009; Campos et al. 2004a; López et al. 2009) and the hydrological model (Booth et al. 2002; Schroeder 1994; Yu 1998; Tsihrintzis and Hamid 1997). Cartographic and thematic information comes from the AEMET, Instituto Cartogrfico Valenciano, LIDAR MDS, Generalitat Valenciana, CORINE LAND COVER 2000 and PNOA.

The modelization has been carried out from a digital terrain model with mesh size of 4 m, that allows to see clearly the behavior of the riverbed and planned urban environment Alginet referrals, the

study area extends from downstream Jucar-Tunia channel to A7 road, comprising a total area of 12.97km^2.

Floods in Alginet (Valencia, Spain) are a recurrent problem in which relevant agencies have worked for years. Several classic hydraulic infrastructures of protection (channeling, cutoff) solve local problems. Alginet has an urban planning which shows a specific urban zone arrangement and kind of buildings. This is the third analysis because it is necessary to detect origin of the problem and to quantify it with high accuracy. Alginet is in foothills of Sierra Falaguera and extent its lands as far as the marsh of Albufera de Valencia. The central area is located at the confluence of four rivers (Seor, Agua, Forca and Belenguera), as well, it collects runoff from Sierra Falaguera (see Fig. 2). The A-7 road makes a border of the city and forms a pool with the infrastructure. Downstream terrain, rice fields slopes are nearly zero.

The municipality had been flooded recently (November 1987, October 1991 or October 2000), some of them of great importance. Alginet, in particular case, has suffered flash-floods and overflow of rivers upstream the city. General causes of problems associated with Alginet floods are listed here:

- Insufficient river drainage in urban zones.
- Limited slope of final river stretch.
- Inappropriate urban development.
- "Pool" effect in infrastructures due insufficient transversal drainage works.

Next photographs (Fig. 3), which were taken on 29th of September of 2009, show previous problems in some important local points in Alginet. Record rainfall for those dates in Alginet environment are lower than 25 years of return period.

To analyze the results, we have selected 55 points or control significant for which have determined the hydraulic parameters of water level, speed and depth of the three simulated flows (see Figs. 4, 5). The choice of these control points was made according to criteria of dangerous black spots detected confluence of canyons or works by important step.

Urban zone in Alginet is in the middle of confluence of several rivers aggravated by the default of drainage in structures. But rainfall in situ is principally responsible for damage in many cases. The

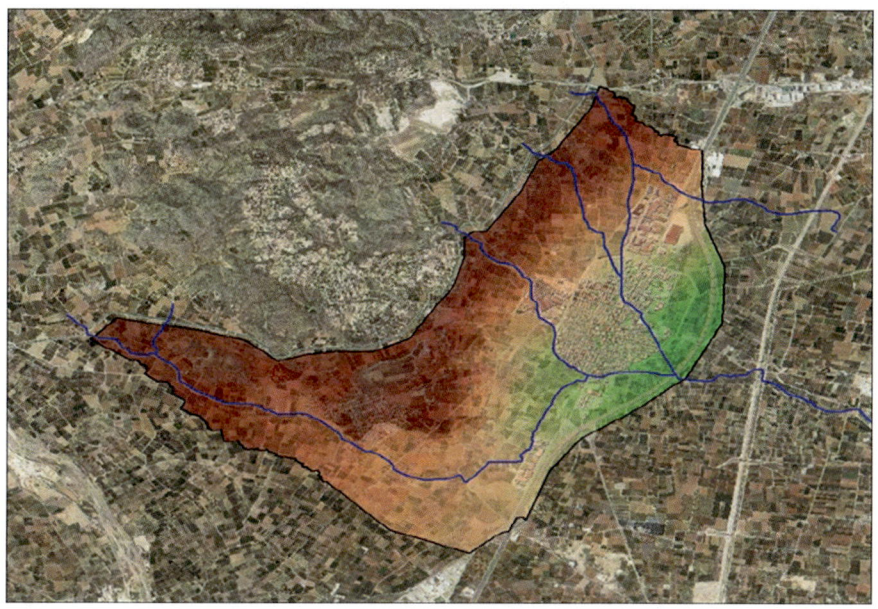

Figure 2
Modeled area. Alginet (lat: 39.263876, lon: −0.47029)

Figure 3
Seor river (Pintor Velzquez st.), Forca river, (Alginet urban zone), A-7 drainage works, (Seor and Forca confluence)

velocities in urban zone are a big problem. Downstream A-7 there is a primitive bedchannel which is a preferential flow zone. In Natural Park of Albufera, flow movements are in a complex net of overflow irrigation channels with low velocities and bidimensional behavior. Agua river flows southern Alginet near to industrial zone and overflows to the urban zone.

To compute the runoff of the modeled area, it is necessary to perform the morphological characterization of the watershed. This has been introduced in the hyetograph model to characterize the study area rainfall and its loss function, defined from the curve number of the study area and the initial abstraction.

The hyetograph is introduced in the model as a rainfall curve (mm) in function of time (s); the hyetograph used for this case we have calculated based on the use of MINISTERIO DE OBRAS PÚBLICAS Y URBANISMO (1990) and is as follows (Fig. 6):

The curve number allows to quantify the interception, retention, infiltration and finally the runoff

Figure 4
Depth in 25, 100 and 500 years return period

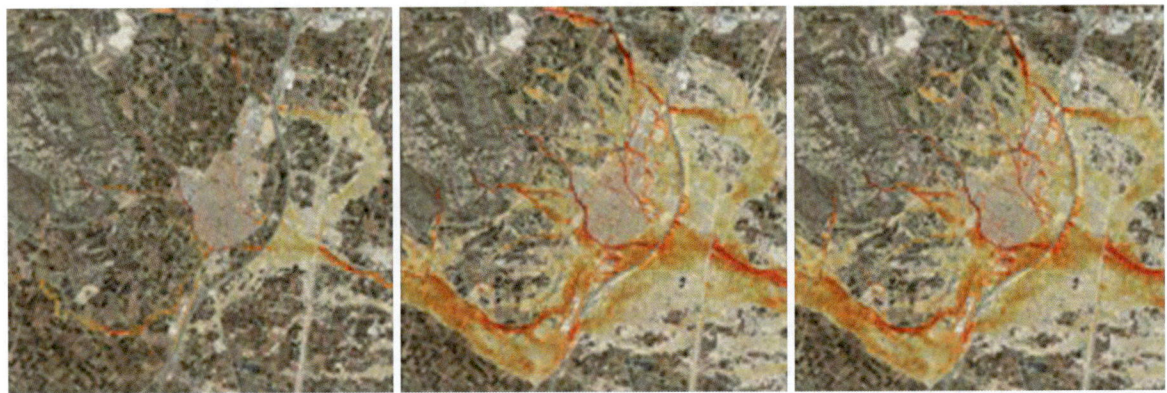

Figure 5
Velocity T = 25, 100 and 500 years return period

terrain from the lithological and soil characteristics, vegetation cover and average slope. In GUAD-2D it is introduced from a raster so that each cell has a curve number and a parameter of initial losses of 20 % (expressed as a percentage of the maximum retention of the soil), common for all the modeled area. When the curve number is high (red in Fig. 7), it involves high runoff or low infiltration. In contrast, when the curve number is low it ensures high infiltration rates and low surface runoff (green in Fig. 7).

To determine the rainfall frequency in the study area, we have set up two scenarios for the Alginet model; one with rain and other without rain. Both models have been modeled with the same calculation parameters (soil friction, bridges, boundary conditions, etc.). The Alginet model without rain uses a hydraulic calculation, while the Alginet model with rain uses a coupled hydrologic-hydraulic calculation.

Figures 8, 9 show the correspondence to the result of both modelings for a return period of 500 years, expressed in depth.

Once the calculations are completed for both scenarios, a comparison of the result obtained for the 500 years of return period is implemented (see Fig. 10), to know the impact of rain in the study area and in particular in the Alginet urban core. This has been quantified to increase the actual depth of rain and locating troubled areas of the urban core due to surface runoff, and thus proposing a solution according to the origin of the damage.

Figures 11, 12 illustrate the differences in outflow of the two models. In these figures the outflow

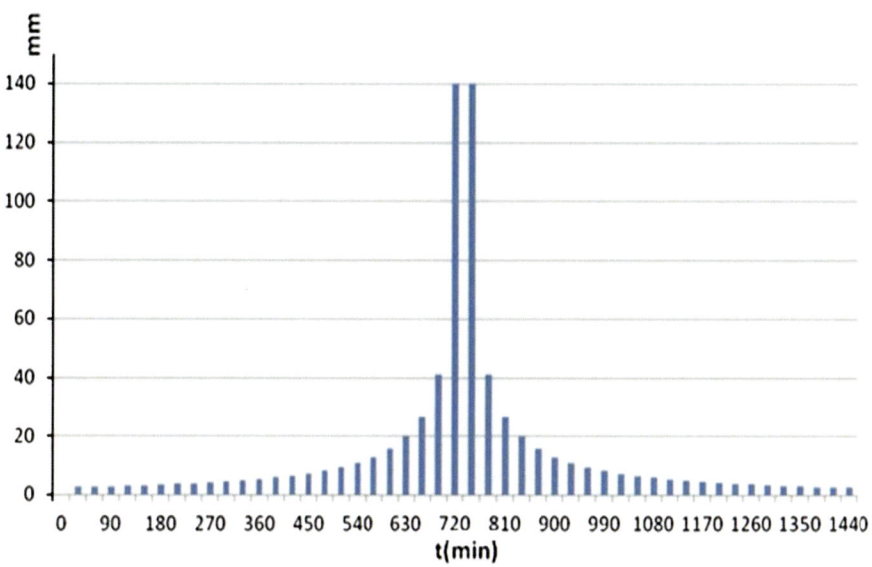

Figure 6
Hyetograph $T = 500$ years

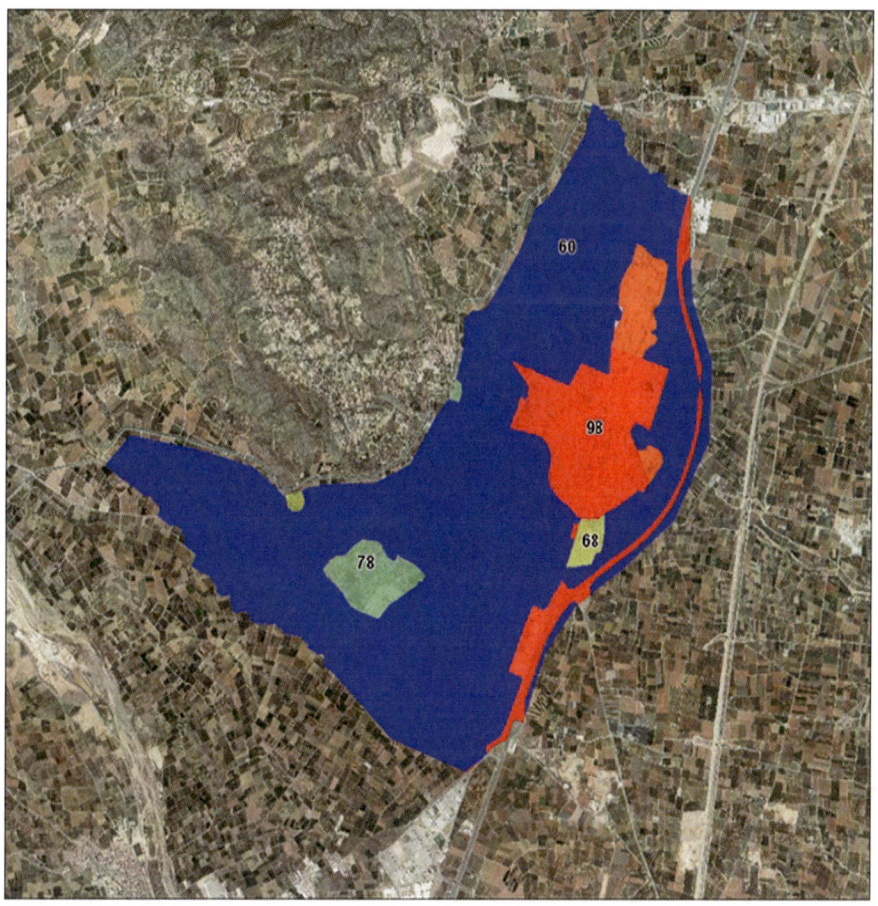

Figure 7
Map of curve number. *Blue* is 60, *yellow* is 68, *green* is 78 and *red* is 98

219

Figure 8
Alginet model without rain (T = 500 years). Legend: *yellow* is for 0.2 m depth, *light blue* is for 0.8 m depth, *red* is for 1 m depth and *dark blue* is for depth greater than 2 m

Figure 9
Alginet model with rain (*T* = 500 years). *Yellow* is for 0.2 m depth, *light blue* is for 0.8 m depth, *red* is for 1 m depth and *dark blue* is for depth greater than 2 m

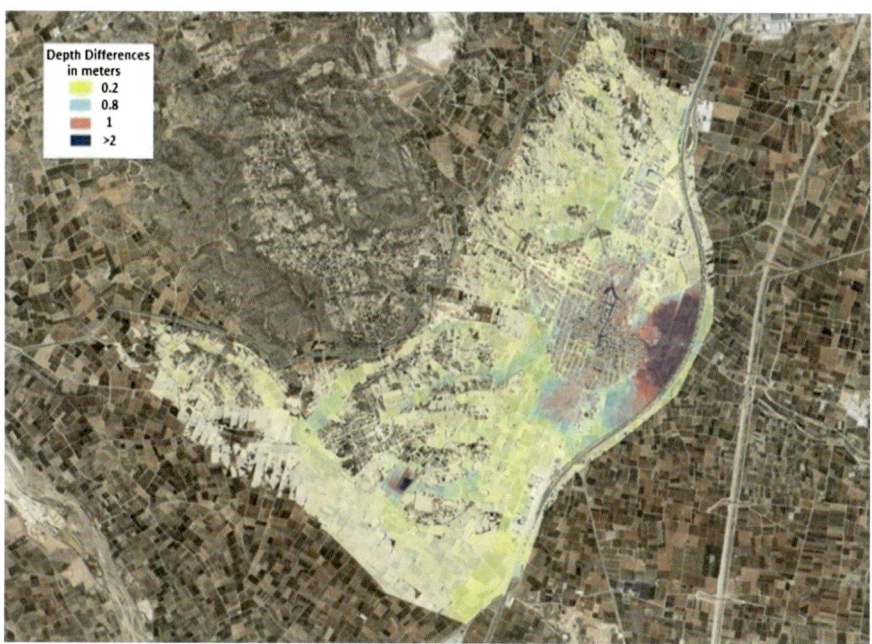

Figure 10

Difference between the two depths of the models ($T = 500$ years). *Yellow* is for 0.2 m depth, *light blue* is for 0.8 m depth, *red* is for 1 m depth and *dark blue* is for depth greater than 2 m

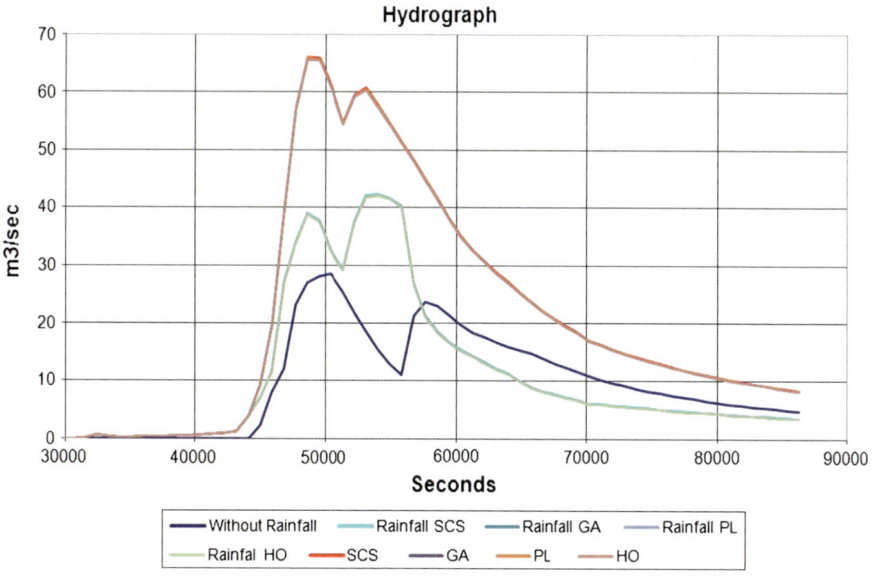

Figure 11
Comparison graph

hydrograph of the Alginet model without rain (dark blue), the hydrograph of the Alginet model with rain (red) and the hydrograph resulting from subtracting both (light blue) have been jointly represented. As it can be seen in the pictures and in the graph, the rainfall has a high frequency in the study area and it should be considered in the calculations.

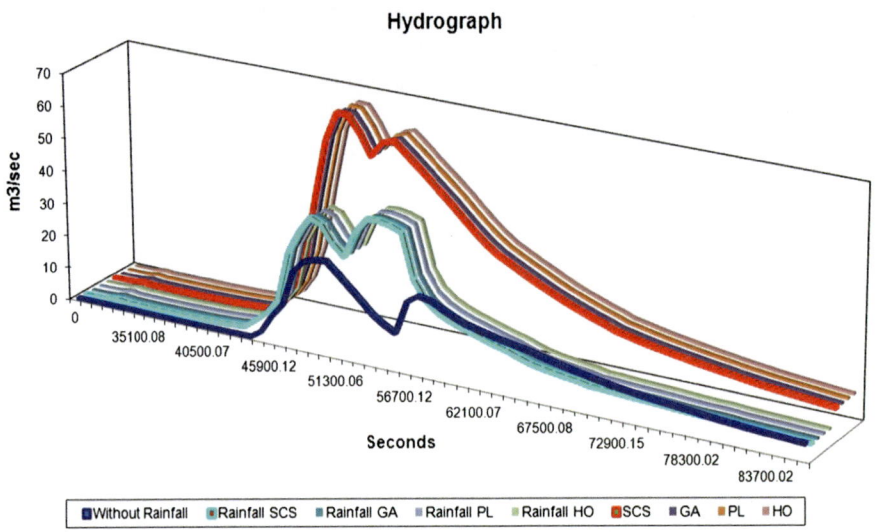

Figure 12
Comparison graph 3d

5. *Conclusions*

The motivation behind this development is to find the best compromise between computational efficiency and physical factors. Surface flow models are completed with simple law and a physical basis for simulating evapotranspiration, infiltration, sub-surface and groundwater flow and trade between sub-surface and underground flows.

The upwind scheme is based on the approximation of the spatial derivatives using the direction and the sign of the advection speed. According to this idea, the flow difference is divided into input and output on each edge of the cell.

The performance of some parts of the model has been evaluated in cases where analytical solutions exist or compared with experimental measurements. For this case there is no real solution with which to compare the solution, but both models are highly tested separately. The models have been presented as an extension to the existing models with a correct definition for marking two-dimensional models.

The analysis of the parameters used to perform the simulations show a strong dependence on the infiltration parameters with the previous state of soil moisture. This implies that for a correct hydrologic response forecast, in addition to the need for information regarding previous event used to calibrate the model parameters, it is necessary to know the pre-event moisture state.

The result shows that the models can predict the water evolution and take into account the results obtained and this can be considered as a promising tool in the rainfall-runoff processes simulation. In the numerical results we can see that if we ignore the rainfall, there are areas of town where flooding does not occur and therefore there are no civil protection measures, therefore unnecessary risks will occur.

Model improvements can come from the collapsing bridges study; in real life bridges might malfunction, which means that the bridge is stuck and will not pass water and result in flooding. A future work on which we are working is to add transport and sediment transport solution; this means that we can study how pollution evolves in lakes or rivers, shedding slopes, or the dangers of floods that transport materials.

Acknowledgments

M. Cea wishes to thank the hospitality of colleagues at ICMAT (CESIC-UAM-UCM-UC3M) in Madrid, Spain, where this work was started during the course 2013–2014. Finally, thanks to Department of Fluid Mechanics the University of Zaragoza.

REFERENCES

E. Abazi. *Modelling Floods in Wide Rivers Using Sobek 1D2D: A Case Study for the Elbe River.* Unesco-IHE, 2005.

A. Bermúdez, A. Dervieux, J-A Desideri, and M. E. Vázquez. *Upwind schemes for the two-dimensional shallow water equations with variable depth using unstructured meshes.* Computer methods in applied mechanics and engineering, *155*(1):49–72, 1998.

A. Bermúdez and M. E. Vázquez. *Upwind methods for hyperbolic conservation laws with source terms.* Computers & Fluids, *23*(8):1049–1071, 1994.

K.J. Beven and Michael J. Kirkby. *A physically based, variable contributing area model of basin hydrology. un modèle à base physique de zone d'appel variable de l'hydrologie du bassin versant.* Hydrological Sciences Journal, *24*(1):43–69, 1979.

D. B. Booth, D. Hartley, and R. Jackson. *Forest cover, impervious-surface area, and the mitigation of stormwater impacts1.* JAWRA Journal of the American Water Resources Association, *38*(3):835–845, 2002.

P. Brufau, M.E. Vázquez-Cendón, and P. García-Navarro. *A numerical model for the flooding and drying of irregular domains.* International Journal for Numerical Methods in Fluids, *39*(3):247–275, 2002.

E. Campos, M. Rodríguez, S. Cordero, S. González, A. Moreno, L. Martínez, V. Bertolín, L. Altarejos, P. Pérez, and E. Martínez. *Análisis de la inundabilidad producida por torrentes y escorrentías en el entorno urbano de alginet y su desagüe a la albufera.* JIA, 2004.

E. Campos, M. Rodríguez, S. Cordero, S. González, A. Moreno, L. Martínez, V. Bertolín, L. Altarejos, P. Pérez, and E. Martínez. *Propuesta de cálculo de la VID en un entorno urbano y plano con un modelo de cálculo simultáneo hidrológico distribuido e hidráulico bidimensional (Dinámica fluvial),(Monográfico: Modelos numéricos en dinámica fluvial).* JIA, 2004.

V. T. Chow, D.R. Maidment, and L.W. Mays. Hidrología aplicada. Mc Graw-Hill Interamericana, 1994.

R. Cronshey. Urban hydrology for small watersheds. Technical report, US Dept. of Agriculture, Soil Conservation Service, Engineering Division, 1986.

J.A. Cunge, F.M. Holly, and A. Verwey. *Practical aspects of computational river hydraulics.* Pitma, London, *xvi*:407–415, 1980.

Ministerio de Obras Públicas y Urbanismo. Instruccin 5.2-IC de Denaje Superficial de Carreteras. 1990.

M. de Saint-Venant. *Théorie du mouvement non permanent des eaux, avec application aux crues des rivières et à l'introduction des marées dans leur lit.* C. R. Acad. Sci. Paris, *73*:147–154, 1871.

M. Ferrer-Juliá, J. Iglesias, A. Cerrillo, A. Vizcaino, E. Martínez, L. Ardiles, and A. López-Peláez. Generación de cartografía geomorfológica con datos lidar para los estudios de inundabilidad. 2009.

L. Garrote, F. Laguna, and J. Lorenzo. Comparación entre distintos modelos comerciales. Jornadas Técnicas sobre Hidráulica Fluvia, CEDEX, p 203, 2008.

S.K. Gudonov. Finite difference methods for the computation of discontinuous solutions of the equations of fluid dynamics. Mat. Sb., *47*:271–306, 1959.

R. E. Horton. *Approach toward a physical interpretation of infiltration capacity.* Soil Sci. Soc. Am. J, *5*, 339–417,1940.

R. J. LeVeque. *Balancing source terms and flux gradients in high-resolution godunov methods: the quasi-steady wave-propagation algorithm.* Journal of computational physics, *146*(1):346–365, 1998.

J. L. López, J. Alavez-Ramírez, and J. L. H. López. *Solución numérica del modelo de saint-venant vía volúmenes finitos.* Revista de Ciencias Básicas UJAT, *8*(2):34–53, 2009.

R. Manning. *On the flow of water in open channels and pipes.* Transactions of the Institution of Civil Engineers of Ireland, *20*:161–207, 1891.

T. Maurer. Catflow, a physically based distributed catchment model, 1997.

E.M. Morris. *Forecasting flood flows in grassy and forested basins using a deterministic distributed mathematical model, 129.* 1980.

J. Murillo, J. Burguete, P. Brufau, and P. García-Navarro. *Coupling between shallow water and solute flow equations: analysis and management of source terms in 2d.* International journal for numerical methods in fluids, *49*(3):267–299, 2005.

J. Murillo, P. García-Navarro, P. Brufau, and J. Burguete. *Extension of an explicit finite volume method to large time steps (cfl >1): application to shallow water flows.* International journal for numerical methods in fluids, *50*(1):63–102, 2006.

F. Olivera, M. Rodriguez, and J. Murillo. Floodplain delineation of the ranillas meander in zaragoza using guad-2d. In: World Environmental and Water Resources Congress 2008@ sAhupuaA, p 1. ASCE, 2008.

O. Palacios-Vélez and B. Cuevas-Renaud. *Transformation of tin data into a kinematic cascade.* Transaction American Geophysical Union, 1989.

J R., Philip *The theory of infiltration: 1. The infiltration equation and its solution.*, Soil science, *83*, 5, 345–358,1957.

P-A. Raviart and E. Godlewsky. Numerical approximation of hyperbolic systems of conservation laws, *118*. Springer, 1996.

J. C. Refsgaard and B. Storm. *Mike-she. vp singh, computer models of watershed hydrology.* Water Res. Publications, *xvi*:809–846, 1995.

P. L. Roe. *Approximate riemann solvers, parameter vectors, and difference schemes.* Journal of computational physics, *43*(2):357–372, 1981.

C.C.M. Rogers, K.J. Beven, E.M. Morris, and M.G. Anderson. *Sensitivity analysis, calibration and predictive uncertainty of the institute of hydrology distributed model.* Journal of hydrology, *81*(1):179–191, 1985.

J.M. Ruiz. Modelo distribuido para la evalućión de recursos hídricos (modelo SIMPA). PhD Thesis, 1999.

S.A. Schroeder. *Reliability of scs curve number method on semi-arid, reclaimed minelands.* International Journal of Surface Mining and Reclamation, *8*(2):41–45, 1994.

J. J. Stoker. Water waves: The mathematical theory with applications, *36*. Wiley, 2011.

V. A. Tsihrintzis and R. Hamid. *Urban stormwater quantity/quality modeling using the scs method and empirical equations1.* JAWRA Journal of the American Water Resources Association, *33*(1):163–176, 1997.

P. Vanderkimpen, E. Melger, and P. Peeters. Flood modeling for risk evaluation: a mike flood vs. sobek 1d2d benchmark study. 2008.

M. E. Vázquez-Cendón. *Improved treatment of source terms in upwind schemes for the shallow water equations in channels with irregular geometry.* Journal of Computational Physics, *148*(2):497–526, 1999.

B. Yu. *Theoretical justification of scs method for runoff estimation.* Journal of irrigation and drainage engineering, *124*(6):306–310, 1998.

(Received March 10, 2014, revised October 6, 2015, accepted October 13, 2015, Published online October 29, 2015)

Pure Appl. Geophys. 173 (2016), 923–935
© 2015 Springer Basel
DOI 10.1007/s00024-015-1124-8

Pure and Applied Geophysics

 CrossMark

On the Effectiveness of Wastewater Cylindrical Reactors: an Analysis Through Steiner Symmetrization

J. I. Díaz[1] and D. Gómez-Castro[1] (iD)

Abstract—The mathematical analysis of the shape of chemical reactors is studied in this paper through the research of the optimization of its effectiveness η such as introduced by R. Aris around 1960. Although our main motivation is the consideration of reactors specially designed for the treatment of wastewaters our results are relevant also in more general frameworks. We simplify the modeling by assuming a single chemical reaction with a monotone kinetics leading to a parabolic equation with a non-necessarily differentiable function. In fact we consider here the case of a single, non-reversible catalysis reaction of chemical order $q, 0 < q < 1$ (i.e., the kinetics is given by $\beta(w) = \lambda w^q$ for some $\lambda > 0$). We assume the chemical reactor of cylindrical shape $\Omega = G \times (0, H)$ with G and open regular set of $\mathbb{R}^2$ not necessarily symmetric. We show that among all the sections G with prescribed area the ball is the set of lowest effectiveness $\eta(t, G)$. The proof uses the notions of Steiner rearrangement. Finally, we show that if the height H is small enough then the effectiveness can be made as close to 1 as desired.

Key words: Wastewater treatment, chemical reactor tanks, effectiveness, Steiner symmetrization.

1. Introduction

One of the most important problems on environment in Geosciences is the treatment of wastewater flows. Most industrial wastewater treatments are carried out in a series of cylindrical-type tanks. In some of them a diffusion-reaction process takes place specially in the trickling filter phase in which wastewater flows downward through a bed of rocks,

[1] Dpto. de Matemática Aplicada, Facultad de Ciencias Matemáticas and Instituto de Matemática Interdisciplinar, Universidad Complutense de Madrid, Madrid, Spain. E-mail: ildefonso.diaz@mat.ucm.es; dgomez91@gmail.com

gravel, slag, peat moss, or plastic media reacting on a layer (or film) of microbial slime covering the bed media. The process (see , e.g., Rodriguez *et al.* 2012; Vicente *et al.* 2011; Rosas *et al.* 2014 and its references) involves adsorption of organic compounds in the wastewater by the microbial slime layer, diffusion of air into the slime layer to provide the oxygen required for the biochemical oxidation of the organic compounds. In this paper, we shall assume that an ideal homogenization process was applied (by passing to the limit $\varepsilon \to 0$ on the porosity of the solid bed) so that the chemical reaction can be assumed as distributed over all the reactor cylinder (see, e.g., Conca *et al.* 2003, 2004 and their references). Simplifying the modeling process we arrive to the consideration of a single, non-reversible catalysis reaction of q-order on a chemical reactor Ω of cylindrical shape

$$\Omega = G \times (0, H),$$

with G an open regular set of $\mathbb{R}^2$ (or more in general $\mathbb{R}^N$) not necessarily symmetric. We point out that, in spite of the abovementioned motivation, our mathematical results can be applied to a larger framework (for instance the own structure of the set Ω can be taken much more in general (see Sect. 3). It is useful to separate the boundary of Ω in its lateral parts $\partial_l \Omega$ and its horizontal parts $\partial_h \Omega$, so that $\partial_l \Omega = \partial G \times (0, H)$ and $\partial_h \Omega$ consists in the union of the top and bottom boundaries: $\partial_h \Omega = (\partial_h \Omega)^H \cup (\partial_h \Omega)_0$ with $(\partial_h \Omega)^H = \Omega \times \{H\}$ and $(\partial_h \Omega)_0 = G \times \{0\}$. We shall use also the notation $\mathbf{x} = (x, y)$ with $x = (x_1, x_2) \in G$ and $y \in (0, H)$. A similar notation can be introduced if $\mathbb{R}^2$ is replaced by $\mathbb{R}^N$ and $(0, H)$ by a set in $\mathbb{R}^m$.

In order to fix ideas we shall consider here the following parabolic model

$$\begin{cases} \frac{\partial w}{\partial t} - \Delta w + \lambda \beta(w) = 0 & \text{in } (0, +\infty) \times \Omega, \\ w = 1 & \text{on } (0, +\infty) \times \partial_l \Omega, \\ \frac{\partial w}{\partial n} = \mu(1 - w) & \text{on } (0, +\infty) \times \partial_h \Omega, \\ w(0, \mathbf{x}) = w_0(\mathbf{x}) & \text{in } \Omega, \end{cases} \tag{1}$$

where

$$\beta(w) = w^q, \quad 0 < q \le 1$$

(q is called reaction order), $\lambda > 0$,

$$w_0 \in L^\infty(\Omega), \quad 0 \le w_0 \le 1, \tag{2}$$

$\mathbf{n}$ denotes the unit normal exterior vector to $\partial_h \Omega$ and the Robin coefficient μ is taken in a generalized way as $\mu \in [0, +\infty]$. In fact, we assume that the value of μ can be different for the top or the bottom surfaces , i.e.,

$$\mu = \begin{cases} \mu_H & \text{on } (\partial_h \Omega)^H = G \times \{H\}, \\ \mu_0 & \text{on } (\partial_h \Omega)_0 = G \times \{0\}. \end{cases}$$

So, very often $\mu_H = 0$ (which corresponds to the case of an open tank) and/or $\mu_0 = +\infty$ (which must be understood as a Dirichlet type boundary condition $w = 1$ on $(0, +\infty) \times (\partial_h \Omega)_0$ and that corresponds to a tank alimented also from the bottom).

The limit case, the case of 0-order reactions, $q = 0$, can also be considered (see Remark 5) with the help of some special multivalued maximal monotone graph of $\mathbb{R}^2$. We also mention that some larger generality can be considered also concerning the differential operator (see Remark 5).

As mentioned before, as proved in CONCA et al. (2004), this is the limit as $\varepsilon \to 0$ of the following models

$$\begin{cases} -\Delta w^\varepsilon = f & \text{in } \Omega^\varepsilon, \\ \frac{\partial w_\varepsilon}{\partial \nu} + \mu(\varepsilon) \beta(w^\varepsilon) = 0 & \text{on } S^\varepsilon, \\ w_\varepsilon = 1 & \text{on } \partial\Omega, \end{cases} \tag{3}$$

where Ω^ε is a domain with fixed obstacles (which due to the chemical implications we will call pellets), where $\varepsilon > 0$ has to do with the size of each obstacle and S^ε represents the boundary of the pellets and $\partial\Omega$ the boundary of the reactor. This kind of problem is an intuitive model of fixed bed reactors. In this sense problem (1) can be seen as a homogenized problem for fixed bed reactors. This can model a large number of process in Geosciences.

Due to one of those "gifts" of interdisciplinarity, problem (1) also models the heat energy stored by the Earth. The so-called Sellers model considers the averaged surface atmospheric climate proposing the equation:

$$\frac{\partial w}{\partial t} - \Delta w + \beta(u) = QH(u).$$

In the above model, the functions H and β are assumed Lipschitz continuous. There is a different model proposed by Budyko in which the function H is assumed to be discontinuous but we shall not pay attention here to this case (see DÍAZ 1996). The techniques presented in this paper (mainly the application of the Trotter–Kato formula and its consequences) are extensible to the Sellers model, since the operator is omega-accretive (see BENILAN et al., unfinished manuscript).

We shall also consider, as by-product of our results concerning the parabolic problem, the associated stationary problem (formally obtained when making $t \to +\infty$)

$$\begin{cases} -\Delta w + \lambda \beta(w) = 0 & \text{in } \Omega, \\ w = 1 & \text{on } \partial_l \Omega, \\ \frac{\partial w}{\partial n} = \mu(1 - w) & \text{on } \partial_h \Omega. \end{cases} \tag{4}$$

The main optimality element in the study of the shape of such chemical reactors is given in terms of a notion introduced in 1957 by R. Aris (see references in STRIEDER and ARIS 1973): the so called *effectiveness factor* which is defined as

$$\eta(t : G, H) := \frac{1}{H|G|} \int_\Omega \beta(w(t, \mathbf{x})) d\mathbf{x}.$$

In a pioneering work, R. Aris presented, in his book (STRIEDER and ARIS 1973), in collaboration with W. Strieder, the study of a linear model ($q = 1$) for a finite number of catalyst particles, which they always consider spherical. Here we will consider cylinders of arbitrary basis and reactions of order less or equal than one, which are much more frequent in practice, but which result in non-linear models requiring delicate mathematical tools. We recall that when $0 < q < 1$ the solutions may give rise to a *dead core*, an interior region where no reaction is taking place. This dead core, which can be defined, for a given $t \ge 0$, as

$$N_w(t) = \{\mathbf{x} \in \Omega : w(t, \mathbf{x}) = 0\}.$$

We shall not give here estimates on the size and location of the dead core regions (see Sect. 4, Remark 4). Obviously, the presence of dead cores affects negatively the global effectiveness, and is to be avoided in the shape optimization process. Intuitively, it represents volume where no catalyst is present, and thus no reaction is taking place.

Although more realistic models may incorporate more complex and sophisticated aspects that the ones here presented, our main goal is to give a conceptual justification of why these reactors are wide and low. In fact, we shall prove here that among all the sections G, with prescribed area, the ball is the set of lowest effectiveness $\eta(t : G, H)$ (Theorem 2.1). Our proof uses the notions of Steiner rearrangement. In contrast to that, we shall also show that if the height of the tank H is small enough then the effectiveness can be made as close to 1 as desired (Theorem 2.2).

The organization of this paper is the following: the above main results are stated in Sect. 2 where some numerical experiences are commented. Section 3 is devoted to the proof of Theorem 2.1. The notion of Steiner rearrangement of a function is introduced and several properties showing the comparison in mass of the Steiner rearrangement of the solution of problem (1) and the solution of the "symmetrized problem" are given. In particular, we show how the so-called Trotter–Kato formula can be applied even under non-autonomous formulation. Finally, Sect. 4 contains the proof of Theorem 2.2 as well as a series of remarks on more general frameworks in which our main results remain valid.

2. Main Results and Some Numerical Experiences

Thanks to the maximum principle, it is clear that the solution w of (1) must satisfy that $0 \leq w(t, \mathbf{x}) \leq 1$ for a.e. $\mathbf{x} \in \Omega$ and for any $t \geq 0$. Then, in which follows, it will be useful to introduce the change of unknown $u = 1 - w$ for which the problem may be rewritten as

$$
\begin{cases}
\frac{\partial u}{\partial t} - \Delta u + \lambda g(u) = \lambda \beta(1) & \text{in } (0, +\infty) \times \Omega, \\
u = 0 & \text{on } (0, +\infty) \times \partial_l \Omega, \\
-\frac{\partial u}{\partial n} = \mu u & \text{on } (0, +\infty) \times \partial_h \Omega, \\
u(0, x) = u_0(x) & \text{on } \Omega,
\end{cases} \tag{5}
$$

where

$$
g(u) = \beta(1) - \beta(1 - u). \tag{6}
$$

Thus, we can assume that g is a continuous increasing function with $g(0) = 0$. We recall that the existence and uniqueness of a weak solution $u \in C([0, +\infty) : L^1(\Omega)) \cap L^\infty((0, +\infty) \times \Omega)$ is today a well-known result. Moreover, it is also known that when $t \to +\infty$ then $u(t, \cdot) \to u_\infty(\cdot)$ in $L^2(\Omega)$ (see e.g., DÍAZ 1994 and its references).

We shall start by giving a rigorous proof of the well-known principle (from an experimental point of view) that among all cylindrical reactors with prescribed volume the one with a circular section is the least effective:

Theorem 2.1 *For fixed basis volume $|G|$effectiveness is least on an circle. That is, let $A > 0$ and let B the ball centered at the origin and let G be any other n-dimensional open regular set such that $|G| = |B| = A$. Then*

$$
\eta(t : B, H) \leq \eta(t : G, H).
$$

Moreover, the same inequality holds for the associated stationary problems.

Remark 1 In contrast to the case in which *the effectiveness is compared with the one on a ball of $\mathbb{R}^3$* having the same volume than Ω, the proof of the above theorem for the stationary case seems quite complicated without proving first the analogous result for the associated parabolic problem. That was one of our motivations not to simplify our formulation to the easier case of the stationary problem.

In order to illustrate the conclusion of Theorem 2.1 we produced a numerical experience concerning a particular (one-parametric) family of elliptic cylinders $G_a \times (0, H)$. The elliptic cylinders are assumed with a prescribed volume V. So, given the lower semiaxis a, the greater semiaxis b_a is given by the identity $\pi a b_a = \frac{V}{H}$. In other words, the ellipse family is defined by the parameter a through the expression

$$
G_a = \left\{ (x_1, x_2) \in \mathbb{R}^2 : \left(\frac{x_1}{a}\right)^2 + \left(\frac{x_2}{b_a}\right)^2 = 1 \right\}, \tag{7}
$$

$$
b_a = \frac{V}{H \pi a}.
$$

The image below shows a minimum of the effectiveness over this one-parametric family of elliptic cylinders $\Omega_a = E_a \times (0, 1)$, in which if we choose $V = \pi H$ and so the value $a = 1$ corresponds to the case of a circular section (Fig. 1).

Our second main result deals with the pure Dirichlet problem ($\mu = +\infty$) and gives a detailed statement of the well-known principle (from an experimental point of view) that among all cylindrical reactors with the prescribed volume low reactors are very effective. We introduce the auxiliary function $\psi \in C^2(\Omega)$ given as the unique solution of

$$\begin{cases} -\Delta\psi = 1 & \text{in } \Omega, \\ \psi = 0 & \text{on } \partial\Omega. \end{cases} \quad (8)$$

Theorem 2.2 *Assume $\mu = +\infty$. Let $V = |\Omega| = |G|H = AH > 0$ be a fixed volume and let B_H be the ball of $\mathbb{R}^N$ centered at the origin such that $|B_H| = A$. Assume also*

$$1 - w_0(\mathbf{x}) \leq \lambda\psi(\mathbf{x}) \quad \text{a.e. } \mathbf{x} \in \Omega.$$

Then

$$\eta(t : G, H) \to 1 \quad \text{as } H \to 0. \quad (9)$$

More precisely, for any $t > 0$ and a.e $\mathbf{x} \in \Omega$

$$1 \geq \beta(w(t, \mathbf{x})) \geq 1$$
$$- \left(\frac{V(4 + 2(N+1))(2N+1)^{-\frac{N+1}{2}}}{\pi^2 \omega_{N+1}} H^2 \right)^{2/(N+3)}.$$
$$(10)$$

The above estimate holds also for the solution of the associated stationary problem (4).

In order to illustrate quantitatively conclusion 2 we produced a numerical experience concerning the family of symmetric cylinder reactors $B_r \times (0, H)$. Motivated by the special case considered in ARIS (1975) (see its Figure 4.5.1) when computing curves for this phenomenon for the linear case $q = 1$, we have taken $H = \gamma^{-2}\left(\frac{16}{3}\right)^{\frac{1}{3}}$ and $r = \gamma\left(\frac{2}{3}\right)^{\frac{1}{3}}$ with γ a variable parameter. In the next figure we can see how $H \to 0$ implies $\eta \to 1$ (Fig. 2). We can also see how, in this case, $\eta \to 1$ as $q \to 0$ (this is because, for this volume, no dead core exists even in the worst case scenario).

Remark 2 The numerical experiences were produced by using a semi-implicit iterative algorithm [see SPIGLER and VIANELLO 1995 for a proof of the convergence in an abstract framework which includes, as a special case, problem [1] under the conditions assumed in this paper). The chosen scheme applies finite differences in time and finite elements in space. The time discretization for time step h is

$$u_{n+1} - h\Delta u_{n+1} = u_n - hg(u_n).$$

The scheme is chosen implicit in time on the diffusion so that the operator in u_{n+1} is coercive, and thus the sequence is uniquely determined in $H_0^1(\Omega)$. However, the method is explicit in the non-linearity,

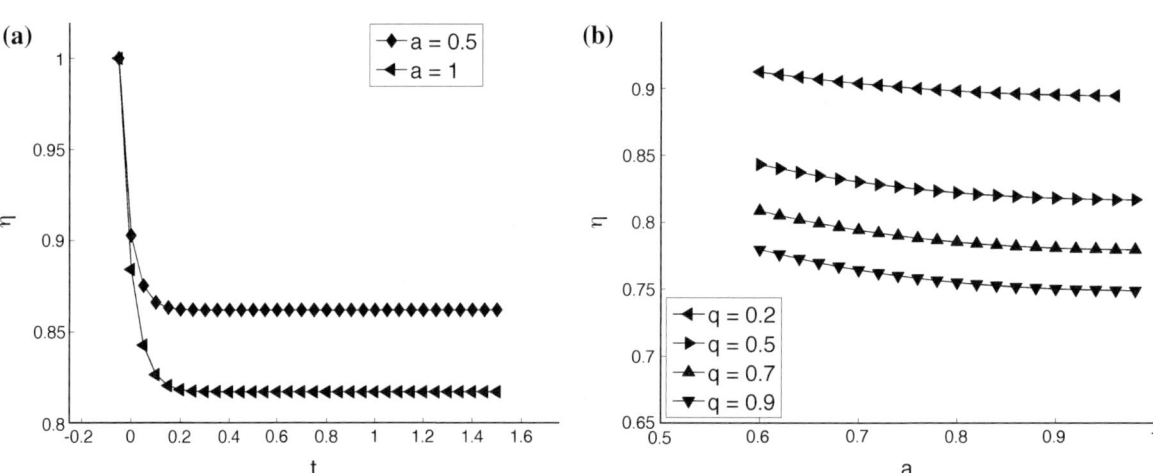

(a)   **(b)**

Figure 1

Effectiveness factor for a family of ellipses with the same area. **a** Time evolution of the effectiveness for two cylinders, one circular $a = 1$ and one elliptical $a = 0.5$ both of the same volume, with initial condition $w_0 = 1$ on Ω. **b** Effectiveness for the elliptic problem for different values of G_a and q

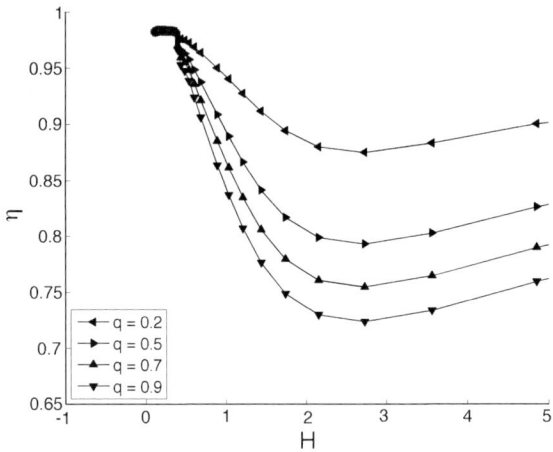

Figure 2
Effectiveness for the elliptic problem on cylinder with varying aspect ratio. Simulation

which makes the problem linear in u_{n+1}, thus allowing for faster simulations. The implementation of the finite element method was performed through the automated library FEniCS, which meshes simple domains in two and three dimensions, constructs the continuous Galerkin finite elements necessary and solves the linear systems.

3. The Circular Section is the Least Effective: Steiner Symmetrization. Proof of Theorem 2.1

The proof of Theorem 2.1 will use some inequalities on Steiner symmetrization obtained in ALVINO et al. (1996). As a matter of fact, we shall improve also a previous result by the authors (DÍAZ and GÓMEZ-CASTRO 2014a) corresponding, essentially, to the case $q \geq 1$. It turns out that our result remains true under a more general setting by replacing the *vertical space* $\mathbb{R}$ by $\mathbb{R}^m$. We start by recalling that given a general measurable function $h : \mathbb{R}^N \times \mathbb{R}^m \to \mathbb{R}$, with $N, m \geq 1$, for a fixed $y \in \mathbb{R}^m$ we can define the Steiner distribution function $\mu_h : \mathbb{R} \times \mathbb{R}^m \to \mathbb{R}$ by means of

$$\mu_h(t, y) = |\{x \in \mathbb{R}^N : |h(x, y)| > t\}|.$$

The Hardy–Littlewood–Polya decreasing rearrangement $h^* : [0, +\infty) \times \mathbb{R}^m \to \mathbb{R}$ is given as

$$h^*(s, y) = \sup\{t > 0 : \mu_h(t, y) > s\}$$
$$= \inf\{t > 0 : \mu_h(t, y) \leq s\}.$$

It is well known that if ω represents a generic measurable subset of $\mathbb{R}^N \times \mathbb{R}^m$ then

$$\int_0^s h^*(\sigma, y) \, d\sigma = \sup_{|\omega| = s} \int_\omega h(\mathbf{x}, y) \, d\mathbf{x}. \qquad (11)$$

Finally, for $y \in \mathbb{R}^m$ prescribed, we define the Steiner symmetrization of h with respect to $\mathbf{x}$ as

$$h^\#(\mathbf{x}, y) = h^*(\omega_N |\mathbf{x}|^N, y),$$

where ω_N is the measure of the N-dimensional ball. The basic idea underlying Steiner symmetrization is to consider the integral of the function over *slices*. Given $s > 0$ and $y \in \mathbb{R}^m$ we take very particular *slices* of the form

$$G(y) = \{\mathbf{x} \in \mathbb{R}^N : u(\mathbf{x}, y) > u^*(s, y)\},$$

where $|G(y)| = s$ (by construction of u^*). Variable s should formally be included in the definition but this will not lead to confusion.

Explicit calculations can be performed in simple cases. The following figure provides and example of the exact distribution function and Steiner rearrangement for the function

$$u(x, y) = \begin{cases} 0, & |(x, y)| > 1, \\ 2(1 - x^2 - y^2), & \frac{1}{2} \leq |(x, y)| \leq 1, \\ 1, & |(x, y)| > 1. \end{cases}$$

In this section we shall use a more general framework. We introduce the following notations (Fig. 3):

$$\Omega = \Omega' \times \Omega''$$

and $(x, y) \in \Omega' \times \Omega''$ for an arbitrary point (note that in our initial framework $\Omega' = G$ and $\Omega'' = (0, H)$). We shall denote by B a ball such that $|B| = |\Omega'|$ and then we introduce

$$\Omega^\# = B \times \Omega''.$$

Remark 3 In the case where we rearrange with respect to all variable, i.e., no y is presented (and, in an abuse of notation $m = 0$) the symmetrization is know as Schwarz symmetrization. Since it will be

(a)   **(b)**   **(c)**

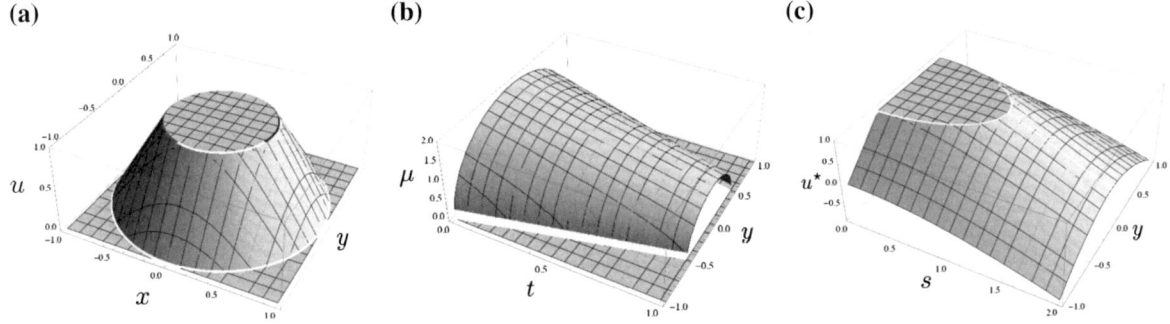

Figure 3
Computation of Steiner symmetrization. **a** Function u. **b** Distribution function μ. **c** Steiner rearrangement u^*

useful to use both symmetrizations, for Schwarz rearrangement we will use the notation $\tilde{u}$. We also introduce the truncation at level $y \in \Omega''$ as

$$u_y(x) = u(x,y), \quad (x,y) \in \Omega' \times \Omega''.$$

Is clear from the definition that,

$$\tilde{u}_y(s) = u^*(s,y).$$

For the case where time is introduced, even though the application is written $u(t,x,y)$ we will never rearrange with respect to t.

The image below (Fig. 4) shows an artistic comparison between Steiner and Schwarz symmetrizations for the function

$$u(x,y,z) = e^{-10x^2 - 5y^2 - 10z^2}(1-x)x(1-y)y(1-z)z,$$

$$(x,y,z) \in [0,1]^3.$$

This function has a single maximum point, and we show cross cuts of symmetrizations.

Our main result leading to the conclusion of Theorem 2.1 is the following:

Theorem 3.1   *Let β be a concave continuous nondecreasing function such that $\beta(0) = 0$. Give $T > 0$ arbitrary and let $f \in L^2(0,T:L^2(\Omega))$ with $f \geq 0$ in $(0,T)$ and let $w_0 \in L^2(\Omega)$ be such that $0 \leq w_0 \leq 1$. Let $\delta > 0$ be fixed and $w \in C([0,T]:L^2(\Omega)) \cap L^2(\delta,T:H_0^1(\Omega))$ and $z \in C([0,T]:L^2(\Omega^\#)) \cap L^2(\delta, T:H_0^1(\Omega^\#))$ be the unique solutions of*

$$(P)\begin{cases} \frac{\partial w}{\partial t} - \Delta w + \lambda\beta(w) = f(t) & \text{in } \Omega \times (0,T), \\ w = 1 & \text{on } \partial\Omega \times (0,T), \\ w(0) = w_0 & \text{on } \Omega, \end{cases}$$

$$(P^\#)\begin{cases} \frac{\partial z}{\partial t} - \Delta z + \lambda\beta(z) = f^\#(t), & \text{in } \Omega^\# \times (0,T), \\ z = 1, & \text{on } \partial\Omega^\# \times (0,T), \\ z(0) = z_0, & \text{on } \Omega^\#, \end{cases}$$

where $z_0 \in L^2(\Omega^\#)$, $0 \leq z_0 \leq 1$ is such that

$$\int_s^{|\Omega'|} z_0^*(\sigma,y)d\sigma \leq \int_s^{|\Omega'|} w_0^*(\sigma,y)d\sigma,$$

$$\forall s \in [0,|\Omega'|] \text{ and a.e. } y \in \Omega''.$$

Then, for any $t \in [0,T]$, $s \in [0,|\Omega'|]$ and a.e. $y \in \Omega''$

$$\int_s^{|\Omega'|} z^*(t,\sigma,y)d\sigma \leq \int_s^{|\Omega'|} w^*(t,\sigma,y)d\sigma. \qquad (12)$$

In terms of the comparison of the effectiveness we have the following consequence (which will be proved in Sect. 3) leading to the proof of Theorem 2.1:

Corollary 3.2   *In the assumptions of Theorem 3.1, for any $t \in [0,+\infty)$ we have*

$$\int_{\Omega^\#} \beta(z(t,\mathbf{x}))d\mathbf{x} \leq \int_\Omega \beta(w(t,\mathbf{x}))d\mathbf{x}. \qquad (13)$$

The interest on the above two results is that the conclusions remains true for the associated stationary problems.

(a) **(b)** **(c)**

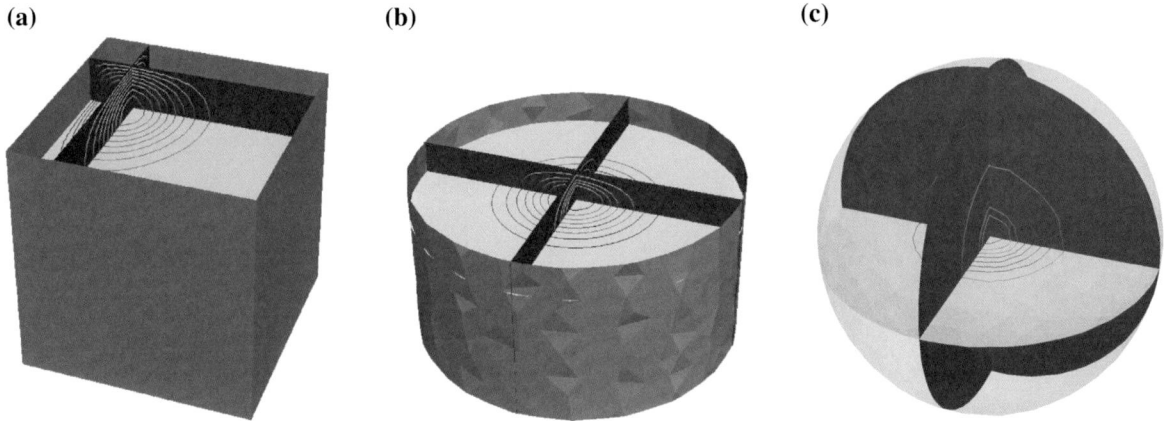

Figure 4

Comparison of Steiner and Schwarz rearrangements of a given function. **a** A given measurable function on $\Omega = [0, 1]^3$, which we choose constant on the boundary. **b** Steiner symmetrization with respect to (x, y). **c** Schwarz symmetrization

Corollary 3.3 *The mass and effectiveness comparison given by* (12) *and* (13), *respectively, remain valid for the solutions of the corresponding stationary problems.*

As mentioned before, Theorem 3.1 extends previous result by the authors (Díaz and Gómez-Castro 2014a). For the proof of this result we apply, essentially, the same techniques as in the cited article, but with some refinements concerning the nature of the non-linear term $\beta(w)$ (i.e., $g(u)$ in the equivalent formulation (5)). In contrast to our work (Díaz and Gómez-Castro 2014a) we shall work with the increasing rearrangement. We start by recalling the following simple property: if $f : [0, |\Omega'|] \to \mathbb{R}$ is a real function such that $0 \leq f \leq L$ then $(L - f)^*(s) = L - f^*(|\Omega'| - s)$ and in particular

$$\int_0^s (L - f(t))^* \mathrm{d}t = L - \int_{|\Omega'|-s}^{|\Omega'|} f^*(t) \mathrm{d}t$$

(the proof can be found, for instance, in Mossino 1984).

As in Díaz and Gómez-Castro (2014a), we shall prove the above theorem by means of the Trotter–Kato formula. So we shall need to consider previously two auxiliary problems. The first problem corresponds to the associated linear diffusion problem:

Proposition 3.4 *Let* $0 \leq w_0, z_0 \leq 1$

$$(A) \begin{cases} \frac{\partial w}{\partial t} - \Delta w = 0, & (0, T) \times \Omega \\ w = 1, & (0, T) \times \partial\Omega \\ w = w_0, & \{0\} \times \Omega \end{cases}$$

$$(A^{\#}) \begin{cases} \frac{\partial z}{\partial t} - \Delta z = 0, & (0, T) \times \Omega^{\#} \\ z = 1, & (0, T) \times \partial\Omega^{\#} \\ z = z_0, & \{0\} \times \Omega^{\#} \end{cases}$$

and

$$\int_s^{|\Omega'|} z_0^*(\sigma, y) \, \mathrm{d}\sigma \leq \int_s^{|\Omega'|} w_0^*(\sigma) \, \mathrm{d}\sigma, \quad s \in [0, |\Omega|].$$

Then

$$\int_s^{|\Omega'|} z^*(t, \sigma, y) \, \mathrm{d}\sigma \leq \int_s^{|\Omega'|} w^*(t, \sigma, y) \, \mathrm{d}\sigma,$$

$$s \in [0, |\Omega|].$$

Proof Let us consider $u = 1 - w$ and $v = 1 - z$. Then u and v are solutions of the problems

$$(B) \begin{cases} \frac{\partial u}{\partial t} - \Delta u = 0, & (0, T) \times \Omega \\ u = 0, & (0, T) \times \partial\Omega \\ u = u_0, & \{0\} \times \Omega \end{cases}$$

$$(B^{\#}) \begin{cases} \frac{\partial v}{\partial t} - \Delta v = 0, & (0, T) \times \Omega^{\#} \\ v = 0, & (0, T) \times \partial\Omega^{\#} \\ v = v_0, & \{0\} \times \Omega^{\#} \end{cases}$$

where now $u_0, v_0 \geq 0$ are given as $u_0 = 1 - w_0$ and $v_0 = 1 - z_0$. Since, for any $\tau \in [0, |\Omega'|]$, we have that

231

$$\int_0^\tau u_0^*(\sigma)\,\mathrm{d}\sigma = L - \int_{|\Omega'|-\tau}^{|\Omega'|} w_0^*(\sigma)\,\mathrm{d}\sigma$$

$$\leq L - \int_{|\Omega-\tau}^{|\Omega'|} z_0^*(\sigma)z_0 \leq \int_0^\tau v_0^*(\sigma)\,\mathrm{d}\sigma$$

then

$$\int_0^\tau u_0^*(\sigma, y)\,\mathrm{d}\sigma \leq \int_0^\tau v_0^*(\sigma, y)\,\mathrm{d}\sigma.$$

Under this conditions, it is proven in CHIACCHIO (2004) that, for any $t \geq 0$ and for any $\tau \in [0, |\Omega'|]$, we have the comparison

$$\int_0^\tau u^*(t, \sigma, y)\mathrm{d}\sigma \leq \int_0^\tau v^*(t, \sigma, y)\mathrm{d}\,\sigma. \qquad (14)$$

The key idea the proof of the result in CHIACCHIO (2004) is to integrate each term of the equation of problem (B) over the sets $\Omega_y(s) = \{x \in \mathbb{R}^N : (x, y) \in \Omega$ and $u(t, x, y) > u^*(t, s, y)\}$ for each $t > 0$ and to use the differentiation formula

$$\left(\frac{\partial^2 F}{\partial y_i \partial y_j}\right)_{i,j} \geq \int_{\Omega_y(s)} \left(\frac{\partial^2 u}{\partial y_i \partial y_j}\right)_{i,j}, \qquad (15)$$

where

$$F(t, s, y) = \int_0^s u^*(t, \sigma, y)\mathrm{d}\sigma.$$

Inequality (15) was proved for the first time in the literature in the paper (ALVINO et al. 1996) (see also an alternative proof in FERONE and MERCALDO 1998). In CHIACCHIO (2004) we find the application of this formula to the parabolic problem (with the additional proof of the comparison with respect the formula obtained for the case of radially symmetric sections).

Applying (14), finally we arrive to the conclusion since

$$\int_{|\Omega'|-\tau}^{|\Omega'|} z^* = L - \int_0^\tau v^* \leq L - \int_0^\tau u^* = \int_{|\Omega'|-\tau}^{|\Omega'|} w^*.$$

which concludes the proof.   □

The second auxiliary problem corresponds to a distributed non-linear ordinary differential equation.

Proposition 3.5   *Let β be a concave continuous non-decreasing function such that $\beta(0) = 0$. Let u, v satisfy*

$$(B)\begin{cases} w_t + \lambda\beta(w) = 0, & \Omega \times (0, T), \\ w = w_0, & \Omega \times \{0\}, \end{cases}$$

$$(B^{\#})\begin{cases} z_t + \lambda\beta(z) = 0, & \Omega^{\#} \times (0, T), \\ z = z_0, & \Omega^{\#} \times \{0\}. \end{cases}$$

Assume

$$\int_s^{|\Omega'|} z_0^*(\sigma, y)\mathrm{d}\sigma \leq \int_s^{|\Omega'|} w_0^*(\sigma, y)\mathrm{d}\sigma,$$

$$\forall s \in [0, |\Omega'|], \quad \text{a.e.} y \in \Omega''.$$

Then we have

$$\int_s^{|\Omega'|} z^*(t, \sigma, y)\mathrm{d}\sigma \leq \int_s^{|\Omega'|} w^*(t, \sigma, y)\mathrm{d}\sigma$$

$$\forall t > 0, s \in [0, |\Omega'|], \quad \text{a.e.} \; y \in \Omega''.$$

Proof   For any $\varepsilon > 0$ and $y \in \Omega''$ prescribed, let $w_{\varepsilon,y}(t, x)$, $z_{\varepsilon,y}(t, x)$ be the solutions of the (ε, y)-parametric family of semilinear parabolic problems

$$(P(\varepsilon, y))\begin{cases} \frac{\partial w}{\partial t} - \varepsilon\Delta_x w + \lambda\beta(w) = f_y(t) & \text{in } \Omega' \times (0, T), \\ w = 1 & \text{on } \partial\Omega' \times (0, T), \\ w(0) = (w_0)_y & \text{on } \Omega', \end{cases}$$

$$(P^{\#}(\varepsilon, y))\begin{cases} \frac{\partial z}{\partial t} - \varepsilon\Delta z + \lambda\beta(z) = f_y^{\#}(t) & \text{in } B \times (0, T), \\ z = 1 & \text{on } \partial B \times (0, T), \\ z(0) = (z_0)_y & \text{on } B. \end{cases}$$

Notice that the diffusion operator is only dependent of the x-variables. Then, by Theorem 1 of DÍAZ (1991) we know that, for any $\varepsilon > 0$ and $y \in \Omega''$ prescribed,

$$\int_s^{|\Omega'|} \widetilde{z_{\varepsilon,y}}(t, \sigma)\mathrm{d}\sigma \leq \int_s^{|\Omega'|} \widetilde{w_{\varepsilon,y}}(t, \sigma)\mathrm{d}\sigma \qquad (16)$$

$$\forall t > 0, s \in [0, |\Omega'|].$$

Moreover, we can know apply Theorem 3.16 on BREZIS (1973)

$$z_{\varepsilon,y} \to z_y \quad \text{as } \varepsilon \to 0 \quad \text{in } C([0, T] : L^2(B)),$$

$$w_{\varepsilon,y} \to w_y \quad \text{as } \varepsilon \to 0 \quad \text{in } C([0, T] : L^2(G)).$$

Then, passing to the limit in (16) we get

$$\int_s^{|\Omega'|} \widetilde{z_y}(t, \sigma)\mathrm{d}\sigma \leq \int_s^{|\Omega'|} \widetilde{w_y}(t, \sigma)\mathrm{d}\sigma$$

$$\forall t > 0, s \in [0, |\Omega'|].$$

Finally, it is enough to observe that since $y \in \Omega''$ is prescribed then the Schwarz rearrangement $\widetilde{w}_y(t, \sigma)$ coincides with the Steiner rearrangement $w^*(t, \sigma, y)$ (see Remark 3) and the result holds. □

3.1. Proof of Theorem 3.1.

Proof of Theorem 3.1 The special case $f = 0$ is easier. Since we know

$$\int_\tau^{|\Omega'|} z_0^*(\sigma, y)\, d\sigma \leq \int_\tau^{|\Omega'|} w_0^*(\sigma, y)\, d\sigma, \quad \forall s, \forall y$$

applying Propositions 3.4 and 3.5 inductively we get

$$\int_\tau^{|\Omega'|} \left[\left(S_A\left(\frac{t}{n}\right) S_B\left(\frac{t}{n}\right) \right)^n z_0 \right]^* (\sigma, y)\, d\sigma$$

$$\leq \int_\tau^{|\Omega'|} \left[\left(S_{A\#}\left(\frac{t}{n}\right) S_{B\#}\left(\frac{t}{n}\right) \right)^n w_0 \right]^* (\sigma, y)\, d\sigma$$

where S_A is the semigroup associated to problem (A) and analogously for S_B, S_P, $S_{A\#}$, $S_{B\#}$, and $S_{P\#}$. Taking limits, applying the Trotter–Kato formula (see Proposition 4.3 BREZIS 1973) and applying convergence under the integral sign we get

$$\int_0^s [S_P(t)z_0]^*(\sigma, y)\, d\sigma \leq \int_0^s [S_{P\#}(t)w_0]^*(\sigma, y)\, d\sigma$$

for any $t \in [0, T]$, for any $s \in [0, |\Omega'|]$ and a.e. $y \in \Omega''$.

For the case $f \neq 0$ and time dependent, the Trotter–Kato formula can be also applied (see, e.g., VUILLERMOT *et al.* 2008). In fact, to deal with the affine case $f(t) \neq 0$ we shall use a "reduction of order technique" argument which can be found on BENILAN *et al.* (unfinished manuscript). We point out that by an approximation argument and then passing to the limit process we can assume, without loss of generality, that in fact $f \in H^1(0, T; L^2(\Omega))$. We shall argue by using the formulation of the problem with homogeneous Dirichlet condition, that is $u = 1 - w$ as unknown, for the case of the general set Ω and with $v = 1 - z$ as unknown for the ball $\Omega^\#$. We also introduce the following notations:

$$\hat{f}(t) = \lambda\beta(1) - f(t).$$

and given any function $\theta \in H^1(0, T; L^2(\Omega))$, for a.e. $t \in (0, T)$ we define the function $\theta(t + \cdot) \in H^1(0, T; L^2(\Omega))$ by the application $s \mapsto \theta(t + s)$. We also introduce the vectorial function $U(t) =$

$(u(t), f(t + \cdot)) \in L^2(\Omega) \times H^1(0, T; L^2(\Omega))$. We proceed in a similar way for the case of the domain $\Omega^\#$: we define $V(t) = (v(t), f^\#(t + \cdot)) \in L^2(\Omega^\#) \times H^1(0, T; L^2(\Omega^\#))$. Then, it is easy to see that U, V are the respective unique solutions of the "autonomous vectorial problems"

$$\begin{cases} \frac{\partial U}{\partial t} + \hat{L}U = 0, & t \in (0, T) \\ U(0) = (u_0, \hat{f}) \end{cases}$$

$$\begin{cases} \frac{\partial V}{\partial t} + \hat{L}V = 0, & t \in (0, T) \\ V(0) = (v_0, \hat{f}^\#) \end{cases}$$

where

$$\hat{L}(u, \xi) = (-\Delta u + g(u) - \xi(0 + \cdot), \xi').$$

Here ξ' represents simply the derivative of ξ. We can use a decomposition $\hat{L} = \hat{L}_1 + \hat{L}_2$ in the following way:

$$\hat{L}_1(u, \xi) = (-\Delta u + h(t)g(u), 0),$$
$$\hat{L}_2(u, \xi) = (-\xi(0 + \cdot), \xi').$$

Let us define the problems

$$(C)\begin{cases} \frac{\partial U}{\partial t} + \hat{L}_1 U = 0, \\ U(0) = (u_0, \hat{f}), \end{cases} \quad (C^\#)\begin{cases} \frac{\partial V}{\partial t} + \hat{L}_1 V = 0, \\ V(0) = (v_0, \hat{f}^\#), \end{cases}$$

$$(D)\begin{cases} \frac{\partial U}{\partial t} + \hat{L}_2 U = 0, \\ U(0) = (u_0, \hat{f}), \end{cases} \quad (D^\#)\begin{cases} \frac{\partial V}{\partial t} + \hat{L}_2 V = 0, \\ V(0) = (v_0, \hat{f}^\#), \end{cases}$$

and the correspondent solution operators

$$S_C(t)(u_0, \hat{f}) = (S_P(t)u_0, \hat{f}),$$
$$S_{C\#}(t)(v_0, \hat{f}^\#) = (S_P(t)u_0, \hat{f}^\#),$$

$$S_D(t)(u_0, \hat{f}) = \left(u_0 + \int_0^t \hat{f}(s)\,ds, \hat{f} \right),$$

$$S_{D\#}(t)(v_0, f^\#) = \left(v_0 + \int_0^t \hat{f}^\#(s)\,ds, \hat{f}^\# \right).$$

Let Q be the projection operator such that $u(t) = QU(t)$. Let us study QS_C and QS_D. Since, for any $t \in [0, T]$, for any $s \in [0, |\Omega'|]$ and a.e. $y \in \Omega''$,

$$\int_0^s u_0^*(\sigma, y)\, d\sigma \leq \int_0^s v_0^*(\sigma, y)\, d\sigma,$$

we have, by the above explicit formulas (for the first component we apply the similar proof as in the case $f = 0$)

$$\int_0^s [Q \, S_C(t)(u_0, f)]^*(\sigma, y)\mathrm{d}\sigma$$

$$\leq \int_0^s [Q \, S_{C^\#}(t)(v_0, f^\#)]^*(\sigma, y)\mathrm{d}\sigma,$$

$$\int_0^s [Q \, S_D(t)(u_0, f)]^*(\sigma, y)\mathrm{d}\sigma$$

$$\leq \int_0^s [Q \, S_{D^\#}(t)(v_0, f^\#)]^*(\sigma, y)\mathrm{d}\sigma.$$

By applying an induction argument again we get

$$\int_0^s \left[Q\left(S_C\left(\frac{t}{n}\right)S_D\left(\frac{t}{n}\right)\right)^n(u_0, f)\right]^*(\sigma, y)\mathrm{d}\sigma$$

$$\leq \int_0^s \left[Q\left(S_{C^\#}\left(\frac{t}{n}\right)S_{D^\#}\left(\frac{t}{n}\right)\right)^n(v_0, f^\#)\right]^*(\sigma, y)\mathrm{d}\sigma.$$

Finally, since all the operators are maximal monotone operators on their respective Hilbert spaces, we can take limits by applying the Trotter–Kato formula (which justify the convergence of the limits) and the result holds. □

3.2. Proof of Corollary 3.2: End of the Proof of Theorem 2.1

For the proof we shall need a classical result.

Lemma 3.6 (HARDY *et al.* 1929) *Let* $\tilde{y}, \tilde{z} \in L^1(0, M)$, $\tilde{y}, \tilde{z} \geq 0$ *a.e.. Suppose y is non-increasing and*

$$\int_0^s \tilde{y}(\sigma) \, \mathrm{d}\sigma \leq \int_0^s \tilde{z}(\sigma) \, \mathrm{d}\sigma, \quad \forall s \in [0, M].$$

Then, for every continuous non-decreasing convex function Φ *we have*

$$\int_0^s \Phi(\tilde{y}(\sigma)) \, \mathrm{d}\sigma \leq \int_0^s \Phi(\tilde{z}(\sigma)) \, \mathrm{d}\sigma \quad \forall s \in [0, M].$$

Proof of Corollary 3.2 Applying the theorem and Lemma 3.6 to

$$\Phi(s) = \beta(1) - \beta(1 - s),$$
$$\tilde{y}(\sigma) = 1 - z^*(|\Omega'| - \sigma, y),$$
$$\tilde{z}(\sigma) = 1 - w^*(|\Omega'| - \sigma, y)$$

we get that

$$\int_s^{|\Omega'|} \beta(z^*(t, \sigma, y))\mathrm{d}\sigma \leq \int_s^{|\Omega'|} \beta(w^*(t, \sigma, y))\mathrm{d}\sigma.$$

It is a classical result (see MOSSINO 1984) that for F Borel and u measurable it holds that

$$\int_{\Omega'} F(u) = \int_0^{|\Omega'|} F(u^*)$$

In particular, the comparison holds between w and z. All that remains is to integrate on Ω'', apply Fubini's theorem and the result follows. □

3.3. The Elliptic Case

Proof of Corollary 3.3 Since there is uniqueness of solutions for the stationary problem (4) then, by applying Corollary 3 of DÍAZ (1994) we get that $w(t) \to w$ in $H^1(\Omega)$, as $t \to +\infty$ (with w the unique solution of problem (4) with $\mu = +\infty$, i.e., the Dirichlet problem $w = 1$ on $\partial\Omega$. Moreover, since the application $u \mapsto u^*$ is continuous with respect to the convergence in L^1 (see e.g., MOSSINO 1984) we get that the mass comparison is stable by passing to the limit as $t \to +\infty$ and the result holds.

4. Proof of Theorem 2.2 and Further Remarks

We shall use the function $\bar{u} = \lambda\psi$ is a supersolution (we recall that ψ is given by (8)). We shall apply the following previous result in the literature due to BANDLE (1985):

Theorem 4.1 *Let* $\Omega \subset \mathbb{R}^n$ *be an open bounded set of measure* $V = |\Omega|$ *such that* Ω *is contained between two parallel* $(n - 1)$-*dimensional hyperplanes at distance* 2ρ *and let* ψ *be the solution of problem* (8). *Then*

$$\|\psi\|_\infty^{1+\frac{n}{2}} \leq CV\rho^2$$

with

$$C = \frac{(4 + 2n)(2n)^{-\frac{n}{2}}}{\pi^2 \omega_n}. \tag{17}$$

Proof of Theorem 2.2 Thanks to the assumption on the initial datum, since we are dealing with the Dirichlet problem $[\mu = +\infty$ in (5)] and $0 \leq u = 1 - w \leq 1$, $0 \leq g(u) \leq 1$, we get that $\bar{u} = \lambda\psi$ is a supersolution of problem (5). Then, applying

Theorem 4.1 to $\Omega = G \times (0, H)$, i.e., with $n = N + 1$ and $2\rho = H$, we get that

$$\|u\|_{L^\infty(0,T;L^\infty(\Omega))} \to 0, \quad \text{as } H \to 0,$$

and, in particular

$$\underset{(0,T)\times\Omega}{\operatorname{essinf}} \beta(w) \to 1, \quad \text{as } H \to 0.$$

More precisely, for any $t \geq 0$ and a.e $\mathbf{x} \in \Omega$

$$1 \geq \beta(w(\mathbf{x}, t)) \geq 1 - \left(\frac{V(2N + 6)(2N + 1)^{-\frac{N+1}{2}}}{\pi^2 \omega_{N+1}} H^2 \right)^{2/(N+3)}, \tag{18}$$

which proves the assertion for the case of the parabolic problem (even if $V = |\Omega| = |G|H$ is prescribed). In the case of the associated stationary problem, since we know that $w(t) \to w$ in $H^1(\Omega)$, as $t \to +\infty$ (see the proof of Corollary 3.3) then, by the dominated Lebesgue theorem we know that $\beta(w(t)) \to \beta(w)$ in $L^\infty(\Omega)$, as $t \to +\infty$ and thus the estimate (18) remains valid replacing $\beta(w(t))$ by $\beta(w)$ (since the bounds are independent of t). $\square$

Remark 4 We shall not enter in this paper in the study of the free boundary (the boundary of the dead core) associated to the solutions $w(t)$ and w of the parabolic and elliptic problems (1) and (4), respectively. We recall that the key assumption for the formation of such free boundary is the condition $0 < q < 1$. We send the reader to the monographs DÍAZ (1985) and ANTONTSEV et al. (2001) for an extensive treatment with numerous references.

Remark 5 All the results of this paper can be extended to more general frameworks according different point of views. For instance, with respect to the diffusion operator it is possible to replace the Laplacian operator $-\Delta w$ by a general second-order elliptic operator of the type

$$Lu = -\sum_{i,j=1}^{N} \frac{\partial}{\partial x_j} \left(a_{ij}(x, y) \frac{\partial u}{\partial x_j} \right) - \sum_{h,k=1}^{m} \frac{\partial}{\partial y_k} \left(b_{hk}(y) \frac{\partial u}{\partial y_h} \right)$$
$$- \sum_{i=1}^{N}\sum_{h=1}^{m} \frac{\partial}{\partial y_h} \left(c_{ih}(y) \frac{\partial u}{\partial x_i} \right) - \sum_{i=1}^{N}\sum_{h=1}^{m} \frac{\partial}{\partial x_i} \left(d_{hi}(y) \frac{\partial u}{\partial y_h} \right) \tag{19}$$

with bounded coefficients (here we followed the notation of Sect. 3). In that case, the comparison via

Steiner symmetrization is made with respect the solution (on a cylinder of symmetric section) associated to the operator

$$L^\# v = -\Delta_x v - \sum_{h,k=1}^{m} \frac{\partial}{\partial y_k} \left(b_{hk}(y) \frac{\partial v}{\partial y_h} \right).$$

No special change in the statements arises if the operator Lu involves transport first-order terms of the type

$$\sum_{k=1}^{m} b_k(y) \frac{\partial u}{\partial y_k}.$$

Quasilinear terms with respect to the x-variable (that is, those in which Steiner symmetrization is performed) can be allowed, too. The presence of transport terms in the x-variable can also be considered, but then the expression of the rearranged operator $L^\# v$ must be modified (see,e.g., CHIACCHIO and MONETTI 2001 and its references). On the other hand, it is still an open problem how to deal with quasilinear terms in the y-variables. We point out that Theorem 4.1 (which play a fundamental role in the proof of Theorem 2.2) was obtained in BANDLE (1985) for the case of a general second-order elliptic operator of the type (19). Concerning the reaction term $\beta(w) = w^q$, the results of this paper can be extended also to the case $q = 0$ by means of the consideration of the maximal monotone graph of $\mathbb{R}^2$ given by

$$\beta(w) = 0 \text{ if } w < 0, \ \beta(w) = 1 \text{ if } w > 0 \quad \text{and} \quad \beta(0) = [0, 1]. \tag{20}$$

(see, e.g., DÍAZ 1985, Chapter 2). As a matter of fact, the proof of Proposition 3.5 (an thus Theorem 3.1) remains valid under the same assumptions on β that Theorem 1 on DÍAZ (1991), i.e., β non-decreasing function with $\beta(0) = 0$ and such that

$$\beta = \beta_1 + \beta_2 \tag{21}$$

where β_1 is concave and β_2 is convex. The results can be extended also to the "enthalpy formulation" of some porous media type equations (associated to a linear operator Lu) in the spirit of the framework presented in DÍAZ (1991, 1992, 2001). It is also possible to extend the results to the more realistic case of suitable coupled systems of the type

$$\begin{cases} \frac{\partial w}{\partial t} - d_w \Delta w + R_1(w, u) = 0 & \text{in } (0, +\infty) \times \Omega, \\ \frac{\partial u}{\partial t} - d_u \Delta u + R_2(w, u) = 0 & \text{in } (0, +\infty) \times \Omega, \end{cases}$$

under suitable structural assumptions on the coupling reaction terms $R_1(w, u)$ and $R_2(w, u)$ (see Theorem 3 of DíAz 1991 for $d_w, d_u > 0$ and DíAz and STAK-GOLD 1994 for $d_w > 0$ and $d_u = 0$). Some results on the Steiner rearrangement for the case of Neumann boundary conditions can be found in FERONE and MERCALDO (2005) and CHIACCHIO (2004).

Remark 6 It can be shown (see BANDLE and VER-NIER-PIRO 2003) that, in spite of Theorem 2.2, domains Ω of optimal effectiveness do not exist for reactions $\beta(w) = w^q$ with $0 < q < 1$. Nevertheless, for the limit case of zero order reactions [with $\beta(w)$ given by (20)] any result proving that there is no dead core for a concrete Ω shows that the effectiveness attains its maximum value for this domain Ω (several criteria for the non-formation of the dead core were given in Chapter 2 of DíAz 1985).

Remark 7 The study of the optimality of the effectiveness factor in terms of shape differentiation on Ω is the main object of the paper (DíAz and GÓMEZ-CASTRO 2014b).

Acknowledgments

We thank Professor A. Romero, from the Chemical Engineering Department of the UCM for the useful conversations we held on industrial wastewater treatment tanks and the references with which he supplied us on the subject. The first author's research was partially supported by the project ref. MTM2011-26119 of the DGISPI (Spain), the UCM Research Group MOMAT (Ref. 910480), and the ITN *FIRST* of the Seventh Framework Program of the European Community's (Grant agreement number 238702).

REFERENCES

A. ALVINO, G.TROMBETTI, J.I. DíAz, P.L. LIONS (1996) *Elliptic Equations and Steiner Symmetrization*, Communications on Pure and Applied Mathematics, Vol. *XLIX*, 217–236, John Wiley and Sons

S.N. ANTONTSEV, J.I, DIAZ AND S.I. SHMAREV *Energy Methods for Free Boundary Problems: Applications to Nonlinear PDEs and Fluid Mechanics*, (Birkhäuser, Boston 2001)

R. ARIS, *The Mathematical Theory of Diffusion and Reaction in Permeable Catalysts* (Oxford University Press, 1975)

C. BANDLE, (1985) *A note on Optimal Domains in a Reaction-Diffusion ProblemIsoperimetric Inequalities and Applications*, Zeitschrift für Analysis unhd ihre Answendungen, B.d. *4*, (3), 207–213

C. BANDLE, S. VERNIER-PIRO (2003) *Estimates for solutions of quasilinear problems with dead cores*, Z. angew. Math. Phys. *54*, 815–821

P. BENILAN, M. CRANDALL, A. PAZY, *Nonlinear evolution equations in Banach spaces* (unfinished manuscript)

H. BREZIS, *Operateurs Maximaux Monotones et Semi-groupes de Contractions dans les Espaces de Hilbert*. Notes de Matematica, 5, (North-Holland, Amsterdam. 1973)

F. CHIACCHIO (2004), *Steiner symmetrization for an elliptic problem with lower-order terms*. Ricerche di Matematica, vol *53* n.1, 87–106

F. CHIACCHIO, V.M. MONETTI (2001), *Comparison results for solutions of elliptic problems via Steiner symmetrization*. Differential and Integral Equations *14* (11), 1351–1366.

C. CONCA, J. I. DíAz, C. TIMOFTE (2003) *Effective Chemical Process in Porous Media*. Mathematical Models and Methods in Applied Sciences, *13*, 1437–1462.

C. CONCA, J. I. DíAz, A. LIÑAN, C. TIMOFTE (2004) *Homogeneization in Chemical Reactive Flows*, Electr. J. Diff. Eqns. 2004 (No.40), 1–22

J. I. DíAz, *Nonlinear Partial Differential Equations and Free Boundaries* (Pitman, London 1985)

J. I. DíAz (1991), Simetrización de problemas parabó licos no lineales: Aplicación a ecuaciones de reacción - difusi ón. Memorias de la Real Acad. de Ciencias Exactas, Físicas y Naturales, Tomo XXVII

J. I. DíAz (1992). Symmetrization of nonlinear elliptic and parabolic problems and applications: a particular overview. In Progress in partial differential equations.elliptic and parabolic problems (ed. C. Bandle et al.), (Pitman Research Notes in Mathematics No 266, Longman, Harlow, Essex) pp. 1–16

J. I. DíAz, *The Mathematics of Models in Climatology and Environment*, ASI NATO Global Change Series I, no. 48 (Springer-Verlag, Heidelberg, Germany, 1996)

J. I. DíAz (2001). *Qualitative Study of Nonlinear Parabolic Equations: an Introduction*. Extracta Mathematicae, *16*, no. 2, 303–341,

J. I. DíAz AND D. GÓMEZ-CASTRO (2014a), *Steiner symmetrization for concave semilinear elliptic and parabolic equations and the obstacle problem*, accepted in Discrete and Continuous Dynamical Systems-S

J. I. DíAz AND D. GÓMEZ-CASTRO, *On the effectiveness of chemical reactors: an analysis through shape differentiation*, (2014b). To appear in Electronic Journal of Differential Equations

J. I. DíAz AND I. STAKGOLD (1994) *Mathematical aspects of the combustion of a solid by distributed isothermal gas reaction*. SIAM. Journal of Mathematical Analysis, Vol *26*, No2, 305–328,

J. I. DíAz (1994), *F. de Thelin. On a nonlinear parabolic problems arising in some models related to turbulent flows*. SIAM Journal of Mathematical Analysis, Vol *25*, No 4, 1085–1111,

V. FERONE AND A. MERCALDO (1998), *A second order derivation formula for functions defined integrals*, C. R. Acad. Sci. Paris, t. *326*, Serie I, 549–554.

V. FERONE AND A. MERCALDO (2005), *Neumann Problems and Steiner Symmetrization*, Communications in Partial Differential Equations, Volume *30*, Issue 10, 1537–1553

G.H. HARDY, J.E. LITTLEWOOD, G. PÓLYA (1929). *Some simple inequalities satisfied by convex functions*. Messenger Math., *58*, pp. 145–152,

J. MOSSINO, *Inegalités Isoperimetriques et Applications en Physique* (Hermann, Paris 1984).

S. RODRIGUEZ, A. SANTOS, A. ROMERO, F. VICENTE (2012), *Kinetic of oxidation and mineralization of priority and emerging pollutants by activated persulfate*, Chemical Engineering Journal *213*, 225–234

J. M. ROSAS, F. VICENTE, E. G. SAGUILLO, A. SANTOS, A. ROMERO (2014) *Remediation of soil polluted with herbicides by Fenton-like reaction: Kinetic model of diuron degradation*, Applied Catalysis B: Environmental *144*, 252–260

R. SPIGLER AND M. VIANELLO (1995), *Convergence analysis of the semi-implicit Euler method for abstract evolution equations*, Numer. Funct. Anal. Optim. *16*, 785–803,

W. STRIEDER, R. ARIS *Variational Methods Applied to Problems of Diffusion and Reaction*, (Springer-Verlag, Berlin 1973)

F. VICENTE, J.M. ROSAS, A. SANTOS, A. ROMERO (2011) *Improvement soil remediation by using stabilizers and chelating agents in a Fenton-like process* Chemical Engineering Journal *172*, 689–697

P.-A. VUILLERMOT, W.F. WRESZINSKI, V.A. ZAGREBNOV (2008), *A Trotter–Kato Product Formula for a Class of Non-Autonomous Evolution Equations, Trends in Nonlinear Analysis: in Honour of Professor V. Lakshmikantham, Nonlinear Analysis, Theory*, Methods and Applications *69*, 1067–1072.

(Received November 20, 2014, revised June 7, 2015, accepted June 9, 2015, Published online July 11, 2015)

Pure Appl. Geophys. 173 (2016), 937–944
© 2015 Springer Basel
DOI 10.1007/s00024-015-1112-z

Defining Geodetic Reference Frame using Matlab®: PlatEMotion 2.0

FLAVIO CANNAVÒ[1] ⓘ and MIMMO PALANO[1]

Abstract—We describe the main features of the developed software tool, namely PlatE-Motion 2.0 (PEM2), which allows inferring the Euler pole parameters by inverting the observed velocities at a set of sites located on a rigid block (inverse problem). PEM2 allows also calculating the expected velocity value for any point located on the Earth providing an Euler pole (direct problem). PEM2 is the updated version of a previous software tool initially developed for easy-to-use file exchange with the GAMIT/GLOBK software package. The software tool is developed in Matlab® framework and, as the previous version, includes a set of MATLAB functions (m-files), GUIs (fig-files), map data files (mat-files) and user's manual as well as some example input files. New changes in PEM2 include (1) some bugs fixed, (2) improvements in the code, (3) improvements in statistical analysis, (4) new input/output file formats. In addition, PEM2 can be now run under the majority of operating systems. The tool is open source and freely available for the scientific community.

Key words: Euler pole, matlab, geodetic reference frame.

1. Introduction and Mathematical Formulation

Kinematic models of plate tectonics rely on the assumption that a rigid displacement confined to the surface of a sphere can be described by a rotation about a virtual axis passing through the centre of the sphere (the so-called Euler's theorem by Leonid Euler in 1776). This implies that an angular velocity vector originating at the centre of the Earth can describe motions of plates. BULLARD et al. (1965) first applied this parameterization of the Euler's theorem to plate tectonics as a means of quantifying and evaluating the WEGENER's (1927) postulated fit of the eastern margin of South America to the western margin of Africa. Since then, subsequent elaboration

of plate tectonics included derivation of rotation poles and angles and rates to describe plate reconstructions of magnetic anomalies (VINE 1966), transform fault azimuths, earthquake slip vectors and spreading rates at mid-ocean ridges (MINSTER and JORDAN 1978), hotspots (e.g., DEMETS et al. 1990; GRIPP and GORDON 1990) and lastly, space geodesy data (SELLA et al. 2002). The most widespread parameterization of the angular velocity vector in plate kinematics is using latitude and longitude to describe the location where the rotation axis cuts the surface of the Earth, and a rotation rate that corresponds to the magnitude of the angular velocity. Formally, the latitude and longitude of the angular velocity vector constitute the so-called 'Euler pole'.

Here we describe the main features of a software tool, called PEM2, which allows solving the direct and inverse problems of plate tectonics Euler pole. PEM2 represents the updated version of a previous software tool described in CANNAVÒ and PALANO (2011). Because the estimation of Euler pole parameters represents an important issue in global tectonics and geodynamics studies, we developed PEM2 in order to give to the scientific community as well as to practitioners an easy-to-use tool. New improvements of PEM2 with respect to the previous version include (1) changes in the code to improve the Euler pole parameters computation, (2) improvement in statistical analysis to properly evaluate the quality of estimated parameters, (3) new input/output file formats.

We have chosen to work in Earth-Centred Earth-Fixed (ECEF) Cartesian coordinates because the plate motion calculations are much easier in this frame using rotation vectors. The mathematical formulation is straightforward: in an ECEF frame, at a given location P, on the Earth's surface, defined by its Cartesian coordinates X, Y and Z (in meters), for a plate rotation defined by the rotation vector Ω (ω_x,

---
[1] Istituto Nazionale di Geofisica e Vulcanologia, Osservatorio Etneo - Sezione di Catania Piazza Roma 2, 95123 Catania, Italy.
E-mail: flavio.cannavo@ingv.it

ω_y, ω_z) (in rad/year), the velocity vector V (v_x, v_y, v_z) (in m/year) is given by the Euler's fixed point theorem (GREINER 1999; PALAIS and PALAIS 2007) which can be expressed as the cross-product:

$$\bar{V} = (\bar{\Omega} \times \bar{P}) = \begin{pmatrix} Z\omega_y - Y\omega_z \\ Z\omega_z - Z\omega_x \\ Y\omega_x - X\omega_y \end{pmatrix} \qquad (1)$$

Thus, for the direct problem, the resulting site velocities are in ECEF frame so they need to be transformed to local NEU frame. A well-known valid velocity conversion at the corresponding geodetic location, defined with geodetic latitude, longitude and height (above ellipsoid) respectively (λ, φ, h), is given by a combination of the three rotations needed to align the ECEF frame with the NEU frame:

$$\bar{L} = \overline{RV} \qquad (2)$$

where R is the rotation matrix:

$$\bar{R} = \begin{bmatrix} -\sin\lambda\cos\varphi & -\sin\lambda\sin\varphi & \cos\lambda \\ -\sin\varphi & \cos\varphi & 0 \\ \cos\lambda\cos\varphi & \cos\lambda\sin\varphi & \sin\lambda \end{bmatrix} \qquad (3)$$

and L (N, E, U) are the components of the velocity in the local frame (VERMEILLE 2002).

The inverse problem consists in calculating the pole parameters by knowing the velocities of a given number of sites. To this aim, the direct problem shown in Eq. (1) can be rewritten in matrix form. In particular for a given site, and taking into account the measurement errors, the matrix for of the equation is:

$$\begin{bmatrix} v_x \\ v_y \\ v_z \end{bmatrix} = A\bar{\Omega} + \varepsilon = \begin{bmatrix} 0 & Z & -Y \\ -Z & 0 & X \\ Y & -X & 0 \end{bmatrix} \begin{bmatrix} \omega_x \\ \omega_y \\ \omega_z \end{bmatrix} + \varepsilon$$

$$(4)$$

where A is the design matrix and ε is the vector of measurement errors. With three unknowns and three equations per site, the problem is over-determined with just two sites, and can be solved by minimizing the residual model errors by a least squares approach (TARANTOLA and VALLETTE 1982):

$$\Omega = (A'WA)^{-1}A'WV \qquad (5)$$

where W is the data weight matrix that is typically chosen as the inverse of the measurement covariance

matrix and A' is the transpose of the matrix A. The resulting coordinate rotations Ω is in ECEF Cartesian reference system. The corresponding Euler pole expressed in geographic coordinates is basically the intersection of Ω with the WGS-84 ellipsoid, and the software adopts the formulas reported in BOWRING (1976, 1985) to make the conversion to latitude and longitude coordinates. The angular velocity is calculated as the strength of the rotations Ω expressed in degrees/Myear.

In order to take into account all the possible uncertainty sources and the nonlinearities of geo-centric-geographic conversion (BOWRING 1976, 1985), a Monte Carlo resampling step has been implemented in PEM2 for propagating the uncertainties (ANDERSON 1976) through the whole inversion process and obtaining robust solutions with valid associated covariance matrix. The Monte Carlo simulations are performed by randomly sampling Gaussian distributions with the observed velocities as mean values and their associated uncertainties as standard deviations. From this set of resampled data, all three-component (i.e., latitude, longitude and angular velocity) solutions from Eq. (5) and subsequent coordinate conversion are collected, and the expected values (means) together with the full covariance matrix are calculated to supply the estimated pole solution with its associated uncertainties.

As a new feature, PEM2 allows the user to calculate the equivalent rotation coordinates in geocentric system and to estimate the parameter variances.

2. Goodness-Of-Fit

In the Euler pole estimation, selecting the right model (CANNAVÒ 2012; CANNAVÒ et al. 2015) and assessing the quality of a model solution becomes a fundamental step to evaluate the quality of the whole process of plate-frame identification. The choice of the right model is always a trial and error process leading to a satisfied solution for some kinds of criteria. To aim the user in this process, different relevant tests can be carried out in PEM2 for an easy model evaluation.

2.1. Chi-Square Test

The quality of the estimated Euler pole can be evaluated statistically by using the Chi-square statistic. The Chi-square statistic is defined by the sum:

$$\chi^2 = E'WE \qquad (6)$$

where E is the vector of residuals $E = (V_x - v_x, V_y - v_y, V_z - v_z)$ and W is the same data weight matrix used for the Monte Carlo inversions [see Eq. (5)]. The method of least-squares is built on the hypothesis that the optimum description of a set of data is one which minimizes the weighted sum of squares of deviations, between the data, V, and the fitting model v (TARANTOLA and VALLETTE 1982). Hence the smaller the Chi-square value, the better the model.

The ratio between the estimated variance in the residues and the theoretical variance of the hypothetic distribution can be estimated by χ^2/v, where, for an Euler pole estimation, $v = 3N - 3$ is the degree of freedom, N is the number of sites. χ^2/v is called the reduced Chi-square statistic and for a good model it assumes values around 1.

In PEM2, the reduced Chi-square of the residues can easily be calculated, allowing to compare different models with different degrees of freedom.

PEM2 allows also evaluating the Chi-square test (SNEDECOR and COCHRAN 1989) to easily assess the goodness-of-fit of the estimated Euler pole based on the site velocity errors. The Chi-square test (χ^2 test) considers as null hypothesis the Chi-square distribution of the asymptotic (by increasing the samples) sampling distribution of residues. We therefore test $\chi^2 \leq \chi^2_{\alpha,n}$ where $\chi^2_{\alpha,n}$ is the inverse of a Chi-square distribution (SNEDECOR and COCHRAN 1989) with n degree of freedom calculated at a probability of α (with 1-α level of confidence). In this new version of the software, it is possible for the user to set the level of confidence of the test.

The reasons for failing the test are mainly two: (1) the weights used in the system are not correct, or (2) there is at least one point containing a gross error (see LEICK 2004, for details). The user can play with the data directly into PEM2 to improve the pole estimation.

2.2. F-test

This test allows comparing the results from two least-square estimations in order to detect which model best fits the data, according to the degree of freedom (i.e., the number of observations minus number of unknowns; see NOCQUET et al. 2001, for more details). As described in the previous section, the least-squares estimation of an Euler pole minimizes the term in Eq. (6).

For a given set of observations, adding unknown parameters in least-squares inversion makes χ^2 decrease. The F-ratio evaluates if the decrease of χ^2 from a model with $n2$ versus $n1$ ($n1 > n2$) degrees of freedom is significant. Since the statistic in (6) is χ^2 distributed with n degrees of freedom, the ratio

$$F = \frac{[\chi^2(n1) - \chi^2(n2)]/(n1 - n2)}{\chi^2(n2)/n2} \qquad (7)$$

follows a Fisher-Snedecor distribution (FORBES et al. 2010) with $n1$ versus $n2$ degrees of freedom. This experimental ratio is compared to the expected value of a $F(n1 - n2, n2)$ distribution, for a given risk level α (corresponding to $100 - \alpha$ % confidence level).

For our objectives, the null hypothesis, following the rigid plate tectonics assumption, is that all the site velocities can be modelled with the same Euler pole. We therefore test

$$F \leq f_\alpha^{n1 - n2, n2} \qquad (8)$$

where $f_{1-\alpha}^{n1n2}$ is the α % fractile for a Fischer-Snedecor law with $(n1, n2)$ degree of freedom. We use this test to verify if the velocity of an individual site is consistent with a set of sites. In particular, we compare the two estimations of the Euler pole with and without an individual site. The degree of freedom for an Euler pole estimation is $n1 = 3 \times N - 3$ for N site velocities, and it becomes $n2 = 3 \times N - 3 - 3$ by subtracting a site. If (8) is not verified, we can assert, with a risk of α % of being wrong that this site velocity is not consistent with the velocity of the other sites.

PEM2 allows estimating the F-test, also showing the sites that are not consistent with all the other velocities. The site check is done simultaneously and separately for each of the sites. It is possible for the user to choose an arbitrary confidence level for the test. A good way to proceed should be to delete the worst sites one by one and performing the test after re-estimating the pole at each step.

2.3. Correlation Coefficient

To estimate the reliability of the results, the calculation of correlation coefficient r has also been implemented on PEM2. Correlation coefficient r is defined as:

$$r = \frac{\sum_{j=1}^{3} \sum_{i=1}^{N} \left(V_i^{(j)} - \mu_{V^{(j)}} \right) \left(v_i^{(j)} - \mu_{v^{(j)}} \right)}{\sqrt{\sum_{j=1}^{3} \sum_{i=1}^{N} \left(V_i^{(j)} - \mu_{V^{(j)}} \right)^2} \sqrt{\sum_{j=1}^{3} \sum_{i=1}^{N} \left(v_i^{(j)} - \mu_{v^{(j)}} \right)^2}} \tag{9}$$

where V_i and v_i are the observed and modelled values respectively, j is the vector component, μ indicates the expected value and N represents the number of sites. The r absolute value ranges between 0 and 1, with a value closer to 1 indicating a better fit. The significance of r can be associated to the coefficient of determination, r^2, which indicates the percentage of data that can be explained by the model. According to statistical decision theory, $r^2 > 70\%$ ($r = 0.837$) is an acceptable test for prediction (MULARGIA and GASPERINI 1995).

3. Software Features

PEM2 is an application that runs in Matlab 7.1 and above. This software tool can be run under the majority of operating systems, namely Windows, Macintosh and Unix/Linux. Since PEM2 has a GUI, an X-Window environment should also be installed on UNIX-like operating systems, and MATLAB should be started with the Java Virtual Machine (JVM) enabled. PEM2 requires the basic MATLAB and the Statistics and Mapping toolboxes. PEM2 includes a set of MATLAB functions (m-files), GUIs (fig-files), map data files (mat-files) and user's manual as well as some example input files. To run the software, add the software folder to the MATLAB search path and type PEM2 in the MATLAB command prompt. The current version is 2.0 but the tool is still being developed and future versions are expected. An example of the PEM2 screenshot is shown in Fig. 1 with some of the main features.

3.1. Loading and Editing Data

The first version of the software (CANNAVÒ and PALANO 2011) was initially developed for easy-to-use file exchange with the GAMIT/GLOBK software package (HERRING et al. 2010). In PEM2, we added the possibility to load and save ASCII files directly formatted for the Generic Mapping Tools (GMT; WESSEL and SMITH 1998). The formats of the input data files accepted by PEM2 are below described:

- *org_file* this file represents the output from the run of the GLORG module of GAMIT/GLOBK. In particular, this file contains a summary of the final solution, including estimates and uncertainties of the parameters and various representations of these parameters (e.g., baseline components). For more details about this file format, please refer to the GAMIT/GLOBK reference manual (http://www-gpsg.mit.edu/~simon/gtgk/docs.htm).

- *apr_file* this file has the form as the Cartesian version of the GAMIT L-file with a blank character in the first column and contains the 8-character site name, *XYZ* position (m), *XYZ* velocity (m/year) in ECEF Cartesian coordinates of the site at a specified epoch (e.g., 2007.239) and associated uncertainties (m/year). There are 11 columns in the format. Each release of the GAMIT/GLOBK software package contains several apr_file, including the version of the latest ITRF solutions and additional files rotated to plate-specific frames. However, after an initial GLORG solution, each user can create a specific apr_file, containing selected sites by extracting them from the org_file output of the GLORG routine. Please refer to the GAMIT/GLOBK reference manual for more details.

- *vel_file* this file must contain the coordinates and velocities of selected sites. It should have an extension vel and should have the "psvelo" format used in GMT (WESSEL and SMITH 1998). Entries should be separated by a space. The first line of the data file is a header which is in free format. *E* and *N* describe the East and North velocities component expressed in mm/year, E_σ and N_σ the

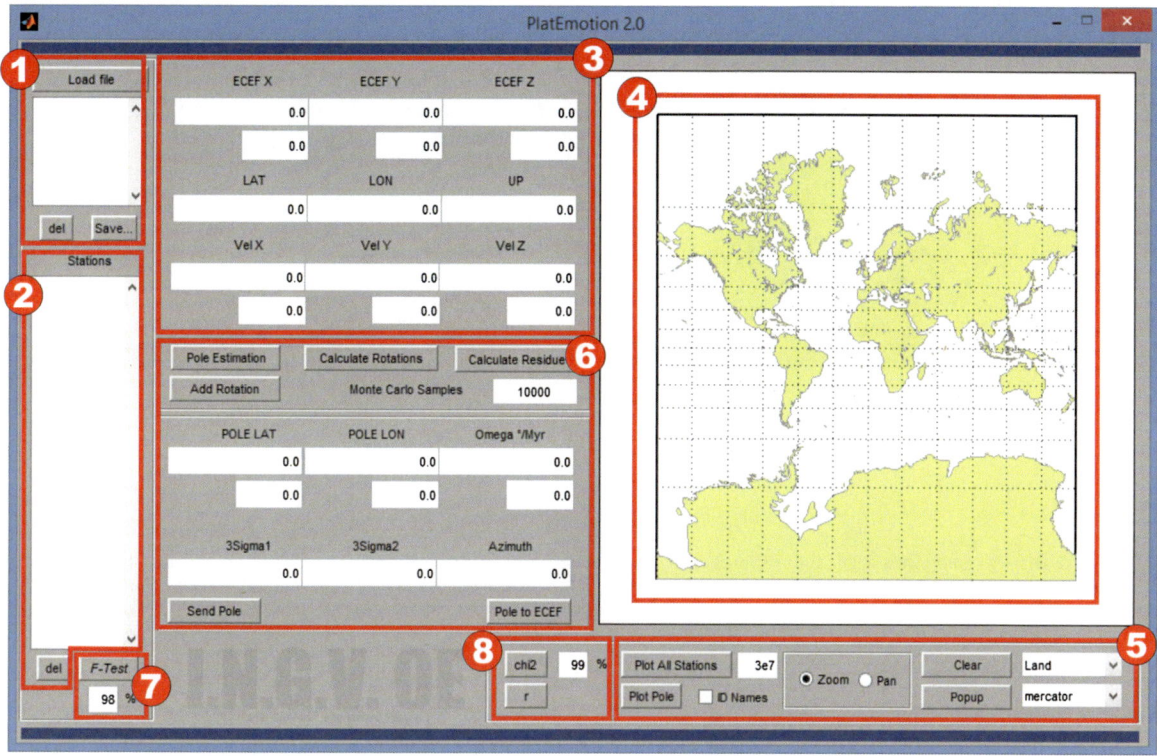

Figure 1

PEM2 screen view showing main functions: *1* panel for loading input data files (see Sect. 3.1 for details about the format of input files). *2* panel for loaded stations. Once an input data file is loaded, all stations contained in it will be listed here and unwanted stations can be selected and deleted from further computations. *3* panel for station parameters. *4* panel for displaying site locations, velocity vectors and Euler pole location in a simple map projection. *5* panel for managing plots on panel 4. *6* panel for Euler pole estimation. The "pole estimation" button performs the pole estimation from the selected loaded data file (inverse problem). PEM2 allows the user to change the number of inversions (e.g., the number of Monte Carlo samples) for errors estimation. The "calculate rotation" button runs the expected velocity vector from a given (or computed) Euler pole at sites contained in a selected loaded data file (direct problem). The button "add rotation" adds the velocities due to a given Euler pole to the sites. The "calculate residue" button computes the residual velocity vector at selected sites from a given (or computed) Euler pole. Once a pole is estimated, it can be plotted on the map projection ("plot pole" button on *panel 5*). The "send pole" button allows transferring the current Euler pole solution to other loaded data frame in the list 1. *7* "F-test" button. It calculates an F-ratio test (NOCQUET *et al.* 2001) in order to test if the velocities of one or more sites are consistent with the other set of velocities. *8* "chi2" and "r" buttons. They are useful for estimating the consistency of the results as described in the previous sections. It is possible for the user to set the level of confidence of the χ^2 test

associated errors and RHO is the correlation between E_σ and N_σ. The entries on each line are site longitude (positive east, decimal degrees), latitude (decimal degrees), east velocity, north velocity, standard deviations for the east and north velocities, RHO and the 4-character site name. All units for the velocities are mm/year. There are eight columns in the format.

As mentioned above, the current PEM2 version contains some example input files.

Output data files are in the *vel_file* or *apr_file* format. This last can be used as input directly in the GLOBK runs.

4. A Case of Study

In the following, as an example of application we estimated new Euler vector components (latitude and longitude of pole, rotation rate) for the Eurasian plate

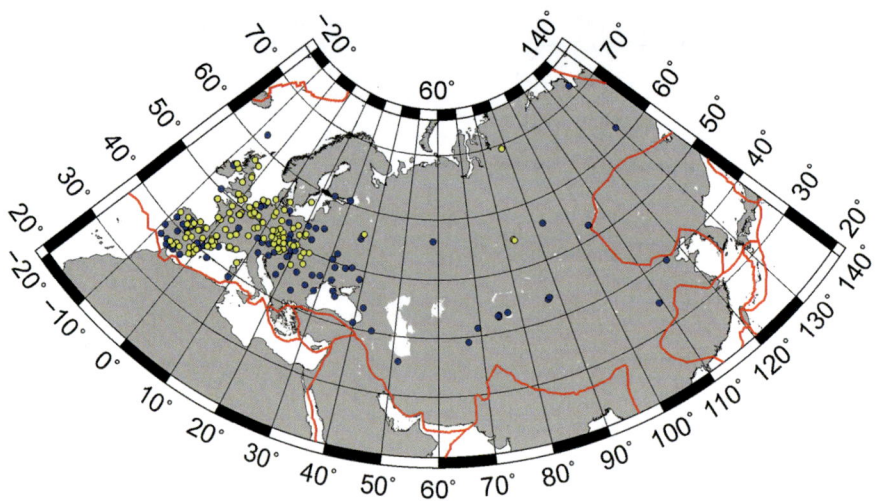

Figure 2
Locations and velocity components of the 216 GNSS sites used to estimate the Euler vector components (Table 1) for the Eurasian Plate. The 132 GNSS sites allowing the final Euler vector components estimation are reported as *yellow dots*. Plate boundaries are reported as *red lines* (BIRD 2003)

by considering a starting dataset of 216 GNSS continuous stations (Fig. 2). These GNSS stations are distributed over most of the Eurasian plate and far from actively deforming boundaries and from Fennoscandia, where horizontal velocities may be significantly affected by glacial isostatic adjustment (e.g., NOCQUET *et al.* 2005). However, in order to show the quality of the tool, this starting dataset includes a number of GNSS stations (~20 %) which, because affected by some local effects (subsidence, fault motion, etc.), have been "a priori" excluded in recent Euler vector components estimation for the Eurasian plate (please see ALTAMIMI *et al.* 2012 and MARQUES *et al.* 2013 for details). The GNSS dataset, spanning the 1995.00–2014.55 time interval, was processed by using GAMIT/GLOBK (HERRING *et al.* 2010) following the strategy described in PALANO (2015). The estimated GNSS velocities were referred to the ITRF2008 reference frame (ALTAMIMI *et al.*

2012). In a first step, after loading the file, we estimated the Euler vector parameters, whose details are reported in Table 1 under the "Strategy 1" section. Although characterized by low uncertainties, the obtained solution does not fit properly the data as evidenced from the large χ^2 value (25.26) resulting in a failure of the χ^2 test at 99 % of confidence. As previously reported, the PEM2 software allows minimizing, with a weighted least squares inversion, the adjustments to observed horizontal velocities of the selected sites, to objectively determine the set of GNSS stations that best defines a rigid plate/block. By using the *F*-ratio test, PEM2 allows also comparing the results from various estimations in order to detect which model best fits the data. For our objectives, the null hypothesis, following the rigid plate assumption, is that all the site velocities can be modeled with the same Euler vector. Considering the 216 GNSS sites previously selected, we therefore

Table 1

Parameters of the Euler vector parameters and associated statistical tests as estimated for the two adopted different strategies (please see text for details) and the relative outcomes of χ^2 test

	Latitude N° (deg)	Longitude E° (deg)	Ω (deg/Myear)	χ^2	χ^2 test at 99 %
Strategy 1	56.623 ± 0.072	−95.099 ± 0.130	0.263 ± 0.001	25.26	Failed
Strategy 2	55.050 ± 0.160	−99.455 ± 0.320	0.258 ± 0.001	1.16	Passed

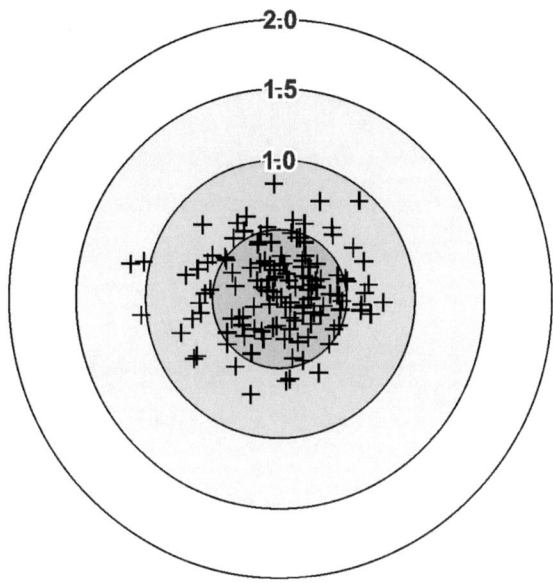

Figure 3
East and North component of the residuals (in mm/year) of the 132
GNSS sites allowing the final Euler vector components estimation
following Strategy 2 (please see text for details)

performed some F-ratio tests in order to detect those
sites not satisfying the F-ratio criteria. As mentioned
above, for each test, all sites not satisfying the F-ratio
criteria were excluded and a subsequent test was
performed after a re-estimation of the pole.

Finally, a total of 132 stations passed the F-ratio
test allowing the Euler vector components estimation
as reported in Table 1 under the "Strategy 2" section.
The reduced χ^2 value of 1.16 clearly indicates a statistically better quality result, as evidenced also by
the χ^2 test that results passed at 99 % of confidence.
Therefore, no significant residuals remain in this
reference frame computation, and as shown in Fig. 3,
most of the residuals are lower than 1 mm/year. In
addition, our estimated Euler vector components for
the Eurasian plate (Table 1) well agree with the
estimations reported in ALTAMIMI *et al.* (2012) and
MARQUES *et al.* (2013).

5. Conclusions

We presented the developed software tool, called
PEM2, for solving the direct and inverse problems of
plate tectonics Euler pole. PEM2 is developed in

Matlab® framework and shows a user-friendly
graphical user interface (GUI). It is designed for
easy-to-use file exchange also with the GAMIT/
GLOBK software package. We showed the quality of
the tool by presenting an example of application.
Other examples of application can be found in
PALANO *et al.* (2013a, b). The tool is open source and
freely available for the scientific community at
https://sourceforge.net/projects/platemotion/. Further
information and documentation on PEM2 may be
obtained from the developers.

Acknowledgments

We thank the Guest Editor, María Charco, and three
anonymous reviewers for their critical reviews,
constructive suggestions and useful comments that
improved the paper.

REFERENCES

ANDERSON, G. M. (1976). *Error propagation by the Monte Carlo
method in geochemical calculations*. Geochimica et Cos-
mochimica Acta, *40*(12), 1533–1538, doi:10.1016/0016-
7037(76)90092-2.

ALTAMIMI, Z., METIVIER, L. and COLLILIEUX, X. (2012). *ITRF2008
plate motion model*. J. Geophys. Res., *117*, B07402, doi:10.1029/
2011JB008930.

BIRD, P. (2003). *An updated digital model of plate boundaries*.
Geochemistry Geophysics Geosystems, *4*(3), 1027, doi:10.1029/
2001GC000252.

BULLARD, E. C., EVERETT, J. E. and SMITH, A. G. (1965). *Fit of
continents around the atlantic*. Roy. Soc. London, Phil. Trans.
Ser. A, V., *258*.

BOWRING, B. R. (1976). *Transformation from spatial to geographical coordinates*. Survey review, *23*(181), 323–327.

BOWRING, B. R. (1985). *The accuracy of geodetic latitude and
height equations*. Survey Review, *28*(218), 202–206.

CANNAVÓ, F. (2012). *Sensitivity analysis for volcanic source modeling quality assessment and model selection*. Computers &
Geosciences, *44*, 52–59, doi:10.1016/j.cageo.2012.03.008.

CANNAVÓ, F. and PALANO, M. (2011). *PlatEMotion: a Matlab® Tool
for geodetic reference frame definition*. Rapporti Tecnici INGV,
201, 1–12.

CANNAVÓ, F., ARENA A. and MONACO, C. (2015). *Local geodetic and
seismic energy balance for shallow earthquake prediction*.
J. Seismol. *19*(1), 1–8.

DEMETS, C., GORDON, R. G., ARGUS, D. F. and STEIN, S. (1990).
Current plate motions. Geophys. J. Int., *101*, 425–478.

FORBES, C., EVANS, M., HASTINGS, N. and PEACOCK, B. (2010). *F
(Variance Ratio) or Fisher–Snedecor Distribution*, in Statistical
Distributions, Fourth Edition, John Wiley & Sons, Inc., Hobo-
ken, NJ, USA. doi: 10.1002/9780470627242.ch20.

GREINER, B. (1999). *Euler rotations in plate-tectonic reconstructions*. Computers & Geosciences, *25*(3), 209–216.

GRIPP, A. E. and GORDON, R. G. (1990). *Current plate velocities relative to the hotspots incorporating NUVEL-1 global plate motion*. Geophys. Res. Lett., *17*, 1109–1112, doi:10.1029/GL017i008p01109.

HERRING, T.A., KING, R.W. and MCCLUSKY, S.C. (2010). *Introduction to GAMIT/GLOBK, Release 10.4*, Department of Earth, Atmospheric and Planetary Sciences, Massachusetts Institute of Technology, Cambridge, MA.

LEICK, A. (2004). *GPS Satellite Surveying*. John Wiley & Sons, ISBN 0471059307.

MARQUES F.O., CATALAO J.C., DEMETS C., COSTA A.C.G. and HILDENBRAND A. (2013). *GPS and tectonic evidence for a diffuse plate boundary at the Azores Triple Junction*. Earth Planet. Sci. Lett. *381*, 177–187, doi:10.1016/j.epsl.2013.08.051.

MINSTER, J.B. and JORDAN T.H. (1978). *Present-day plate motions*. J. Geophys. Res. *83*(11), 53, doi: 10.1029/JB083iB11p05331.

MULARGIA, F. and GASPERINI, P. (1995). *Evaluation of the applicability of the time and slip-predictable earthquake recurrence models to Italian seismicity*. Geophys. J. Int., *120*, 453–473, doi:10.1111/j.1365-246X.1995.tb01832.x.

NOCQUET, J.-M., CALAIS, E, ALTAMIMI, Z, SILLARD, P. and BOUCHER, C. (2001). *Intraplate deformation in the western Europe deduced from an analysis of the International Terestrial Reference Frame 1997 (ITRF97) velocity field*. J. Geophys. Res., *106*, B6, 11239–11257, doi:10.1029/2000JB900410.

NOCQUET, J.-M., CALAIS, E. and PARSONS, B. (2005). *Geodetic constraints on glacial isostatic adjustment in Europe*. Geophys. Res. Lett., *32*, L06308, doi:10.1029/2004GL022174.

PALAIS, B., and PALAIS, R. (2007). *Euler's fixed point theorem: The axis of a rotation*. Journal of Fixed Point Theory and Applications, *2*(2), 215–220.

PALANO, M., GONZÁLEZ, P. J. and FERNÁNDEZ, J. (2013a). *Strain and stress fields along the Gibraltar Orogenic Arc: Constraints on active geodynamics*. Gondwana Research, *23*(2), 1071–1088, doi:10.1016/j.gr.2012.05.021.

PALANO, M., IMPRESCIA, P. and GRESTA, S. (2013b). *Current stress and strain-rate fields across the Dead Sea Fault System: Constraints from seismological data and GPS observations*. Earth and Planetary Science Letters *369–370*, 305–316, doi:10.1016/j.epsl.2013.03.043.

PALANO, M. (2015). *On the present-day crustal stress, strain-rate fields and mantle anisotropy pattern of Italy*. Geophys. J. Int., *200*(2), 969–985, doi:10.1093/gji/ggu451.

SELLA, G.F., DIXON, T.H. and MAO, A. (2002). *REVEL: A model for recent plate velocity from space geodesy*. J. Geophys. Res., *107*, doi:10.1029/2000JB000033.

SNEDECOR, G.W. and COCHRAN, W.G. (1989). *Statistical Methods, Eighth Edition*. Iowa State University Press.

TARANTOLA, A. and VALLETTE, B. (1982). *Generalized nonlinear inverse problems solved using the least squares criterion*. Rev. Geophys. Space Phys., *20*, 219–232, doi:10.1029/RG020i002p00219.

VERMEILLE, H. (2002). *Direct Transformation from Geocentric to Geodetic Coordinates*. J. Geod. *76*(8), 451–454, doi:10.1007/s00190-002-0273-6.

VINE, F.J. (1966). *Spreading of the ocean floor: new evidence*. Science, *154*, 1405–1415, doi: 10.1126/science.154.3755.1405.

WEGENER, A. (1927). *Die geophysikalischen Grundlagen der Theorie der Kontinentverschiebung*. Scientia, Milan, *41*, 103–116.

WESSEL, P. and SMITH, W.H.F. (1998). *New improved version of the Generic Mapping Tools released*. EOS Trans. AGU, *79*, 579, doi:10.1029/98EO00426.

(Received December 7, 2014, revised May 5, 2015, accepted May 27, 2015, Published online June 9, 2015)

Pure Appl. Geophys. 173 (2016), 945–961
© 2015 Springer Basel
DOI 10.1007/s00024-015-1130-x

Identification of Bedrock Lithology using Fractal Dimensions of Drainage Networks extracted from Medium Resolution LiDAR Digital Terrain Models

Joaquín Cámara,[1] Vicente Gómez-Miguel,[1] and Miguel Ángel Martín[2]

Abstract—Geologists know that drainage networks can exhibit different drainage patterns depending on the hydrogeological properties of the underlying materials. Geographic Information System (GIS) technologies and the increasing availability and resolution of digital elevation data have greatly facilitated the delineation, quantification, and study of drainage networks. This study investigates the possibility of inferring geological information of the underlying material from fractal and linear parameters describing drainage networks automatically extracted from 5-m-resolution LiDAR digital terrain model (DTM) data. According to the lithological information (scale 1:25,000), the study area is comprised of 30 homogeneous bedrock lithologies, the lithological map units (LMUs). These are mostly igneous and metamorphic rocks, but also include some sedimentary rocks. A statistical classification model of the LMUs by rock type has been proposed based on both the fractal dimension and drainage density of the overlying drainage networks. The classification model has been built using 16 LMUs, and it has correctly classified 13 of the 14 LMUs used for its validation. Results for the study area show that LMUs, with areas ranging from 177.83 ± 0.01 to 3.16 ± 0.01 km^2, can be successfully classified by rock type using the fractal dimension and the drainage density of the drainage networks derived from medium resolution LiDAR DTM data with different flow support areas. These results imply that the information included in a 5-m-resolution LiDAR DTM and the appropriate techniques employed to manage it are the only inputs required to identify the underlying geological materials.

1. Introduction

Drainage networks indicating paths of overland flow concentration are the geomorphological result of climate, tectonics, geology and time (Lifton and Chase 1992). Drainage networks are present everywhere, even in remote places such as the largest of Saturn's moons, Titan, where the climatic conditions are hugely different to those currently found on Earth (Cartwright *et al.* 2011). Furthermore, the evidence of their existence are still present today in places like Mars where the tectonic activity and the overland fluid disappeared long time ago (Hynek and Phillips 2003).

For many decades of the past century the study of a drainage network entailed the accomplishment of a huge and thorough delineation work. Nowadays, the evolution of Geographic Information Systems (GIS), the increasing resolution and availability of digital elevation data, the development of specific algorithms (O'Callaghan and Mark 1984; Tarboton 1997), and the increasing computer power available in personal computers have simplified the delineation of drainage networks and facilitated their use in hydrological studies. However, the proper delineation of drainage networks is still a complex problem due to the scale dependence of digital terrain models (DTMs) and drainage networks (Helmlinger *et al.* 1993). The answer to the question where do channels begin (Montgomery and Dietrich 1988, 1992) is still being investigated (Tesfa *et al.* 2011; Schwanghart *et al.* 2013).

Geologists know that drainage networks show different drainage patterns depending on the geological properties of the underlying materials. This knowledge represents one of the fundamental aspects of photo-geology, which is the use of aerial photographs to interpret the geology of a region and to compile geological maps from them. In fact, the first step of geological and soil mapping, after the proper preparation of the aerial photographs, is the delineation of drainage networks at the finer scales allowed by the aerial photographs (Allum 1966).

[1] Departamento de Edafología, Universidad Politécnica de Madrid, E.T.S.I. Agrónomos, Avda. Puerta de Hierro 2, 28040 Madrid, Spain. E-mail: joaquincamaragajate@yahoo.es

[2] Departamento de Matemática Aplicada, Universidad Politécnica de Madrid, E.T.S.I. Agrónomos, Avda. Puerta de Hierro 2, 28040 Madrid, Spain.

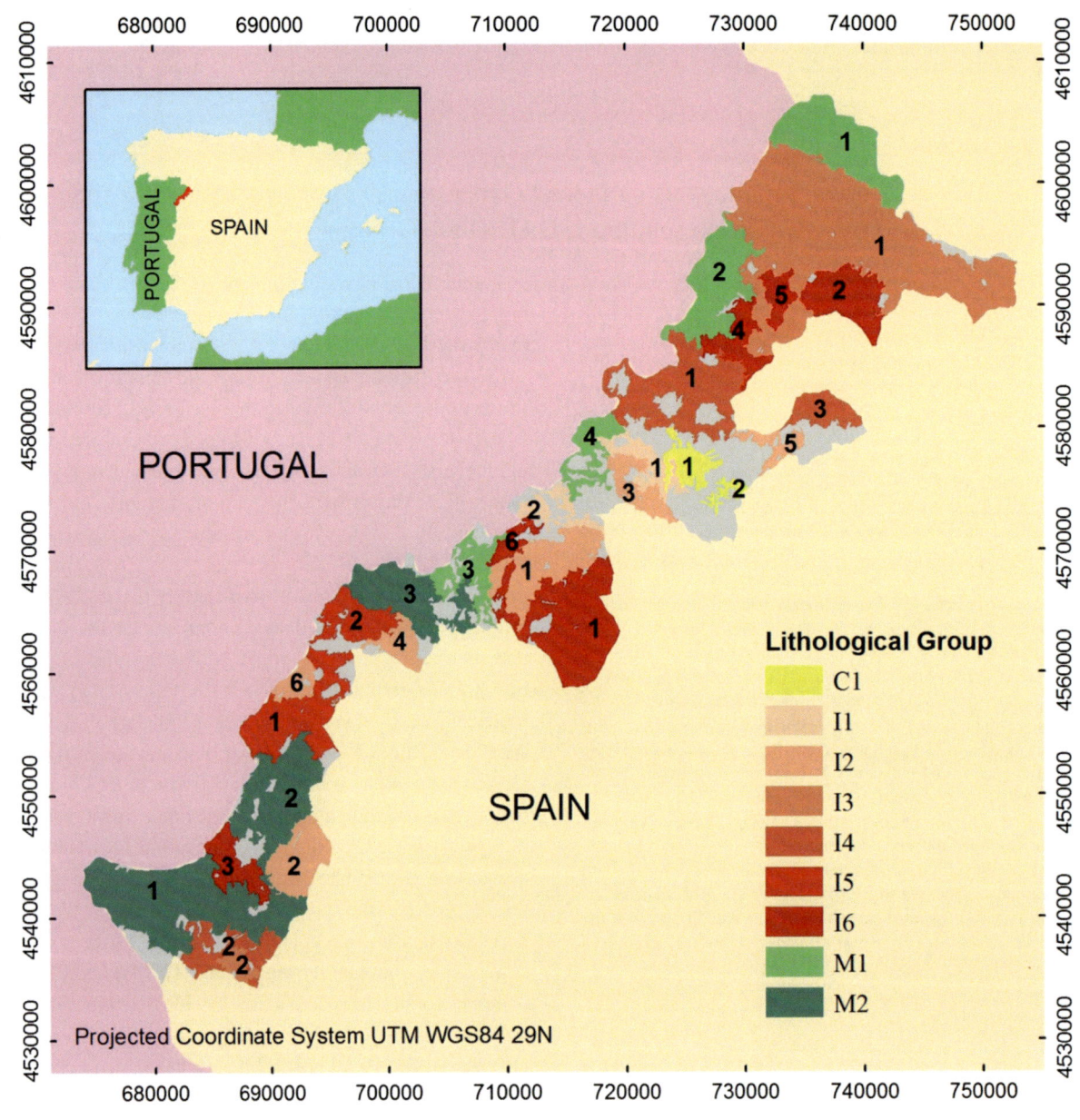

Figure 1
Overview of the study area showing the 30 lithological map units under study

Almost 50 years ago, HOWARD (1967) proposed a qualitative classification of drainage patterns that is reproduced nowadays in most geology textbooks. The different classes of drainage patterns are dendritic, parallel, pinnate, trellis, rectangular, radial, and annular. Each of these classes is related to different landforms, rocks, or slope steepness in a qualitative way. Fortunately, some authors have since decided to quantify the complexity involved in the architecture of these natural structures. The first works quantitatively ordering and describing the drainage networks (HORTON 1932, 1945; STRAHLER 1957) revealed important relationships between their elements and were used to establish a series of ratios known as Hortonian laws that still today are commonly used in hydrological studies. These works provided the first evidence of the scaling properties of drainage networks. Research

Table 1

Lithological groups included in the lithological map of the region

Code	Geological time		Lithological group		MAGNA50	
	Era	Period	Type of rock	Brief description	Sheet no	Unit
T1	Cenozoic	Miocene	Sedimentary	Conglomerates with heterometric pebbles	449	25
S1	Cenozoic	Paleogene	Sedimentary	Lutites and silts	423	16
A1	Paleozoic	–	Filonian	Quartz dikes and pegmatites	449	1
I1	Paleozoic	Ordovician	Igneous	Ortoneisses metagranitic, and granite	423	3
I2	Paleozoic	Devonian	Igneous	Granite: medium grained, two-mica with abundant quartz dikes.	423	11
I3	Paleozoic	Devonian	Igneous	Granite: medium to coarse grained, two-mica, porphyritic	475	3
I4	Paleozoic	Devonian	Igneous	Granite: coarse grained, two-mica, not porphyritic	475	10
I5	Paleozoic	Devonian	Igneous	Granite: coarse to very coarse grained, biotitic, porphyritic	449	11
I6	Paleozoic	Devonian	Igneous	Granite: fine to medium grained, two-mica, leucocratic	449	17
I7	Paleozoic	Devonian	Igneous	Inhomogeneous granitoids, migmatitics, microporphyritic	423	7
M1	Paleozoic	Precambrian	Metamorphic	Pelitic paragneiss interbedded with quartzite	422	16
M2	Paleozoic	Cambrian	Metamorphic	Pelitic and psammitic metasediments interbedded with quartzite	449	19
M3	Paleozoic	Ordovician	Metamorphic	Alternating level of shales and quartzitic shales	449	23

Table 2

Participation of the lithological groups in the study area, and in the 30 selected LMUs

Code	Study area			30 LMUs selected		
	Area		LMU	Area		LMU
	km^2	%	Count	km^2	%	Count
T1	4.50 ± 0.01	0.5	14	–	–	0
S1	15.20 ± 0.01	1.5	6	13.93 ± 0.01	1.6	2
A1	1.07 ± 0.01	0.1	23	–	–	0
I1	35.58 ± 0.01	3.6	19	23.41 ± 0.01	2.7	2
I2	100.82 ± 0.01	10.2	18	95.03 ± 0.01	11.1	6
I3	197.18 ± 0.01	20.0	16	183.22 ± 0.01	21.4	2
I4	88.07 ± 0.01	8.9	18	83.45 ± 0.01	9.8	3
I5	77.24 ± 0.01	7.8	38	56.42 ± 0.01	6.6	2
I6	148.24 ± 0.01	15.0	34	127.72 ± 0.01	14.9	6
I7	17.21 ± 0.01	1.7	26	–	–	0
M1	132.47 ± 0.01	13.4	27	111.79 ± 0.01	13.1	4
M2	166.55 ± 0.01	16.9	16	159.88 ± 0.01	18.7	3
M3	3.90 ± 0.01	0.4	2	–	–	0
	988.04 ± 0.01		257	854.85 ± 0.01		30

on these scaling properties led hydrologists to the recognition of the fractal and multifractal nature of drainage networks (RODRÍGUEZ-ITURBE and RINALDO 1997; DE BARTOLO *et al.* 2000, 2004, 2006; GAUDIO *et al.* 2004; ARIZA-VILLAVERDE *et al.* 2013) that is changing the way drainage networks are investigated.

The question follows that if drainage networks are fractals, and drainage patterns are controlled by geology, can fractal measures of drainage networks

be used to help answer the question of what is the underlying lithology?

Some authors have explored the lithologic control on drainage networks concluding that fractal measures are related to the lithologic characteristics of the source rocks (GAUDIO *et al.* 2006), and are more influenced by tectonic uplift than by lithology (DOMBRADI *et al.* 2007). However, the study units in these works were entire basins (DOMBRADI *et al.* 2007) or large areas of 100s of

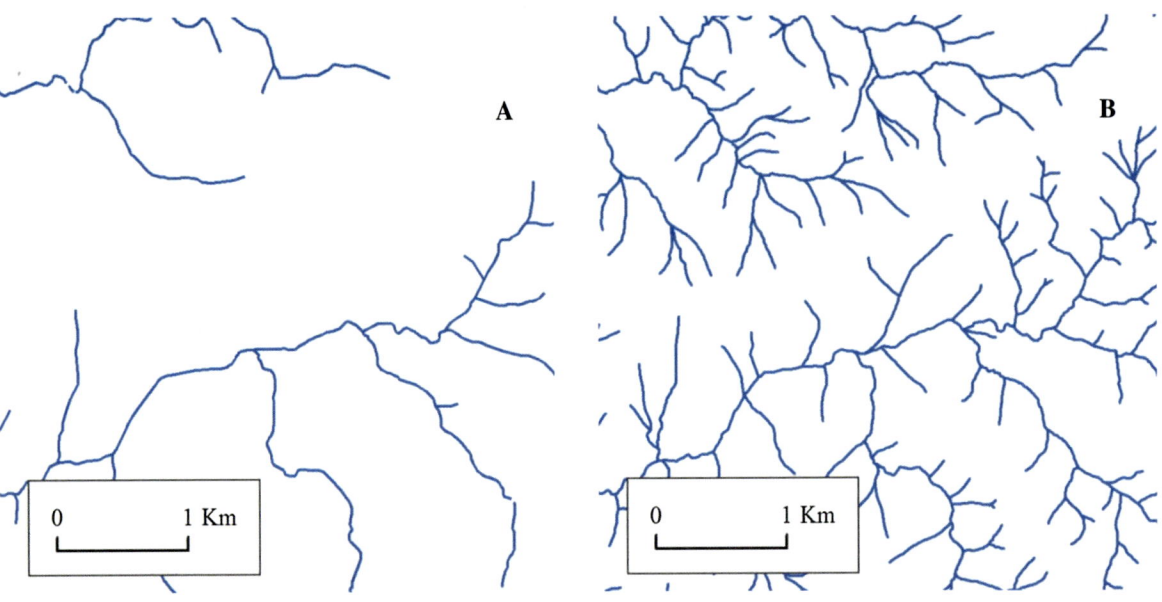

Figure 2
Drainage networks included in the National Topographic Map of Spain (**a**), scale 1:25,000 (http://www.ign.es), and the digitized drainage network obtained by photogrammetric methods (**b**)

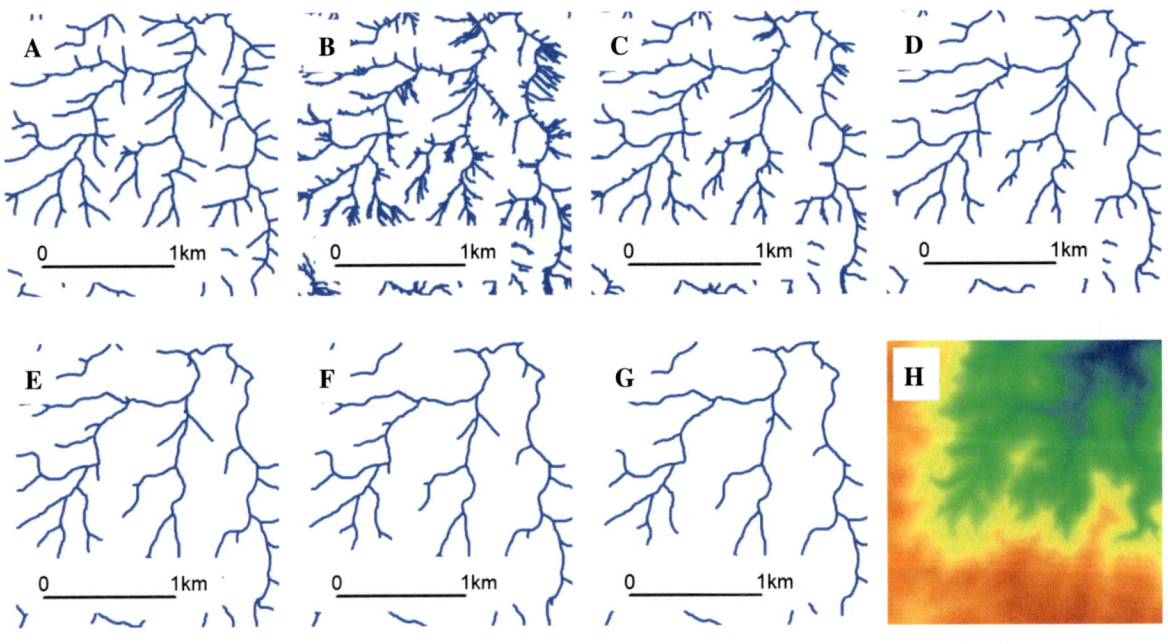

Figure 3
Details of the digitized drainage network of the LMU M2_01 (**a**), details of the drainage networks automatically extracted with the different flow support areas; 2500 m² (**b**), 5000 m² (**c**), 10,000 m² (**d**), 20,000 m² (**e**), 30,000 m² (**f**), 40,000 m² (**g**), and details of the 5-m-resolution LiDAR DTM

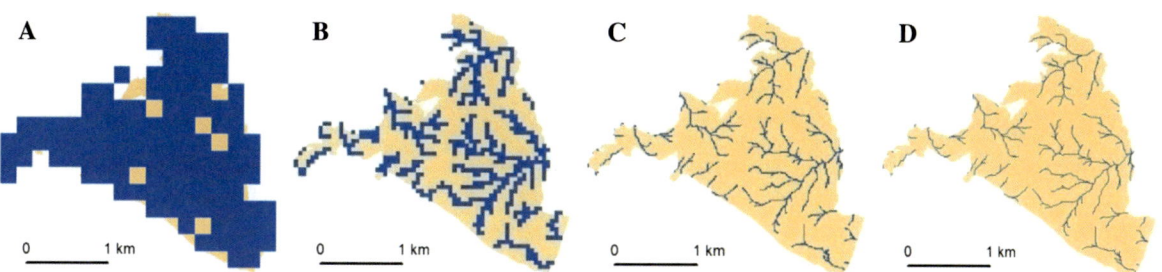

Figure 4

Box counting method applied to the drainage network of the LMU I2_04 extracted with the FAT value of 20,000 m². The boxes of size 320 m (**a**), 80 m (**b**), 20 m (**c**), and 5 m (**d**) occupied by the drainage network

Table 3

Quantitative variables describing the LMUs used to perform the statistical analysis

Abbrev.	Description
HI	Hypsometric integral
E mean	Weighted mean elevation of the LMU expressed in 'meters above sea level'
Area	Area of the lithological map units expressed in km²
Perimeter	Perimeter of the Lithological Map Units expressed in m
Ratio A/P	Ratio area-perimeter of the lithological map unit expressed in m
DD_{2500}	Drainage density of the drainage network obtained with FAT = 2500 m²
DD_{5000}	Drainage density of the drainage network obtained with FAT = 5000 m²
$DD_{10,000}$	Drainage density of the drainage network obtained with FAT = 10,000 m²
$DD_{20,000}$	Drainage density of the drainage network obtained with FAT = 20,000 m²
$DD_{30,000}$	Drainage density of the drainage network obtained with FAT = 30,000 m²
$DD_{40,000}$	Drainage density of the drainage network obtained with FAT = 40,000 m²
$DD_{DIGITIZED}$	Drainage density of the digitized drainage network
FAT_{DD}	Theoretical FAT obtained for the digitized drainage network based on $DD_{DIGITIZED}$
FD_{2500}	Fractal dimension of the drainage network obtained with FAT = 2500 m²
FD_{5000}	Fractal dimension of the drainage network obtained with FAT = 5000 m²
$FD_{10,000}$	Fractal dimension of the drainage network obtained with FAT = 10,000 m²
FD_{20000}	Fractal dimension of the drainage network obtained with FAT = 20,000 m²
$FD_{30,000}$	Fractal dimension of the drainage network obtained with FAT = 30,000 m²
$FD_{40,000}$	Fractal dimension of the drainage network obtained with FAT = 40,000 m²
$FD_{DIGITIZED}$	Fractal dimension of the digitized drainage network
FAT_{FD}	Theoretical FAT obtained for the digitized drainage network based on $FD_{DIGITIZED}$

km² (GAUDIO *et al.* 2006), where more than one homogeneous bedrock lithology would be expected. The use of study units defined by semi-detailed scale lithological maps has been recently suggested (BLOOMFIELD *et al.* 2011; CÁMARA *et al.* 2013) and could enlighten the problem and provide further conclusive results.

The objective of this study was to investigate if the fractal dimension of the drainage networks automatically extracted from medium resolution DTM data could quantitatively identify geological characteristics of the underlying materials. One aspect that the authors want to emphasize is that the drainage networks derived with different flow support areas from LiDAR DTMs were used to capture the terrain irregularity resulting from erosion and fragmentation processes caused by current or past climatic conditions on different geological materials, instead of to solve the problem of perfect delineation of the current drainage streams.

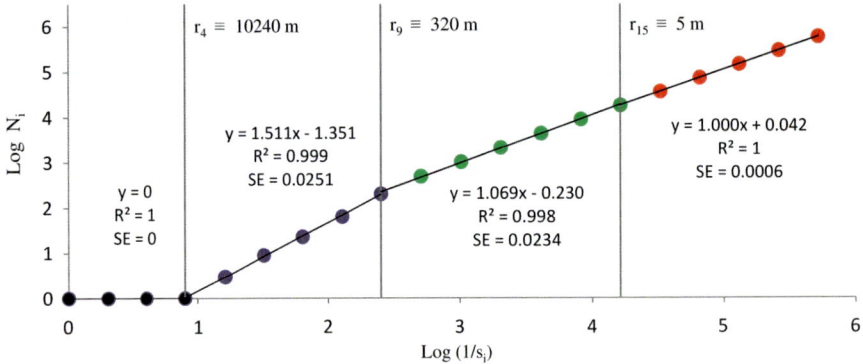

Figure 5
Log–log plot of the number of boxes that intersect the drainage network (N_i) versus the size factor of the intersected boxes ($1/s_i$). The numbers reflect the scaling behavior, for pixel sizes from 81.92 km to 15.625 cm, of the drainage network of the LMU I5_02 extracted with the FAT value of 10,000 m^2

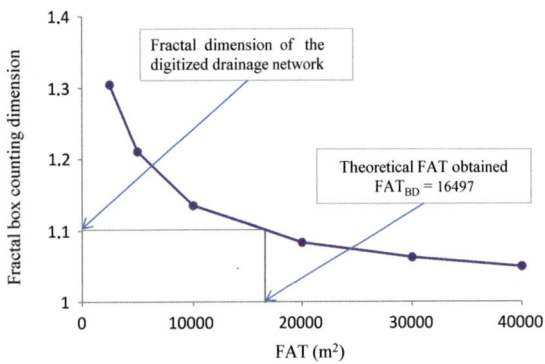

Figure 6
Method applied to calculate the flow accumulation threshold value that theoretically would generate a drainage network with the same fractal dimension as the digitized drainage network

2. Materials and Methods

2.1. Study Area

The study area is located in Northwest Spain (Fig. 1), and it corresponds to the wine growing region called 'Arribes'. This area is entirely placed on the Douro River Basin. The Douro River is the third longest river in the Iberian Peninsula. It flows East to West crossing Spain and Portugal to the Atlantic Ocean.

The Arribes landscape is characterized by deep canyons formed by the Douro River and the main rivers of its left-side tributary system, i.e. the Tormes, Uces, Huebra, and Águeda Rivers. These canyons are carved into the surrounding peneplain, which was formed by igneous and metamorphic rocks from the Hercinian orogeny. Sedimentary materials, that filled the Douro Basin during the Cenozoic Era, surface the highest zones of the peneplain.

The Arribes climate shows a xeric soil moisture regime, which is the typical moisture regime in areas of Mediterranean climate (Soil Survey Staff 2014), and a mesic soil temperature regime with a mean annual soil temperature between 8 and 15 °C. The annual rainfall is close to 600 mm all across the region (GÓMEZ-MIGUEL et al. 2011). The main meso-climatic differences inside the region are determined by the landscape. Thus, the lower parts of the canyons, which are protected against the dominant Atlantic winds, show higher mean annual temperatures than those recorded on the peneplain. According to the Köppen-Geiger climate classification scheme (KÖPPEN 1918; KÖPPEN and GEIGER 1928), the climate of the canyons is classified as Csa (temperate climate with dry and hot summer) while the climate of the peneplain is classified as Bsk (arid climate steppe cold).

2.2. Data Set

Three sources of information of the study area were used for this paper: the soil map, scale 1:25,000, the 5-m-resolution DTM, and the drainage network obtained by photogrammetric methods.

The lithological information contained in the soil map units has been used to define the lithological map. It is worth noting that during the creation of the soil map, at a scale of 1:25,000, the boundaries between different lithologies were delineated by refining the National Geological Map of Spain, at a scale of 1:50,000, using photographic interpretation of aerial photographs (23 x 23 cm), scale 1:18,000.

The soil map units with the same lithological information were dissolved using ArcGIS 10.1 to generate the Lithological Map Units (LMUs). The lithological map contains 257 LMUs of 13 lithological groups. Table 1 shows a brief description of the lithological groups included in the lithological map, and also the geological map unit number and the sheet number of the National Geological Map of Spain, scale 1:50,000 (MAGNA50) matching the lithological group. An exhaustive description of the geological map units can be consulted online at the website of the Instituto Geológico y Minero de España (http://info.igme.es/cartografia/magna50.asp).

Thirty LMUs of the lithological map were chosen to carry out the study (Table 2). The initial criterion for the choice was the selection of the 30 biggest LMUs. In this selection there was only one LMU of sedimentary material. Therefore, the smallest LMU of the initial group of 30 was replaced by the sedimentary LMU S1_02. Nine different lithological groups, which cover 97.3 % of the total area of the studied region, are represented within these 30 LMUs. The nine lithological groups belong to the three types of rock mentioned above: igneous, metamorphic, and sedimentary rocks. Table 4, which is placed in the "Results" section, shows a quantitative geomorphological description of the 30 LMUs.

Table 4

Morphological properties of the LMUs

LMU	HI	Elevations of DTM data				Area (km²)	Perimeter (m)	Ratio A/P (m)
		E_{min} (masl)	E_{max} (masl)	E_{mean} (masl)	Std. Desv.			
S1_01	0.621 ± 0.001	684 ± 1	809 ± 1	762 ± 1	19.1	10.778 ± 0.001	42,671 ± 1	252.6 ± 0.1
S1_02	0.593 ± 0.001	727 ± 1	763 ± 1	745 ± 1	6.8	3.155 ± 0.001	28,000 ± 1	112.7 ± 0.1
M1_01	0.703 ± 0.001	625 ± 1	789 ± 1	728 ± 1	28.8	42.057 ± 0.001	45,653 ± 1	921.2 ± 0.1
M1_02	0.782 ± 0.001	325 ± 1	713 ± 1	522 ± 1	89.4	37.594 ± 0.001	62,732 ± 1	599.3 ± 0.1
M1_03	0.684 ± 0.001	521 ± 1	785 ± 1	727 ± 1	27.8	16.944 ± 0.001	82,385 ± 1	205.7 ± 0.1
M1_04	0.647 ± 0.001	229 ± 1	696 ± 1	560 ± 1	108.4	15.195 ± 0.001	60,058 ± 1	253.0 ± 0.1
M2_01	0.613 ± 0.001	415 ± 1	768 ± 1	674 ± 1	60.3	81.212 ± 0.001	90,623 ± 1	896.2 ± 0.1
M2_02	0.631 ± 0.001	545 ± 1	743 ± 1	706 ± 1	26.2	44.393 ± 0.001	67,369 ± 1	658.9 ± 0.1
M2_03	0.703 ± 0.001	725 ± 1	786 ± 1	764 ± 1	10.1	34.275 ± 0.001	67,024 ± 1	511.4 ± 0.1
I1_01	0.629 ± 0.001	187 ± 1	690 ± 1	537 ± 1	137.9	13.279 ± 0.001	59,576 ± 1	222.9 ± 0.1
I1_02	0.508 ± 0.001	512 ± 1	806 ± 1	745 ± 1	31.2	10.134 ± 0.001	53,241 ± 1	190.3 ± 0.1
I2_01	0.780 ± 0.001	594 ± 1	677 ± 1	635 ± 1	17.8	32.984 ± 0.001	58,462 ± 1	564.2 ± 0.1
I2_02	0.709 ± 0.001	383 ± 1	786 ± 1	694 ± 1	71.0	22.758 ± 0.001	30,500 ± 1	746.2 ± 0.1
I2_03	0.734 ± 0.001	373 ± 1	683 ± 1	606 ± 1	38.2	13.507 ± 0.001	35,033 ± 1	385.5 ± 0.1
I2_04	0.812 ± 0.001	750 ± 1	803 ± 1	781 ± 1	10.6	10.422 ± 0.001	25,306 ± 1	411.8 ± 0.1
I2_05	0.639 ± 0.001	187 ± 1	730 ± 1	586 ± 1	118.5	7.934 ± 0.001	31,974 ± 1	248.1 ± 0.1
I2_06	0.696 ± 0.001	188 ± 1	765 ± 1	674 ± 1	81.4	7.423 ± 0.001	20,304 ± 1	365.6 ± 0.1
I3_01	0.791 ± 0.001	594 ± 1	804 ± 1	746 ± 1	20.5	177.831 ± 0.001	163,445 ± 1	1,088.0 ± 0.1
I3_02	0.489 ± 0.001	741 ± 1	820 ± 1	784 ± 1	15.7	5.385 ± 0.001	15,607 ± 1	345.0 ± 0.1
I4_01	0.773 ± 0.001	122 ± 1	630 ± 1	414 ± 1	126.2	52.089 ± 0.001	97,180 ± 1	536.0 ± 0.1
I4_02	0.753 ± 0.001	667 ± 1	779 ± 1	728 ± 1	25.4	18.565 ± 0.001	67,233 ± 1	276.1 ± 0.1
I4_03	0.578 ± 0.001	698 ± 1	768 ± 1	738 ± 1	11.8	12.797 ± 0.001	22,721 ± 1	563.2 ± 0.1
I5_01	0.734 ± 0.001	324 ± 1	753 ± 1	632 ± 1	94.4	38.127 ± 0.001	70,706 ± 1	539.2 ± 0.1
I5_02	0.842 ± 0.001	516 ± 1	817 ± 1	728 ± 1	50.4	18.291 ± 0.001	43,748 ± 1	418.1 ± 0.1
I6_01	0.721 ± 0.001	429 ± 1	747 ± 1	678 ± 1	55.8	57.474 ± 0.001	77,833 ± 1	738.4 ± 0.1
I6_02	0.544 ± 0.001	323 ± 1	750 ± 1	615 ± 1	92.9	25.627 ± 0.001	33,896 ± 1	756.1 ± 0.1
I6_03	0.575 ± 0.001	382 ± 1	707 ± 1	592 ± 1	76.4	15.168 ± 0.001	37,720 ± 1	402.1 ± 0.1
I6_04	0.548 ± 0.001	122 ± 1	662 ± 1	453 ± 1	130.6	15.127 ± 0.001	42,215 ± 1	358.3 ± 0.1
I6_05	0.570 ± 0.001	187 ± 1	723 ± 1	525 ± 1	122.8	8.426 ± 0.001	18,989 ± 1	443.8 ± 0.1
I6_06	0.718 ± 0.001	324 ± 1	753 ± 1	632 ± 1	94.4	5.901 ± 0.001	28,670 ± 1	205.8 ± 0.1

The DTM used in this study is a 5-m-resolution DTM generated from LiDAR data with a density of 0.5 point/m^2. The altimetric accuracy of the medium resolution DTM has a root mean square deviation equal to or less than 0.5 meters. The DTM data source was the Instituto Geográfico Nacional de España© (IGN). These data are freely available under registration on the "Download Centre" of the IGN's website (http://www.ign.es). According to VAZE et al. (2010), medium resolution LiDAR DTMs are quality data to capture the nature and complexity of drainage networks, even more than high-resolution LiDAR DTMs, with grid sizes equal to or less than 1 m^2, which generate a greater noise effect (ZANDBERGEN 2010).

The drainage network of the studied area was delineated in previous works by the authors (GÓMEZ-MIGUEL et al. 2011). The drainage network is a vector file obtained by stereoscopic view of the aerial photographs cited above, which was delineated over the 25-cm-resolution orthophotos of the region, considering all the evidence of continuous or intermittent watercourses on the terrain. No comparable similar scale hydrological data are available. Figure 2 shows a visual comparison between the finer scale of the available hydrological data and the digitized drainage network used in this study.

2.3. Method of Drainage Network Extraction

The cartographic operations were carried out with the Hydrology Toolbox included in the Spatial Analyst extension of the software ArcGIS 10.1, which was developed by the University of Texas (MAIDMENT 2002). The method of extraction of drainage networks included in ArcGIS is based on the maximum gradient method, also known as the D8 algorithm, which was proposed by O'Callaghan and Mark (O'CALLAGHAN and MARK 1984).

In this study, six drainage networks were generated from the LiDAR-DTMs by applying six different values of the flow accumulation threshold (FAT). The six FAT values used correspond to flow support areas of 2,500, 5000, 10,000, 20,000, 30,000, and 40,000 m^2 (Fig. 3).

The drainage network obtained by photogrammetric restitution was compared in the 30 LMUs with the automatically extracted drainage network aiming to explore which FAT value generates the drainage network most similar to the delineated one.

2.4. Geomorphological Approach to the LMUs

The medium resolution LiDAR DTM was also used to calculate a topographic index of the LMUs. The calculated index was the hypsometric integral (HI) which was introduced by STRAHLER (1952). This index has been traditionally used in the analysis of river basins as a classic measure of erosional stage. In this study, the HI was not applied to defined watersheds, but to LMUs.

The method selected for the estimation of the HI was the elevation-relief ratio, which was proposed by Pike and Wilson (PIKE and WILSON 1971). Following this method, the HI is expressed as

$$HI = \frac{E_{mean} - E_{min}}{E_{max} - E_{min}}, \qquad (1)$$

where E_{mean} is the weighted mean value of the 5-m DTM pixel values belonging to the LMU, E_{max} is the maximum elevation value of the LMU, and E_{min} is the minimum elevation value. The appropriateness of this method has been recently reaffirmed (SINGH et al. 2008).

Some basic characteristics of the LMUs such as the area, the perimeter, and the area-perimeter ratio were also quantified.

A classic measure in hydrology, the drainage density, was calculated on the LMUs for the seven drainage networks under investigation. A theoretical FAT value of the digitized drainage network (FAT$_{DD}$) was calculated for each LMU by relating the drainage density of the digitized drainage network to the decreasing evolution with the FAT increments of the drainage density of the automatically extracted drainage networks.

2.5. Fractal Analysis: Box Counting Method

The fractal dimension of the drainage networks was estimated applying a fixed-size algorithm known as the box counting method, which was introduced by RUSSEL et al. (1980). The box counting method has been widely used in fractal analysis since fractals were brought onto the scientific scene by Mandelbrot (MANDELBROT 1983).

Table 5

Drainage densities of the drainage networks obtained with different FAT values (DD$_{FAT}$)

LMU	DD$_{2500}$	DD$_{5000}$	DD$_{10,000}$	DD$_{20,000}$	DD$_{30,000}$	DD$_{40,000}$	DD$_{DIGITIZED}$	FAT$_{FD}$
S1_01	12.19	6.37	3.91	2.63	2.20	1.88	3.12	16202 ± 1
S1_02	8.92	4.02	2.00	0.95	0.68	0.54	1.65	13332 ± 1
M1_01	11.64	7.18	4.79	3.63	3.10	2.76	4.03	16519 ± 1
M1_02	18.84	10.71	6.34	4.17	3.30	2.86	3.80	24287 ± 1
M1_03	17.97	9.90	6.01	4.23	3.63	3.18	5.36	13647 ± 1
M1_04	15.40	8.32	5.15	3.64	3.15	2.82	4.85	11977 ± 1
M2_01	13.52	7.63	4.94	3.57	3.06	2.73	5.17	9568 ± 1
M2_02	16.94	9.56	5.91	4.21	3.59	3.24	4.60	17708 ± 1
M2_03	16.82	9.09	5.53	3.88	3.27	2.94	4.70	15009 ± 1
I1_01	19.49	11.08	6.78	4.54	3.56	3.13	4.49	20563 ± 1
I1_02	15.66	9.07	5.88	4.20	3.53	3.20	6.02	9790 ± 1
I2_01	17.69	10.37	6.57	4.47	3.63	3.13	5.17	16642 ± 1
I2_02	17.11	9.74	6.21	4.33	3.57	3.11	5.41	14271 ± 1
I2_03	15.40	8.84	5.88	4.05	3.23	2.79	4.44	17854 ± 1
I2_04	16.36	9.47	6.14	4.04	3.24	2.81	4.49	17872 ± 1
I2_05	20.56	12.66	7.28	4.97	3.74	3.12	3.83	33420 ± 1
I2_06	18.50	10.42	6.42	4.27	3.39	2.87	4.51	18877 ± 1
I3_01	15.79	9.94	6.44	4.33	3.50	3.01	3.56	29233 ± 1
I3_02	13.18	8.47	5.71	4.08	3.39	2.88	3.85	23338 ± 1
I4_01	17.16	10.10	6.43	4.41	3.62	3.18	4.35	20760 ± 1
I4_02	15.37	10.10	6.89	5.04	4.25	3.86	5.41	18007 ± 1
I4_03	18.18	11.35	7.19	4.70	3.62	3.02	2.99	–
I5_01	17.58	10.19	6.37	4.32	3.48	2.99	4.87	17305 ± 1
I5_02	17.47	10.27	6.47	4.22	3.27	2.74	4.47	18866 ± 1
I6_01	18.51	11.12	7.00	4.69	3.75	3.19	3.90	28442 ± 1
I6_02	18.11	11.34	7.29	4.73	3.67	3.07	2.98	–
I6_03	19.91	11.28	6.69	4.43	3.65	3.22	5.23	16474 ± 1
I6_04	20.53	11.64	6.88	4.32	3.31	2.79	3.55	27625 ± 1
I6_05	19.80	11.83	7.43	4.71	3.60	2.95	3.07	38133 ± 1
I6_06	21.13	11.66	6.53	4.22	3.54	3.08	3.97	23579 ± 1

The corresponding errors for the DD$_{FAT}$ and DD$_{DIGITIZED}$ values shown in this table are ±0.01

As is described by Rodriguez-Iturbe and Rinaldo (RODRIGUEZ-ITURBE and RINALDO 1997), the box counting method consists of a two-dimensional analysis of an object. The object is covered by a regular grid of side *r*. Then, the number of grid boxes containing the object is counted, obtaining the value $N(r)$. The *r* value is halved repeatedly and the series of N_i values is obtained by counting the occupied boxes. Taking a reference box size (r_1), the size factor (*s*) is calculated as $s_i = r_i/r_1$. As *r* decreases, the following relation is obtained:

$$\frac{\log N_i}{\log\left(\frac{1}{s_i}\right)}, \qquad (2)$$

which converges to a finite value defined as the box counting dimension.

The entire study area has been covered by a square of 81.92 × 81.92 km. This square has been sequentially subdivided until it formed a grid of 15.625 cm. The reference box size has been established at 320 m because bigger box sizes capture the scaling properties of the contour and shape of the LMU and do not provide further hydrological information. Figures 4 and 5 illustrate the box counting method and its results.

The minimum box size considered for the estimation of the fractal box counting dimension was set by the scale of the LiDAR-DTM, which means a box size of 5 m. The box counting dimension was estimated using the scale range between $r_1 = 320$ m and $r_7 = 5$ m for the seven drainage networks in the 30 LMUs. The coefficients of

Table 6

Fractal dimensions of the drainage networks obtained with different FAT values (FD$_{FAT}$)

LMU	FD$_{2500}$	FD$_{5000}$	FD$_{10,000}$	FD$_{20,000}$	FD$_{30,000}$	FD$_{40,000}$	FD$_{DIGITIZED}$	FAT$_{FD}$
S1_01	1.2523	1.1151	1.0381	0.9905	0.9750	0.9684	1.0279	12141 ± 1
S1_02	1.1300	1.0088	0.9113	0.8525	0.8160	0.7954	0.8787	15551 ± 1
M1_01	1.2885	1.1786	1.0961	1.0498	1.0306	1.0150	1.0810	13263 ± 1
M1_02	1.3998	1.2629	1.1496	1.0794	1.0496	1.0357	1.0713	22709 ± 1
M1_03	1.3403	1.1996	1.0998	1.0491	1.0254	1.0164	1.0816	13593 ± 1
M1_04	1.3056	1.1608	1.0676	1.0225	1.0020	0.9924	1.0680	9977 ± 1
M2_01	1.3292	1.1947	1.1075	1.0541	1.0390	1.0283	1.1193	9325 ± 1
M2_02	1.3849	1.2450	1.1417	1.0833	1.0597	1.0459	1.1005	17057 ± 1
M2_03	1.3658	1.2178	1.1119	1.0486	1.0289	1.0234	1.0907	13347 ± 1
I1_01	1.3557	1.2263	1.1266	1.0688	1.0440	1.0316	1.0603	23426 ± 1
I1_02	1.2904	1.1638	1.0830	1.0385	1.0170	1.0125	1.0748	11848 ± 1
I2_01	1.3774	1.2495	1.1564	1.0883	1.0628	1.0470	1.1171	15775 ± 1
I2_02	1.3920	1.2559	1.1561	1.0975	1.0744	1.0614	1.1280	14806 ± 1
I2_03	1.3290	1.1987	1.1140	1.0486	1.0253	1.0177	1.0685	16952 ± 1
I2_04	1.3429	1.2165	1.1262	1.0586	1.0248	1.0059	1.0752	17551 ± 1
I2_05	1.3762	1.2694	1.1719	1.1099	1.0606	1.0348	1.0313	–
I2_06	1.3784	1.2423	1.1386	1.0737	1.0496	1.0308	1.0873	17910 ± 1
I3_01	1.3719	1.2642	1.1720	1.1033	1.0746	1.0618	1.0785	28642 ± 1
I3_02	1.2770	1.1946	1.1119	1.0539	1.0419	1.0198	1.0585	19210 ± 1
I4_01	1.3710	1.2425	1.1460	1.0794	1.0536	1.0437	1.0809	19771 ± 1
I4_02	1.3047	1.2106	1.1350	1.0833	1.0627	1.0497	1.1014	16497 ± 1
I4_03	1.3848	1.2748	1.1835	1.1020	1.0717	1.0447	1.0446	–
I5_01	1.3845	1.2537	1.1562	1.0867	1.0578	1.0400	1.1082	16901 ± 1
I5_02	1.3705	1.2475	1.1493	1.0697	1.0428	1.0272	1.0888	17602 ± 1
I6_01	1.3988	1.2778	1.1812	1.1165	1.0874	1.0723	1.0755	37939 ± 1
I6_02	1.3992	1.2907	1.1983	1.1244	1.0948	1.0809	1.0452	–
I6_03	1.4093	1.2715	1.1578	1.0904	1.0611	1.0541	1.1089	17261 ± 1
I6_04	1.3971	1.2678	1.1636	1.0869	1.0556	1.0431	1.0584	29103 ± 1
I6_05	1.3929	1.2714	1.1706	1.1057	1.0691	1.0486	1.0426	–
I6_06	1.3630	1.2186	1.1025	1.0380	1.0133	1.0042	1.0154	29146 ± 1

The corresponding errors for the FD$_{FAT}$ and FD$_{DIGITIZED}$ values shown in this table are ±0.0001

determination (R^2) for these estimations were higher than 0.990 in all of the 210 cases investigated.

The box counting dimension of the digitized drainage network was used within each LMU to estimate the theoretical FAT value (FAT$_{FD}$) that would generate a drainage network with the same value of the box counting dimension as the digitized drainage network. Figure 6 illustrates the procedure employed to obtain the FAT$_{FD}$.

2.6. Statistical Analysis

The 21 quantitative variables describing the LMUs, initially considered to perform the statistical analyses, are shown in Table 3.

An ANOVA test was performed to investigate the variance behavior of each variable, grouping the LMUs by type of rock and by lithological group.

Unfortunately, each lithological group was represented by a number of LMUs too small, so that the results obtained for these groups were not statistically significant.

A statistical classification model of the LMUs by type of rock was proposed based on the discriminant analysis of the linear and fractal parameters that describe the drainage networks.

Discriminant analysis is multivariate analysis of variance which is used to determine which continuous variables discriminate between different naturally occurring groups. In discriminant analysis, the independent variables are the predictors and the dependent variables are the groups (POULSEN and FRENCH 2004).

The 16 LMUs used to generate the classification model were S1_01, I1_01, I2_01, I2_03, I2_05, I3_01, I4_01, I4_03, I5_01, I6_01, I6_03, I6_05,

Table 7

Results of the variables calculated for the LMUs grouped by rock type

| | Type of rock | | | |
	Sedimentary	Metamorphic	Igneous	All
Code	S	M	I	–
Count	2	7	21	30
Variable				
HI	0.607 ± 0.028a	0.680 ± 0.042a	0.673 ± 0.046a	0.671 ± 0.034
Area (km^2)	6.97 ± 7.47a	38.81 ± 16.28a	27.11 ± 16.02a	28.50 ± 12.01
Ratio A/P	182.7 ± 137.1a	578.0 ± 208.4b	466.9 ± 95.3ab	473.9 ± 87.2
DD_{2500}	10.56 ± 3.20a	15.88 ± 1.89b	17.79 ± 0.87b	16.86 ± 1.01
DD_{5000}	5.19 ± 2.30a	8.91 ± 0.94b	10.52 ± 0.46c	9.79 ± 0.65
$DD_{10,000}$	2.95 ± 1.88a	5.52 ± 0.44b	6.62 ± 0.23c	6.12 ± 0.41
$DD_{20,000}$	1.79 ± 1.64a	3.91 ± 0.22b	4.43 ± 0.12c	4.13 ± 0.27
$DD_{30,000}$	1.44 ± 1.48a	3.30 ± 0.17b	3.55 ± 0.10b	3.35 ± 0.22
$DD_{40,000}$	1.21 ± 1.31a	2.93 ± 0.15b	3.05 ± 0.10b	2.90 ± 0.19
$DD_{DIGITIZED}$	2.38 ± 1.44a	4.64 ± 0.42b	4.30 ± 0.37b	4.25 ± 0.34
FAT_{DD}	14767 ± 2813ab	15531 ± 3506a	23475 ± 3775b	21041 ± 3054
FD_{2500}	1.191 ± 0.120a	1.345 ± 0.030b	1.365 ± 0.016b	1.349 ± 0.021
FD_{5000}	1.062 ± 0.104a	1.208 ± 0.027b	1.243 ± 0.014c	1.223 ± 0.020
$FD_{10,000}$	0.975 ± 0.124a	1.111 ± 0.021b	1.148 ± 0.012c	1.127 ± 0.018
$FD_{20,000}$	0.921 ± 0.135a	1.055 ± 0.015b	1.082 ± 0.011b	1.065 ± 0.018
$FD_{30,000}$	0.896 ± 0.156a	1.034 ± 0.014b	1.055 ± 0.009b	1.039 ± 0.018
$FD_{40,000}$	0.882 ± 0.170a	1.022 ± 0.013b	1.040 ± 0.009b	1.025 ± 0.018
$FD_{DIGITIZED}$	0.953 ± 0.146a	1.087 ± 0.013b	1.074 ± 0.012b	1.069 ± 0.016
FAT_{FD}	13846 ± 3342ab	14182 ± 3370a	25736 ± 5683b	22247 ± 4464

Different letters in a row indicate significant differences ($P < 0.05$). LSD test

M1_01, M1_03, M2_01, and M2_03. The remaining 14 LMUs under investigation were used to validate the classification model within the study area. All of the confidence intervals are shown at the 95 % confidence level.

3. Results

3.1. Morphological Properties of the Lithological Map Units

A quantitative geomorphological approach to the LMUs is shown in Table 4. The results reflect the morphological heterogeneity of the studied LMUs. These 30 LMUs together covered 86.5 ± 0.1 % of the wine growing region. The area of the LMUs ranged from the 177.83 ± 0.01 km^2 of the biggest LMU (I3_01) to the 3.15 ± 0.01 km^2 of the smallest LMU (S1_02). The area/perimeter ratio was significantly related to the logarithm of the area, with a coefficient of determination (R^2) equal to 0.694.

The variability of the HI values in our study agrees with HURTREZ and LUCAZEAU (1999) who concluded that relief characteristics at small and meso-scales are independent of the type of rock. However, the geological information managed by Hurtrez was scale 1:1,000,000, and it is logical to assume that this information scale was too coarse for small scale relief studies.

The HI values were higher than 0.500, except for the LMU I3_02 unit. It is significant that this unit gave the highest E_{mean} value, which means that this LMU was placed on the peneplain. Although the HI has been calculated in areas of homogeneous lithology instead of in defined watersheds, the HI values could be interpreted as an indication of the erosional stage of the LMU. Thus, in this study, the lower HI values are present in the LMUs located in the highest or the lowest zones of the region. This is reasonable because the peneplain, which is by definition an erosional surface, is placed in the higher parts of the region, while the lower parts of the canyons are actually the most eroded ones.

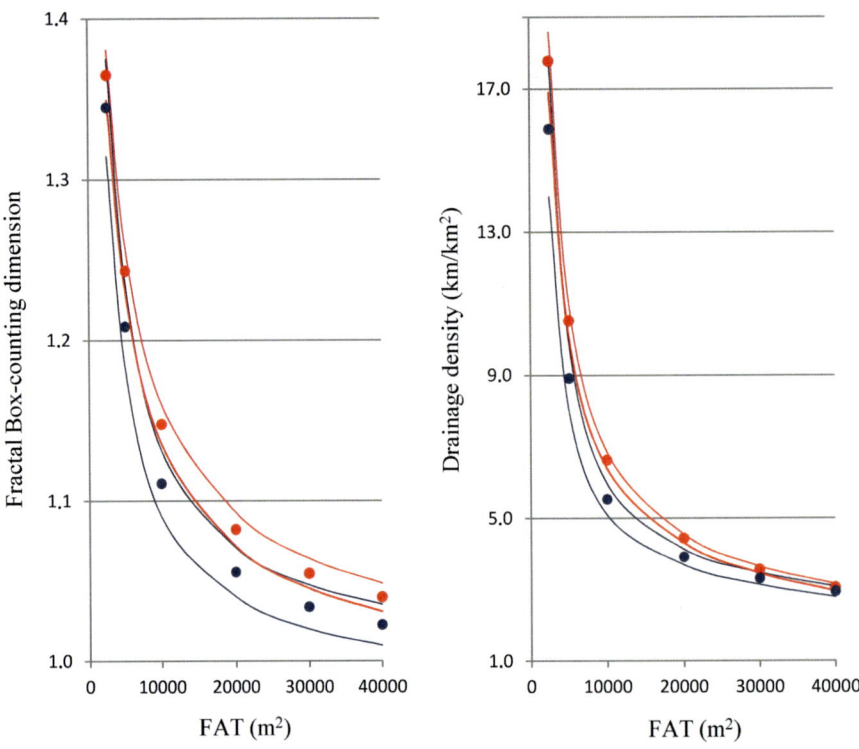

Figure 7
Fractal dimension and drainage density of the drainage networks automatically extracted with different FAT values. Data are grouped by type of rock: Igneous rocks (*red*) and Metamorphic rocks (*blue*). The points represent the mean value of the group and the lines define the 95 % confidence intervals (color figure online)

3.2. Drainage Density

This traditional measure in hydrology has been calculated for all of the LMUs, for the digitized drainage network, and also for the six drainage networks automatically extracted with different FAT values. The results of these calculations are shown in Table 5.

As expected, the drainage density of the drainage networks decreased as the FAT values used to generate them increased. If more area is needed to begin channelization, channels are shorter, and the drainage density is lower. This decrease of the drainage density with increasing FAT values matches a potential curve. The coefficients of determination for this fit were higher than 0.970 for all of the 30 LMUs. The theoretical FAT values (FAT_{DD}) ranged from 9568 ± 1 to 38133 ± 1 m^2. Two of the 30 LMUs produced $DD_{DIGITIZED}$ values lower than the $DD_{40,000}$ values.

3.3. Fractal Box Counting Dimension

The fractal box counting dimension of the drainage networks was estimated using the box size range between 320 and 5 m (Table 6). The coefficients of determination (R^2) of these estimations were higher than 0.990 in all of the 210 cases studied.

As was described for the drainage density results, the decrease of the fractal dimension with increasing FAT values matches a potential curve. The quality of the fitting was slightly lower but the R^2 values were still higher than 0.930 for all of the 30 LMUs.

The theoretical FAT values (FAT_{FD}) ranged from 9325 ± 1 to 37939 ± 1 m^2; four of the 30 LMUs produced $FD_{DIGITIZED}$ values lower than the $FD_{40,000}$ values.

3.4. LMUs Results Grouped by Type of Rock

The results for the LMUs of the 21 variables described in Table 2 have been grouped by type of

Table 8

Coefficients of the classification functions for each type of rock

	Type of rock		
	Sedimentary S	Igneous I	Metamorphic M
FD_{5000}	6116.76	12075.5	12084.3
$FD_{10,000}$	−7708.09	−13332.5	−13610.9
$FD_{20,000}$	424.089	370.423	937.564
$FD_{40,000}$	7909.55	7351.37	7027.38
DD_{5000}	64.2871	−98.4981	−82.0603
$DD_{10,000}$	−752.441	−607.533	−656.47
$DD_{20,000}$	1550.62	1841.16	1862.3
$DD_{40,000}$	−951.595	−955.07	−970.832
K value	−3325.38	−3998.12	−3869.93

rock (Table 7). The results for the sedimentary LMUs have to be taken into account carefully because the study population is very small.

According to the ANOVA test, 5 of the 21 variables showed significant differences in their means by type of rock at the 95 % confidence level. These variables were DD_{5000}, $DD_{10,000}$, $DD_{20,000}$, FD_{5000}, and $FD_{10,000}$. Figure 7 graphically shows the results of the LMUs grouped by rock type, showing the mean values and confidence intervals of the drainage networks and the fractal dimension results of the groups of metamorphic and igneous LMUs. The sedimentary LMUs have been removed from the graph because, due to the small number and the variability of the sedimentary LMUs, the confidence intervals for this group were physically inconsistent.

The HI values reported by HURTREZ and LUCAZEAU (1999) for granite (0.45 ± 0.01) and schist (0.29 ± 0.01) in the Herault basin are lower than those obtained in our study.

3.5. Classification Model of LMUs in Different Types of Rock

The statistical classification model of the LMUs by type of rock proposed in this study was determined by discriminant factorial analysis. The classification model was built using the 16 LMUs of sedimentary, metamorphic, and igneous rock indicated in the Sect. 2. Beginning with the 21 variables describing the LMUs, the discriminant analysis determined the 8 variables selected for the statistical classification

model. These variables were DD_{5000}, $DD_{10,000}$, $DD_{20,000}$, $DD_{40,000}$, FD_{5000}, $FD_{10,000}$, $FD_{20,000}$, and $FD_{40,000}$. This classification model was validated using the remaining 14 LMUs. Thirteen of these 14 LMUs were successfully classified by type of rock. The visual analysis of the aerial orthophotos of the wrongly classified LMU I6_06 revealed that within the third smallest LMU (5.39 ± 0.01 km^2), and probably affecting the results, there is a town, called Villarino de los Aires with about a thousand inhabitants, and an hydroelectric power station with an installed capacity of 810 MW. It is noteworthy that only fractal and linear parameters describing the drainage networks automatically extracted from 5-m-resolution LiDAR DTMs were needed to classify the LMUs by their type of rock in the study area.

The coefficients of the classification functions for each type of rock are provided in Table 8.

According to this table, the classification function for the sedimentary rock (S) is

$$S \text{ value} \equiv 6116.76 \cdot FD_{5000} - 7708.09 \cdot FD_{10,000}$$
$$+ 424.089 \cdot FD_{20,000} + 7909.55 \cdot FD_{40,000}$$
$$+ 64.2871 \cdot DD_{5000} - 752.441 \cdot DD_{10,000}$$
$$+ 1550.62 \cdot DD_{20,000} - 951.595 \cdot DD_{40,000}$$
$$- 3325.38$$

$$(3)$$

In order to classify a new LMU, the three classification functions must be calculated, and the type of rock assigned to the LMU would be the type of rock of the classification function obtaining the highest value of the three classification functions.

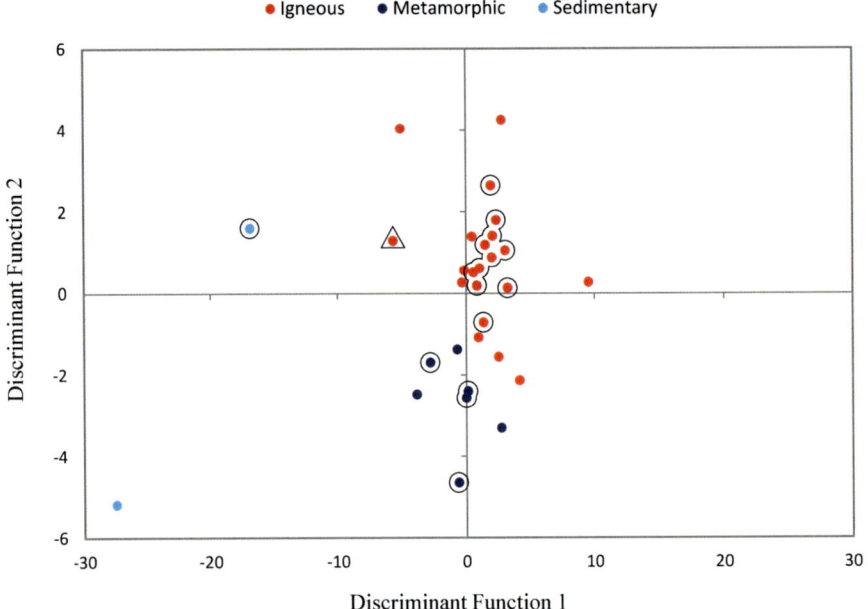

Figure 8
Representation of the LMUs on the two first discriminant functions. The circled LMUs were used to develop the classification model, and the rest of LMUs were use to validate it. The igneous LMU I6_06 within the triangle was the only LMU that was wrongly classified as metamorphic rock

The representation of the 30 LMUs on the two first discriminant functions (Fig. 8) shows considerable dispersion of the LMUs belonging to the same class, which was more evident for the sedimentary LMUs. This means that this classification model is not extremely solid, and also that it needs more replications on different materials and probably substantial modifications to be able to be applied over large scales. However, the existence of this classification model itself represents evidence of the suitability of our approach.

The results strongly suggest that the information included in a 5-m-resolution LiDAR DTM and the appropriate techniques to manage it could be the only inputs required to identify the underlying type of rock.

4. Discussion

The complexity and the irregularity of the topographical relief are caused by the landscape forming factors, but the different effects or expressions of these causes are determined by the underlying geological materials and their responses to erosion and fragmentation processes.

In HOWARD's (1967) work on drainage patterns, the author mentioned that in any small area, where all other factors are held constant, drainage texture may provide information on underlying materials and indirectly on structure.

In our analysis, drainage networks automatically extracted from the DTM were not considered properly defined hydrological features. Instead, they were considered fractal structures capturing the topographical relief roughness produced by the landscape forming factors on different geological materials. Topographic expression, fault situation, and drainage patterns are considered intrinsic characteristics of the lithological material that photo-geologic interpretation techniques can qualitatively identify (ALLUM 1966), and that this study aims to quantify.

This perspective makes it possible to investigate the underlying geological material, independently of the hydrologic factors such as groundwater inputs or the current seasonality of precipitation, even in remote places where hydrologic factors have not acted for a long time. The results of the statistical classification model proposed in this study seem to validate our approach and the applied techniques.

However, the proposed classification method only resulted in a coarse classification of the geological materials which was only validated for the study area. The small number of LMUs included in the study precludes statistical analysis of the lithological groups. More replications are needed on each lithological group and on different lithological materials to enable a finer classification by lithological group instead of the proposed classification by type of rock.

The inclusion in the analysis of other elements strongly linked to the bedrock lithology, such as the particle-size distribution of the soils developed on it, is currently being investigated (CÁMARA *et al.* 2015) and could make it possible to perform a finer discrimination between lithological materials. These kinds of studies, addressing the heterogeneity and process complexity of watershed hydrology at all scales, are being vigorously pursued to advance watershed and hydropedology science (McDONELL *et al.* 2007; LIN 2003).

The consolidation of robust statistical classification models to identify the underlying lithological material from DTM could represent the basis to implement tools for the automatic delineation of different lithological materials.

5. Conclusions

This paper has investigated the role of drainage networks extracted automatically from medium resolution DTMs in identifying the underlying lithological materials. The investigation was carried out by applying fractal analysis techniques to the drainage networks of 30 areas of homogeneous bedrock lithology located in Northwest Spain. These areas of homogeneous bedrock lithology are lithological map units (LMUs) defined by the available lithological data, at a scale 1:25,000. The types of rock included in these 30 areas were mostly igneous and metamorphic, but there were also some sedimentary rocks present.

Six different flow accumulation threshold (FAT) values were used for the extraction of the drainage

networks from the 5-m-resolution LiDAR DTM, and the six resulting drainage networks were compared with the drainage network obtained using photogrammetric methods.

The decreasing magnitudes of the fractal dimension and drainage density of the automatically extracted drainage networks with increasing FAT values matched potential curves for all of the LMUs. The coefficients of determination of the fits of all these curves were higher than 0.970 for the drainage density values and higher than 0.930 for the fractal dimension values.

A statistical classification model of the LMUs by type of rock has been proposed based on both fractal dimension and drainage density of the drainage networks.

The classification model was built using 16 LMUs of sedimentary, metamorphic, and igneous rocks. This model correctly classified 13 of the 14 LMUs used for its validation. The applicability of this model was not tested outside of the study area.

More replications of this procedure on different regions with different geological materials are needed to generalize our findings, but the results in the study area show that LMUs, with areas ranging from 177.83 ± 0.01 to 3.16 ± 0.01 km^2, can be successfully classified by type of rock, using the fractal dimension and the drainage density of the drainage networks derived from medium resolution LiDAR DTM with different flow support areas.

The results suggest that the information included in a 5-m-resolution LiDAR DTM and the appropriate techniques to manage it could be the only inputs required to identify the underlying geological material.

Acknowledgments

This study was partially funded by the European Regional Development Fund (ERDF) through the research project "SUVIDUR: Sustainability of viticulture in the area of the Douro River". Miguel Ángel Martín was supported in part by the Plan Nacional de Investigación Científica y Técnica (Spain) under ref. AGL-2011-25175.

REFERENCES

ALLUM, J.A.E., Photogeology and regional mapping (Pergamon Press, Oxford 1966).

ARIZA-VILLAVERDE, A.B., JIMÉNEZ-HORNERO, F.J., and GUTIÉRREZ DE RAVÉ, E. (2013), *Multifractal Analysis Applied to the Study of the Accuracy of DEM-Based Stream Derivation*, Geomorphology *197*, 85–95.

BLOOMFIELD, J.P., BRICKER, S.H., and NEWELL, A.J. (2011), *Some Relationships between Lithology, Basin Form and Hydrology: A Case Study from the Thames Basin, UK*. Hydrol. Process. *25*, 2518–2530.

CÁMARA, J., GÓMEZ-MIGUEL, V., and MARTÍN, M.A. (2013), *Lithologic Control on the Scaling Properties of the First-Order Streams of Drainage Networks: A Monofractal Analysis*, Vadose Zone J. *12*, 1–8.

CÁMARA, J., GÓMEZ-MIGUEL, V., and MARTÍN, M.A. (2015), *Quantifying the Relationship Between Drainage Networks at Hillslope Scale and Particle Size Distribution at Pedon Scale*, Fractals *23*, 1540007.

CARTWRIGHT, R., CLAYTON, J.A., and KIRK, R.L. (2011), *Channel Morphometry, Sediment Transport, and Implications for Tectonic Activity and Surficial Ages of Titan Basins*, Icarus *214*, 561–570.

DE BARTOLO, S.G., GABRIELE, S., and GAUDIO, R. (2000), *Multifractal Behaviour of River Networks*. Hydrol. Earth Syst. Sci. *4*, 105–112.

DE BARTOLO, S.G., GAUDIO, R., and GABRIELE, S. (2004), *Multifractal Analysis of River Networks: Sandbox Approach*. Water Resour. Res. *40*, W02201.

DE BARTOLO, S.G., VELTRI, M., and PRIMAVERA, L. (2006), *Estimated Generalized Dimensions of River Networks*. J. Hydrol. *322*, 181–191.

DOMBRADI, E., TIMÁR, G., BADA, G., CLOETINGH, S., and HORVÁTH, F. (2007), *Fractal Dimension Estimations of Drainage Network in the Carpathian-Pannonian System, Global Planet*. Change *58*, 197–213.

GAUDIO, R., DE BARTOLO, S.G., PRIMAVERA, L., VELTRI, M., and GABRIELE, S. (2004), *Procedures in Multifractal Analysis of River Networks: A State of the Art Review*, IAHS Publ. *286*, 228–237.

GAUDIO, R., DE BARTOLO, S.G., PRIMAVERA, L., GABRIELE, S., and VELTRI, M. (2006), *Lithologic Control on the Multifractal Spectrum of River Networks*, J. Hydrol. *327*, 365–375.

GÓMEZ-MIGUEL, V., SOTÉS, V., CÁMARA, J., and HERNANDO, G., Delimitación de las Zonas Vitícolas de la DO Arribes. Informe Técnico no Publicado (Univ. Politécnica de Madrid, Madrid 2011).

HELMLINGER, K.R., KUMAR, P., and FOUFOULA-GEORGIOU, E. (1993), *On the Use of Digital Elevation Model Data for Hortonian and Fractal Analyses of Channel Networks*, Water Resour. Res. *29*, 2599–2613.

HORTON, R.E. (1932), *Drainage Basin Characteristics*, Trans. Am. Geophys. Union *13*, 350–361.

HORTON, R.E. (1945), *Erosional Development of Streams and Their Drainage Basins: Hydrophysical Approach to Quantitative Morphology*, Geol. Soc. Am. Bull. *56*, 275–370.

HOWARD, A.D. (1967), *Drainage Analysis in Geologic Interpretation: A Summation*, AAPG Bulletin *51*, 2246–2259.

HURTREZ, J., and LUCAZEAU, F. (1999), *Lithologic Control on Relief and Hypsometry in the Herault Drainage Basin (France)*, Comptes Rendus de l'Académie des Sciences-Series 11A-Earth and Planetary Science *328*, 687–694.

HYNEK, B.M., and PHILLIPS, R.J. (2003), *New Data Reveal Mature, Integrated Drainage Systems on Mars Indicative of Past Precipitation*, Geology *31*, 757–760.

KÖPPEN, W. (1918), *Klassification der Klimate nach Temperatur, Niederschlag und Jahreslauf*, Petermanns Geogr. Mitt. *64*, 193–203.

KÖPPEN, W., and GEIGER, R., Die Klimate der Erde (Verlag Justus Perthes, Gotha 1928).

LIFTON, N.A., and CHASE, C.G. (1992), *Tectonic, Climatic, and Lithologic Influences on Landscape Fractal Dimension and Hypsometry: Implications for Landscape Evolution in the San Gabriel Mountains, California*, Geomorphology *5*, 77–114.

LIN, H. (2003), *Hydropedology: Bridging Disciplines, Scales, and Data*, Vadose Zone J. *2*, 1–11.

MAIDMENT, D.R., ARCHYDRO: GIS for Water Resources (ESRI Press, Redlands 2002).

MANDELBROT, B.B., The Fractal Geometry of Nature (W.H. Freeman, New York 1983).

MCDONELL, J.J., SIVAPALAN, M., VACHÉ, K., DUNN, S., GRANT, G., HAGGERTY, R., HINZ, C., HOOPER, R., KIRCHNER, J., RODERICK, M.L., SELKER, J., and WEILER, M. (2007), *Moving Beyond Heterogeneity and Process Complexity: A New Vision for Watershed Hydrology*, Water Resour. Res. *43*, W07301.

MONTGOMERY, D.R., and DIETRICH, W.E. (1988), *Where Do Channels Begin?* Nature *336*, 232–234.

MONTGOMERY, D.R., and DIETRICH, W.E. (1992), *Channel Initiation and the Problem of Landscape Scale*, Science *255*, 826–830.

O'CALLAGHAN, J., and MARK, D.M. (1984), *The Extraction of Drainage Networks from Digital Elevation Data*, Comput. Vision Graph. *3*, 323–344.

PIKE, R.J., and WILSON, S.E. (1971), *Elevation-Relief Ratio, Hypsometric Integral and Geomorphic Area-Altitude Analysis*, Geol. Soc. Amer. Bull. *82*, 1079–1084.

POULSEN, J., and FRENCH, A. (2004), Discriminant function analysis (DA). Retrieved April 20, 2015, from http://userwww.sfsu.edu/efc/classes/biol710/discrim/discrim.pdf.

RODRÍGUEZ-ITURBE, I., and RINALDO, A., Fractal River Basins: Chance and Self-Organization (Cambridge Univ. Press, Cambridge 1997).

RUSSEL, D., HANSON, J., AND OTTO, E. (1980), *Dimension of Strange Attractors*, Physical Review Letters *45*, 1175–1178.

SCHWANGHART, W., GROOM, G., KUHN, N.J., and HECKRATH, G. (2013), *Flow Network Derivation from a High Resolution DEM in a Low Relief, Agrarian Landscape*, Earth Surf. Process. Landforms *38*, 1576–1586.

SINGH, O., SARANGI, A., and SHARMA, M.C. (2008), *Hypsometric Integral Estimation Methods and Its Relevance on Erosion Status of North-Western Lesser Himalayan Watersheds*, Water Resour. Manage. *22*, 1545–1560.

Soil Survey Staff, Keys to Soil Taxonomy, 12th ed. (USDA, National Resources Conservation Service, National Soil Survey Center, Lincoln 2014).

STRAHLER, A.N. (1952), *Hypsometric (Area-Altitude) Analysis of Erosional Topography*, Geol. Soc. Am. Bull. *63*, 1117–1142.

STRAHLER, A.N. (1957), *Quantitative Analysis of Watershed Geomorphology*, Trans. Am. Geophys. Union *38*, 913–920.

TARBOTON, D.G. (1997), *A New Method for the Determination of Flow Directions and Upslope Area in Grid Digital Elevation Models*, Water Resour. Res. *33*, 309–319.

TESFA, T.K., TARBOTON, D.G., WATSON, D.W., SCHREUDERS, K.A.T., BAKER, M.E., and WALLACE, R.M. (2011), *Extraction of*

Hydrological Proximity Measures from DEMs Using Parallel Processing, Environ. Modell. Softw. *26*, 1696–1709.

VAZE, J., TENG, J., and SPENCER, G. (2010), *Impact of DEM Accuracy and Resolution on Topographic Indices*, Environ. Modell. Softw. *25*, 1086–1098.

ZANDBERGEN, P.A. (2010), *Accuracy Considerations in the Analysis of Depressions in Medium-Resolution LIDAR DEMs*, GISci. Remote Sens. *47*, 187–207.

(Received March 29, 2014, revised June 21, 2015, accepted June 23, 2015, Published online July 17, 2015)

Pure Appl. Geophys. 173 (2016), 963–982
© 2015 Springer Basel
DOI 10.1007/s00024-015-1162-2

Pure and Applied Geophysics

CrossMark

Integrated GPR and ERT as Enhanced Detection for Subsurface Historical Structures Inside Babylonian Houses Site, Uruk City, Southern Iraq

EMAD H. AL-KHERSAN,[1] JASSIM M. T. AL-ANI,[2] and SALAH N. ABRAHEM[3]

Abstract—Uruk archaeological site, which located in Al-Muthanna Governorate southern Iraq, was investigated by integrated geophysical methods, ground penetration radar (GPR) and electric resistivity tomography (ERT) to image the historical buried structures. The GPR images show large radar attributes characterized by its continuous reflections having different widths. GPR attributes at shallower depth are mainly representing the upper part of Babylonian Houses that can often be found throughout the study area. In addition, radargrams characterized objects such as buried items, buried trenches and pits which were mainly concentrated near the surface. The ERT results show the presence of several anomalies at different depths generally having low resistivities. It is clear that the first upper zone can be found throughout the whole area and it may represent the top zone of the Babylonian houses. This zone is characterized by its dry clay and sandy soil containing surface broken bricks and slag mixed with core boulders. The second one underneath the top shows a prominent lower resistivity zone. It is probably caused by the moisture content that reduces the resistivity. The thickness of this zone is not equal at all parts of the site. The third deeper zone typically represents the archaeological walls. Most of the main anomalies perhaps referred to the buried clay brick walls. The map of the archaeological anomalies distribution and 3D view of the foundations at the study area using GPR and ERT techniques clearly show the characteristics of the Babylonian remains. A contour map and 3D view of Uruk show that the archaeological anomalies are concentrated mainly at the NE part of the district with higher values of wall height that range between 6 and 8 m and reach to more than 10 m. At the other directions, there are fewer walls with lower heights of 4–6 m and reach in some places the wall foot.

Key words: Uruk, Al-Muthanna, GPR and 2D-imaging, Babylon, remains, Emad Al-Khersan, Gilgamesh.

1. Introduction

Uruk represents Iraq's ancient site, which includes the vestiges of some of the world's earliest cities. It thrived from the beginning of the fifth millennium BC until the end of the third century AD, when it finally declined and was abandoned. Its ruins are covered by tens centimetres of sand (POLLOCK *et al.* 1996). Uruk is the first largest Sumerian settlements and the most important religious centres in Mesopotamia. The Sumerian traders have moved Uruk culture and surroundings. Uruk people gradually evolved their own economies comparing with other cultures (BEAULIEU 2003). It is famous as the capital city of Gilgamesh, hero of the Epic of Gilgamesh. He built the city wall around Uruk and he was the king of it. Uruk citizens have designed very well canal system through the city that has been described as "Venicein the desert". This canal system flowed throughout the city connecting it with the maritime trade on the ancient Euphrates River as well as the surrounding agricultural belt (FASSBINDER *et al.* 2003).

The studied archaeological site of this research is located between the longitudes 45°37′28″E–45°39′7.3″E and latitudes 31°18′34.5″N–31°20′14.5″N. It is situated about 30 km east of Al-Muthanna Governorate, southern Iraq near the boundary between the Mesopotamia and the southern desert. Moreover, HRITZ (2012) had been drawn subsoils column at the marshes region nearby the Uruk site (Fig. 1). The maximum extent of this site is 3 km N–S and 2.5 km E–W covering an area of about 5.5 km². It lies within the lower parts of Mesopotamia that is characterized by its flat topography (BUDAY 1980). Hills of ancient civilization inside the investigated site represent the historical buildings such as, houses and temples or

---

[1] Geology Department, College of Science, Basrah University, Basrah, Iraq. E-mail: emreal60@yahoo.com

[2] Geology Department, College of Science, University of Baghdad, Baghdad, Iraq. E-mail: jassimthabit@yahoo.com

[3] Physics Department, College of Science, University of Al-Muthanna, Al-Muthana, Iraq. E-mail: zircon200036@yahoo.com

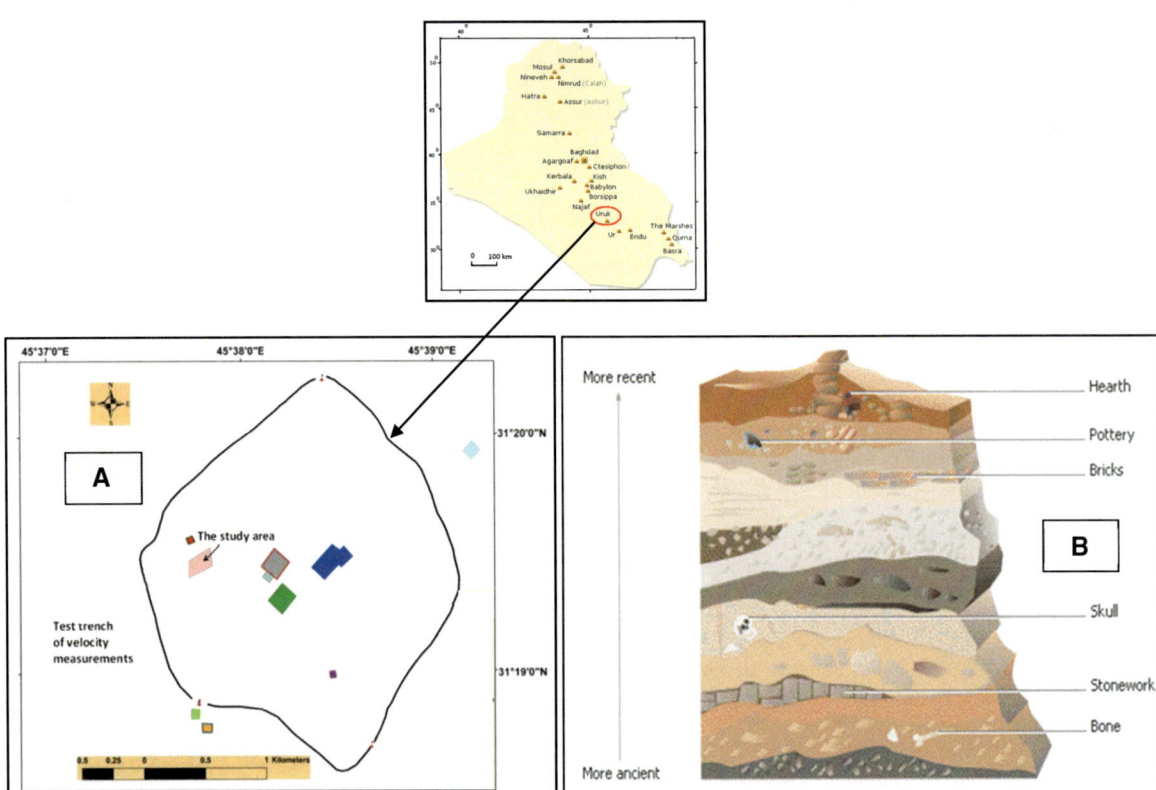

Figure 1
a Location map of the study area (Babylon Houses); and **b** subsoil column inside URUK site (HRITZ 2012)

ziggurats (BAKER 2002). The study area was covered by the Quaternary alluvium deposits. It mainly consists of clay, silt and sand sediments (AL-HASHIMI 1974). Different materials such as bricks, debris, sediment from channels, undisturbed soil and sunbaked bricks are scattered on the ground surface of the area. There are many exposed buildings such as walls, temples, arced gates and tombs. Most of the exposed walls have widths ranging from approximately 1–2 m. The dimensions and design of clay bricks are different. The important point for geophysical surveying, some 5000 years later, is the fact that Gilgamesh used baked (burnt) bricks for the walls he built. The enrichment of ferromagnetic minerals such as Magnetite in the archaeological structures is owing to the use of fire and magneto tactic bacteria in organic debris (FASSBINDER and BECKER 2001). The early wall of Uruk had a mantle of baked bricks, filled with cheaper mud bricks, which would give an ideal base

for electrical prospecting because of the conductivity of burnt clay. Since any drilling in the archaeological sites is not allowed, the data about the groundwater are obtained from an artesian well located at a close village with coordinates of 45°15′E and 31°26′N about 500 m west of the study area. The well is drilled by the General Commission for Groundwater (GCGW)/Al-Muthanna branch. The static water level in this well is 3 m from the natural ground surface; moreover, the mobile water depth is at 4.7 m. The electrical conductivity of groundwater sample is measured and found to be 2000 microMohs/cm (high salinity media).

The main objectives of this study are:

1. Comparison between the common electrical arrays of the ERT technique to find the most suitable array for the study area that leading to the best results.

2. Using GPR and ERT techniques to determine the positions and depths of underneath archaeological remains inside Uruk site. Therefore, a map for the buried remains can be deduced.

3. Comparison between both GPR and ERT surveys in delineating the accurate depth and shape of the subsurface.

2. Field Work

To define an accurate coordinate information for the study area, we made use of Garmin GPS, total station, and Arc GIS 9.3. The result of this step produced a topographic map to the area (Fig. 2). The locations of GPR and ERT profiles were determined according to this map as grids of GPR and ERT surface lines.

2.1. GPR Data Acquisition

We have used Sweden MALA Geosciences RAMAC/GPR field equipment of mono-static 250 MHz ground-coupled shielded antenna. Before starting the data production, we considered equipment capability to resolve the target archaeological feature. Thus, we performed a man-made testing facility. The followed steps are described below:

- A location was selected outside the western gate of Uruk city in contact with the city wall of the archaeological site, and a trench was dug at this location with dimensions of 1.75 × 1.1 m and depth 1.8 m.

- Different archaeological materials such as bricks, debris, sediments from channels, undisturbed soil and sun-baked bricks scattered here and there in settlement, are collected, buried and are covered by

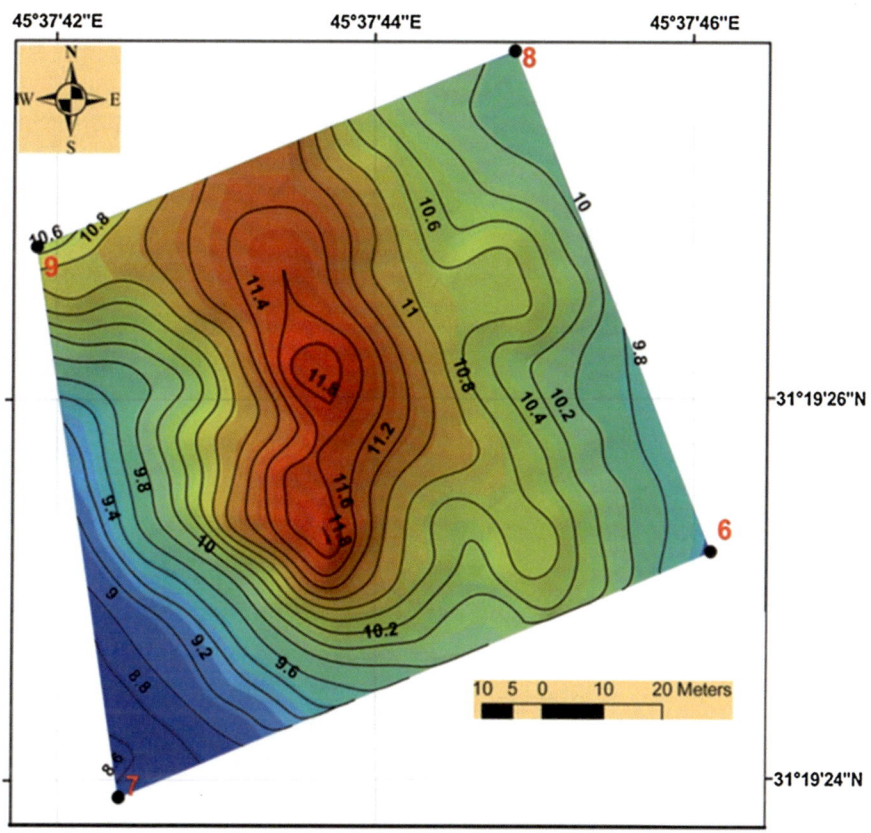

Figure 2
The topographic map of the Babylonian houses and the surroundings

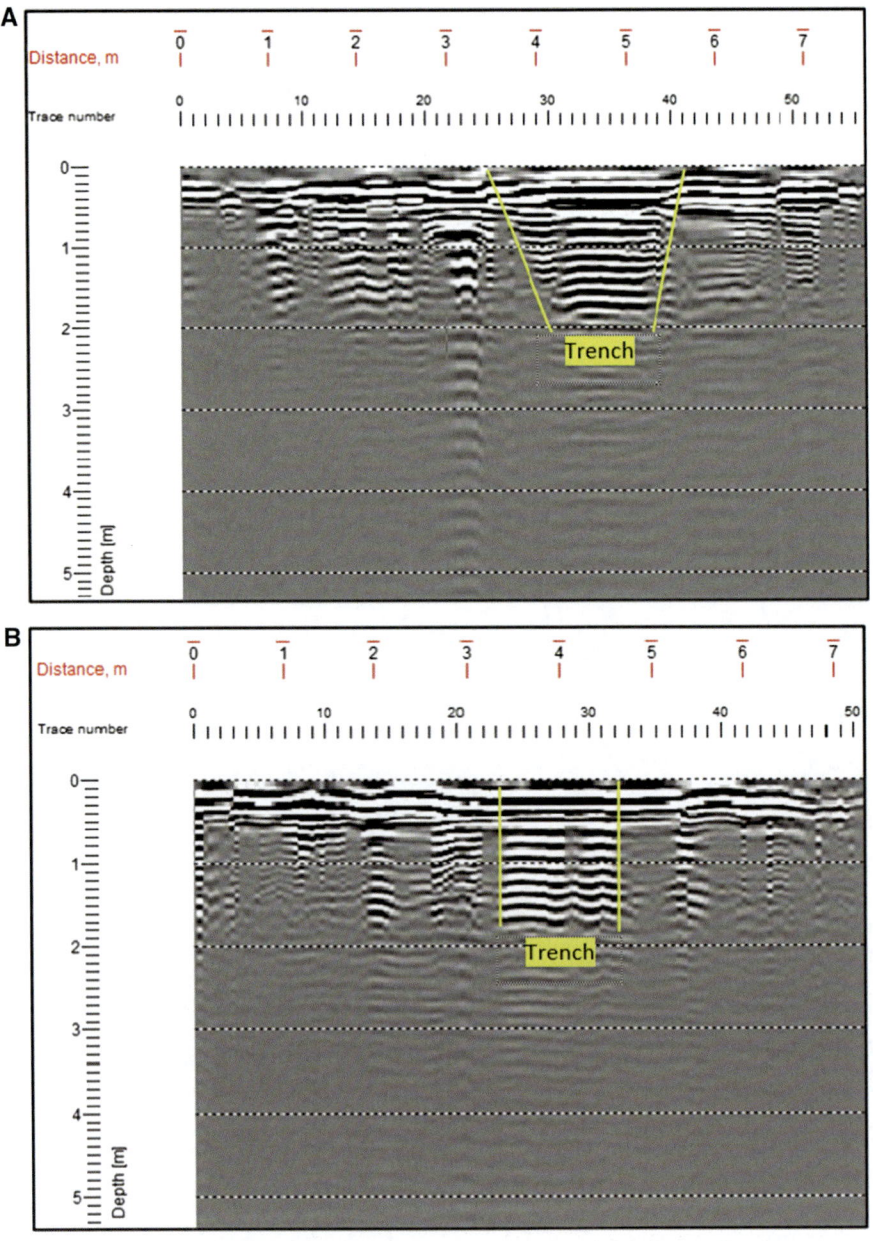

Figure 3
GPR tested profiles **a** *above* and **b** *below*

thin veneer of soil. These materials were returned back to their original positions after GPR test.

- Several GPR settings were used until the suitable one is chosen and then applied for all selected profiles. These settings are: number of trace stack is 4 and sometimes autostack, antenna spacing from ground surface is 0.1 m, sampling interval 0.4 m, time window 203 ns, sampling frequency

2607 MHz and time sampling interval is 0.7 ns. Therefore, two perpendicular GPR profiles were conducted over this artificial buried wall, the first profile (A) trends SE–NW, and the other (B) trends NE–SW (Fig. 3).

- The two-way travel times from the radargrams were measured; and then the depth of the artificial wall is known, the velocity is measured using the travel time from the GPR record and the known reflector depth (d). The average velocity (v) of the radar signal can be determined from the formula: (OSWIN 2009)

$$V = \frac{2d}{t} \qquad \therefore V = \frac{2d}{t} = \frac{2 \times 1.8}{36} = 0.1 \, \text{m/ns}$$

t, total time and the antenna is directly over the known target.

Therefore, the average velocity of 0.10 m/ns is applied for all sections on the basis of known depth to reflector. The topographic variation inside the studied area was surveyed precisely, so that the GPR profiles can display with correct topography. A grid pattern parallel GPR survey system was carried out in the area and ninety SW–NE parallel profiles with 1 m spacing between each other were conducted in October 2011.

2.2. ERT Data Acquisition

Before starting the data production, we considered the best-fit electrode configuration tests to resolve the target archaeological feature. In other words, we tested the best suitable spreading geometry at the site. This step is represented by deploying different electrode configurations at the same profile location. We compared the results of three common electrical resistivity configurations (i.e. Wenner, Wenner–Schlumberger and dipole–dipole). The comparison focused on the resolution efficiency at the site (LOKE 2010). To investigate the imaging capabilities of these electrode configurations, three test survey profiles, which are URUK-TEST-WEN, URUK-TEST-WEN-SCH and URUK-TEST-DIPDIP corresponding to Wenner, Wenner–Schlumberger and dipole–dipole configurations, were carried out using the Sweden ABEM Terrameter SAS-4000 Lund imaging multi-electrode system. The electrode layout of these profiles is ranged from 0 to 60 m and the electrode basic spacing is 1.5 m. It is known that each of the electrode arrays has its own advantages and limitations in fieldwork. The image created by means of the ERT for the same structure will be different for each array. For these reasons, choosing the right one for the resistivity surveys is important (LOKE 2010).

For resistivity imaging, the electrode arrays might have different imaging abilities for a model, i.e. differences in spatial resolution, tendency for artefacts in the images, deviation from the true model resistivity and interpretable maximum depth. The sensitivity patterns play important roles for the resolving capability in the inversion of the data. To obtain a high resolution and reliable image, the electrode array used should ideally give data with the maximum anomaly information and reasonable data coverage.

The choice of the best array for a field survey depends on the type of structure to be mapped, the sensitivity of the resistivity metre and the background noise level. Among the characteristics of an array that should be considered are (1) sensitivity of the array to vertical and lateral changes in the subsurface resistivity, (2) depth of investigation, (3) horizontal data coverage (4) signal strength (LOKE 2004) (5) resolution for the different models (6) imaging quality with different data densities and (7) sensitivity to noise levels (DAHLIN and ZHOU 2004). The first two characteristics can be determined from the sensitivity function of the array for a homogeneous earth model. The sensitivity function basically tells us the degree to which a change in the resistivity of a section of the subsurface will influence the potential measured by the array. The higher the value of the sensitivity function is the greater of the influence of the subsurface region on the measurement. Here, we try to discuss some of them as follows:

2.2.1 Sensitivity of the Array to Vertical and Horizontal Changes

For the Wenner array, the sensitivity sections (Fig. 4) show large values near the surface between C1 and P1 electrodes, as well as between C2 and P2 electrodes. This means that if a small body with a higher resistivity than the background medium is placed in these zones, the measured apparent resistivity value will decrease. This phenomenon is also known as an "anomaly inversion". In comparison, if the high resistivity body is placed between the P1 and P2 electrodes where there are large sensitivity values, the measured apparent resistivity will increase. This is the basis of the offset Wenner method by BARKER

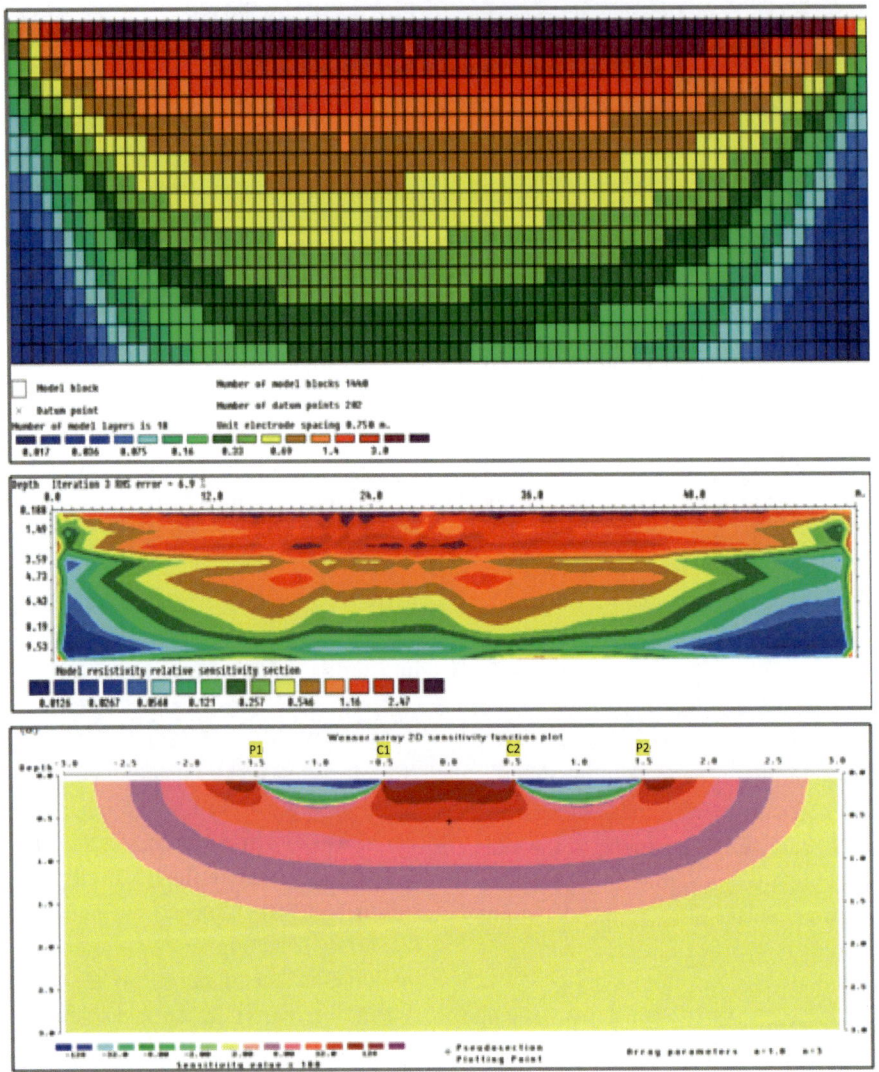

Figure 4
The subsurface sensitivity (*above*), the blocks sensitivity (*middle*) and the sensitivity (*below*) sections of the test survey Profile URUK-TEST-WEN for Wenner array

(1992) to reduce the effects of lateral variations in resistivity sounding surveys. In these sections, the sensitivity plot for the Wenner array has almost horizontal contours beneath the centre of the array. Because of this property, the Wenner array is relatively sensitive to vertical changes in the resistivity below the centre of the array. However, it is less sensitive to the horizontal changes. In general, Wenner array is good in resolving vertical changes (i.e. horizontal structures), but relatively poor in detecting horizontal changes (i.e. narrow vertical structures).

Figure 5 shows the sensitivity sections for the Wenner–Schlumberger array. The sensitivity contours of this array have a slight vertical curvature below the centre of the array. At high "n" values, the high sensitivity lobe beneath P1 and P2 electrodes becomes more separated from the high sensitivity values near C1 and C2 electrodes. This means that this array is moderately sensitive to both horizontal (for low "n" values) and vertical structures (for high "n" values).

Dipole–dipole array has been, and is still, widely used in resistivity because of the low EM coupling

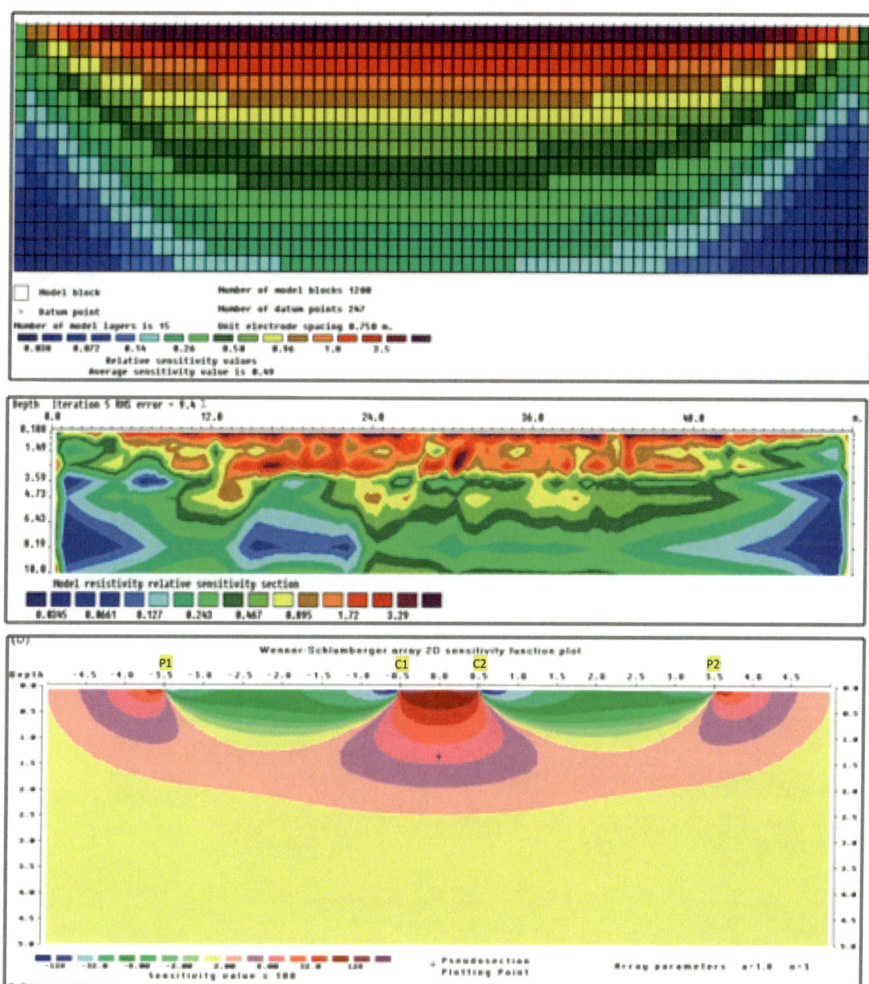

Figure 5
The subsurface sensitivity (*above*), the blocks sensitivity (*middle*) and the sensitivity (*below*) sections of the test survey Profile URUK-TEST-WEN-SCH for Wenner–Schlumberger array

between the current and potential circuits. The sensitivity contour pattern becomes almost vertical for "n" values greater than 2. Thus, the dipole–dipole array is very sensitive to horizontal changes in resistivity, but relatively insensitive to vertical changes in the resistivity. That means it is good in mapping vertical structures, such as walls, archaeological groove and cavities, but relatively poor in mapping horizontal structures (Fig. 6).

2.2.2 The Inverse Model and Depth Investigation

Another three tests on the same profile were conducted to examine the imaging capabilities of these arrays for the inversion models (Fig. 7). The depth of investigation of the Wenner, Wenner–Schlumberger and dipole–dipole arrays at this test survey is equal to 11, 8.68 and 6.13 m, respectively. A median depth of investigation means that the upper section of the earth above the "median depth of investigation" has the same influence on the measured potential as the lower section. This tells us roughly what depth we can see with an array. This depth does not depend on the measured apparent resistivity or the resistivity of the homogeneous earth model. It should be noted that the depths are strictly only valid for a homogeneous earth model, but they are probably good enough for planning field surveys.

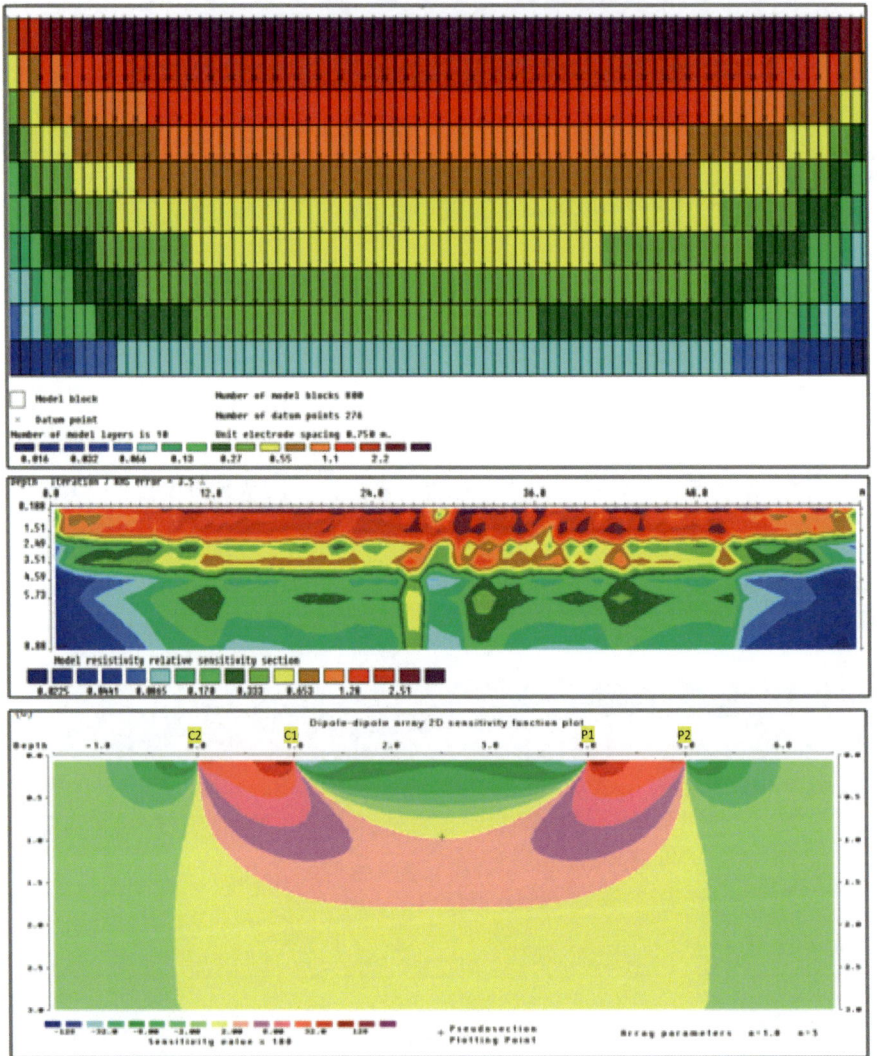

Figure 6

Shows the subsurface sensitivity (*above*), the blocks sensitivity (*middle*) and the sensitivity (*below*) sections of the test survey Profile URUK-TEST-DIPDIP for dipole–dipole array

If there are large resistivity contrasts near the surface, the actual depth of investigation could be somewhat different. In general, the dipole–dipole array has a shallower depth of investigation compared to the Wenner and Wenner–Schlumberger arrays for ERT survey.

From these models, one can see that the dipole–dipole array measurement yields the highest resolution and the best image for vertical anomalies. The Wenner and Wenner–Schlumberger arrays have similar behaviour of imaging ability due to the resemblance of their electric field and measurements,

Figure 7 ▶

The measured and the inverse model of the test survey Profile URUK-TEST-WEN for Wenner, Wenner–Schlumberger and dipole–dipole configurations

with their main strength in the depth determination, which is good in relation to the dipole–dipole array. However, the spatial resolution of the Wenner array is poorer than the dipole–dipole and Wenner–Schlumberger arrays. The imaging resolution of the dipole–dipole is better than others, particularly for the location of vertical structures. Accordingly, the

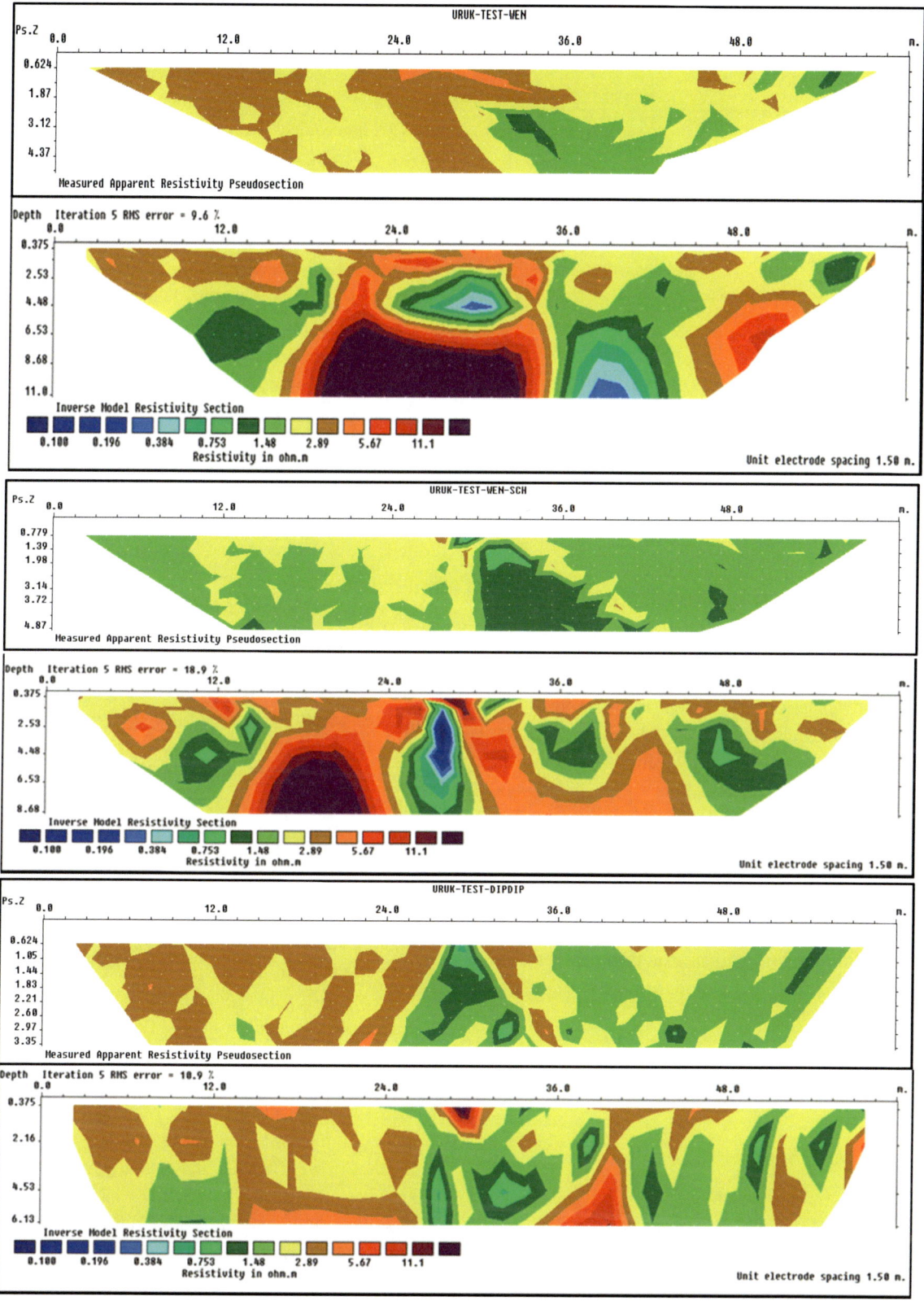

Table 1

The densities of data points of the three test survey profiles

Configuration	Density of data points
Wenner	202
Wenner–Schlumberger	247
Dipole–dipole	276

dipole–dipole array was chosen for future ERT surveys within the study area.

Finally, to obtain a high resolution and reliable image, the electrode array used should ideally give data with maximum anomaly information and reasonable data coverage (AIZEBEOKHAI 2010). Consequently, resistivity acquisition included a 32 parallel ERT northeast profiles each being 100 m long, comprising 41 electrodes were collected in February, 2012 over an area of 9300 m^2, with 2.5 m electrode spacing using the suitable dipole–dipole array. The spacing between these profiles equals 3 m to give systematic information about the area understudy.

2.2.3 Horizontal Data Coverage

One disadvantage of the Wenner array for 2D-resistivity imaging surveys is the relatively poor horizontal coverage as the electrode spacing is increased and the dipole–dipole array has better horizontal data coverage than the Wenner. The horizontal data coverage for Wenner–Schlumberger array is slightly wider than the Wenner array, but narrower than that obtained with the dipole–dipole array. Table 1 illustrates that the data points of the dipole–dipole survey are more than the other surveys. This means that the survey time for the dipole–dipole array is longer than the others.

3. Results and Discussion

3.1. GPR Interpretation

The greater the amplitude of wave reflections through a medium is, the greater the difference in chemical and physical characteristics of the buried material. The change in contrast on the GPR profile is isolated regions of high and/or low contrast.

Amplitude may be analysed to understand the possible material compositions of buried targets. It is important to note that the unique reflections created by covered features may change from site to site based on geological factors such as soil saturation levels or complications attributed to ground coupling. However, with the knowledge of what to look for, data interpretation, while not easy, will in time become much less confusing. Although there is some consistency in the identification of the major target types, there is rarely a way to identify exactly what material the target is made of. Hypothetically, the compact sediments may not display a visual change upon excavation, but the GPR may give a change in amplitude in the region, due to the physical change (resistivity) in relation to the surrounding earth. Waveforms often signify the boundaries of subsurface changes. It is made of wavelets, which is a positive or negative shift in amplitude. These are formed as a result of the change in physical or chemical property of the targets that the signal encounters, and they often signify the top and bottom of buried items. When combined, the wavelets create a waveform, which is then compiled with the other waveforms from any given GPR sample to create a composite amplitude trace, or image of what lies beneath the GPR unit. Among many other uses, the analysis of these waveforms allows to understand subsurface of the survey region. Areas of low-amplitude waves usually indicate uniform matrix material or soils while those of high amplitude denote areas of high subsurface contrast such as buried archaeological features, voids or important changes. Lateral changes in amplitude, phase or reflection patterns in the radar record can be caused by changes in rock, soil type and moisture content.

RadExplorer, Ver.1.4-GX software (2005) was used to process and interpret the collected GPR data. The GPR raw data processing includes Dc removal, background removal, bandpass filter and predictive deconvolution. Time adjustment and topography have been applied for all the profiles in this study and processed with the same range of filter values, because the study area contains subsurface targets with the same original lithological characters (clay and silt). The presence of clay layer has an influence which causes high adsorption of the electromagnetic

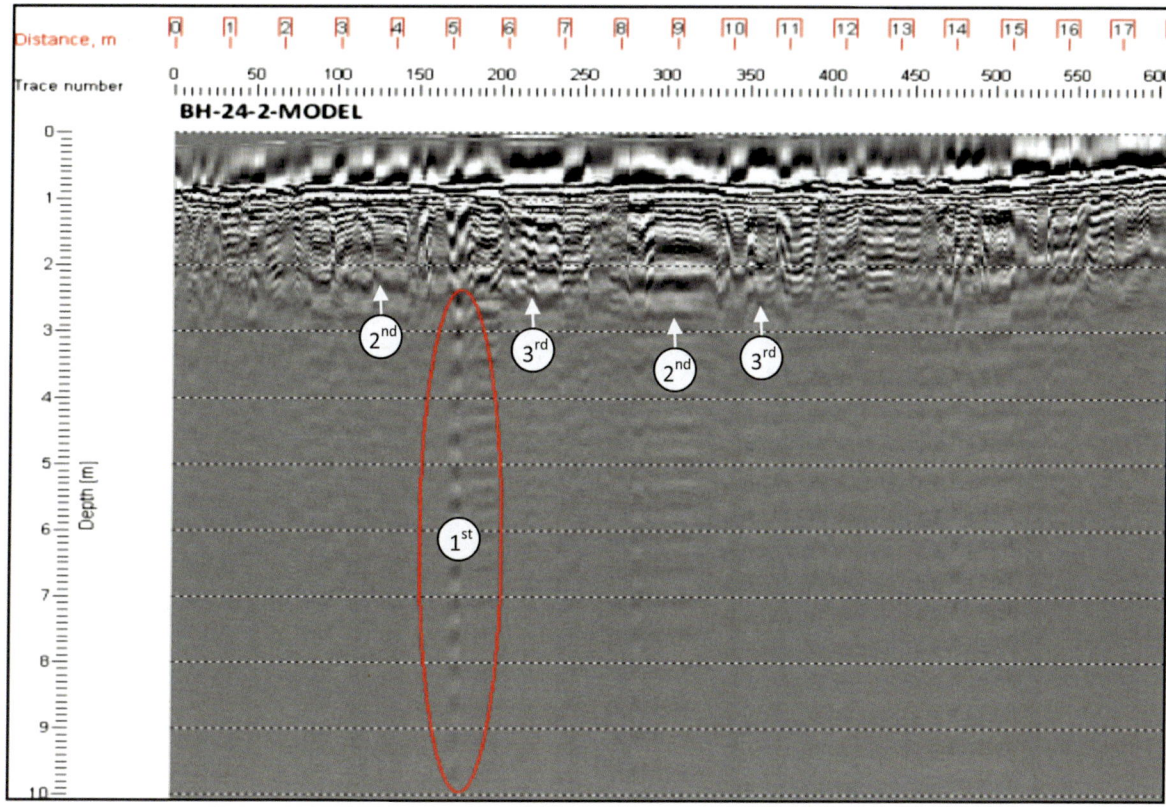

Figure 8
GPR model of profile (BH-24-2-MODEL) shows dense or reflective (1st) sub-horizontal buried items

energy that limits its penetration to not deeper than 2–3 m, although dielectric permittivity of the dried subsoil may have low value due to water table that exceeds nearly 5 m below surface. For this reason, the GPR data show maximum depth not more than 5 m. According to the above facts and results, and from reviewing the models of the Babylonian houses district, it is clear that:

- The first type of the image features is present with shallow depth of attributes with sub-horizontal, partly wavy reflections; however, it can often be found throughout the whole area representing the top part of the Babylonian houses district. This zone is characterized by dried clay and sandy soil including broken and weathered different archaeological materials such as broken brick and slag mixed with core boulders. These processes produce shallow layers of clay and sand deposits with different archaeological materials such as broken brick and slag.

- The second type of reflections shows deep attributes. These are characterized by continuous reflections with different widths. These reflections are typical feature for the archaeological walls. Most of the main attributes perhaps refer to buried remains of clay brick walls. The archaeological walls have approximated real depth between 1 and 6 m, and width from 1 to 2 m.

- The third type of reflections represents the point reflection attributes (small hyperbola) which are presented at the upper and/or the lower parts of the images. They occur either within the first or second type of reflections. The third type of reflections includes four subtypes of reflections, these are:

  – The 1st subtype is dense or reflective buried items, which are capable of creating echoes. The GPR signals cannot penetrate these objects. This phenomenon creates high-amplitude signals of repeating or echoing bands upon the screen. The electromagnetic wave bounces back and forth

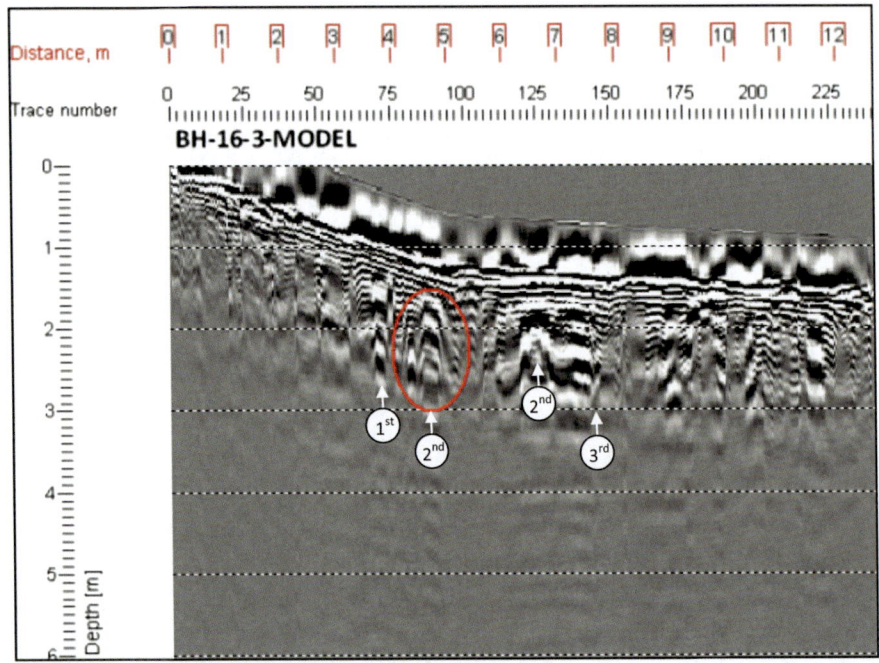

Figure 9
GPR model of profile (BH-16-3-MODEL) shows parabola (2nd) represented an object which is located at apex of an arc

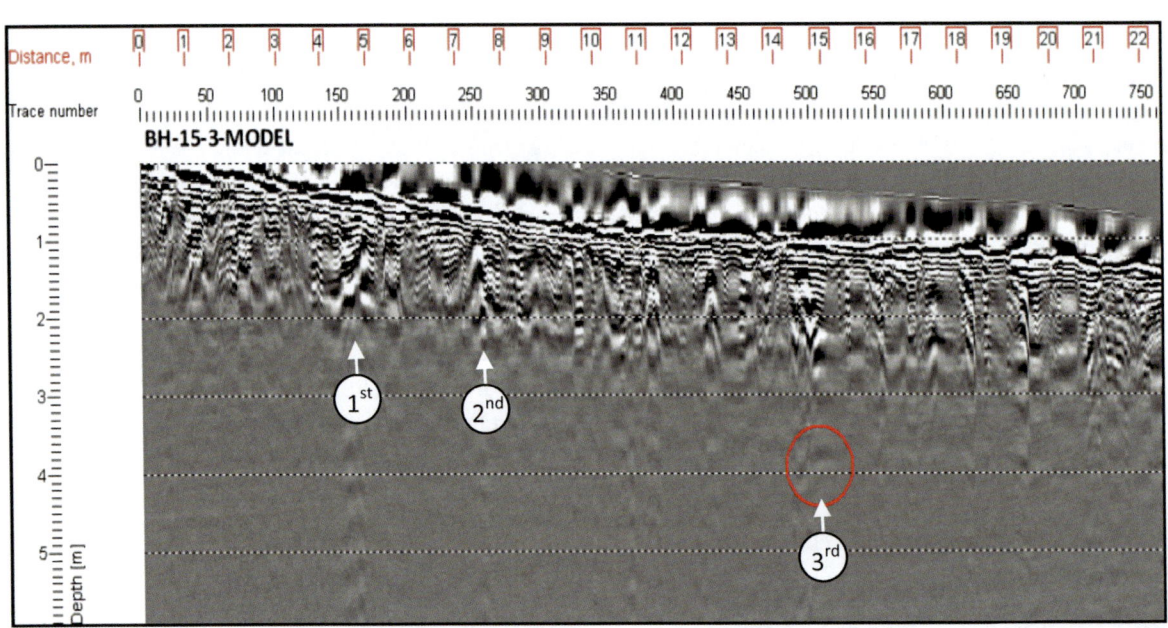

Figure 10
GPR model of profile (BH-15-3-MODEL) shows planar reflection (3rd)

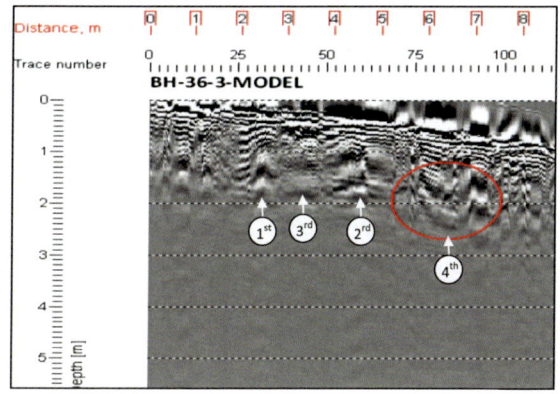

Figure 11
GPR model of profile (BH-36-3-MODEL) shows buried trenches and pits create cup-shaped (4th)

from the highly reflective object to the GPR device over and over again. The reflected signature from the dense or reflective buried item is at most times unmistakable. The buried features are capable of creating parabolic signatures such as coffins, ceramics, bricks or boulders (Fig. 8).

– The 2nd subtype is a parabola which represents an object located at the apex of the arc. Small hyperbola is created as the GPR footprint moves across rounded object. Reflections of this type are distinct and relatively easy to identify (Fig. 9).

– The 3rd subtype is the planar reflections. These reflections are occurring from buried targets and take the form of planar reflections. They appear as horizontal attributes and exhibit higher amplitude in relation to the surrounding matrix. These reflections are often the result of a physical discontinuity or a horizontal feature of archaeological interest. They may identify the layering of the soil. In general, it gives regular reflections along the GPR profile with some distinguished features here and there may refer to the small archaeological remains (Fig. 10).

– The 4th subtype is presented as buried trenches and pits. Covered pits and V-shaped trenches are identifiable features. The signatures indicate pit-like features and trenches. This creates high-amplitude cup-shaped or V-shaped signature on the display screen (Fig. 11). Covered pits may indicate the presence of grave pits.

Figure 12 is an example of the interpreted three segments of the GPR profile lines of Babylonian houses district showing the mentioned different types of radargram attributes at the study area.

3.2. ERT Interpretation

Also, the RES2DINV software is used to interpret the measured raw apparent resistivity data. This software uses the rapid least-squares inversion method to model the final resistivity sections. The optimization method basically tries to reduce the difference between the calculated and measured apparent resistivity values by adjusting the resistivity of the model blocks. A measure of this difference is given by the root-mean-squared (RMS) error. However, the model with the lowest possible RMS error can sometimes show large and unrealistic variations in the model resistivity values and might not always be the best model from a geological perspective. In general, the most prudent approach is to choose the model at the iteration after which the RMS error does not change significantly. This usually occurs between 3rd and 5th iterations (LOKE and DAHLIN 2002).

Figure 13 displays the ERT model of profiles 12, 15 and 32 as examples of the inverse model in the study area. The 2D resistivity imaging profiles have maximum depth of investigation of 13.5 m. A comparison of the results from adjacent profiles shows almost similar anomalies at similar depths. The resistivity of subsurface material is affected by porosity (shape, size and connection of pores), moisture (water) content, dissolved electrolytes, temperature of pore water, conductivity of minerals, the chemistry of the groundwater and other fluids trapped in the pore spaces within the soil matrix and/or the presence or absence of buried debris and structures. For a typical site, fine materials such as clay and silt are generally less resistive while coarse sand and gravel are generally more resistive. The soil (clay or sand) will appear more resistive when it is dry and less resistive when it is wet. The presence of subsurface walls often appears as a vertically oriented anomaly, and may be either conductive or resistive

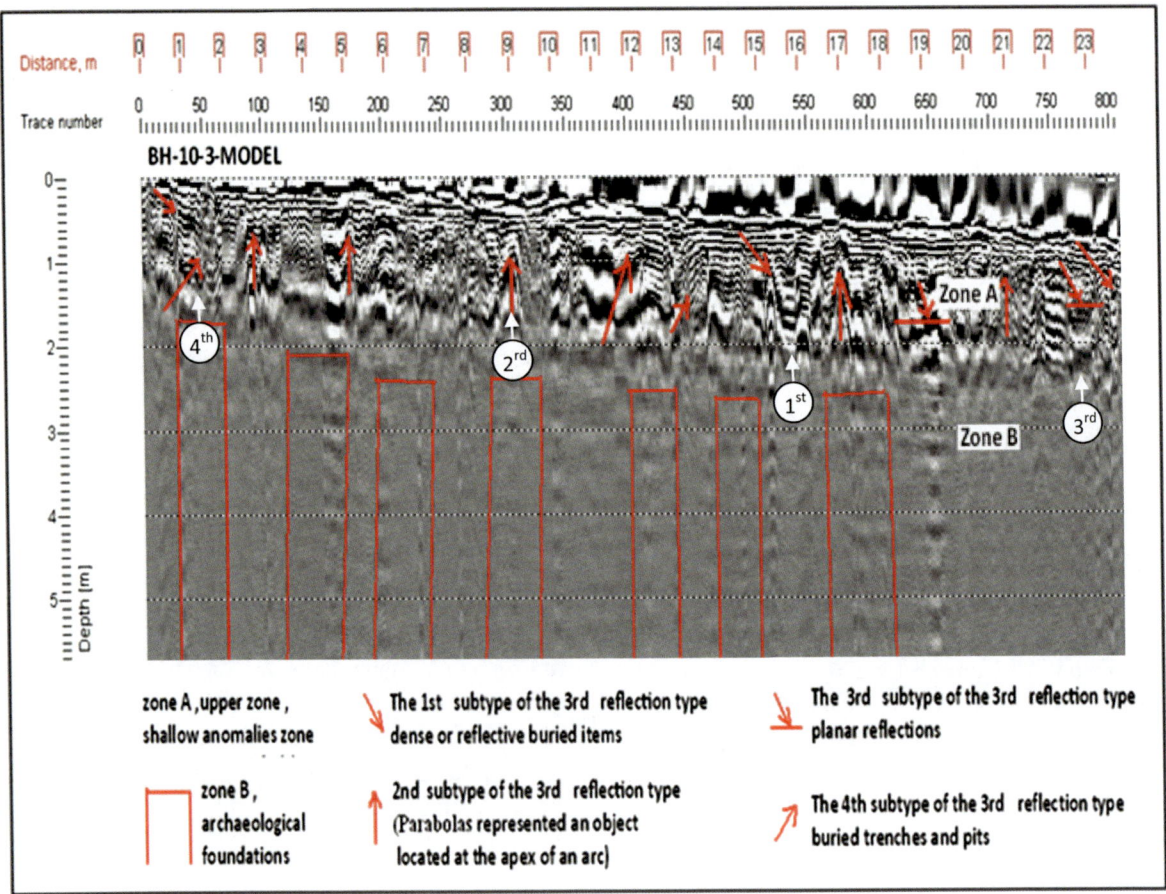

Figure 12
Interpreted example of three segments of (BH-10-3-.MODEL) GPR profile showing the legend of the different types of radargram attributes at the study area

depending on what type of fluid present (e.g. clean groundwater and/or unweathered/weathered contamination). Resistivity is linked to moisture content or porosity. Features such as wall foundations will give a relatively high resistivity response, while ditches and pits that retain moisture give a lower one. The archaeological graves can be considered as a lateral anomaly in a homogenous medium similar to the tunnel and cavity. An anomalous zone of the water- or clay-filled archaeological grave is distinguishably as very low or lowest resistivity zone, surrounding with the higher background. High resistivity zones are correlated with air-filled archaeological structures such as graves, caves, voids or holes in the overburden that had formed over archaeological walls. In general, in the archaeological structures, competent

structures have high resistivity. Weathered rocks would show a much lower resistivity than the competent one. High-conductivity zone in the resistivity images may be due to infilling by clay, or water within the pores. Generally, all the resistivity values are very low to low values not exceeding more than 11.1 Ohm m due to corresponding site subsoils or layers: The interpretation of all models basically declares the presence of three zones underneath Uruk site as mentioned below:

The upper one with a resistivity value (2.89–5.67 Ohm m) is interpreted as an alluvium soil consisting of sand and clay. In Uruk, weathering of archaeological structures produces clayed and sandy soil with core boulders and other partially weathered material because these structures are constructed

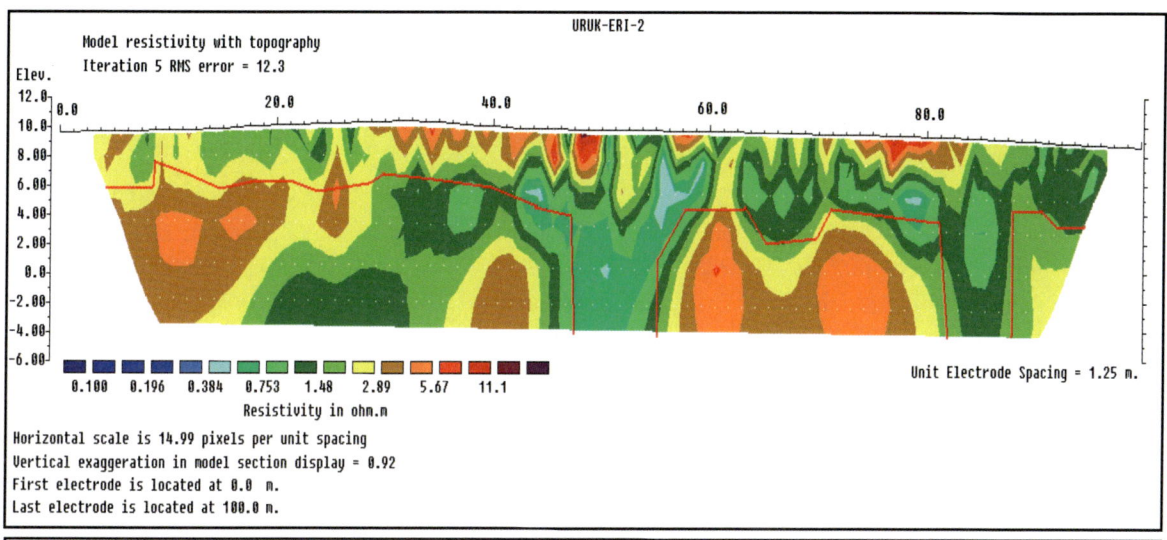

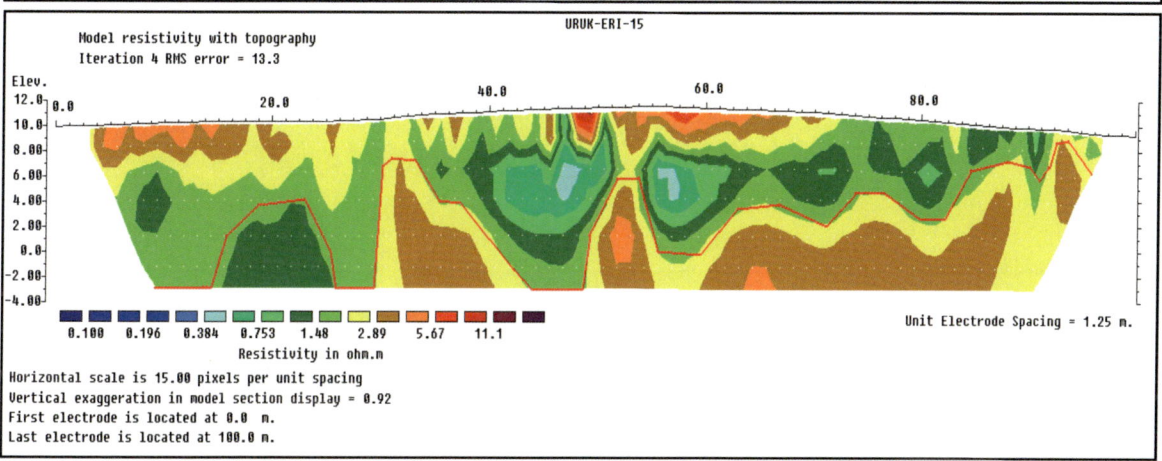

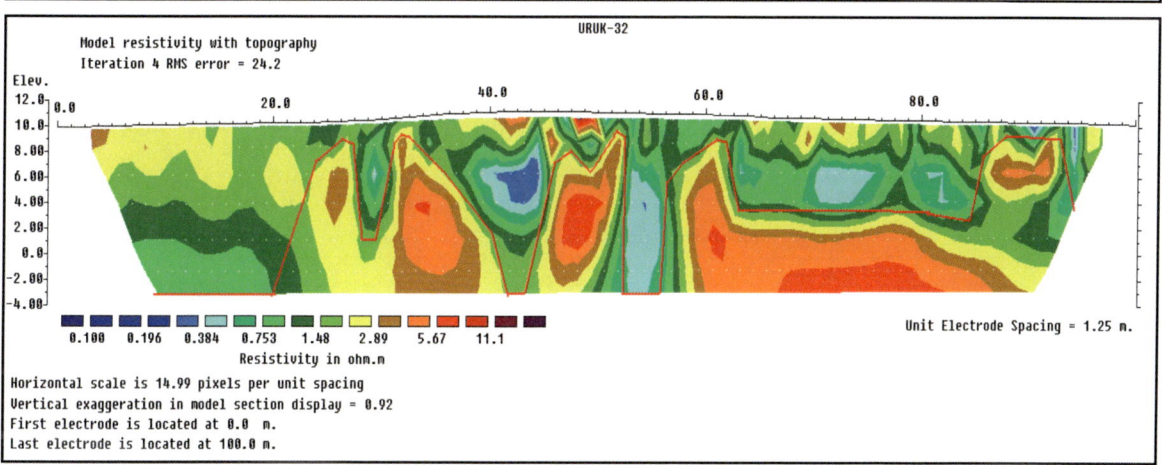

Figure 13
Interpreted 2D inverse model of profiles 12, 15 and 32 as some examples from the area under study

mainly of clay. The highest resistive areas nearby the surface are caused by the dried upper layers. In this zone, some anomalies caused by boulders with resistivity equal to 5.67 Ohm m can be seen at depth 0.5 m from the surface. The second zone shows a prominent lower resistive (0.753 Ohm m) zone below the first one. This is probably caused by the moisture in this region that reduces the resistivity. Thickness of this layer differs from other parts of the site. In addition, some anomalous archaeological materials (5.67 Ohm m) related to core boulders, air cavities or graves presented at this zone can clearly be observed. Underneath this intermediate zone is a more resistive zone which indicates more intact zone of archaeological structures down to the bottom of the image. The images indicate that an archaeological structure extended vertically through the images is most probably related to the buried remains and ruins of old buildings and walls that have resistivity values which are either low (0.384 Ohm m) or resistive

(5.67–11 Ohm m) at depth ranging from 6 m to 13.5 m related to the archaeological wall.

3.3. Comparison Between GPR and ERT

Figures 14 and 15 show examples of both GPR and ERT results at the Babylonian houses district. The reflections existed within the radargrams are described in terms of reflection continuity, shape, amplitude, internal reflection configuration and external form (pattern of reflections). It seems that the GPR profile has a maximum depth of investigation of 1–6 m depending on the subsoil case. The GPR images demonstrate deeper anomalies characterized by continuous reflections with different widths. These deeper reflections correspond to a zone on the 2D-resistivity models with the higher resistive archaeological walls of the third zone extending vertically through the images. The GPR images show reflections of attributes which are presented at the

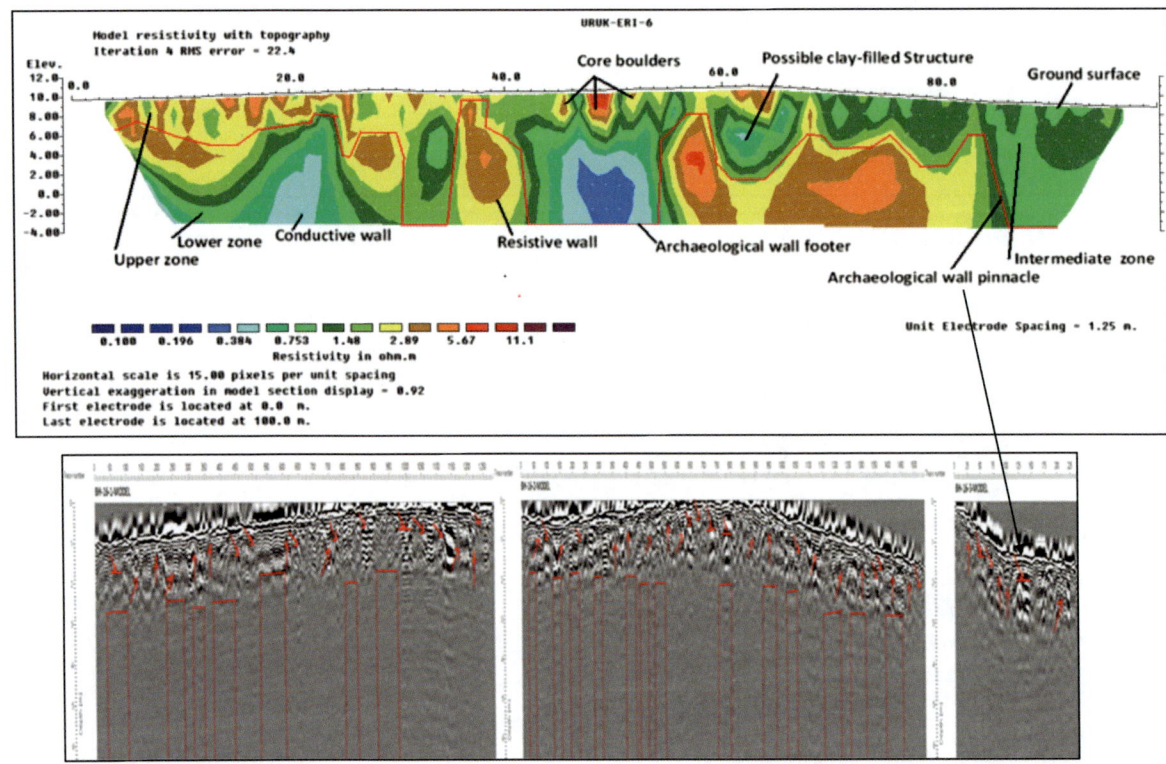

Figure 14
Interpreted inverse model of ERT (profile-6) and its corresponding GPR segment profiles within the area

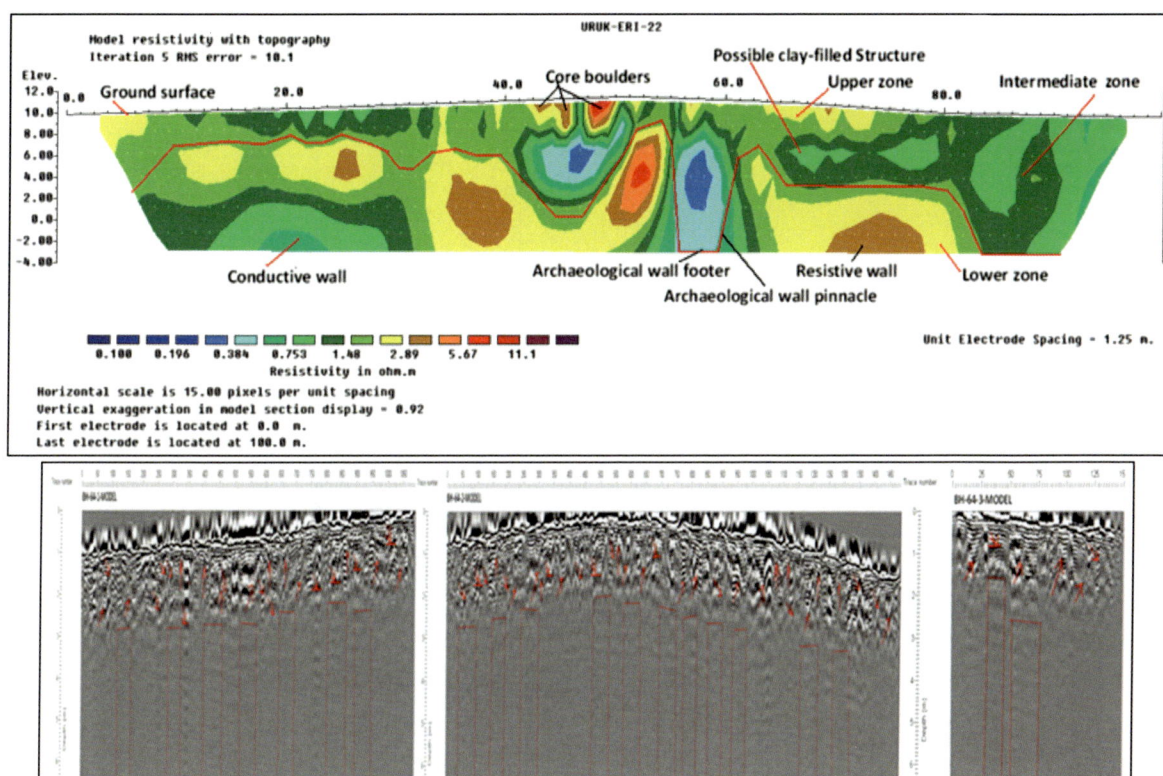

Figure 15
Interpreted inverse model of ERT (profile-22) and its corresponding GPR segment profiles within the area

shallower depth and characteristic for the upper part of the district and can often be found throughout the area and represent the top part of the Babylonian houses. However, the ERT sections also show the upper zone near the surface of the earth with a highly resistivity value. There is intermediate lower resistive layer with thickness differing from parts to others within the study site. Also, the GPR and ERT anomalies are characterized by the existence of dense and buried items, buried trenches and pits or object.

The GPR profiles have maximum depth of investigation of 6 m and ERT profiles have maximum depth of investigation of 13.5 m. These results establish whether the information from GPR and ERT is complementary in delineating the subsurface. According to the previous results, it is clear that the two methods are integrated and used for completing the description of the area understudy.

A contour map (Fig. 16) and 3D-view (Fig. 17) of the distribution of the archaeological anomalies at the

study area were plotted using data of the ERT transects (from the 2D-resistivity line 1–32) and GPR transects (from the GPR line 1 to GPR line 90) with the assistance of Surfer and GIS programs, respectively. These data represent the anomalies at the study area; therefore, high anomaly means high elevation of the old wall present at the certain location and vice versa. The wall footer indicates the absence of the archaeological wall at this location. The values of the elevations of the walls are taken as guide for the archaeological anomalies. These 2D and 3D maps of the archaeological foundations of Uruk site show that the archaeological anomalies are concentrated mainly at the NE part of the district. Additionally, high values of anomalies are concentrated along this direction. In this part of the district, the walls are higher than that at the SW, SE and NW parts ranging mainly from 6 to 8 m and exceed 10 m in some locations at the NE area. At the remaining parts trending SW, SE and NW, the walls are lower in

Reprinted from the journal

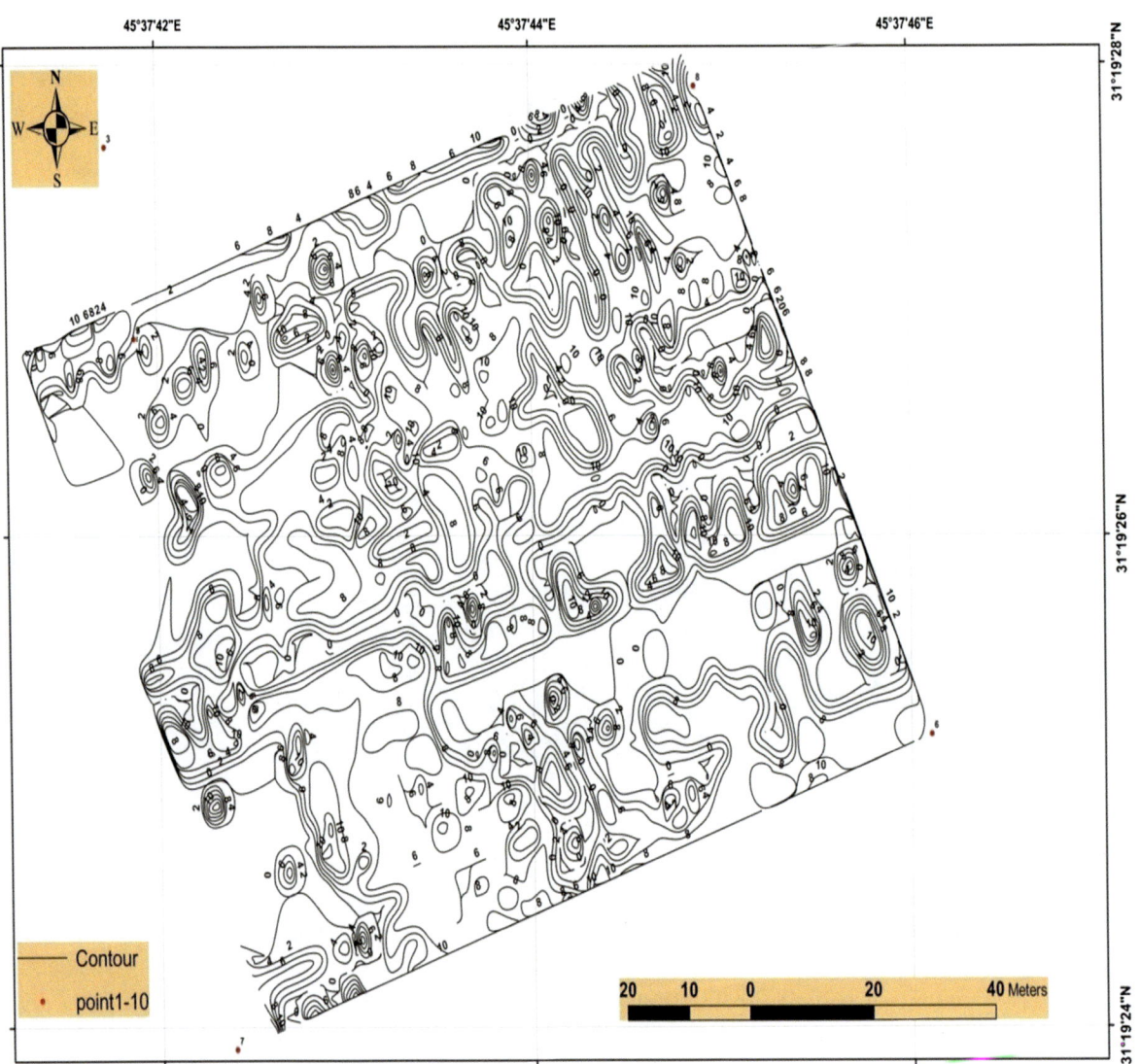

Figure 16
Map shows anomalies of the archaeological foundations of Uruk site from GPR and 2D-imaging data

heights ranging mainly between 4 and 6 m. In some places (especially at the west part) of district, the height of the wall reaches the wall foot.

4. Conclusions

According to the GPR and ERT profiles' interpretation of the study area, it seems that the acquired GPR radargrams show large and continuous signal attributes with different widths. In radargrams, at the shallower depth, GPR attributes strongly represent the upper part of Babylonian Houses throughout the study area. Besides, some objects such as buried items, buried trenches and pits were mainly concentrated near the surface.

ERT dipole–dipole array results show the presence of several anomalies at different depths mostly having low resistivities. Thus, it is clear that the first upper zone can often be found throughout the whole area and it may represent the top zone of the Babylonian houses that consists of dry clay and sandy soil including some broken bricks and slag mixed with core boulders distributed here and there at the

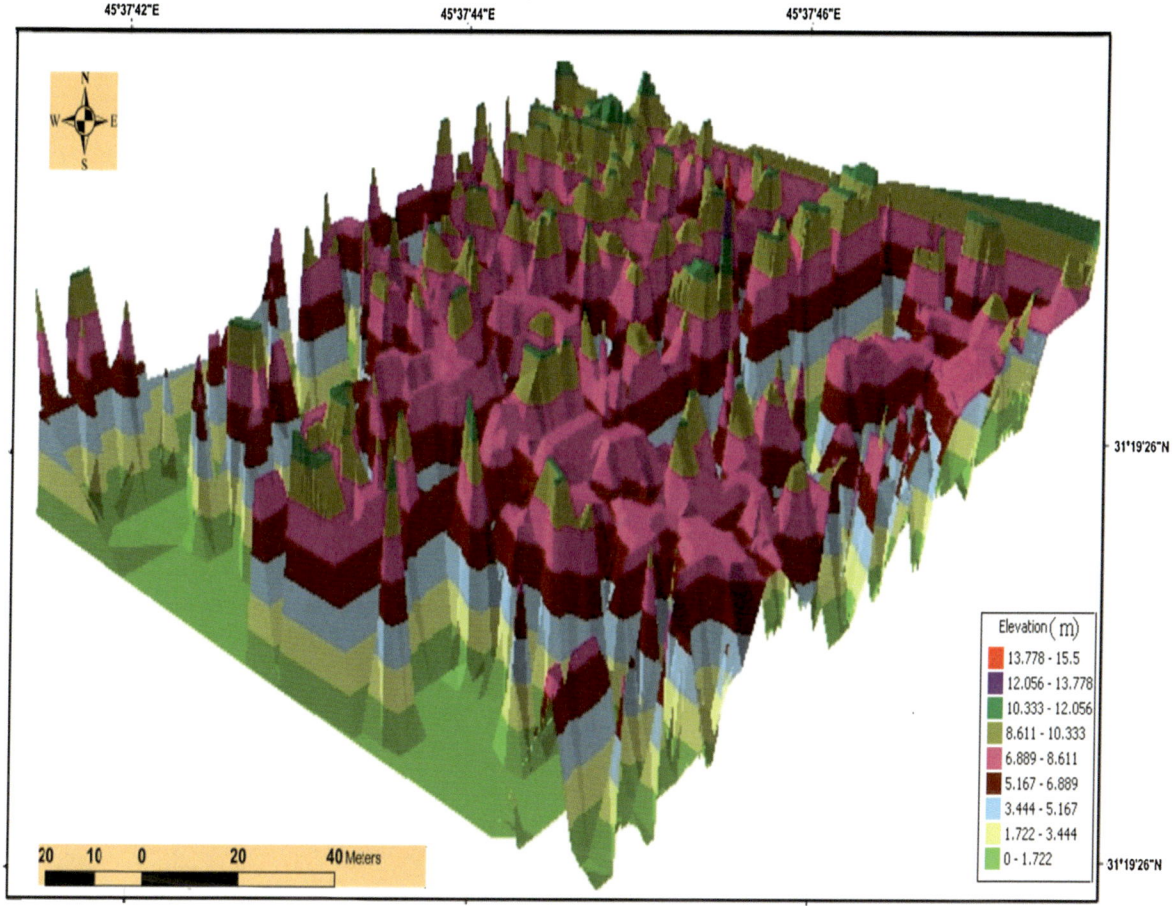

Figure 17
3D view of the archaeological foundations of Babylonian houses data

surface. The second zone clarifies a prominent lower resistivity zone. It is probably presented by the moisture content that reduces the resistivity. The thickness of this zone is not equal at all parts of the site. Finally, the third deeper zone typically represents the archaeological walls. Most of the distinct anomalies perhaps referred to the buried clay brick walls.

The constructed map of the archaeological anomalies distribution and 3D view of the foundations at the study area using GPR and ERT methods clearly demonstrate the characteristics of the Babylonian remains. A contour map and 3D view of Uruk show that the archaeological anomalies are concentrated at the NE part of the area understudy having

highly walls height that ranging between 6–8 m and up to more than 10 m in some places. At the other directions, there are fewer walls with lower heights of 4–6 m and seem in some places closer to the wall foot.

Acknowledgments

The authors are highly grateful to Dr. Firas H. Al-Menshed and Dr. Ahmed S. Al-Zubedi for their sincere assistance in the field works. Thanks also extend to Dr. Ali Z. Al-Khashan for reading the manuscript of the present research.

REFERENCES

AIZEBEOKHAI, A. P., (2010), *2D and 3D geoelectrical resistivity imaging, Theory and field design*, Scientific Research and Essays Vol. *5*, 14p.

AL-HASHIMI, H., (1974), Stratigraphy and palaeontology of the subsurface rocks of Samawa Area, SOM, Iraq.

BAKER, H. D., (2002), The Urban landscape in first millennium BC Babylonia, University of Vienna.

BARKER, R.D., (1992), *A simple algorithm for electrical imaging of the subsurface*, First Break, Vol. *10*, No. 2, 10p.

BEAULIEU, P., (2003), The Pantheon of Uruk during the Neo-Babylonian period, 424p.

BUDAY, T., (1980), Regional geology of Iraq, Stratigraphy and Palaeogeography, GEOSURV, Baghdad, Iraq, Vol. 1, 445p.

DAHLIN, T. and ZHOU, B., (2004), *A numerical comparison of 2D resistivity imaging with ten electrode arrays*, Geophysical Prospecting, Vol. *52*, 20 p.

FASSBINDER, J. and BECKER, H., (2001), "Uruk-City of Gilgamesh (Iraq) First tests in 2001 for magnetic prospecting", dans Becker, H., Fassbinder, J. W.E, Magnetic prospecting in Archaeological Sites, Vol. VI, 5p.

FASSBINDER, J., BECKER, H. and VAN ESS, M. (2003), Prospections magnetiques a Uruk (Warka): La cite du roi Gilgamesh (Irak).

HRITZ, C., (2012), History and Archaeology in Southern Mesopotamia, Department of Anthropology, Pennsylvania State University Basrah.

LOKE, M.H. and DAHLIN, T., (2002), *A comparison of Gauss-Newton and quasi-Newton methods in resistivity imaging inversion*. Journal of applied geophysics, vol. *49*, 19 p.

LOKE, M.H., (2004), Tutorial: 2D and 3D electrical imaging surveys, 127p.

LOKE, M.H., (2010), Tutorial: 2D and 3D electrical imaging surveys, 154p.

OSWIN, J., (2009), A Field guide to geophysics in archaeology, Praxis Publishing, No 2009925774, Chic Ester, UK, 243p.

POLLOCK, S., POPE, M. and COURSE, Y., (1996), *Household production at the Uruk Mound, Abu Salabikh, Iraq*, American Journal of Archaeology, Vol. *100*, No. 4, 16p.

(Received June 26, 2014, revised July 29, 2015, accepted August 1, 2015, Published online August 14, 2015)

Pure Appl. Geophys. 173 (2016), 983–993
© 2014 Springer Basel
DOI 10.1007/s00024-014-0959-8

| **Pure and Applied Geophysics**

CrossMark

A Coupled Vegetation–Crust Model for Patchy Landscapes

Shai Kinast,[1] Yosef Ashkenazy,[1] and Ehud Meron[1,2]

Abstract—A new model for patchy landscapes in drylands is introduced. The model captures the dynamics of biogenic soil crusts and their mutual interactions with vegetation growth. The model is used to identify spatially uniform and spatially periodic solutions that represent different vegetation-crust states, and map them along the rainfall gradient. The results are consistent extensions of the vegetation states found in earlier models. A significant difference between the current and earlier models of patchy landscapes is found in the bistability range of vegetated and unvegetated states; the incorporation of crust dynamics shifts the onset of vegetation patterns to a higher precipitation value and increases the biomass amplitude. These results can shed new light on the involvement of biogenic crusts in desertification processes that involve vegetation loss.

1. Introduction

Water-limited vegetation landscapes are usually patchy (Valentin *et al.* 1999; Deblauwe *et al.* 2008). Vegetation patch formation is a means by which dryland vegetation copes with water stress. The formation of patches devoid of vegetation provides additional sources of water to adjacent vegetation patches through various mechanisms of water transport, which help the vegetation to sustain itself. Self-organized vegetation patchiness of this kind is currently viewed as a symmetry-breaking pattern formation phenomenon driven by positive feedbacks between two main processes: local vegetation growth and water transport toward the growing vegetation. Several water-transport forms have been identified, including overland water flow, water conduction by

laterally extended root zones, soil-water diffusion, and fog advection (Rietkerk and van de Koppel 2008; Meron 2012; Kinast *et al.* 2014; Borthagaray *et al.* 2010; Borgogno *et al.* 2009). The mechanism by which local vegetation growth enhances water transport toward patches of growing vegetation depends on the type of water transport. We focus here on overland water flow as a major type of water transport in dryland landscapes.

Soil areas devoid of vegetation are often covered by thin biogenic soil crusts (West 1990). Depending on the precipitation regime, soil characteristics, and disturbances, these crusts may consist of one or more organisms, including cyanobacteria, green algae, fungi, lichens, and mosses. Soil crusts reduce soil erosion by water and wind. They also provide a source of carbon and nitrogen for vascular plants. Most important to our discussion here is their capability to induce overland water flow (runoff) by changing the rate of surface-water infiltration into the soil. Crusts that are dominated by cyanobacteria, for example, can absorb water several times their dry weight in only a few seconds (Campbell 1979). This results in crust swelling and soil-pore blocking and, consequently, in significant reduction of water infiltration shortly after rain starts (Verrecchia *et al.* 1995; Eldridge *et al.* 2012). Because cyanobacteria are photosynthetic organisms, their growth is hindered by vegetation, which limits exposure to sunlight. The reduced infiltration in crusted areas and the absence of crusts in vegetated areas result in an infiltration contrast: low infiltration rates in sparsely vegetated areas and high rates in densely vegetated areas. Additional factors contributing to this outcome include soil mounds generated by dust deposition (Shachak and Lovett 1998) and higher soil porosity in vegetation patches (Puigdefábregas 2003; Stavi *et al.* 2009). The infiltration contrast induces overland

[1] Department of Solar Energy and Environmental Physics, Blaustein Institutes for Desert Research, Ben-Gurion University of the Negev, 84990 Sede Boqer Campus, Israel. E-mail: shaikinast@gmail.com
[2] Department of Physics, Ben-Gurion University of the Negev, 84105 Beer Sheva, Israel.

water flow toward densely vegetated areas, which accounts for the enhancement of water transport by local vegetation growth, and closes the positive feedback loop (vegetation growth → water transport → vegetation growth) that drives vegetation pattern formation (hereafter the "infiltration feedback").

Mathematical models that incorporate infiltration feedback into the model's equations (RIETKERK *et al.* 2002; GILAD *et al.* 2004, 2007) indeed capture a nonuniform stationary instability of uniform vegetation that gives rise to periodic vegetation patterns. These models capture the effect of biogenic crusts on vegetation pattern formation implicitly by introducing an infiltration-contrast parameter that quantifies the differences between infiltration rates in vegetated and unvegetated areas. This modeling approach, however, ignores the properties and dynamics of the biogenic crust, which limits the applicability of the models in two main respects. First, different types of biogenic crust are present in nature; ignoring their properties severely limits the ability to distinguish between the effects of different crust types on vegetation pattern formation (YAIR *et al.* 2011). The second respect is related to the absence of competition for space between biogenic crusts and vegetation. It is well established (PRASSE and BORN-KAMM 2000) that crusts can suppress vegetation growth by preventing seed germination. This effect may be important in desertification processes,[1] for example shrubland–crustland transitions, an example of which is shown in Fig. 1; rapid soil coverage by crusts after degradation of the woody vegetation may delay or even prevent vegetation regrowth.

In the work discussed in this paper we studied the effect of biogenic soil crusts on vegetation pattern formation by adding an equation for crust dynamics to an earlier vegetation model (GILAD *et al.* 2007), and modifying the remaining model equations to take into account the coupled crust–vegetation dynamics and crust–water dynamics. We note that models of crust dynamics have been proposed and studied elsewhere (BÄR *et al.* 2002; MANZONI *et al.* 2014; KINAST *et al.* 2013). To the best of our knowledge,

Figure 1
Degraded landscape in the Northern Negev (Israel) after a series of droughts. The *white patches* consist of shells of dead snails that used to feed on dead branches of living shrubs, and constitute a "ghost pattern" of a former vegetation spot pattern. From SHACHAK (2011) (with permission)

however, the coupling between vegetation dynamics and crust dynamics in a spatial context has not yet been studied.

We first show that the new crust–vegetation model reproduces the sequence of vegetation states along the rainfall gradient that have been predicted by earlier vegetation models. We then present additional predictions that emphasize the effect of crust dynamics in the development of vegetation.

2. A Crust–Vegetation Model

The new crust–vegetation model we propose consists of four dynamic variables, representing the areal densities of vegetation biomass (B), crust biomass (C), soil water (W), and surface water (H), all having the dimensions mass per unit area. The model is based on a simplified version (KINAST *et al.* 2014) of the vegetation model introduced by GILAD *et al.* (2004, 2007). The model's equations are:

$$B_T = G_B B(1 - B/K_B) - M_B B - \frac{\phi_B BC}{(B + B_0)^m} + D_B \nabla^2 B,$$
(1a)

$$C_T = G_C C(1 - C/K_C) - M_C C - \phi_C CB + D_C \nabla^2 C,$$
(1b)

$$W_T = IH - N(1 - RB/K_B)W - G_W W + D_W \nabla^2 W,$$
(1c)

---

[1]   Desertification is defined as an irreversible reduction in biological productivity (biomass production rate) as a result of climate fluctuations or anthropogenic disturbances.

$$H_T = P - IH - G_H H - \nabla \cdot \mathbf{J}, \qquad (1d)$$

where,

$$G_B = \Lambda_B W (1 + EB)^2, \qquad (2a)$$

$$G_C = \Lambda_{CW} W + \Lambda_{CH} H, \qquad (2b)$$

$$G_W = \Gamma_B B (1 + EB)^2 + \Gamma_{CW} C, \qquad (2c)$$

$$G_H = \Gamma_{CH} C, \qquad (2d)$$

$$I = A \left(\frac{f_C C + Q_C}{C + Q_C} \right) \left(\frac{B + Q_B f_B}{B + Q_B} \right), \qquad (2e)$$

$$\mathbf{J} = -D_H H^\beta \nabla H. \qquad (2f)$$

We assume in this model that the vegetation and the biogenic crust are both characterized by logistic growth and linear mortality. The "carrying capacity" K_B represents genetic constraints, such as stem architecture and strength, whereas K_C mainly represents constraints of exposure to sunlight. The growth rates G_B and G_C both depend on water availability but assume different functional forms. Plants exploit below-ground water (W) through water uptake by their roots. This is accounted for by Eq. (2a), where E is a measure for the root-to-shoot ratio and relates root size to above-ground biomass B. The particular biomass dependence of the growth rate G_B follows from the assumption of confined root zones (ZELNIK et al. 2013). The crust exploits both below-ground and above-ground water, hence the form of Eq. (2b).[2] For the same reason the equations for W and H both contain terms describing water uptake by the crust ($\Gamma_{CW} CW$ in G_W and $\Gamma_{CH} CH$ in G_H, respectively).

The vegetation and the biogenic crust compete indirectly by consumption of the common water resource, and directly by competition for space. Plants can suppress the growth of biogenic crusts by spreading litter that limits sunlight. Plants can also destroy biogenic crusts if the litter is toxic (BOEKEN AND ORENSTEIN 2001). These effects are represented by the parameter ϕ_C in Eq. (1b). Biogenic crusts suppress the growth of vegetation by preventing seed seeding and germination (PRASSE and BORNKAMM 2000). The suppression effect, however, applies only to the seed germination phase; once germination occurs, suppression fades out. To account for this biomass-dependent effect we model the vegetation-decay rate as $\phi_B C / (B + B_0)^m$, where the parameter ϕ_B quantifies the suppression, B_0 represents the biomass of a seedling, and the exponent m represents the rate at which the suppression effect decays as the vegetation grows.

Vegetation patches often spread in space by local seed dispersal or by clonal growth. These processes are described in the model by a linear diffusion term in the biomass equation (Eq. 1a). Long-distance seed dispersal can be capture by replacing the diffusion term by an integral over a kernel function (THOMPSON et al. 2008). The spatial spread of biogenic crusts is also a local process [crust fronts may propagate as fast as few centimeters per day (DODY et al. 2011)], which we model by a linear diffusion term in Eq. (1b).

A major component of the infiltration feedback is the development of an infiltration contrast between crusted and vegetated soil, which is modeled by the infiltration function (Eq. 2e). It is common to distinguish between physical soil crust and biogenic soil crust. The physical crust consists of a dense layer of soil particles formed by the effect of rainfall after the soil dries out (SELA et al. 2012). The biogenic crust consists of microorganisms such as cyanobacteria, microfungi, lichens, and mosses (BELNAP AND LANGE 2001). The effects of the two crust types are captured by the monotonic dependencies of infiltration rate on B and C, as Fig. 2 illustrates.

The dimensionless parameters f_B and f_C in Eq. (2e) quantify the infiltration contrasts induced by physical and biogenic soil crusts, respectively (no contrast for $f_B = 1$ or $f_C = 1$, and high contrast for $f_B \ll 1$ or $f_C \ll 1$). Increased infiltration under the plants canopy, because of, e.g., soil mounding (DUNNE et al. 1991), is also represented by f_B. The values of these parameters enable us to control the strength of these two independent properties.

The infiltration contrast induces surface water gradients, which generate an overland water flux $\mathbf{J}$ toward vegetation patches. This is another component of the infiltration feedback which we model by Eq. (2f). In previous studies (MERON 2011), the value $\beta = 1$ was used in the expression for $\mathbf{J}$. This choice leads

---

[2] In distinguishing between below-ground soil water, W, and above-ground surface water, H, we consider the ground level to represent the upper surface of the few-millimeters thick crust.

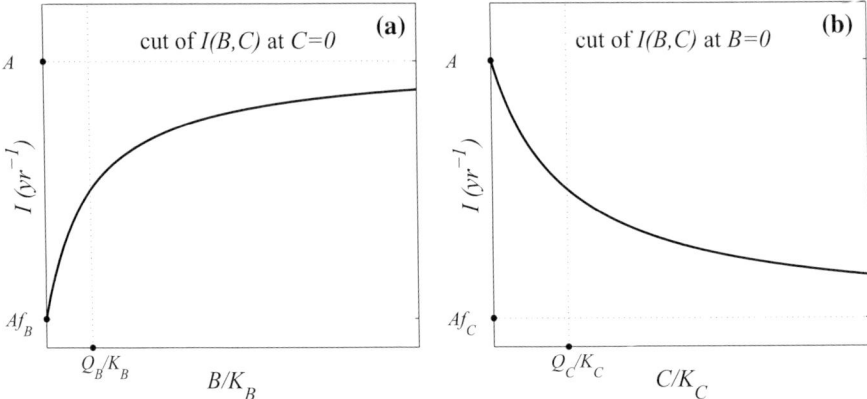

Figure 2

Cuts of the infiltration rate function $I = I(B,C)$ (Eq. 2e) at $C = 0$ (**a**), showing the dependence of the infiltration rate on the proportional vegetation biomass B/K, and at $B = 0$ (**b**), showing the dependence of the infiltration rate on the proportional crust biomass C/K. The infiltration contrast between bare and vegetated soil (because of the physical crust) is quantified by f_B, and the contrast between bare and crusted soil is quantified by f_C, where $0 \leq \{f_B, f_C\} \leq 1$. *Low* values of f_B or f_C represent high infiltration contrasts

to a nonlinear diffusion term in Eq. (1d) proportional to $\nabla^2 H^2$ (GILAD *et al.* 2007). Here we choose the value $\beta = 0$, which leads to a linear diffusion term, $D_H \nabla^2 H$, and simplifies numerical studies of the model's equations. Linear diffusion does not capture the compact nature of overland water flow, but we verified that our results do not depend crucially on this detail. This is in accordance with VAN DER STELT *et al.* (2013), who observed that nonlinear water diffusion does not have a crucial qualitative effect on the results of a similar model of patterned vegetation. Assuming unsaturated soil, water transport below ground level is also considered to be a linear diffusion process ($D_W \nabla^2 W$). Fast soil-water transport in comparison with vegetation spread, with strong uptake, constitute another type of pattern-forming feedback, which can induce vegetation patterns by a Turing instability (KINAST *et al.* 2014).

The remaining factors affecting water dynamics are rainfall, represented by the precipitation rate P, and evaporation of soil water at a rate $N(1 - RB/K_B)$, which takes into account reduced evaporation by shading. The numerical values we use for all model parameters are given in Table 1.

It proves beneficial to study the model equations using non-dimensional variables and parameters, which enables us to eliminate redundant parameters. The non-dimensional quantities we use are defined in Table 2. The non-dimensional model equations read:

$$b_t = g_b b(1 - b) - b - \frac{\varphi_b bc}{(b + b_0)^m} + \nabla^2 b \quad (3a)$$

$$c_t = g_c c(1 - c) - \mu c - \varphi_c cb + \delta_c \nabla^2 c \quad (3b)$$

$$w_t = \mathcal{I}h - v(1 - rb)w - g_w w + \delta_w \nabla^2 w \quad (3c)$$

$$h_t = p - \mathcal{I}h - g_h h + \delta_h \nabla^2 h \quad (3d)$$

where:

$$g_b = vw(1 + \eta b)^2 \quad (4a)$$

$$g_c = v(\lambda_{cw} w + \lambda_{ch} h) \quad (4b)$$

$$g_w = \gamma_b b(1 + \eta b)^2 + \gamma_{cw} c \quad (4c)$$

$$g_h = \gamma_{ch} c \quad (4d)$$

$$\mathcal{I} = \alpha \left(\frac{f_c c + q_c}{c + q_c} \right) \left(\frac{b + q_b f_b}{b + q_b} \right) \quad (4e)$$

3. Vegetation-Crust States Along the Rainfall Gradient

Earlier vegetation models that capture overland water flow (RIETKERK *et al.* 2002; GILAD *et al.* 2004, 2007), predict five basic vegetation states along the rainfall gradient; uniform vegetation, gap patterns, stripe or labyrinthine patterns, spot patterns, and bare soil. The models further predict bistability ranges

Table 1

Model parameters: their symbols, descriptions, units, and values used in this study

Parameters	Description	Units	Value
Λ_B	Vegetation growth rate	$(kg/m^2)^{-1}\ year^{-1}$	0.032
E	Root augmentation per unit biomass of vegetation	$(kg/m^2)^{-1}$	1.5
K_B	Maximum standing biomass of vegetation (carrying capacity)	kg/m^2	1
M_B	Vegetation mortality rate	$year^{-1}$	1.2
ϕ_B	Vegetation suppression by crust	$year^{-1}$	1
B_0	Vegetation biomass reference value beyond which the suppression by crust approaches its minimum	kg/m^2	0.05
m	Steepness of suppression of competition term by vegetation	–	1
Λ_{CW}	Crust growth rate as a result of uptake of soil water	$(kg/m^2)^{-1}\ year^{-1}$	0.035
Λ_{CH}	Crust growth rate as a result of uptake of surface water	$(kg/m^2)^{-1}\ year^{-1}$	0.01
K_C	Maximum crust biomass (carrying capacity)	kg/m^2	0.003
M_C	Crust mortality rate	$year^{-1}$	0.2
ϕ_C	Crust suppression by vegetation	$year^{-1}$	20
A	Maximum infiltration rate in uncrusted soil	$year^{-1}$	10
Q_B	Vegetation biomass reference value beyond which infiltration rate under a vegetation patch approaches its maximum	kg/m^2	0.05
Q_C	Crust biomass reference value beyond which infiltration rate under a crust patch approaches its minimum	kg/m^2	0.0006
f_B	Infiltration contrast between bare soil and vegetated soil	–	1
f_C	Infiltration contrast between bare soil and crusted soil	–	0.1
N	Soil water evaporation rate	$year^{-1}$	4
R	Evaporation reduction due to shading	–	0.95
P	Mean annual precipitation rate	$kg/m^2 year^{-1}$	(0, 500)
Γ_B	Soil water consumption rate per unit vegetation biomass	$(kg/m^2)^{-1} year^{-1}$	30
$\Gamma_C W$	Soil water consumption rate per unit crust biomass	$(kg/m^2)^{-1} year^{-1}$	0.1
$\Gamma_C H$	Surface water consumption rate per unit crust biomass	$(kg/m^2)^{-1} year^{-1}$	0.02
D_B	Vegetation seed dispersal coefficient	$m^2/year$	6.25×10^{-4}
D_C	Crust spores dispersal coefficient	$m^2/year$	6.25×10^{-3}
D_W	Transport coefficient for soil water	$m^2/year$	6.25×10^{-2}
D_H	Bottom friction coefficient between surface water and ground surface	$m^2/year$	5

The values of the parameters appearing in the equations for vegetation biomass (B), the soil water (W), and the surface water (H) are taken from GILAD *et al.* (2007). The values for the crust (C) equation are based on BELNAP and LANGE (2001), GARCIA-PICHEL *et al.* (2003), PRASSE and BORNKAMM (2000), ZAADY and SHACHAK (1994), and BOEKEN and ORENSTEIN (2001). The units of mean annual precipitation rate (P) are equivalent to mm/year

between any pair of consecutive vegetation states, which result in a wide variety of non-periodic patterned states (MERON 2012). These predictions agree well with observations (DEBLAUWE *et al.* 2008), and therefore provide an important test for the new vegetation-crust model proposed here. To discover what states along the rainfall gradient the model equations (Eqs. 3a–3d) predict, we studied stationary solutions in one spatial dimension, using a numerical continuation method and linear stability analysis, and complemented this analysis with direct numerical integration of the model's equations in one and two spatial dimensions, as described below. In both cases we assumed that the development of an infiltration contrast between vegetated and unvegetated areas is

because of biogenic crusts only, by choosing $f_b = 1, f_c = 0.1$.

Figure 3 shows bifurcation diagrams for stationary solutions of Eqs. (3a–3d) in one spatial dimension, and displays the maximum values of the vegetation biomass (Fig. 3a) and of the crust biomass (Fig. 3b) as functions of precipitation rate.

Four types of stable uniform solutions can be distinguished. The first is a constant solution that describes bare-soil devoid of vegetation and crust, $\mathcal{B}$ ($b = 0, c = 0$). It exists for all precipitation values but is stable only for $0 < p < p_0$. At $p = p_0$ the bare-soil solution loses stability to another constant solution devoid of vegetation that describes uniform crust, $\mathcal{C}$ ($b = 0, c \neq 0$). This solution is stable up to $p = p_2$

Table 2

Relations between non-dimensional variables and parameters and dimensional variables and parameters appearing in the dimensional form of the model Eqs. (1a–1d)

Quantity	Scaling	Quantity	Scaling
b	B/K_B	α	A/M_B
c	C/K_C	q_c	Q_C/K_C
w	$\Lambda_B W/N$	q_b	Q_B/K_B
h	$\Lambda_B H/N$	f_c	f_C
v	N/M_B	f_b	f_B
η	EK_B	r	R
b_0	B_0/K_B	γ_b	$\Gamma_B K_B/M_B$
φ_b	$\phi_B K_C/M_B K_B^m$	γ_{cw}	$\Gamma_{CW} K_C/M_B$
φ_c	$\phi_C K_B/M_B$	γ_{ch}	$\Gamma_{CH} K_C/M_B$
λ_{CW}	Λ_{CW}/Λ_B	p	$\Lambda_B P/N M_B$
λ_{CH}	Λ_{CH}/Λ_B	δ_c	D_C/D_B
μ	M_C/M_B	δ_w	D_W/D_B
t	$M_B T$	δ_h	D_H/D_B
$\mathbf{x}$	$\mathbf{X}\sqrt{M_B/D_B}$		

where it bifurcates to a constant mixed vegetation–crust solution, $\mathcal{M}$ ($b \neq 0, c \neq 0$). The mixed solution branch terminates (as a physical solution) at $p = p_3$ on a constant solution branch that describes uniform vegetation devoid of crust, $\mathcal{V}$ ($b \neq 0, c = 0$). The mixed solution $\mathcal{M}$, however, is unstable to the growth of nonuniform perturbations, which leads to a nonuniform solution branch, $\mathcal{P}$ ($b \neq 0, c \neq 0$), describing a periodic mixed pattern of vegetation and crust (Fig.

4). The periodic solution branch $\mathcal{P}$ emanates from the constant solution $\mathcal{M}$ very close to $p = p_2$ and returns to $\mathcal{M}$ very close to $p = p_3$. Because both bifurcations are subcritical, the stable part of the periodic solution branch $\mathcal{P}$ occupies a wider precipitation range bounded by two fold bifurcations at p_1 and at p_4. This range includes a bistability subrange, $p_1 < p < p_2$, with the uniform crust solution $\mathcal{C}$, and a bistability subrange, $p_3 < p < p_4$, with the uniform vegetation solution $\mathcal{V}$.

Altogether the following sequence of stable states has been found along the rainfall gradient in one spatial dimension: uniform vegetation $\mathcal{V}$ ($p > p_3$), periodic spatial pattern $\mathcal{P}$ ($p_1 < p < p_4$), uniform crust $\mathcal{C}$ ($p_0 < p < p_2$) and bare soil $\mathcal{B}$ ($0 < p < p_0$). An additional finding is the existence of two bistability ranges:

1 uniform vegetation and periodic patterns; and
2 uniform crust and periodic patterns.

These results are consistent with those obtained in the earlier models when associating the bare-soil solution $\mathcal{B}$ and the crust solution $\mathcal{C}$ with the "bare-soil state" of the earlier models.

Figure 4 shows typical spatial profiles of periodic solutions, obtained by numerical integration of Eqs. (3a–3d) in one spatial dimension, at two precipitation values, $p = 1$ and $p = 2$, located near the low-precipitation and high-precipitation edges of the periodic

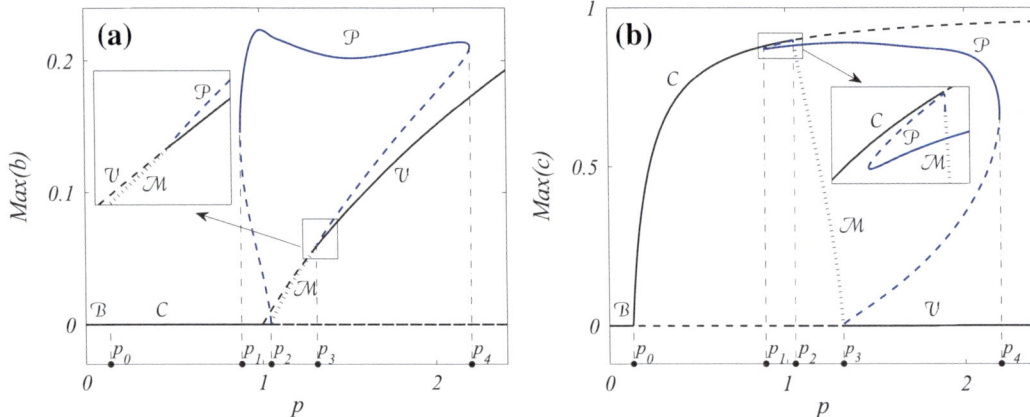

Figure 3

Bifurcation diagram for stationary solutions of the vegetation-crust model. Shown are the maximum values of the vegetation biomass b (**a**) and of the crust biomass c (**b**) as functions of precipitation rate p. *Solid lines* represent stable solutions and *dashed (dotted) lines* represent unstable solutions to uniform (nonuniform) perturbations. Five distinct solutions are denoted: bare soil $\mathcal{B}$, uniform crust $\mathcal{C}$, uniform mixture of vegetation and crust $\mathcal{M}$, uniform vegetation $\mathcal{V}$, and periodic vegetation-crust pattern $\mathcal{P}$. The latter emanates from $\mathcal{M}$ and returns to $\mathcal{M}$ very close to the bifurcation points where $\mathcal{M}$ connects to $\mathcal{C}$ ($p = p_2$) and to $\mathcal{V}$ ($p = p_3$). The *insets* show magnifications of the neighborhoods of these bifurcation points. Not shown in the bifurcation diagrams are negative solutions, which represent unphysical states. Parameter values are as given in Table 1

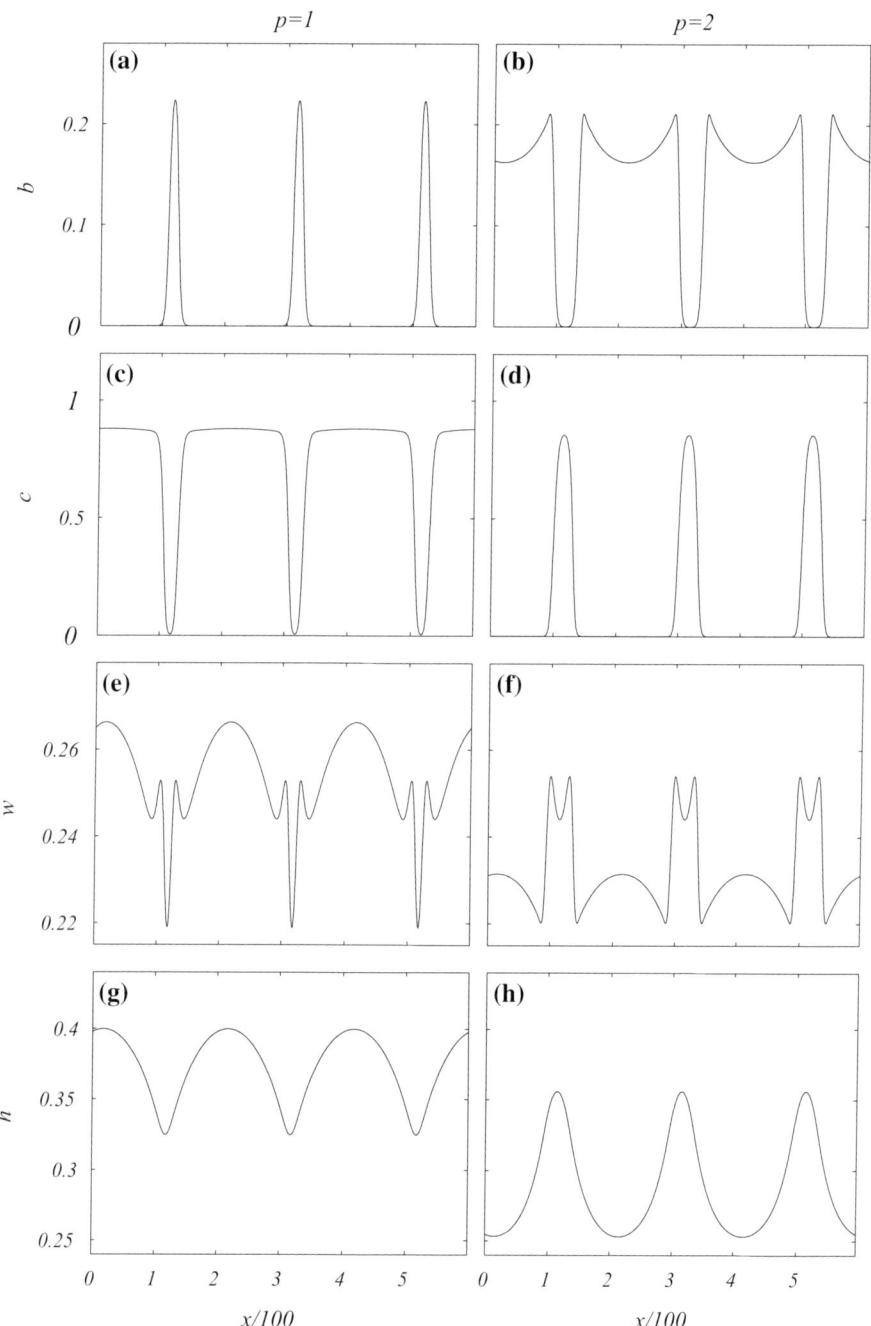

Figure 4
Spatially periodic solutions of Eqs. (3a–3d) in one spatial dimension. Shown are the spatial profiles of all dynamic variables at low precipitation (**a, c, e, g**) and at high precipitation (**b, d, f, h**). Parameter values are as given in Table 1

solution branch depicted in Fig. 3. At $p = 1$ the solution appears as vegetation spots surrounded by crusted soil (Fig. 4a, c), whereas at $p = 2$ the solution appears as crust gaps in vegetated soil (Fig. 4b, d). The remaining parts of this figure show the associated spatial profiles of the soil–water and surface–water variables. As Fig. 4e, f show, the minima of soil water content (w) coincide with maxima of vegetation biomass (b) because of the high uptake rate g_w. Figure 4g, h show that the maxima of surface-water height (h) coincide with the maxima of crust biomass (c). This can be understood from Eq. (4e), because infiltration of surface water is a monotonically decreasing function of crust biomass, as shown by Fig. 2b.

In two spatial dimensions, previous models predicted three basic types of periodic solutions, representing hexagonal spot patterns at relatively low precipitation, stripes (labyrinthine) patterns at intermediate precipitation, and hexagonal gap patterns at relatively high precipitation. These three patterned vegetation states are also found by numerical integration of Eqs. (3a–3d), as shown by Fig. 5. Shown are the time evolution of the same initial conditions at increasing precipitation values and the asymptotic approach to hexagonal-spot, stripe, and hexagonal-gap patterns.

4. The Significance of Modeling Crust Dynamics

The vegetation-crust model (Eqs. 3a–3d) not only reproduces the main qualitative behaviors found in previous models, but also provides new insights. Figure 6 shows a comparison of the bifurcation diagram presented in Fig. 3a with the corresponding diagram of a reduced model, obtained by setting the growth rate of the crust variable to zero ($g_c = 0$), choosing $f_b = 0.1$, and leaving all other parameters unchanged. The right hand side of the crust equation (Eq. 3b) includes then the negative terms only, which drive the crust biomass to zero and reduce the four-variable vegetation-crust model (Eqs. 3a–3d) to a three-variable vegetation model. The latter coincides with a simplified version of the Gilad et al. model (2007) studied earlier (ZELNIK et al. 2013).

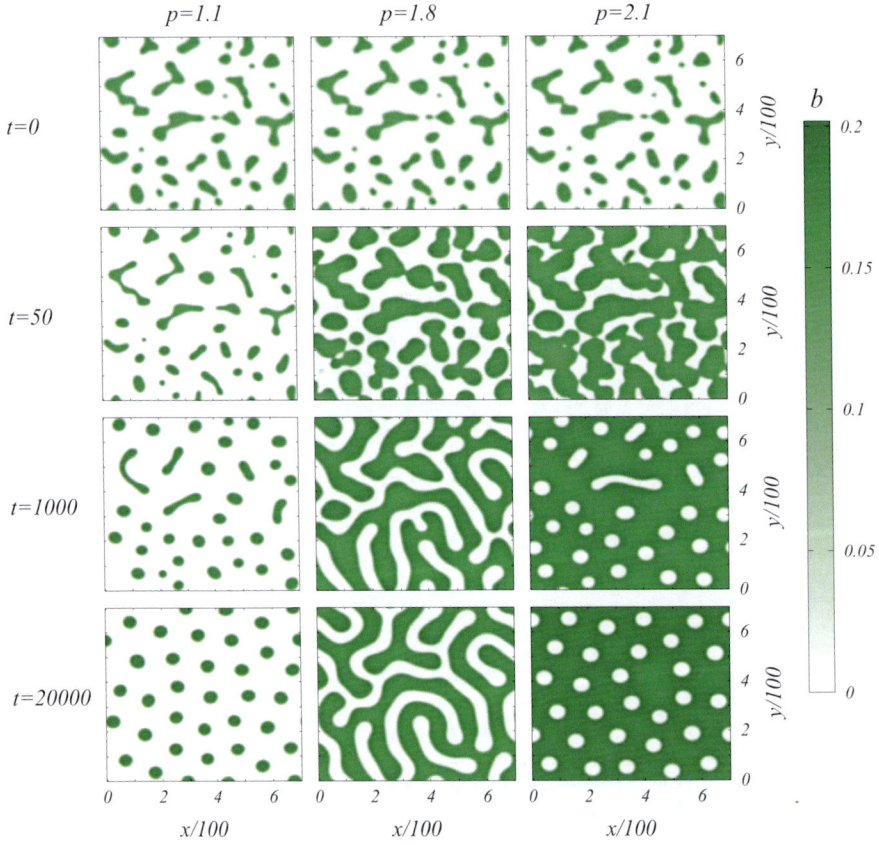

Figure 5

Three basic types of asymptotic vegetation–crust patterns. Shown are snapshots of three simulations of Eqs. (3a–3d) at increasing precipitation values, starting from the same initial condition. At $p = 1.1$ the dynamics converge to a spot pattern, at $p = 1.8$ to stripe patterns, and at $p = 2.1$ to a gap pattern. *Darker shades* represent higher vegetation biomass (b). The crust biomass forms an anti-phase pattern, occupying the *light-shade areas* that are devoid of vegetation. Parameter values are as given in Table 1

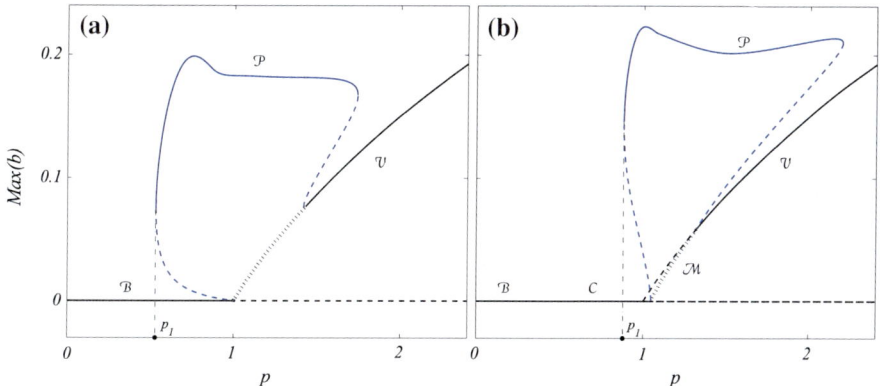

Figure 6

Comparison between bifurcation diagrams with (**b**) and without (**a**) a dynamic crust, both showing the vegetation biomass (*b*) as a function of precipitation (*p*). *Solid* (*dashed*) *lines* represent stable (unstable) solutions. Addition of a dynamic crust shifts the fold bifurcation point, $p = p_1$, to higher precipitation and biomass values. Along with these shifts the unstable branch of the periodic vegetation solution $\mathcal{P}$ shifts upward. Parameter values are as given in Table 1, except for $f_b = 0.1, f_c = 1, \lambda_{CW} = \lambda_{CH} = 0$ in (**a**)

The bifurcation diagram of the vegetation-crust model (Fig. 6b) differs from that of the vegetation model (Fig. 6a) in several structural respects:

1 it contains the additional solution branches $\mathcal{C}$ and $\mathcal{M}$;

2 the periodic solution $\mathcal{P}$ is a mixed vegetation-crust solution; and

3 the periodic solution emanates from and returns to the (unstable) uniform mixed state solution branch $\mathcal{M}$ (rather than the uniform vegetation solution).

Note that the spatial profiles of the vegetation (*b*) and of the crust (*c*) along the periodic solution branch are anti-phase, as shown by Fig. 4; that is, maxima of *b* correspond to minima of *c* and vice versa.

More significant from an ecological perspective are two quantitative differences related to the fold bifurcation at $p = p_1$ at which the stable periodic state $\mathcal{P}$ appears. Including crust dynamics shifts this bifurcation point to higher precipitation and biomass values. The shift to a higher precipitation rate increases the range of the unproductive state, $\mathcal{B}$ or $\mathcal{C}$, at the expense of the productive vegetation-pattern state $\mathcal{P}$, whereas the shift to higher biomass values increases the attraction basin of the unproductive state within its bistability range with the productive vegetation-pattern state. These results suggest that incorporating crust dynamics in vegetation models can be highly significant for studying state transitions involving vegetation loss or vegetation recovery. The inclusion of dynamic crust also shifts $p = p_4$ to a higher value, implying the persistence of vegetation gap patterns at higher precipitation rates and a wider bistability range of vegetation gap patterns and uniform vegetation.

5. Conclusions

A new model for patchy water-limited landscapes has been introduced. Unlike earlier models, in which soil-crust effects are considered in a parametric way, through a biomass-dependent infiltration rate, the new model captures the actual dynamics of biogenic soil crusts and their mutual interactions with vegetation growth. Using the model we mapped the vegetation-crust states along the precipitation axis and found them to be consistent extensions of the results of previous models. We further emphasized significant differences between the new and earlier models in the bistability range of productive and unproductive states; taking into account crust dynamics shifts the fold-bifurcation point at which stable vegetation patterns appear to higher precipitation and biomass values.

These differences can be attributed in part to the competition term in Eq. (3a), which models the suppression effects that biogenic crusts exert on vegetation growth by slowing seed germination. The suppression effect and its fadeout as vegetation grows, depend on the parameter ϕ_B, *m* and B_0. Further studies are needed to clarify how these parameters affect the stable and unstable branches of the periodic solution $\mathcal{P}$.

The vegetation-crust model (Eqs. 3a–3d) may shed new light on the effects of biogenic crusts on the response of dryland ecosystems to rainfall variability, and may improve understanding of desertification processes, such as that shown in Fig. 1, and of means to facilitate recovery to the original state. To this end model studies with periodic or stochastic precipitation to simulate successive droughts should be conducted.

Acknowledgments

We wish to thank Golan Bel, Jost von-Hardenberg, Eli Zaady and Yuval Zelnik for helpful discussions. The research leading to these results has received funding from the Israel Science Foundation (Grant Numbers 75/12 and 305/13).

REFERENCES

A. Ana I. Borthagaray, M. A. Fuentes, and P. A. Marquet. *Vegetation pattern formation in a fog-dependent ecosystem.* Journal of Theoretical Biology, 265:18–26, 2010.

M. Bär, J. Hardenberg, E. Meron, and A. Provenzale. *Modelling the survival of bacteria in drylands: the advantage of being dormant.* Proceedings of the Royal Society of London. Series B: Biological Sciences, 269(1494):937–942, 2002.

J. Belnap and O. L. Lange. Biological Soil Crusts: Structure, Function, and Management. Springer, 2001.

B. Boeken and D. Orenstein. *The effect of plant litter on ecosystem properties in a Mediterranean semi-arid shrubland.* J. Veg. Sci., 12:825–832, 2001.

F. Borgogno, P. D'Odorico, F. Laio, and L. Ridolfi. *Mathematical models of vegetation pattern formation in ecohydrology.* Reviews of Geophysics, 47:RG1005, 2009.

S.E. Campbell. *Soil stabilization by a prokaryotic desert crust: implications for precambrian land biota.* Origins of Life, 9:335–348, 1979.

V Deblauwe, N Barbier, P Couteron, Olivier Lejeune, and Jan Bogaert. *The global biogeography of semi-arid periodic vegetation patterns.* Glob. Ecol. Biogeogr., 17:715–723, 2008.

Avraham Dody, Roni Hakmon, Boaz Asaf, and Eli Zaady. *Indices to monitor biological soil crust growth rate-lab and field experiments.* Natural Science, 3(6), 2011.

Thomas Dunne, Weihua Zhang, and Brian F. Aubry. *Effects of rainfall, vegetation, and microtopography on infiltration and runoff.* Water Resources Research, 27(9):2271–2285, 1991.

Sjors van der Stelt, Arjen Doelman, Geertje Hek, and Jens D.M. Rademacher. *Rise and fall of periodic patterns for a generalized klausmeier-gray-scott model.* Journal of Nonlinear Science, 23(1):39–95, 2013.

D. J. Eldridge, E. Zaady, and Shachak M. *Infiltration through three contrasting biological soil crusts in patterned landscapes in the negev, israel.* J Stat Phys, 148:723–739, 2012.

Ferran Garcia-Pichel, Jayne Belnap, Susanne Neuer, and Ferdinand Schanz. *Estimates of global cyanobacterial biomass and its distribution.* Algological Studies, 109(1):213–227, 2003.

E. Gilad, J. von Hardenberg, A. Provenzale, M. Shachak, and E. Meron. *Ecosystem Engineers: From Pattern Formation to Habitat Creation.* Phys. Rev. Lett., 93(9):098105, Aug 2004.

E. Gilad, J. von Hardenberg, A. Provenzale, M. Shachak, and E. Meron. *A mathematical model of plants as ecosystem engineers.* Journal of Theoretical Biology, 244(4):680–691, 2007.

Shai Kinast, Yuval R. Zelnik, Golan Bel, and Ehud Meron. *Interplay between Turing Mechanisms can Increase Pattern Diversity.* Phys. Rev. Lett., 112:078701, Feb 2014.

Shai Kinast, Ehud Meron, Hezi Yizhaq, and Yosef Ashkenazy. *Biogenic crust dynamics on sand dunes.* Phys. Rev. E, 87:020701, Feb 2013.

S. Manzoni, S.M. Schaeffer, G. Katul, A. Porporato, and J.P. Schimel. *A theoretical analysis of microbial eco-physiological and diffusion limitations to carbon cycling in drying soils.* Soil Biology and Biochemistry, 73(0):69–83, 2014.

E. Meron. *Pattern-formation approach to modelling spatially extended ecosystems.* Ecological Modelling, 234:70–82, 2012.

E. Meron. *Modeling dryland landscapes.* Math. Model. Nat. Phenom., 6:163–187, 2011.

J. Puigdefábregas. *The role of vegetation patterns in structuring runoff and sediment fluxes in drylands.* Earth Surface Processes and Landforms, 30:133–147, 2003.

R. Prasse and R. Bornkamm. *Effect of microbiotic soil surface crusts on emergence of vascular plants.* Plant Ecol., 150:65–75, 2000.

M. Rietkerk and J. van de Koppel. *Regular pattern formation in real ecosystems.* Trends in Ecology and evolution, 23(3):169–175, 2008.

M. Rietkerk, M.C. Boerlijst, F. van Langevelde, R. HilleRis-Lambers, J. van de Koppel, L. Kumar, H.H.T. Prins, and A.M. De Roos. *Self-organization of vegetation in arid ecosystems.* Am, Nat., 160:524–530, 2002.

M. Shachak and G. M. Lovett. *Atmospheric deposition to a desert ecosystem and its implication for management.* Ecological Applications, 8:455–463, 1998.

M. Shachak. *Ecological systems in northern negev (in hebrew).* Ecology and Environment, 1:18–29, 2011.

Shai Sela, Tal Svoray, and Shmuel Assouline. *Soil water content variability at the hillslope scale: Impact of surface sealing.* Water Resources Research, 48(3), 2012.

I. Stavi, H. Lavee, E. D. Ungar, and P. Sarah. *Ecogeomorphic feedbacks in semiarid rangelands: A review.* Pedosphere, 19(2):217–229, 2009.

S. Thompson, G. Katul, and S. M. McMahon. *Role of biomass spread in vegetation pattern formation within arid ecosystems.* Water Resour. Res., 44:W10421, 2008.

C Valentin, J.M Herbés, and J Poesen. *Soil and water components of banded vegetation patterns.* CATENA, 37:1–24, 1999.

E. Verrecchia, A. Yair, G. J. Kidron, and K. Verrecchia. *Physical properties of the psammophile cryptogamic crust and their consequences to the water regime of sandy softs, north-western negev desert, israel.* J Arid Environments, 29:427–437, 1995.

N. E. West. *Structure and Function of Microphytic Soil Crusts in Wildland Ecosystems of Arid to Semi-arid Regions.* Advances in Ecological Research, *20*:179–223, 1990.

A. Yair, R. Almog, and M. Veste. *Differential hydrological response of biological topsoil crusts along a rainfall gradient in a sandy arid area: Northern negev desert, israel.* CATENA, 87(3):326–333, 2011.

E. Zaady and M. Shachak. *Microphytic soil crust and ecosystem leakage in the Negev Desert.* Am. J. Bot., *81*:109, 1994.

Y. R. Zelnik, S. Kinast, H. Yizhaq, G. Bel, and E. Meron. *Regime shifts in models of dryland vegetation.* Philosophical Transactions R. Soc. A, *371*:20120358, 2013.

(Received May 4, 2014, revised September 25, 2014, accepted October 10, 2014, Published online November 27, 2014)

Pure Appl. Geophys. 173 (2016), 995–1009
© 2014 Springer Basel
DOI 10.1007/s00024-014-0928-2

Pure and Applied Geophysics

Morphological Functions with Parallel Sets for the Pore Space of X-ray CT Images of Soil Columns

F. San José Martínez,[1] F. J. Muñoz Ortega,[1] F. J. Caniego Monreal,[1] and F. Peregrina[2]

Abstract—During the last few decades, new imaging techniques like X-ray computed tomography have made available rich and detailed information of the spatial arrangement of soil constituents, usually referred to as soil structure. Mathematical morphology provides a plethora of mathematical techniques to analyze and parameterize the geometry of soil structure. They provide a guide to design the process from image analysis to the generation of synthetic models of soil structure in order to investigate key features of flow and transport phenomena in soil. In this work, we explore the ability of morphological functions built over Minkowski functionals with parallel sets of the pore space to characterize and quantify pore space geometry of columns of intact soil. These morphological functions seem to discriminate the effects on soil pore space geometry of contrasting management practices in a Mediterranean vineyard, and they provide the first step toward identifying the statistical significance of the observed differences.

1. Introduction

One of the most pervasive features of natural soils is its structure as expressed by the size, shape, and arrangement of the soil particles and voids, including both the primary particles to form compound particles (i.e. soil aggregates) and the compound particles themselves (BREWER, 1964). Soil structure plays a major role in soil functioning, including its contribution to accumulation and protection of soil organic matter, to optimization of soil water and air regimes, and to storage and availability of plant nutrients (BOSSUYT et al., 2002; VON LÜTZOW et al., 2006). Performance of many of these functions strongly depends on pore space geometry. For example, it has been shown that gradients of a number of soil characteristics exist inside soil. Among them are gradients in oxygen concentrations of the soil air (SEXSTONE et al., 1985), gradients in concentrations of a variety of elements, including Ca, Mg, K, Na, Mn, K, Al, and Fe (SANTOS et al., 1997; JASINSKA et al., 2006), and in organic matter compositions (ELLERBROCK and GERKE, 2004; URBANEK et al., 2007). These differences in turn influence soil structure that is of particular importance for processes such as soil carbon sequestration (SIX et al., 2000; DENEF et al., 2001; CHENU and PLANTE, 2006).

In this work, we propose a quantitative description of geometrical characteristics of soil pore space as volume, surface, shape, and connectivity within the unified framework that provides mathematical morphology (SERRA, 1982). Mathematical morphology includes a plethora of mathematical techniques to analyze and parameterize the geometry of different features of soil structure. These techniques belong to well-established mathematical fields such as integral geometry (SANTALÓ, 1976), stochastic geometry (MATHERON, 1975), or digital topology and geometry (KLETTE and ROSENFELD 2004). They make available a sound mathematical background that guides the process from image acquisition and analysis to the generation of synthetic models of soil structure (ARNS et al., 2004) to investigate key features of flow and transport phenomena in soil (LEHMANN, 2005; MECKE and ARNS, 2005).

X-ray computed tomography (CT) provides a direct and non-destructive procedure to use three-dimensional information to quantify geometrical features of soil pore space (Peyton et al. 1994; Perret

---
[1] Department of Applied Mathematics, E.T.S.I. Agrónomos, Universidad Politécnica de Madrid, 28040 Madrid, Spain. E-mail: fernando.sanjose@upm.es; f.j.munoz.ortega@upm.es; j.caniego@upm.es
[2] Servicio de Investigación y Desarrollo Tecnológico Agro-alimentario, Instituto de Ciencias de la Vid y el Vino, CSIC-Universidad de La Rioja-Gobierno de la Rioja, La Rioja, 26076 Logroño, Spain. E-mail: fernandoperegrina@hotmail.com

et al., 1999; Pierret *et al.*, 2002; Mees *et al.*, 2003; LEHMANN *et al.*2006; SAN JOSÉ MARTÍNEZ *et al.*, 2010; ZHOU *et al.*, 2013). During the last few decades, mathematical morphology has been successfully used to analyze different characteristics of the rich three-dimensional geometrical information gained through X-ray CT (Banhart, 2008). Among the tools of mathematical morphology, Minkowski functionals (ARNS *et al.*, 2002; LEHMANN *et al.*, 2006), which belong to the mathematical theory of integral geometry (SANTALÓ, 1976), are particularly worthy of consideration since they provide computationally efficient means to measure four fundamental geometrical properties of three-dimensional geometrical objects such as soil pore space. These properties are the volume, the boundary surface, the integral mean curvature, and the connectivity of the object of interest. Hadwiger's theorem (SANTALÓ, 1976) states that any functional that assigns a number to any three-dimensional object and meets some self-evident and natural geometrical restrictions is a linear combination of these Minkowski functionals. Then, these functionals are powerful tools to describe quantitatively 3D geometry. MECKE (1998) and ROTH *et al.* (2005) made use of Minkowski functions based on threshold variation of Minkowski functionals to characterize two-dimensional porous structures. San José Martínez *et al.* (2013) used the same methodology with the pore space of columns of intact soil. Also, two-dimensional porous structures were investigated by MECKE (2002) and VOGEL *et al.* (2005) with Minkowski functions based on dilations and erosions. ARNS *et al.* (2002, 2004) considered the evolution of Minkowski functionals with dilations and erosions to characterize 3D images of Fontainebleau sandstone. Renard and Allard (San José Martínez 2013) used the Euler number as a function of erosion/dilation to explore the role of connectivity for the characterization of heterogeneous aquifers with 2D models.

In this work, we introduced two morphological transformations, namely erosion and dilations, and morphological functions built over Minkowski functionals. These morphological functions take account of the evolution of Minkowski functionals as dilations and erosions are performed on the object of interest, the pore space of soil columns imaged

with X-ray CT. In this way, different geometrical objects are provided that can be seen as parallel sets of the pore space. Then, the Minkowski functionals of the new objects are computed and represented as a function of the radius of the ball of the structuring element of the corresponding dilation/erosion. We observed that morphological functions of dilation/erosion seem to discriminate between two pore structures in a Mediterranean vineyard subjected to contrasting management practices: conventional tillage and permanent cover crop of resident vegetation.

2. Morphology of Pore Space Volume

Morphological analysis mimics other scientific procedures, and in some instances it can be seen as a two-step process. To illustrate this point, let us consider, for instance, the procedure to determine particle size distributions by sieving. This technique first generates a series of subsets of primary mineral particles, the oversize sets corresponding to each sieve size; then, these oversize sets are weighted. In morphological analysis, first, geometrical transformations are applied to the object of interest in an image, and then measurements are carried out. When the granulometry of an image of grains of different sizes shall be determined, successive morphological operations are performed on the image. These operations consist on the elimination of grains smaller than a certain size with a suitable morphological transformation (Fig. 1). Each one of these operations is followed by the measurement of the area for 2D images or the volume for 3D images, of the grains left (SERRA, 1982). Figure 1 illustrates this procedure in a CT image of a packing of sand particles. Now we are going to describe the basic morphological operations, i.e. dilations and erosions. Finally, the notions of Minkowski functionals and morphological functions will be presented.

3. Morphological Operations

Grains or pore space in a 3D CT image of soil will be idealized as sets of points in three-dimensional

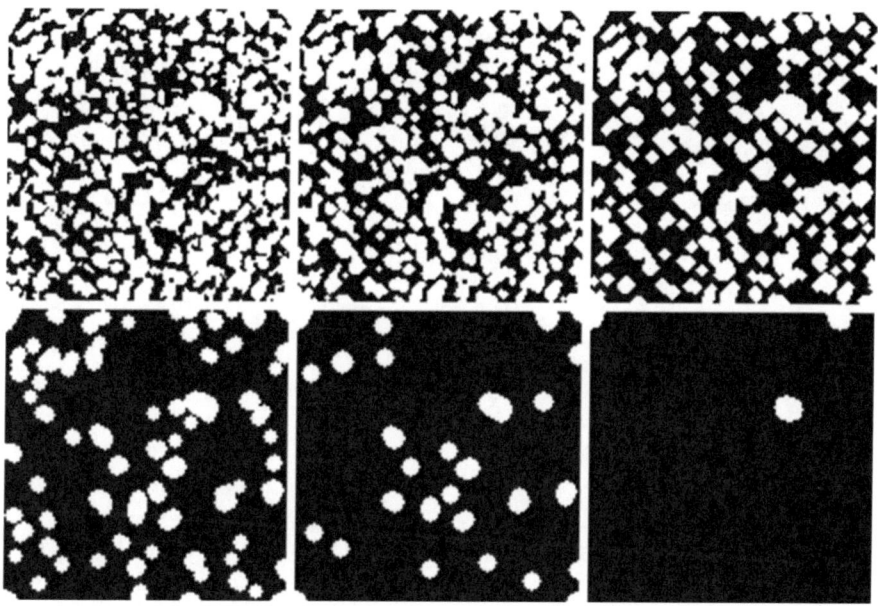

Figure 1
Granulometric analysis of a section of a CT image of 15.4 mm side of a packing of sand particles by successive morphological operations

space. These types of geometrical objects will be the mathematical objects of interest. In this work we will focus on soil pore space as the geometrical object of interest. Mathematically, an object is a closed and bounded set. A ball is a closed set if it contains the points of the spherical surface that defines its boundary. And it is a bounded set because it is contained in a sphere of finite radius. Dilation of an object expands it. This new object can be thought of as being the union of all balls with a given radius r centered at points of the original object. If the original object is a ball of radius r_0, the dilated object by balls of radius r will be a new ball of radius $r_0 + r$.

We consider a generic object K and a ball B of radius 1 whose center is located at the origin of coordinates. Both K and B are objects, closed and bounded sets, but K is the object of interest or simply an object that we scrutinize with the object B that is called the structuring element. A ball of radius r centered at the origin, rB, is obtained by multiplying the coordinates of the points of B by r. In a ball of radius 1, centered at point x, B_x, is obtained adding x to every point of B. Scalar multiplication by a positive number r produces an expansion with scaling factor r when $r > 1$, and a contraction with scaling factor r when $r < 1$. Addition with a vector x

produces a translation in the direction of the vector x at a distance equal to the "length" of this vector, its modulus. Then, we have the following mathematical expressions that define the sets rB and B_x (OSHER and MÜCKLICH 2000):

$$rB = \{ry : y \in B\} \quad \text{and} \quad B_x = \{y + x : y \in B\} \tag{1}$$

That is to say, rB is the set of points ry when y belongs to B, and B_x is the set of points $y + x$ when y belongs to B. In these expressions, ry stands for the scalar multiplication of the scalar r and the vector y, and $y + x$ represents the sum of two vectors, y and x. Thus, the dilation (Fig. 2) of the object K by balls of radius r, that is the union of all balls rB_x of radius r centered at points x of K, will be another object K_r defined as

$$K_r = \bigcup_{x \in K} rB_x. \tag{2}$$

The set K_r is also called the parallel body of K at a distance r or r-parallel body to K. This is the set of all points within a distance smaller than r from the object K. In this work, the structuring element will be a ball centered at the origin. Then, the dilation of an object by a ball of radius r is equivalent to the r-

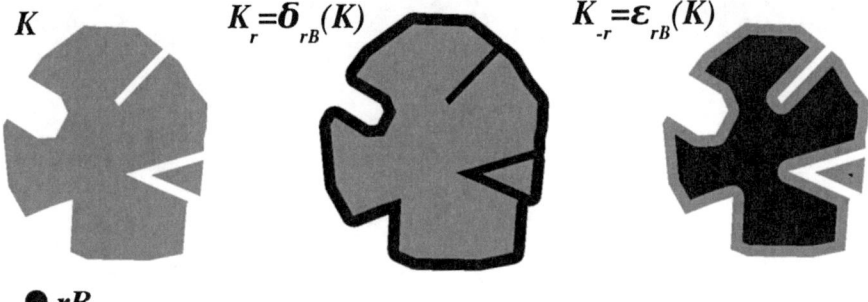

K $K_r = \delta_{rB}(K)$ $K_{-r} = \varepsilon_{rB}(K)$

● rB

Figure 2
Effect of dilation $K_r = \delta_{rB}(K)$ (grey plus black) and erosion $K_{-r} = \varepsilon_{rB}(K)$ (black) of object K by the structuring element rB

parallel body to K. Roughly speaking, it is like a "skin" of thickness r is added to K.

We will analyze binary (black and white) images of soil. They contain two complementary phases: the phase of voids (pores) and the phase of soil matrix (mineral particles). As we said previously, in this study, the pore space is the object of interest and it will be white, while the mineral matrix will form the background and it will be black, as is customary in image analysis. Then, the erosion of one phase is equivalent to the dilation of the complementary phase. Erosion of the pore space is dilation of the soil matrix, and erosion of the soil matrix is dilation of pore space. For an object K, the erosion by a ball of radius r is defined as (ARNS et al., 2002).

$$K_{-r} = \{x : rB_x \subset K\} \qquad (3)$$

Consequently, the erosion of an object K by a ball rB corresponds to the set of all positions of their centers within K where the structuring element rB fits completely into K (Fig. 2). Roughly speaking, it is like a "layer" of thickness r is removed from K. Therefore, we may generalize the notion of r-parallel body so that K_r will be a dilation for $r > 0$, and erosion for $r < 0$ and the original object K for $r = 0$ (ARNS et al., 2002).

4. Measurements: Minkowski Functionals

What is the area of a two-dimensional object or the volume of a three-dimensional one when the object is dilated? Let us consider a simple object like a square or a cube with edges of size a and a disk or a

ball of radius r as a structuring element. In the plane, the area of the dilated object K_r of a square K by a disk rB can easily be computed as (Fig. 3).

$$A(K_r) = A(\delta_{rB}(K)) = a^2 + 4a\,r + \pi\,r^2$$
$$= A(K) + L(K)r + A(B)r^2. \qquad (4)$$

In this expression, A stands for the area and L stands for the length of the perimeter of the square K. Here, B is the disk centered at the origin with radius 1. In the space, we get

$$V(K_r) = V(\delta_{rB}(K)) = a^3 + 6a^2\,r + 3\pi a r^2 + \frac{4}{3}\pi\,r^3$$
$$= V(K) + S(K)r + M(K)r^2 + V(B)r^3$$
$$\qquad (5)$$

Here, V stands for the volume, S for the area of the boundary, and M for the mean breadth multiplied by 2π (it can be shown that the mean breadth of a

Figure 3
Dilation of a square with a disk as structuring element

cube of edge a is $3a/2$ (SANTALÓ, 1976). Here, B is the ball centered at the origin with radius 1.

Now, let us consider a general convex object in d-dimensional linear space; then one has the Steiner formula (OSHER and MÜCKLICH, 2000).

$$V(K_r) = \sum_{i=0}^{d} \binom{d}{i} W_i^{(d)}(K). \tag{6}$$

In this expression, $W_i^{(d)}(K)$ are the Minkowski functionals. There are $d+1$ Minkowski functionals in dimension d.

Minkowski functionals are a complete set of geometrical features as established by Hadwiger's theorem (SANTALÓ, 1976). In simple terms, this theorem states that any functional that assigns a number to any object of interest and fulfills some very natural geometric restrictions is a linear combination of the Minkowski functionals with numbers as scalars of this linear combination.

There are three Minkowski functionals in the plane and four in space. In the plane (the two-dimensional linear space), one has

$$W_0^{(2)}(K) = A(K), \; W_1^{(2)}(K) = L(K) \quad \text{and}$$
$$W_2^{(2)}(K) = A(B)\chi(K). \tag{7}$$

In this expression, A stands for the area, L stands for the length of the perimeter of K, and $\chi(K)$ for its Euler-Poincaré characteristic. Here, B is the disk centered at the origin with radius 1. In space (the three-dimensional, linear space), one has

$$W_0^{(3)}(K) = V(K), \quad W_1^{(3)}(K) = (1/3)S(K),$$
$$W_2^{(3)}(K) = (1/3)M(K) \quad \text{and} \tag{8}$$
$$W_3^{(3)}(K) = V(B)\chi(K).$$

Here, B is the ball centered at the origin with radius one, V stands for the volume, S for the area of the boundary, and M for the mean breadth multiplied by 2π (it can be shown that the mean breadth of a cube of edge a is $3a/2$ (SANTALÓ, 1976). As before, $\chi(K)$ is the Euler-Poincaré characteristic of the spatial object K. See Appendix 2 for more details on interpretation of these functionals.

Another important feature of Minkowski functionals is that they are easy to compute (MICHIELSEN 2001). For computational purposes, points of geometrical objects are considered a voxel of a digital image (i.e. the elements of regular lattice). Taking into account the C-additivity property (see Appendix 1) and the fact that digital images are sets of cubes (or voxels), their computation reduces to the computation of the Minkowski functionals on cubes and their intersections (vertices, edges, and faces) (LIKOS et al., 1995).

5. Morphological Functions

Mathematical morphology offers a powerful description of objects in terms of functions. This technique is similar to the process that provides particle size distributions by morphological analysis of soil images (SERRA, 1982; SOILLE, 2002; VOGEL, 2002).

Consider a 3D binary image of soil where the void phase K is the object of interest. Let K_r be, as before, the dilation of K by balls of radius r when $r > 0$ and the erosion of K by balls of radius r when $r < 0$. Then, consider any Minkowski functional, say M, and the function

$$f(r) = M(K_r) \tag{9}$$

This family of functions built over the Minkowski functionals provides a way to investigate the morphology of the pore space K as it is dilated and eroded with balls of increasing radius r. VOGEL et al. (2005) used this approach on 2D images to describe crack dynamics in clay soil. ROTH et al. (2005) make use of opening (i.e. erosion followed by dilation) to build Minkowski functions to quantifying permafrost patterns with aerial photographs. These functions add new information to that provided by Minkowski functionals as they yield the pore size distribution of the porous structure. ARNS et al. (2004) characterized disordered systems and matched model reconstructions to 3D images of Fontainebleau sandstone with Minkowski functions based on dilations and erosions. VOGEL et al. (2010) took advantage of Minkowski functions based on openings to quantify soil structure of arable soil and of repacked sand using 3D images from X-ray tomography of samples of different sizes recorded at different resolutions.

Mecke (1996) considered a different type of Minkowski function. In this case, the original 2D image is a grayscale image before segmentation. A series of binary images were obtained when the threshold varied from the minimum value of the grayscale to its maximum. Minkowski functionals were evaluated on each binarized image of the series, and four Minkowski functions were defined when the Minkowski functionals evolved as a function of threshold. Roth et al. (2005) also made use of this type of functions to quantify permafrost patterns obtained from aerial 2D photographs.

In this work, we will investigate, in a three-dimensional setting, how Minkowski functions based on parallel sets of binary 3D X-ray CT images of soil columns can be used to characterize soil pore structure of cultivated soil.

6. Materials and Methods

6.1. Soil Columns: Sample Collection

The columns were collected at the experimental farm "Finca La Grajera", a property of La Rioja region government, northern Spain, Latitude, 42°26′34 18″N; longitude 2°30′53 07″W, in December 2010. The field slope was about 10.2 % with west-east orientation. The soil was classified as fine-loamy, mixed, thermic Typic Haploxerepts according to the USDA soil classification (Soil Survey Staff, 2006), and contained 230 g kg^{-1} clay, 433 g kg^{-1} silt, 337 g kg^{-1} sand, 9.3 g kg^{-1} organic matter, and 149 g kg^{-1} carbonates, with pH 8.62 and electrical conductivity 0.17 dS m^{-1} at the Ap horizon (0–20 cm). Climate in the area is semiarid according to the UNESCO aridity index (UNESCO, 1979), with heavy winter rains and summer drought conditions. For the period 2005–2009, the average annual precipitation was 470 mm, average annual temperature was 13 °C, and average annual potential evapotranspiration (FAO-Penman) was 1,132 mm.

In this study, we considered four columns collected between rows of the vineyard that was established in 1996 with *Vitis vinifera* L. "Tempranillo", grafted onto 110-R rootstock. Two types of soil cover management in between rows were undertaken:

(*T*) conventional tillage management between rows, which consisted of a soil tillage of 15-cm depth by cultivator once every 4–6 weeks, as required for weed control during the grapevine growth cycle; (*C*) permanent cover crop of resident vegetation, which was dominated by annual grass and forbs common to La Rioja vineyards (see Peregrina et al., 2010, for more details). Columns were extracted vertically by percussion drilling between rows, within PVC cylinders of 7.5 cm interior diameter and 30 cm height from the upmost part of soil profile. As a consequence, only the upper half of the column was affected by tillage that was undertaken 3 months before the collection of samples.

6.2. Image Acquisition, Filtering, and Segmentation

Soil columns were scanned at Fraunhofer ITWM facilities (Germany) with a PerkinElmer amorphous silicon (a-Si) detector with 2,048 × 2,048 pixels and a Feinfocus FXE 225.51 microfocus beam source tube. It was operated at 190 kV (53 μA) acceleration voltage and 20 W target power. The tube had a tungsten target installed. In addition, a collimator to reduce stray radiation and a 200-μm steel filter in front of the target was used. Only the upper half of the column was scanned to image the tilled part of the columns from tilled soil, and the region between 6.5 and 15 cm was selected to have a resolution of 50 μm. In this way, soil macro-pore structure important for intense renewal of air and serving to transport and distribute water in soil (Brewer, 1964) was imaged.

Raw data from tomography correspond to a stack of 1,706 two-dimensional, 16-bit grayscale images with a pixel size of 50 μm. These horizontal sections are disks of 7.5 cm diameter, 50 μm apart from one another. Thus, the 3D image is made up of voxels of 50 μm. Light values of the grayscale designate voxels corresponding to low densities of the soil column, whereas high values indicate voxels of high density parts of the column. The original 2D projections were filtered by a 3 × 3 median filter before reconstruction in order to reduce random noise from the detector. It is a nonlinear smoothing method used to reduce isolated noise without blurring sharp edges (Wang and Lai, 2009).

The segmentation process provides a way to separate the object of interest from the background, in this case, the pore space from the soil matrix. This process produces binary images when a threshold is selected, and every voxel with a grayscale value lower than the selected threshold is considered part of the pore space and set to 1 (white), while every voxel with a grayscale value higher than the selected threshold is considered part of the soil matrix and set to 0 (black). ImageJ version 1.47v, a public domain program developed at the National Institutes of Health, was used for image processing. We selected a global method as we focused primarily on the analysis of geometrical features evolutions. The modes method of thresholding was chosen to generate binary images (SONKA et al., 1998) for its performance (IASSANOV et al., 2009). In this procedure, the histogram is iteratively smoothed until there are only two local maxima. Then, the threshold is chosen at the midpoint between these local maxima. Figure 4 illustrates image binarization, and Fig. 5 shows the view of 3D reconstruction of pore space in a binary image. The plot of histograms with logarithmic scale on the vertical axis is displayed (Fig. 4) to show the two maxima. Notice the different pore structures that display a typical sample from soil under cover crop of resident vegetation and from soil under conventional tillage (Fig. 5). The homogeneity of the pore space produced by tillage is obvious (T samples) as compared to the much more heterogeneous result of the cover resident vegetation crop (C samples).

6.3. Computing Minkowski Functions for Parallel Sets

We will consider binary images segmented with the modes method procedure. In these images, the pore space will be the object of interest while the soil matrix will be the background. Now, to study pore structure, we will investigate the evolution of Minkowski functionals as successive erosions, and dilations with balls of increasing radius are performed on the binary images (ARNS et al., 2002; VOGEL et al. 2005).

We follow the procedure developed by MECKE (1996) and the code published by MICHIELSEN (2001) to compute Minkowski functionals. For the sake of

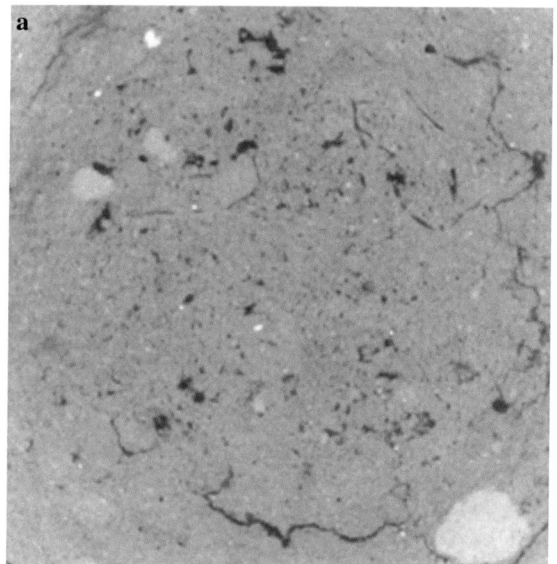

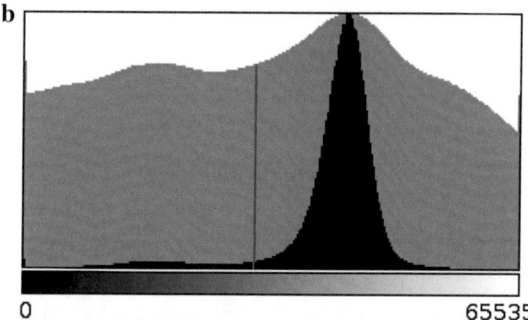

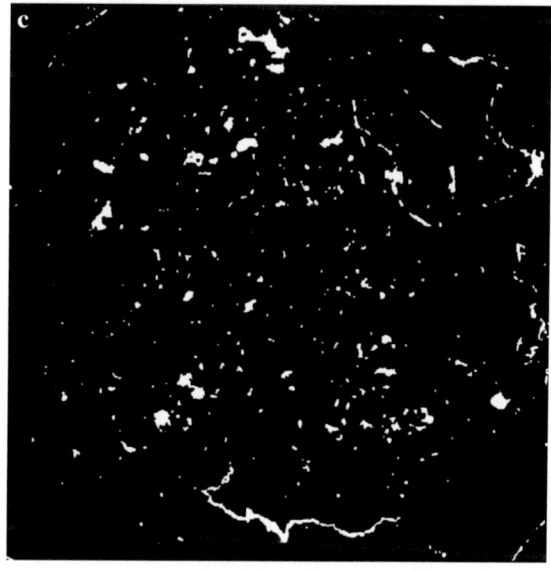

Figure 4

Segmentation process on a horizontal section of 960 × 960 pixels of column C1: **a** gray-scale image, **b** histogram with (*black*) and without (*grey*) logarithmic scales, and the resulting threshold marked with a vertical *red line*, and **c** segmented image (*white voids, black solid*)

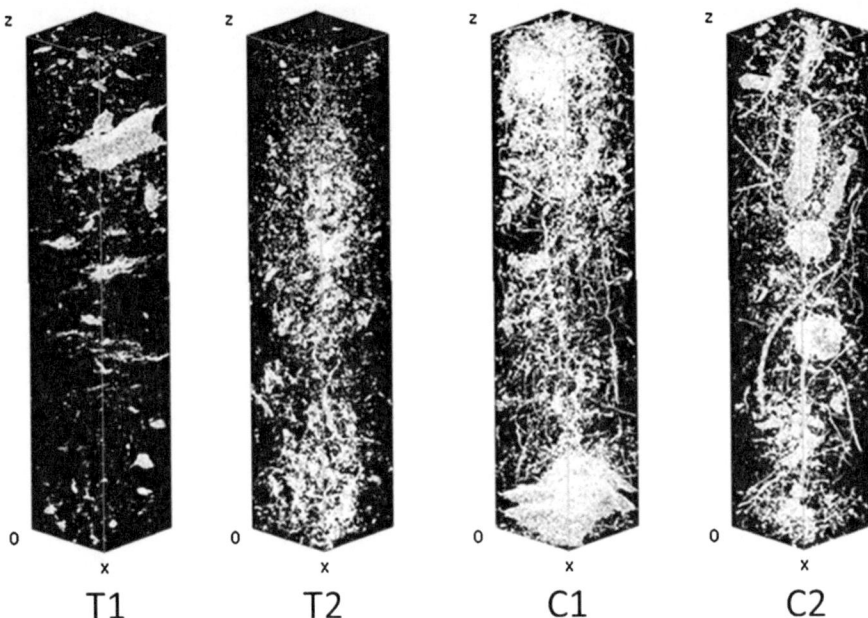

T1   T2   C1   C2

Figure 5
3D reconstructions of the pore geometry (*white*) in each soil column in a box that is 8.5 cm high (*z* axis) and 1.7 cm long (*x* axis) and wide (*y* axis)

clarity, let us illustrate this procedure in 2D images made up of pixels that geometrically are squares. The object of interest, K, is a finite union of squares (compact and convex object). Each square is considered to be decomposed into the four points of its four vertices, the four open segments of its four edges, and the rest of the square, i.e. its interior. Then, the square of each pixel is the union of nine disjoint sets: four points, four open segments, and the interior of the square. As a consequence, we only need to know the Minkowski functional of these three types of sets (a point, an open segment, and an open square), and then use C-additivity extended to the union of an arbitrary amount of sets. If n_s is the number of squares of the object, n_e the number of edges, and n_v the number of vertices of the pixels of the object of interest are counted once, it is easy to verify that (MICHIELSEN 2001).

$$A(K) = n_s, \quad L(K) = -4n_s + 2n_e \quad \text{and} \quad (10)$$
$$\chi(K) = n_s - n_e + n_v.$$

For three-dimensional objects, a similar argument shows that (MICHIELSEN 2001).

$$V(K) = n_c, \quad S(K) = -6n_c + 2n_f,$$
$$\pi^{-1}M(K) = 3n_c - 2n_f + n_e \quad \text{and} \quad (11)$$
$$\chi(K) = -n_c + n_f - n_e + n_v$$

In this expression, n_c is the number of cubes and n_f is the number of faces of the voxels of the object K, counted once.

The Euler-Poincaré characteristic—Euler number, for short—describes the connectivity of an object. In order to reconcile this global topological point of view with the local counterpart that displays the computation of this number in terms of numbers of cubes, faces, edges, and vertices, it is necessary to define when voxels are connected, or equivalently, when are they neighbors. In the plane, a common choice is to consider that two black pixels are connected when they have an edge or a vertex in common. In the three-dimensional space, it is customary to consider two black voxels connected when they have a face, an edge, or a vertex in common. This implies that any voxel is connected to 26 voxels or it has 26 neighbors (MICHIELSEN and DE RAEDT, 2001).

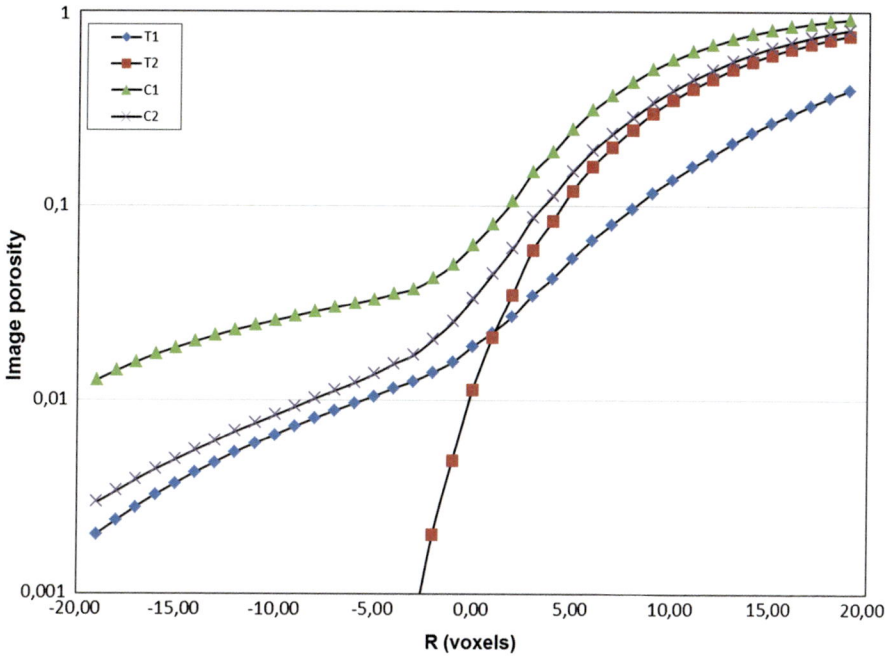

Figure 6
Image porosity as a function of diameter of erosion/dilatation

7. Results and Discussion

To evaluate Minkowski functionals, each column was divided into five consecutive cubes that shared a face, from top to bottom. The cubes had 340 voxels per edge and they were centered on the axes of the column in order to avoid voxels belonging to the container or voxels representing soil near the sampling tube that might have been damaged during sampling. The pore space in each cube was eroded/dilated to yield parallel sets. Diameters of balls took 19 different values for erosions and 19 for dilation, as well; it was incremented from 0 in steps of the voxel size (i.e. 50 µm). As Minkowski functionals are additive, their values for each column were obtained by simply adding the corresponding values of the cubes of the column. We considered densities of Minkowski functionals. Thus, we had volume fraction or image porosity, specific boundary surface area, specific integral of mean curvature, and specific Euler number of the pore space.

Figures 6, 7, 8, 9 display the evolution of these geometrical densities as functions of erosion/dilation diameter (R). As stated above, dilations of pore space produce an increase of its volume. Let us remark that

this effect is more pronounced when there are tunnels of soil materials through voids because dilations reduce them, even if it also depends on the complexity of the pore-solid interface as measured by surface area and integral of mean curvature. Roughly speaking, dilations turn some voxels of the soil matrix into voxels of its pore space. Hence, this morphological operation expands the void part of the sample. Erosion produces the inverse process. Differences between soil samples under natural resident vegetation cover (C) and samples under conventional tillage (T) are noticeable even if samples T2 and C2 have a similar evolution for dilations. Nevertheless, the evolution of image porosity (Fig. 6) and specific boundary surface (Fig. 7) with erosions diverges. This suggests that geometrical features of sample T2 are smaller than three voxels as they vanish with erosions of diameter smaller than that size. The opposite behavior is observed on sample C1. The erosion with the larger ball still left an important amount of porosity in this sample. Overall, samples with natural resident vegetation cover (C) store a greater amount of volume fraction and specific surface at any diameter of the balls used to erode/dilate as compared to samples from tilled soil (T). This is

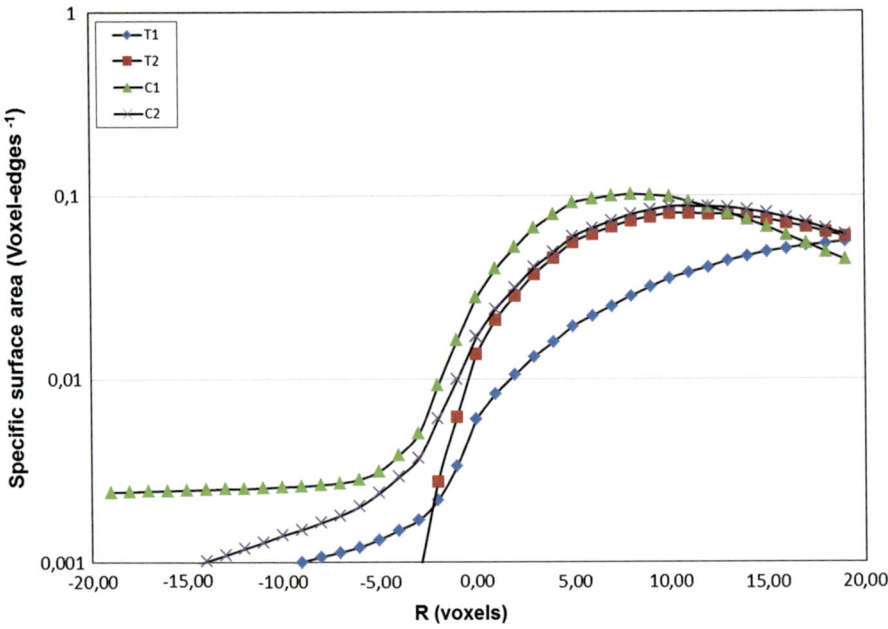

Figure 7
Specific surface area (voxel-edges^{-1}) as a function of diameter of erosion/dilatation

consistent with results reported by PEREGRINA *et al.* (2010).

Figures 8 and 9 depict the evolution of the specific integral of mean curvature—mean curvature, for short—and connectivity. Let us remember that the connectivity is evaluated as the number of connected components of the object of interest minus its tunnels plus its cavities (see Appendix 2). Tunnels are redundant loops or handles, as torus-like holes through the object of interest. As we are dealing with images of a natural soil, we may assume that there are no soil materials completely surrounded by voids and, as a consequence, the Euler number corresponds to the number of connected components of the pores space minus the number of tunnels of solid materials through the pore space. The morphological functions of the specific mean curvature (Fig. 8) and connectivity (Fig. 9) seem to indicate that conventional tillage and resident vegetation cover produces two different pore structures; this difference is especially apparent when comparing samples C1 and T1. Sample C1 yields more specific mean curvature than sample T1 when dilated with balls smaller than nine voxels. In this range of diameters, mostly small voids connecting soil matrix should populate sample C1 as

compared to sample T1, as is apparent from Fig. 5. High Euler numbers of sample C1 at small diameters seem to suggest this behavior. But large diameters decrease specific mean curvature and Euler number of sample C1, producing negative values. Nevertheless, in the case of T1, these geometrical measurements have lower growth. In the case of connectivity, it is negative for the largest diameter of dilations. This suggests that the pore structure of sample C1 contains a great amount of small features as the number of small voids (i.e. connected components) exceeds the number of tunnels of solid materials through them; therefore, high values of the specific mean curvature from these small features of the C1 pore space might be explained by the regularity of the surface that enclosed them, and they are also compatible with their small size. Moreover, C1 seems to display a rich structure as compared to sample T1. Between diameters 8 and 9, the graphs of both samples intersect at a positive specific mean curvature, but sample C1 has negative Euler characteristic. Therefore, it suggests that geometrical features similar in size should dominate sample T1, while the dilations of sample C1 show a more complex structure highly connected with tunnels through it, as it seems to indicate negative

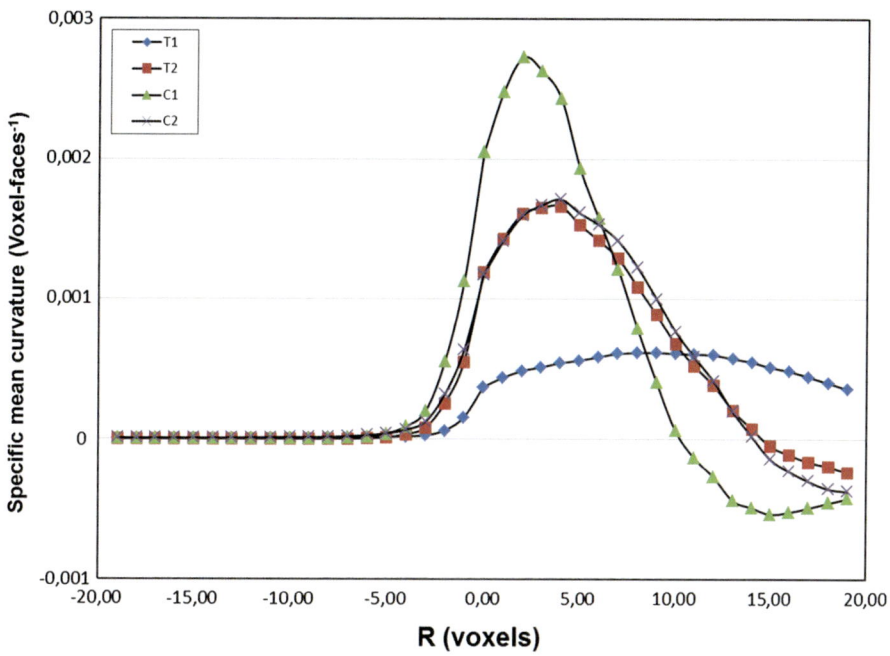

Figure 8
Specific curvature (voxel-faces^{-1}) as a function of diameter of erosion/dilatation

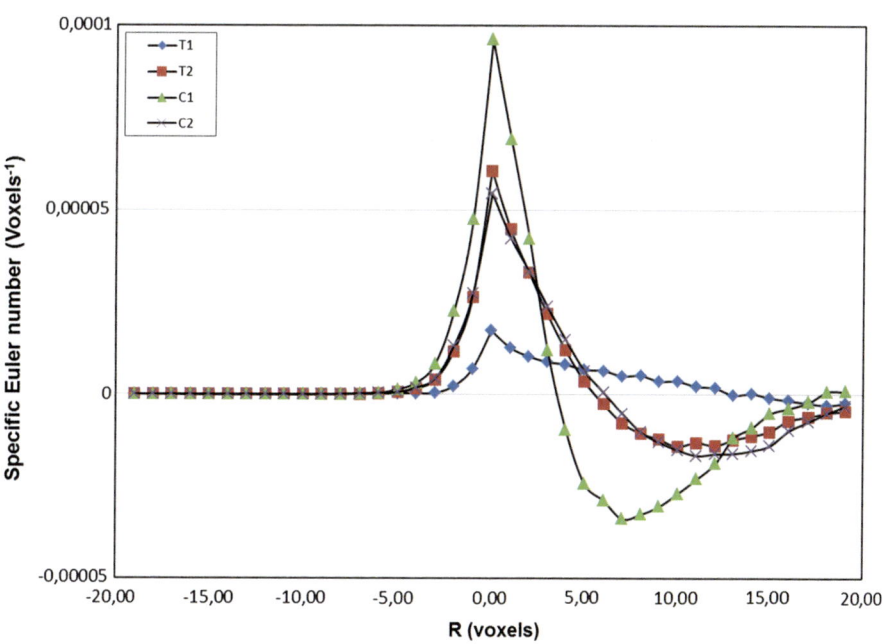

Figure 9
Specific Euler number (voxel^{-1}) as a function of diameter of erosion/dilatation

Euler numbers. The low variation of specific mean curvature and Euler numbers of sample T1 is compatible with a pore structure made up with irregular geometrical features of similar sizes that collapse as diameter of dilation increases and do not generate a complex and highly connected structure.

These results open the door to new investigations to identify statistically significant differences in soil structure due to contrasting management practices. It was the necessary first step towards further research that should include a richer sample. Then, the trends that suggest this study would be the hypothesis of those new investigations. Therefore, this might provide the basis for new projects that are likely to be lengthy and costly, as there is the need for a greater amount of 3D tomograms of large soil columns.

It has been reported that different land use and management practices significantly affect directions and magnitudes of the soil processes by contributing different quantities and qualities of biomass inputs, generating different levels of soil disturbance, influencing soil temperature and moisture regimes. These differences generate notable changes in soil physical and hydraulic properties, including changes in soil organic matter content, soil porosity, hydraulic conductivity, and water retention (Wang et al., 2012; Zhou et al., 2013). Our results suggest that the evolution of morphological features with dilation/erosion is a suitable indicator of soil structure for cultivated soil, and it seems to describe the influence of two different soil management practices (i.e. conventional tillage and natural cover crop) on soil structure in a Spanish Mediterranean vineyard. It is worth noting here how these results reflect the different pore structures as depicted by Fig. 5. The homogeneity of the pore space produced by tillage is obvious as compared to the heterogeneity of samples under resident vegetation cover. Similar geometrical features seem to dominate samples T2 and C2, but big structures discriminate between them and explain the behavior of the morphological functions of image porosity and specific boundary surface when sample T2 is eroded. These results are consistent with previous studies on the impact of land use on soil structure (Kravchenko et al., 2011; Wang et al., 2012) when they remarked on the homogeneity of the pore structure of conventional tillage as compared with no-till.

Soil structure is regarded as one of the main providers of physical protection of soil organic matter and carbon sequestration by soils (Six et al., 2000). One of the mechanisms of such protection is a reduced access of organic material inside soil voids to

decomposing microorganisms. The differences that we are observing in the porosity patterns between C and T samples hint at their potentially different effectiveness for protecting carbon. Clearly, T samples with their network of bigger voids will be offering greater microbial access, thus poorer protection than the C samples that have more porosity connected with smaller features. Observations of Ananyeva et al. (2013) support this hypothesis.

8. Conclusions

In this work, we have introduced the essential tools of mathematical morphology in order to quantify the geometrical morphology of soil structure. We made use of 3D images from X-ray CT of soil columns collected at the experimental farm "Finca La Grajera", property of the La Rioja region government, northern Spain. In this study, we considered four columns collected between rows of the vineyard that was established in 1996 with Vitis vinifera L. "Tempranillo". Two types of soil management in between rows were undertaken: (T) conventional tillage management between rows, which consists of a soil tillage of 15-cm depth by cultivator once every 4–6 weeks, as required for weed control during the grapevine growth cycle; (C) permanent cover crop of resident vegetation, which was dominated by annual grass and forbs common to La Rioja.

We have presented the building blocks of mathematical morphology, the morphological operations of dilation, erosion. We have dealt with the Minkowski functionals (i.e. volume, boundary surface, curvature, and connectivity) and the Minkowski functions that take account of the evolution of the Minkowski functionals as morphological operations are performed on the 3D object of interest with balls of increasing diameter.

Our results suggest that the evolution of morphological features with dilation/erosion is a suitable indicator of soil structure for cultivated soil and it seems to describe the influence of two different soil management practices (i.e. conventional tillage and natural cover crop) on soil structure in a Spanish Mediterranean vineyard. It is worth noting here how these results reflect the different pore structures as depicted by Fig. 5. The homogeneity of the pore space produced by tillage is obvious as compared to the heterogeneity of samples under resident vegetation crop. Similar geometrical features seem to dominate samples T2 and C2, but big structures discriminate between them and explain the behavior of specific image porosity and boundary surface when sample T2 is eroded.

These geometrical descriptors that seem to discriminate between these two types of samples could be used as inputs for morphological models of natural soil structures. But further investigations are needed to establish quantitatively the statistical significance of the observed impact of contrasting management practices on soil structure.

Acknowledgments

This work was partially supported by Plan Nacional de Investigación Científica, Desarrollo e Investigación Tecnológica (I+ D+I) under ref. AGL2011/25175 and DGUI (Comunidad de Madrid) and UPM under ref. QM100245066. We thank the staff of the Servicio de Investigación y Desarrollo Tecnológico Agroalimentario (Gobierno La Rioja) for providing the experimental plots and helping with the field work.

Appendix 1

Let us be more precise and specify the objects of interest and the geometrical conditions of Hadwiger's theorem. A class of objects to which this theorem applies is the class of sets that can be viewed as the union of a finite number of convex objects. An object K is convex when it contains any point of the segment that joins two of its points. The class of objects made up of finite unions of convex sets is worth considering as any three-dimensional binary image can be considered an element of this class. Binary images are sets of voxels which may be thought of as being cubes, and then any geometrical structure of interest in a binary image is a finite union of convex objects, which are the voxels.

There are three geometrical conditions that a functional to which Hadwiger's theorem applies must fulfill. The first one is motion invariance: the number assigned by a functional must be independent of the position of the object in space when the object is translated or rotated. The second one is C-additivity:

$$\mathcal{F}(K_1 \cup K_2) = \mathcal{F}(K_1) + \mathcal{F}(K_2) - \mathcal{F}(K_1 \cap K_2) \quad (12)$$

That is to say, the number assigned by a functional $\mathcal{F}$ to the union of two objects K_1 and K_2 equals the value of the functionals over those two objects minus parts counted twice. And the third condition is continuity. Consider a sequence of objects $\{K_n\}$ that approaches the object K as n tends to infinity. An example of this is the sequence of r-parallel bodies of an object K; it is clear that the sequence of r-parallel bodies $\{K_n\}$ with $r = 1/n$, approaches K as n goes to infinity or, equivalently, as r goes to zero. Then, the continuity condition is fulfilled if $\mathcal{F}(K_n)$ tends to $\mathcal{F}(K)$ as n goes to infinity. Under these conditions there are $d + 1$ numbers c_i such that

$$\mathcal{F}(K) = \sum_{i=0}^{d} c_i W_i^{(d)}(K) \quad (13)$$

where $W_i^{(d)}(K)$ are the Minkowski functionals that assign to any object a number and K belongs to the d-dimensional linear space.

Appendix 2

When the boundary surface of a three-dimensional object is smooth, the third functional, the surface integral of the mean curvature, $M(K)$, may be interpreted as the mean breadth of the object (OSHER and MÜCKLICH, 2000). This functional might also be an indicator of the surface boundary shape. Points on the boundary surface of an object with positive curvatures settle on convex parts (protrusions) while points with negative curvatures belong to concave parts (hollows). Hence, the mean curvature of convex points will be positive while it will be negative for concave points. Taking into account that the surface integral of the mean curvature over a certain boundary region of K may be interpreted as the average of the mean curvature over this surface region, the third functional, $M(K)$, should be positive for convex parts of the boundary surface while it should be negative for concave parts.

When the object of interest K corresponds to the pore space P, the Euler-Poincaré characteristic $\chi(P)$ is an index of the topology of the pore phase and it quantifies pore connectivity (VOGEL and KRETZSCHMAR, 1996). In the plane, Euler-Poincaré can be computed subtracting the number of holes of the object, $H(K)$, from the number of connected components, $CC(K)$ (MECKE, 1998):

$$\chi(K) = CC(K) - H(K) \qquad (14)$$

In this context, a connected component of an object is any part of it whose points are connected to one another by curves of points contained in the object. Then, a disk has Euler-Poincaré characteristic equal to 1 because it has one connected component and no holes. A punctured disk has Euler-Poincaré number equal to 0, a disk punctured twice, -1, and so on. If the object is just the union of n separated grains on an image, the Euler-Poincaré characteristic equals n. This object has n connected components. Similar definitions and relations hold in space though distinction between two kinds of holes must be made. In space, the Euler-Poincaré characteristic can be computed as the sum of the number of connected components, $CC(K)$, and the number of cavities of the object, $C(K)$, subtracted by the number of tunnels, $T(K)$ (MECKE, 1998):

$$\chi(K) = CC(K) - T(K) + C(K) \qquad (15)$$

Cavities are holes completely surrounded by the object, while tunnels are handles or redundant loops as torus-like holes through the object connected with the exterior or background. If the object is just a separate union of n grains of an image, the Euler-Poincaré characteristic equals n. Then, a solid ball has Euler-Poincaré characteristic equal to 1, a ball with a cavity in it, 2, a ball with two cavities, 3, and so on. But, if the ball has a tunnel that goes through it, the Euler-Poincaré characteristic is 0, two tunnels gives a Euler-Poincaré characteristic equal to -1, and so on.

REFERENCES

ANANYEVA, K., W. WANG, A.J.M. SMUCKER, M.L. RIVERS, A.N. KRAVCHENKO. 2013. *Can intra-aggregate pore structures affect*

the aggregate's effectiveness in protecting carbon? Soil Biology and Biochemistry *57*:868–875.

ARNS, C.H., M.A. KNACKSTEDT, and K.R. MECKE. 2002. Characterizing the morphology of disordered materials. In: K.R. MECKE and D. Stoyan (Eds.). Morphology of condensed matter. LNP 600. Springer, Berlin. pp. 37–74.

ARNS, C.H., M.A. KNACKSTEDT, and K.R. MECKE. 2004. *Characterization of irregular spatial structures by parallel sets and integral geometric measures.* Colloids and Surfaces A: Physicochem. Eng. Aspects, *241*:351–372.

BOSSUYT, H., SIX, J., HENDRIX, P. F., 2002. *Aggregate-protected carbon in no-tillage and conventional tillage agroecosystems using carbon-14 labeled plant residue.* Soil Science Society America Journal 66, 1965–1973.

BREWER, R. 1964. Fabric and mineral analysis of soils. Wiley, New York.

CHENU, C., PLANTE, A.F., 2006. *Clay-sized organo-mineral complexes in a cultivation chronosequence: revisiting the concept of the primary organo-mineral complex.* European Journal of Soil Science 57, 596–607.

DENEF, K., SIX, J., BOSSUYT, H., FREY, S.D., ELLIOTT, E.T., MERCKX, R., PAUSTIAN, K., 2001. *Influence of dry–wet cycles on the interrelationship between aggregate, particulate organic matter, and microbial community dynamics.* Soil Biology and Biochemistry *33*, 1599–1611.

ELLERBROCK, R.H., GERKE, H.H., 2004. *Characterizing organic-matter of soil aggregate coatings and biopores by Fourier transform infrared spectroscopy.* European Journal of Soil Science 55, 219–228.

IASSONOV, P., T. GEBREGENUS, AND M. TULLER. 2009. *Segmentation of X-Ray CT Images of Porous Materials: A Crucial Step for Characterization and Quantitative Analysis of Pore Structures.* Water Resour. Res. *45*: W09415, doi:10.1029/2009WR008087.

JASINSKA, E., BAUMGARTL, T., WETZEL, H., HORN, R., 2006. *Heterogeneity of physico–chemical properties in structured soils and its consequences.* Pedosphere 16, 284–296.

KLETTE, R. AND ROSENFELD, A., 2004. Digital geometry. Geometric methods for digital picture analysis. Morgan Kaufmann Series in Computer Graphics and Geometric Modeling, Morgan Kaufmann, San Francisco.

KRAVCHENKO, A.N., WANG, W., SMUCKER, A.J.M., RIVERS, M.L., 2011. *Long-term Differences in Tillage and Land Use Affect Intra-aggregate Pore Heterogeneity.* Soil Science Society America Journal 75, 1658–1666.

LIKOS, C. N., MECKE, K. R. and WAGNER, H., 1995. *Statistical morphology of random interfaces in microemulsions.* J. Chem. Phys., *102*:9350–9360.

LEHMANN, P. 2005. *Pore structures: measurement, characterization and relevance for flow and transport in soils.* Proc Appl Math Mech 5, 39–42.

LEHMANN, P., P. WYSS, A. FLISCH, E. LEHMANN, P. VONTOBEL, M. KRAFCZYK, A. KAESTNER, F. BECKMANN, A. GYGI, and H. FLÜHLER. 2006. *Tomographical imaging and mathematical description of porous media used for the prediction of fluid distribution.* Vadose Zone J. 5:80–97.

MATHERON, G., 1975. Random sets and integral geometry. Wiley, New York

MECKE, K.R. 1996. *Morphological characterization of patterns in reaction–diffusion systems.* Phys. Rev. E. *53*(5): 4794–4800.

MECKE, K. R. 1998. *Integral geometry and statistical physics.* Inter. J. Mod. Phys. B. *12*(9):861–899.

MECKE, K. R. 2002. *The shape of parallel surfaces: porous media, fluctuating interfaces and complex fluids.* Physica A *314*:655–662.

MECKE, K. and C.H. ARNS. 2005. *Fluids in porous media: a morphometric approach.* J. Phys.: Condens. Matter. *17*:S503–S534.

MEES, F., R. SWENNEN, M. VAN GEET, and P. JACOBS. (Eds.), 2003. Applications of X-ray computed tomography in geosciences. Geological Society, London, Special publication, 215, pp. 243.

MICHIELSEN, K., and H. DE RAEDT. 2001. *Integral–geometry morphological image analysis.* Physics Reports, *347*:461–538.

OHSER, J., and F. MÜCKLICH. 2000. Statisitical analysis of microstructure in materials sciences. Wiley, Chichester

PEREGRINA, F., C. LARRIETA, S. IBÁÑEZ, and E. GARCÍA-ESCUDERO. 2010. *Labile organic matter, aggregates, and stratification ratios in a semiarid vineyard with cover crops.* Soil Sci. Soc. Am. J. *74*(6):1–11.

PERRET, J., S.O. PRASHER, A. KANTZAS, and C. LANGFORD. 1999. *Three-dimensional quantification of macropore networks in undisturbed soil cores.* Soil Sci. Soc. Am. J. *63*: 1530–1543.

PEYTON, R.L., C.J. GANTZER, S.H. ANDERSON, B.A. HAEFFNER, and P. PFEIFER. 1994. *Fractal dimension to describe soil macropore structure using X-ray computed tomography.* Water Resour. Res. *30*:691–700.

PIERRET, A., Y. CAPOWIEZ, L. BELZUNCES, and C.J. MORAN. 2002. *3D reconstruction and quantifi cation of macropores using x-ray computed tomography and image analysis.* Geoderma *106*:247–271.

RENARD, P. and A. DENIS. 2013 *Connectivity metrics for subsurface flow and transport,* Advances in Water Resources, *51*(0):168-196.

ROTH, R., J. BOIKE, and H. J. VOGEL. 2005. *Quantifying Permafrost Patterns using Minkowski Densities.* Permafrost and Periglac. Process. *16*:277–290.

SAN JOSÉ MARTÍNEZ. F., M.A. MARTÍN, F.J. CANIEGO, M. TULLER, A. GUBER, Y. PACHEPSKY, C. GARCÍA-GUTIÉRREZ, 2010. *Multifractal analysis of discretized X-ray CT images for the characterization of soil macropore structures.* Geoderma, *156*:32–42.

SAN JOSÉ MARTÍNEZ, F., F.J. MUÑOZ, F.J. CANIEGO, F. PEREGRINA, 2013. *Morphological Functions to Quantify Three-Dimensional Tomograms of Macropore Structure in a Vineyard Soil with Two Different Management Regimes.* Vadose Zone J. Vol. 12, No. 3.

SANTALÓ, L.A. 1976. Integral geometry and geometric probability. Addison-Wesley Publishing Co. Inc. Reading, Massachusetts

SANTOS, D., MURPHY, S.L.S., TAUBNER, H., SMUCKER, A.J.M., HORN R., 1997. *Uniform separation of concentric surface layers from aggregates.* Soil Science Society America Journal *61*, 720–724.

SERRA, J. 1982. Image analysis and mathematical morphology. Academic Press Inc. Orlando, Florida

SEXSTONE, A.J., REVSBECH, N. P., PARKIN, T.B., TIEDJE J.M., 1985. *Direct measurement of oxygen profiles and denitrification rates in soil aggregates.* Soil Science Society America Journal *49*, 645–651.

SIX, J., ELLIOTT, E.T., PAUSTIAN, K., 2000. *Soil macroaggregate turnover and microaggregate formation: a mechanism for C sequestration under no-tillage agriculture.* Soil Biology and Biochemistry *32*, 2099–2103.

SONKA, M., V. HLAVAC, and R. BOYLE. 1998. Image processing, analysis, and machine vision. (2nd ed.) PWS, an Imprint of Brooks and Cole Publishing Inc.

SOILLE, P. 2002. Morphological textural analysis: an introduction. In K.R. MECKE and D. Stoyan (eds.). Morphology of condensed matter. LNP 600, pp. 215–237. Springer, Berlin.

URBANEK E., HALLETT, P., FEENEY D., HORN, R., 2007. *Water repellency and distribution of hydrophilic and hydrophobic compounds in soil aggregates from different tillage systems.* Geoderma *140*, 147–155.

VOGEL, H.J., 2002. Topological characterization of porous media. In K.R. MECKE and D. Stoyan (eds.). Morphology of condensed matter. LNP 600, pp. 75–92. Springer, Gerlin

VOGEL, H.J., H. HOFFMANN, and K. ROTH. 2005. *Studies of crack dynamics in clay soil. I. Experimental methods, results, and morphological quantification.* Geoderma *125*:203–211.

VOGEL, J.H., U. WELLER, and S. SCHLÜTER. 2010. *Quantification of soil structure based on Minkowski functions.* Comput. Geosci. *36*:1236–1245.

VON LÜTZOW, M., KÖGEL-KNABNER, I., EKSCHMITT, K., MATZNER, E., GUGGENBERGER, G., MARSCHNER, B., FLESSA, H., 2006. *Stabilization of organic matter in temperate soils: mechanisms and their relevance under different soil conditions-a review.* European Journal of Soil Science 57, 426–445.

WANG, M., and CH.H LAI. 2009. A Concise Introduction to Image Processing using C++ Chapman & Hall/CRC (Numerical Analysis and Scientific Computing Series).

WANG, W., KRAVCHENKO, A.N., SMUCKER, A.J.M., LIANG, W., RIVERS, M.L., 2012. *Intra-aggregate pore characteristics: X-ray computed microtomography analysis.* Soil Science Society America Journal *76*, 1159–1171.

ZHOU, H, X. PENG, E. PERFECT, T. XIAO, and G. PENG. 2013. *Effects of organic and inorganic fertilization on soil aggregation in an Ultisol as characterized by synchrotron based X-ray microcomputed tomography.* Geoderma *195–196*: 23–30.

(Received March 28, 2014, revised July 31, 2014, accepted August 26, 2014, Published online September 7, 2014)

Pure Appl. Geophys. 173 (2016), 591–606
© 2015 Springer Basel
DOI 10.1007/s00024-015-1088-8

∎Pure and Applied Geophysics

A Comparison of Different-Mode Fields Generated from Grounded-Wire Source Based on the 1D Model

Nan-Nan Zhou,[1] Guo-qiang Xue,[1] Hai Li,[1] M. Younis,[1] Dong-yang Hou,[1] Hua-sen Zhong,[1] Wei-ying Chen,[1]
and Jiang-wei Cui[1]

Abstracts—Traditional TEM study mainly focuses on the generation and application of the TE field using a loop or grounded-wire source; but in recent decades, lots of efforts have been made for implementation of the TM field and even the integration of the TE field with the TM one into anomaly detection in the subsurface. However, no applicable principles have been proposed for selecting the optimal electromagnetic field for various subsurface targets. The transient electromagnetic (TEM) fields generated from grounded-wire source consist of the TE-mode response (current-carrying wire), the TE–TM mode response (grounding ends) and the combined TEM-mode response (current-carrying wire and grounding ends). This study performs a comparison of TE/TE–TM/TEM fields by generating them from grounded-wire source and testing their distribution characteristics, detection depth, and sensitivity to anomalies, using both synthetic 1D model and two field surveys in China. The comparisons demonstrate that, the detection depth of the TE–TM field is smaller than those of both the TE and combined TEM fields. Meanwhile, for electric field, the TE–TM response provides a better detection than the TEM one, but with an uneven distribution. Therefore, the TE–TM electric field requires well-designed arrangements of receiving positions when applied to real projects. For the magnetic field, the TEM response has the best detection capability compared to the TE and TE–TM ones, but is least sensitive to layer thickness and resistivity, especially for an embedded layer with low resistivity.

Key words: TE, TE–TM, sensitivity, detection depth, field distribution.

1. Introduction

Theoretically, electromagnetic fields can be divided into the transverse electric-type (TE) and the transverse magnetic-type (TM) field. For the fixed-direction source moment, the vertical electric field is absent in TE-mode fields and the vertical magnetic field does not exist in the TM-mode fields (Ward and Hohmann 1991).

The dipole source is one of the most popular source types in theoretical research and is the basis of the actual source in the application. Three popular types of dipole include the vertical magnetic dipole for generating the TE field, vertical electric dipole for generating the TM field, and horizontal magnetic or electric dipole for generating both fields. The modes of the fields generated from the actual large-size source can then be analyzed according to calculation means.

The field equations of the rectangular loop are derived as the integral of the corresponding component of electric dipole (Raiche and Spies 1981; Poddar 1982, 1983; Raiche 1987; Xue et al. 2012). Recent study (Zhou and Xue 2014) illustrates that the rectangle loop source generates only the current-carrying wire (TE), since the grounding ends (TE–TM) from each electric dipole could be ignored. Therefore, for generating EM field with both TE (current-carrying wire) and TE–TM (grounding ends), a long grounded-wire source should be used (Kauahikaua 1978; Weir 1980), and this study uses such source so that all types of fields could be generated for comparison. For the convenience of description and comparison, here, we denote the finite-length wire as the TE field, the grounded ends as the TE–TM field, and the total response the combined TEM field.

When configuring the grounded-wire source for real applications, long-offset transient electromagnetic (LOTEM) is the most commonly used grounded-wire TEM configurations. In 1980s, the time-derivative of the vertical magnetic field, which

¹ Key Laboratory of Mineral Resources, Institute of Geology and Geophysics, Chinese Academy of Sciences, Beijing 100029, China. E-mail: znncas@126.com

is related to the current-carrying wire, is often ob-served in LOTEM survey (STRACK 1984; KELLER *et al.* 1984; YANG 1986), and thereby the study on grounded-wire sources at that time focused on only the TE field. However, with the development of new technologies in recent years, research on long grounded-wire source has been extended to generat-ing both the horizontal electric and magnetic fields and applying them to various industry and discipline detections. For example, COX (1980) proposed the collection of the horizontal electric field to study seafloor geology and noted that the electric field was sensitive to the resistive target. VOZOFF *et al.* (1985) used the LOTEM electric field to map resistive horizons in petroleum exploration. HORDT *et al.* (1992) performed joint inversion of LOTEM and MT data from Munsterland, where horizontal electric field and time-derivative of magnetic field from horizontal electric dipole were used. Besides those, the apparent resistivity (CALDWELL and BIBBY 1998), the effect of topographic (HORDT and MULLER 2000; HORDT and SCHOLL 2004), 3-D forward modeling (PODDAR 1996; COMMER and NEWMAN 2004), 3-D in-version (COMMER *et al.* 2005, 2006) of LOTEM have also been widely studied in recent years.

The LOTEM configuration is also the most widely used setup for marine CSEM (EDWARDS and CHAVE 1986; EDWARDS 1997; SCHWALENBERG *et al.* 2005; WEITEMEYER *et al.* 2006; EVANS 2007), but is based on dipole hypothesis. For large-offset LOTEM, the dipole hypothesis is reasonable and can be applied to the modeling and interpretation of LOTEM data. However, for the short-offset TEM, the dipole hy-pothesis will cause great errors, and thereby short-offset TEM has been the research focus recently for resolution improvement. STREICH and BECKEN (2011) analyzed the approximation caused by the dipole hypothesis of finite-length wire from all components in the frequency-domain, and ZHOU *et al.* (2015) analyzed the error caused by dipole hypothesis in time domain. In order to extend the observation re-gion from long-offset to short-offset region with better accuracy, superposing-dipole method has been developed for calculating the response generated from the grounded-wire source, and this study im-plements such method for response computation.

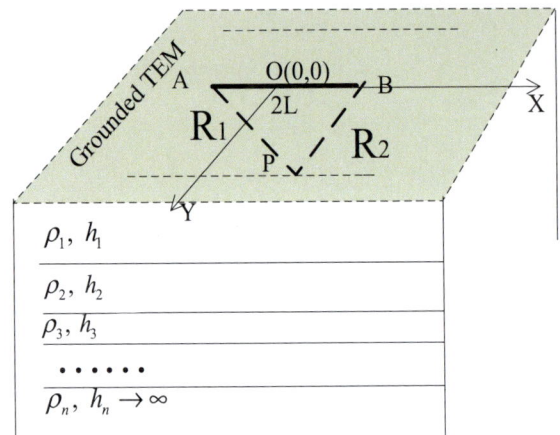

Figure 1
Diagrammatic of the transmitter–receiver geometry above layered earth

In this study, we first generates the various fields (TE, TE–TM, and combined TEM) from grounded-wire source using the superposing-dipole method, and analyzes their distribution characteristic, detection depth and sensitivity in all zone (both large-offset and short-offset). Then, the principles of selecting the most effective configuration and field are discussed for various geo-electric targets in applications. Fi-nally, two field examples in China are conducted to verify our comparisons of the TE, TE–TM, and combined TEM fields.

1.1. Response of the Long Grounded-wire TEM

In Fig. 1, we show the layered Earth model and grounded-wire TEM setup, from which the EM fields could be calculated as (WARD and HOHMANN 1991):

$$E_x = -\frac{I}{4\pi}\left[\frac{x}{r}\int_0^\infty \left[(1-r_{TM})\frac{u_0}{\hat{y}_0} - (1+r_{TE})\frac{\hat{z}_0}{u_0}\right] \times J_1(\lambda r)\mathrm{d}\lambda\right]_{R_1}^{R_2}$$
$$-\frac{\hat{z}_0 I}{4\pi}\int_{-L}^{L}\int_0^\infty (1+r_{TE})\frac{\lambda}{u_0}J_0(\lambda r)\mathrm{d}\lambda\mathrm{d}x'$$

$$\tag{1}$$

$$E_y = -\frac{I}{4\pi}\left[\frac{y}{r}\int_0^\infty \left[(1-r_{TM})\frac{u_0}{\hat{y}_0} - (1+r_{TE})\frac{\hat{z}_0}{u_0}\right]J_1(\lambda r)\mathrm{d}\lambda\right]_{R_1}^{R_2}$$

$$\tag{2}$$

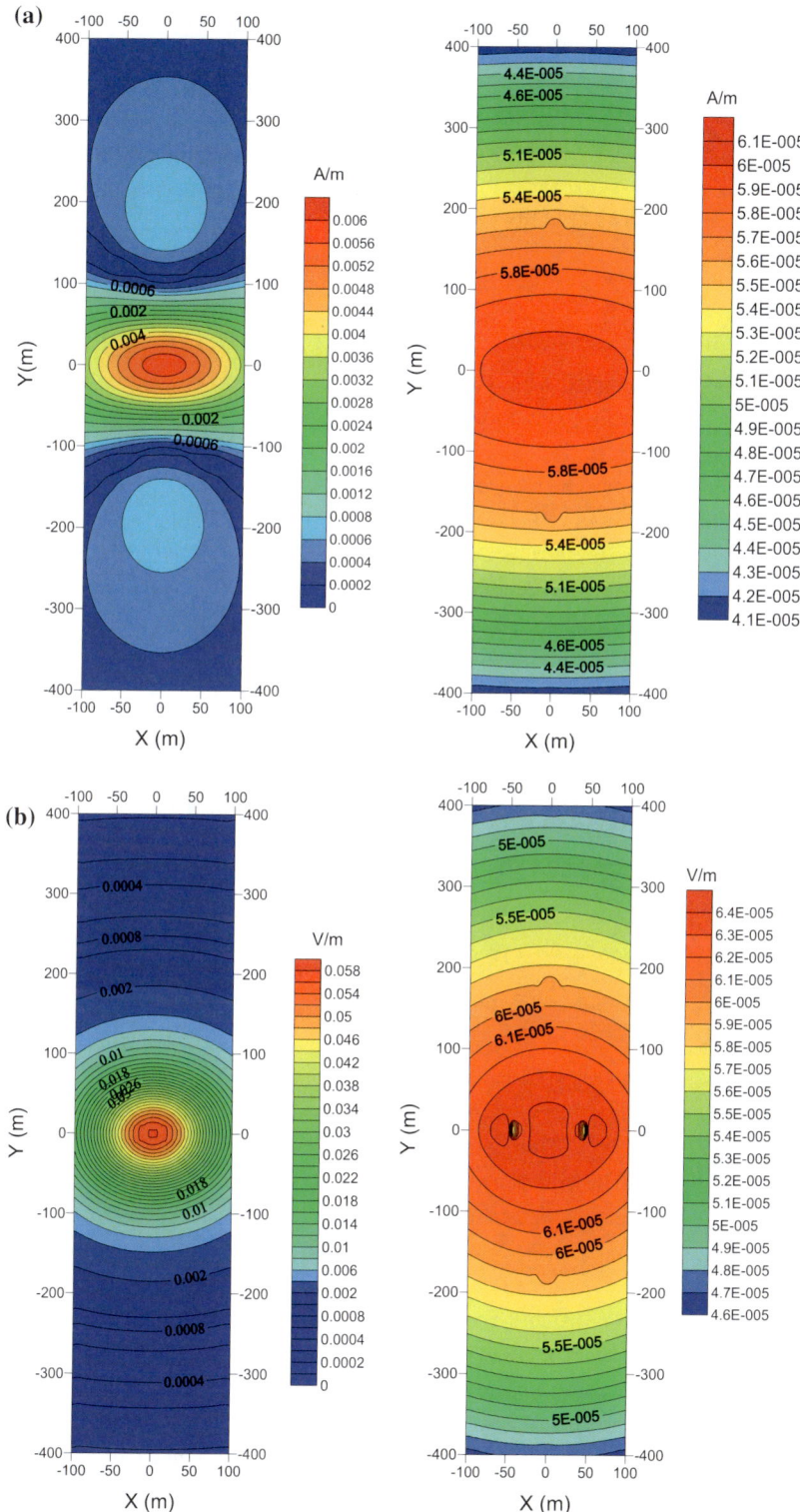

Figure 2
Distribution of combined TEM fields: H_y (**a**) and E_x (**b**) at 1e-5 s (*left*) and 1e-3 s (*right*)

$$H_x = \frac{I}{4\pi}\left[\frac{y}{r}\int_0^\infty (r_{TM} + r_{TE})e^{u_0 z}J_1(\lambda r)d\lambda\right]_{R_1}^{R_2} \quad (3)$$

$$H_y = -\frac{I}{4\pi}\left[\frac{x}{r}\int_0^\infty (r_{TM} + r_{TE})e^{u_0 z}J_1(\lambda r)d\lambda\right]_{R_1}^{R_2}$$

$$-\frac{I}{4\pi}\int_{-L}^{L}\int_0^\infty (1 - r_{TE})e^{u_0 z}\lambda J_0(\lambda r)d\lambda dx' \quad (4)$$

$$H_z = \frac{I}{4\pi}\int_{-L}^{L}\frac{y}{r}\int_0^\infty (1 + r_{TE})e^{u_0 z}\frac{\lambda^2}{u_0}J_1(\lambda r)d\lambda dx', \quad (5)$$

where, $2L$ is the length of the grounded wire; (x, y, z) is the spatial location of the recording station; r is the distance between the recording station and the source; R_1 and R_2 are distances between the recording station and the grounded ends of the wire; $J_1(\lambda r)$ and $J_0(\lambda r)$ are first-order and zero-order Bessel functions of first kind; $r_{TE} = Y_0 - \hat{Y}_1/Y_0 + \hat{Y}_1$, $Y_0 = u_0/\hat{z}_0$, $Z_0 = u_0/\hat{y}_0$, $\hat{z}_0 = i\omega\mu_0$, $\hat{y}_0 = i\omega\varepsilon_0$; ε_0 is the electric permittivity of free space; μ_0 is the magnetic permeability of free space.

For the N-layer model

$$\hat{Y}_j = Y_j\frac{\hat{Y}_{j+1} + Y_j\tanh(u_j h_j)}{Y_j + \hat{Y}_{j+1}\tanh(u_j h_j)}; \quad \hat{Y}_N = Y_N;$$

$$\hat{Z}_j = Z_j\frac{\hat{Z}_{j+1} + Z_j\tanh(u_j h_j)}{Z_j + \hat{Z}_{j+1}\tanh(u_j h_j)}; \quad \hat{Z}_N = Z_N;$$

$$j = N - 1, \ N - 2, \ldots\ldots, 1,$$

where, $Y_j = u_j/\hat{z}_j$, $Z_j = u_j/\hat{y}_j$; $u_j = (k_x^2 + k_y^2 - k_j^2)^{1/2}$ is the vertical wavenumber, $k_j^2 = -\hat{z}_j\hat{y}_j = \omega^2\mu_j\varepsilon_j - i\omega\mu_j\sigma_j$ is the total wavenumber, and $k_x^2 + k_y^2 = \lambda^2$ is the horizontal wavenumber, ε_j is the electric permittivity of subsurface j layer, μ_j is the magnetic permeability of subsurface j layer. The equations are calculated according to NESTOR and ALUMBAUGH (2011).

EM field is dependent on the orientation of grounded wire. Take the x-orientation grounded-wire source as an example. The generated E_x and H_y represent the combined TEM fields, consisting of the TE field (E_{x}^{TE}, H_y^{TE} from current-carrying wire) and TE–TM field (E_x^{TE-TM}, H_y^{TE-TM} from grounding ends). E_y and H_x represent TE–TM fields and are only generated from grounding ends. H_z represents the TE field and is only generated from current-

carrying wire. Therefore, the electromagnetic fields from long grounded-wire contain not only TE and TE–TM fields, but also the combined TEM fields. Compared to the large-loop source and other configurations, the long grounded-wire source is more applicable for our investigation of the distribution characteristics, sensitivity, and detection depth for various fields, such as TE, TE–TM, and TEM.

1.2. Distribution Characteristic of Different-Mode Electromagnetic Fields

Based on the coordinate system shown in Fig. 1, Eqs. (1)–(5) are then used to calculate the responses and analyze the distribution of electromagnetic fields. To be clear, the length of wire source is 100 m, and the transmitting current is 10 A. The parameters of geo-electric model are:

$$\rho_1 = 100 \ \Omega m, \ \rho_2 = 10 \ \Omega m, \ \rho_3 = 100 \ \Omega m,$$
$$h_1 = 400 \ m; \ h_2 = 10 \ m$$

First, we show the distribution of the combined TEM fields (H_y and E_x) in Fig. 2. As shown, the maximum value of the combined TEM fields appears at the center of the source, and also the distribution of the fields is symmetric with the source as the center. The contour lines of E_x are concentric circles, and the value of these lines decreases with increasing distance. For H_y field, the distribution demonstrates similar characteristics with one trap appearing symmetrically at early time. This could be caused by the return current.

Then, we generate the TE fields, E_x^{TE} and H_y^{TE} from combined TEM field. As shown in Fig. 3a, b, the distribution of E_x^{TE} and H_y^{TE} (TE field) is similar to that of E_x and H_y (combined TEM field) shown in Fig. 2. Figure 3c shows the distribution of the TE field (H_z) at different times, in which we notice a symmetric distribution of H_z with the source as the pivot, similar to the extracted E_x^{TE} and H_y^{TE}. But we notice one apparent difference in the maximum value of H_z, which appears at the mid-perpendicular regions. The distance between the position of maximum value and the source becomes even larger with increasing time, and the 'smoke ring' theory

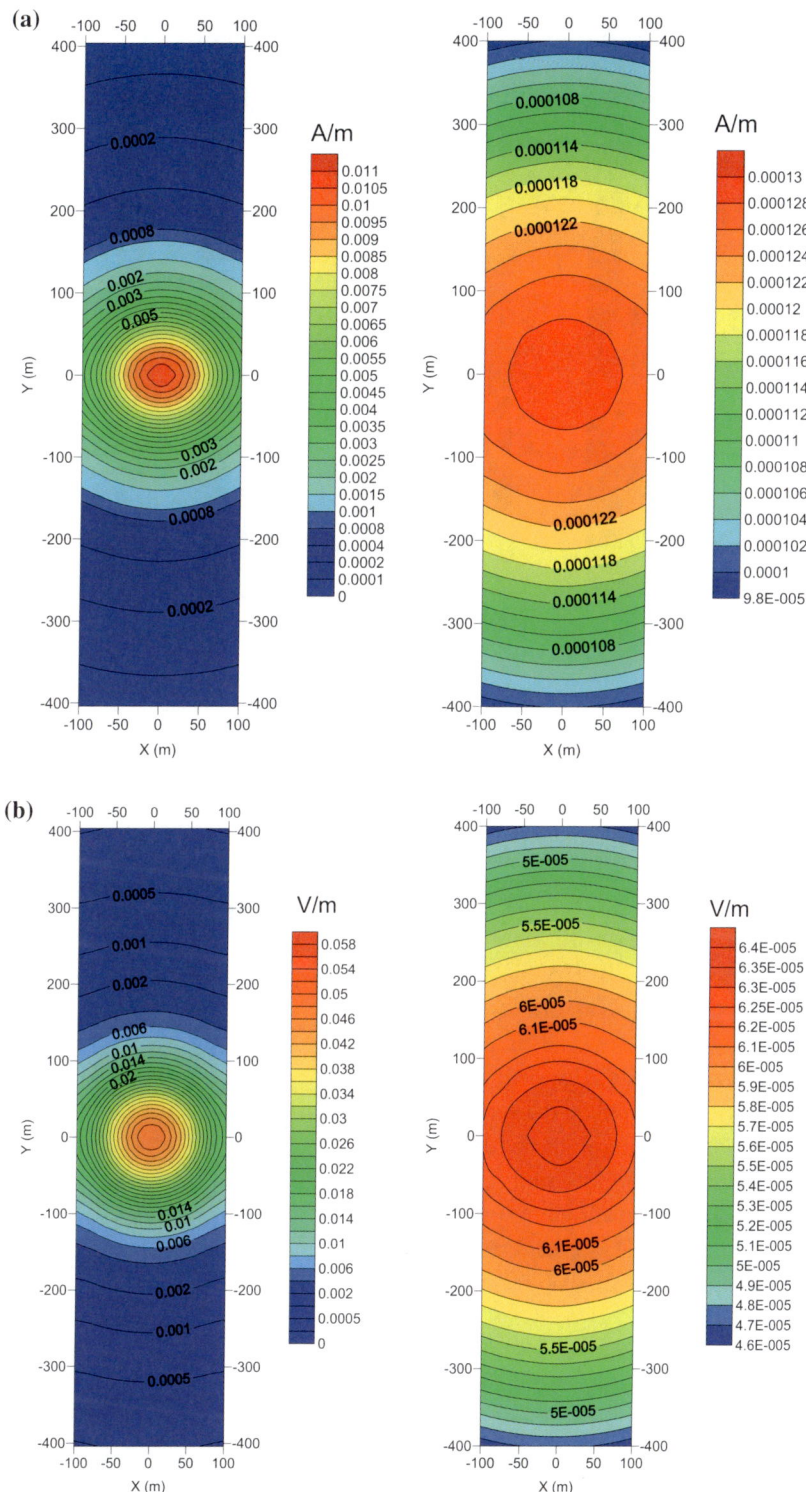

Figure 3
Distribution of TE fields: H_y^{TE} (**a**), E_x^{TE} (**b**) and H_z (**c**) at 1e-5 s (*left*) and 1e-3 s (*right*)

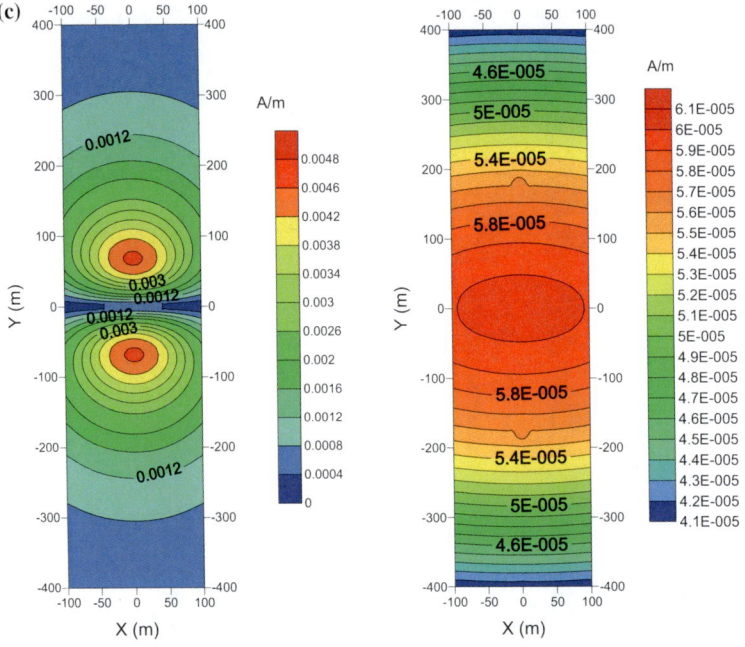

Figure 3
continued

(NABIGHIAN 1979) could be a good explanation for such analysis.

Figure 4 shows the distribution of the TE–TM fields (E_y and H_x) at different times, in which we notice that the distribution of TE–TM field is symmetrical about the central point of the source, due to the fact that these fields (E_y and H_x) are generated from the grounding ends. However, the TE–TM fields are unevenly distributed on the surface of the earth. The maximum values of these fields appear at an angle of approximately 45° from the x-axis, while the minimum values appear along the y-axis mid-perpendicular to the source. For the distance between the position of maximum value and source, it also increases with time.

In conclusion, the distribution of TE and combined TEM fields is symmetrical about the source, but for the TE–TM fields, it is a four-quadrant and symmetrical about the center of the source. Thus, the requirement for selecting the recording points of the TE–TM field is stricter compared to that for the TE and TEM fields.

1.3. Sensitivity of Different-mode Fields to Anomaly

In our study, we evaluate the relative difference to illustrate the sensitivity of different fields to anomaly:

$$\eta = 2 \times \left| \frac{F_{\text{anomaly}} - F_{\text{without}}}{F_{\text{anomaly}} + F_{\text{without}}} \right| \times 100\,\%, \qquad (6)$$

where F_{anomaly} is the response of the geo-electric structure with anomaly and F_{without} is the response without anomaly.

Using Eq. 6, larger differences indicate a higher sensitivity. We design 1-D models to quantify the sensitivity, with H and K representing the geo-electric structure with anomaly and D representing the structure without anomaly.

H: $\rho_1 = 100\ \Omega\text{m}$, $\rho_2 = 10\ \Omega\text{m}$, $\rho_3 = 50\ \Omega\text{m}$, $h_1 = 100\ \text{m}; h_2 = 10\ \text{m}$

K: $\rho_1 = 100\ \Omega\text{m}, \rho_2 = 1000\ \Omega\text{m}, \rho_3 = 50\ \Omega\text{m}$, $h_1 = 100\ \text{m}; h_2 = 10\ \text{m}.$

D: $\rho_1 = 100\ \Omega\text{m}, \rho_2 = 50\ \Omega\text{m}, h_1 = 100\ \text{m}.$

The coordinate of the forward calculation is (200, 600) as shown in Fig. 1. Based on such model, the sensitivity to the conductive anomaly is analyzed

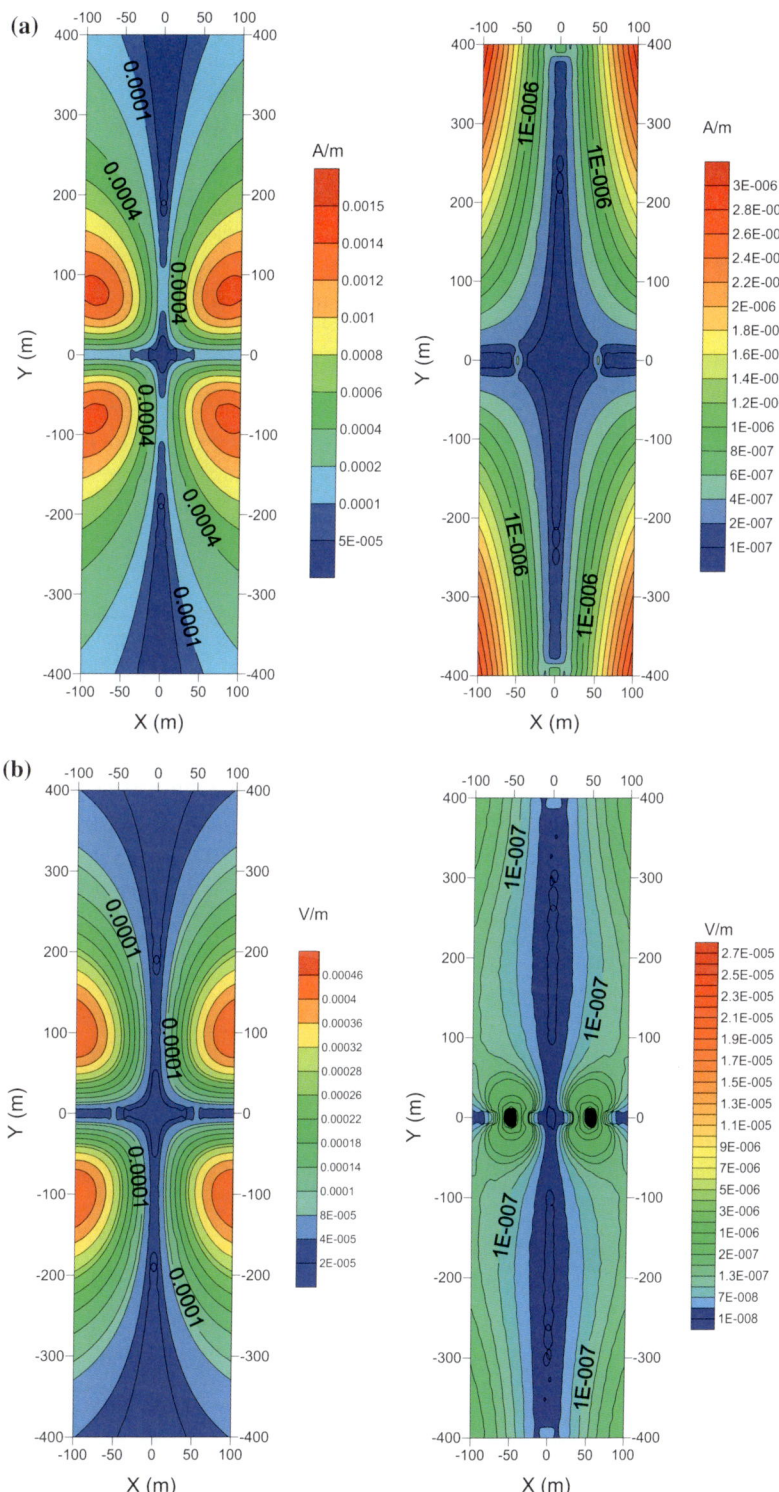

Figure 4
Distribution of TE–TM fields H_x (**a**) and E_y (**b**) at 1e-5 s (*left*) and 1e-3 s (*right*)

through the calculation of difference between H and D models, and the sensitivity to resistive anomalies is analyzed through the calculation of difference between K and D models.

1.3.1 Sensitivity to Low-resistivity Anomaly (H vs D)

Figure 5 shows the relative difference of magnetic fields between H and D models, from which we notice that the difference of H_y is larger than that of H_x and H_z. Such large difference indicates that the detection capability of the TE–TM field (H_x) to conductive anomaly is higher than that of the TE (H_z), but is smaller than that of combined TEM fields (H_y).

Similarly, Fig. 6 shows the relative difference of electric fields between H and D models, from which we notice that the difference of E_y is much larger than that of E_x. Therefore, the detection capability of the TE–TM electric field to conductive anomaly is also higher than that of the combined TEM electric field.

Then, in order to better understand the sensitivity of different-mode fields to the thickness of the embedded layer, we calculate the maximum values of relative difference for H models with varying thickness of embedded layer, and show the maximum values of relative difference between H and D models with different thickness of embedded layer (h_2) in Fig. 7. As shown in Fig. 7a, the maximum values of relative difference for H_x and H_z increase faster than H_y with increasing thickness of embedded layer, which indicates that the TE magnetic field (H_z) and TE–TM magnetic field (H_x) are more sensitive to the thickness of the embedded layer than TEM field (H_y). Similar features are also noticed from Fig. 7b. The TEM electric field (E_x) is more sensitive to the thickness of the embedded layer than TE–TM field (E_y).

In order to better understand the sensitivity of different-mode fields to the resistivity of the embedded layer, we calculate the maximum values of relative difference for H models with varying resistivity of embedded layer. Figure 8 shows the maximum values of relative difference between H and D models with increasing resistivity of embedded layer. As shown in Fig. 8a, the maximum values of

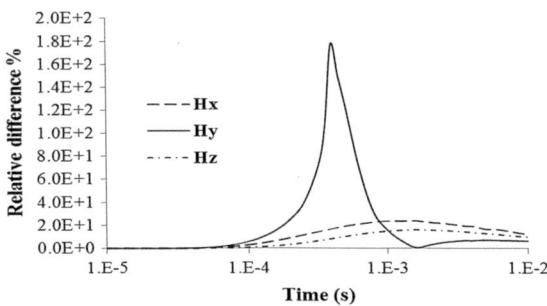

Figure 5
Difference *curves* of magnetic fields between H and D models vs time at point (200, 600)

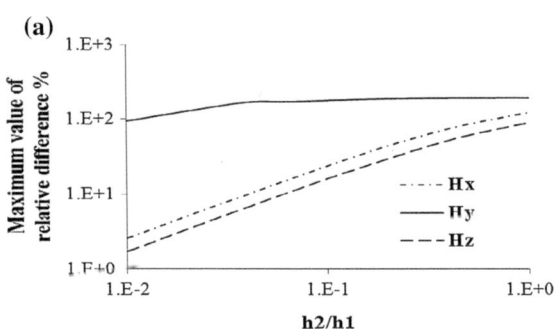

Figure 7
Maximum values of relative difference between H and D models vs thickness of embedded layer. **a** Magnetic field, **b** electric field

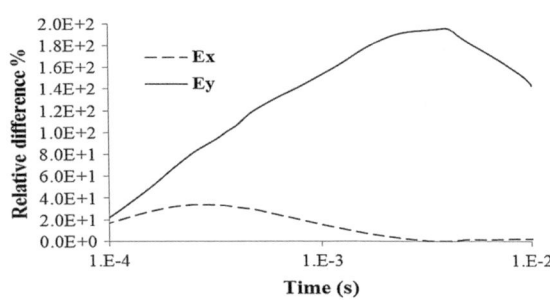

Figure 6
Difference *curves* of electric fields between H and D models vs time at point (200, 600)

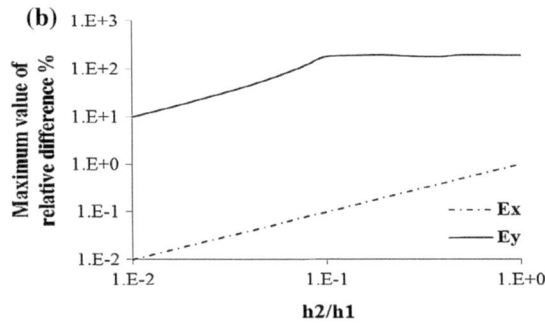

(a)

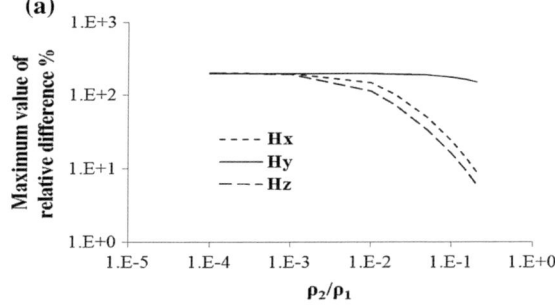

(b)

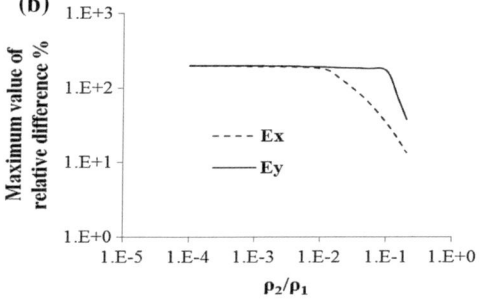

Figure 8
Maximum values of relative difference between H and D models vs resistivity of embedded layer. **a** Magnetic field, **b** electric field

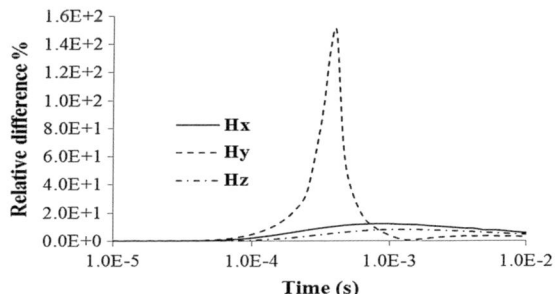

Figure 9
Difference *curves* of magnetic fields between K and D models vs time at point (200, 600)

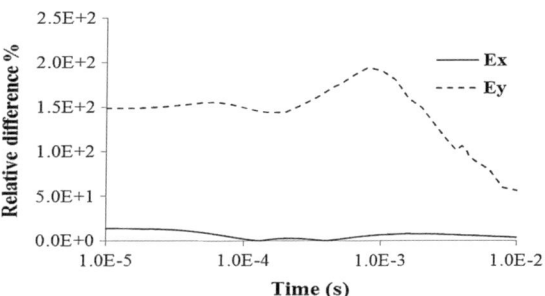

Figure 10
Difference *curves* of electric fields between K and D models vs time at point (200, 600)

relative difference for H_x and H_z decrease faster than H_y with increasing resistivity of embedded layer, which indicates that the TE magnetic field (H_z) and TE–TM magnetic field (H_x) are more sensitive to the resistivity of the embedded layer than TEM field (H_y). Similar features are also noticed from Fig. 8b. The TEM electric field (E_x) is more sensitive to the resistivity of the embedded layer than TE–TM field (E_y).

1.3.2 Sensitivity to High-resistivity Anomaly (K vs D)

Figure 9 shows the relative difference of magnetic fields between K and D models, from which we notice that the difference of H_x is higher than that of H_z, but is smaller than that of H_y. Therefore, the detection capability of the TE–TM magnetic field to resistive anomaly is higher than that of TE field, but is smaller than that of combined TEM field.

Figure 10 shows the relative difference of electric fields between K and D models, from which we notice that the difference of E_y is far larger than that of E_x. This indicate that the detection capability of the TE–TM field to resistive anomalies is higher than that of the combined TEM field.

In order to better understand the sensitivity of different fields to the thickness of the embedded layer, we change the thickness of the embedded layer, i.e., h_2 and calculate the associated response. The maximum values of relative difference between K and D models with different thickness of embedded layer are shown in Fig. 11. As shown in Fig. 11a, when the thickness of the embedded layer is smaller than one-tenth of the first layer, the maximum values of relative difference demonstrates the same variation trend for different magnetic fields, i.e., TE (H_z), TE–TM (H_x) and TEM-mode (H_y). Therefore, those magnetic fields have the same sensitivity to resistive anomaly when the thickness of the embedded layer is small. However, when the thickness of the embedded layer is larger than one-tenth of the first layer, the maximum values of relative difference for H_x and H_z increase faster than that for H_y, so that the TE magnetic field (H_z) and TE–TM magnetic field (H_x) are more sensitive to the thickness of the embedded layer than TEM field (H_y). In Fig. 11b, we show the

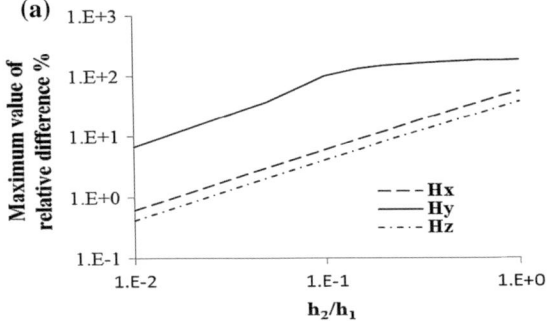

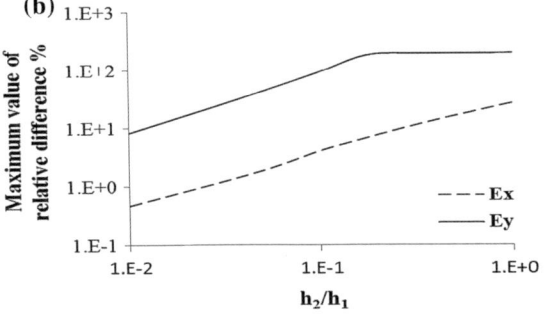

Figure 11
Maximum values of relative difference between K and D models vs thickness of embedded layer. **a** Magnetic field, **b** electric field

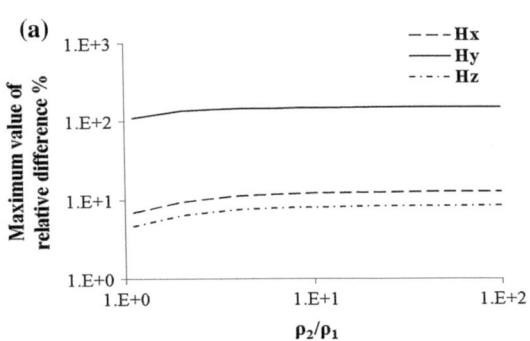

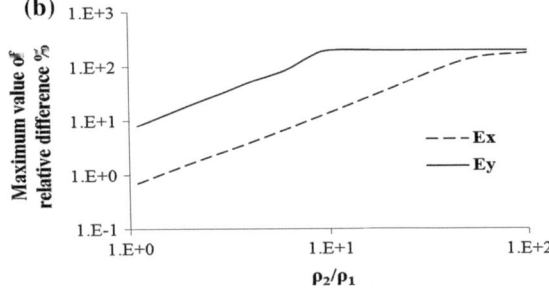

Figure 12
Maximum values of relative difference between K and D models vs resistivity of embedded layer. **a** Magnetic field, **b** electric field

difference curve for different electric fields, from which the TEM electric field (E_x) is more sensitive to the thickness of the embedded layer than TE–TM field (E_y) when the thickness of the embedded layer is large.

In order to better understand the sensitivity of different fields to the resistivity of the embedded layer, we change the resistivity of the embedded layer, i.e., ρ_2 and calculate the associated response. The maximum values of relative difference between K and D models with increasing resistivity of embedded layer are shown in Fig. 12. As shown in Fig. 12a, the maximum values of relative difference demonstrate the same variation trend for different magnetic fields, i.e., TE (H_z), TE–TM (H_x) and TEM-mode (H_y). Therefore, those magnetic fields have the same sensitivity to high resistivity of the embedded layer. For the different electric fields shown in Fig. 12b, when the sensitivity of the embedded layer is smaller than tenfold resistivity of the first layer, the

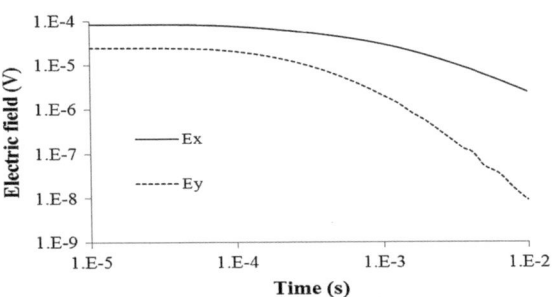

Figure 13
Response *curves* of the TE–TM electric fields and combined TEM field vs time at point (200, 600)

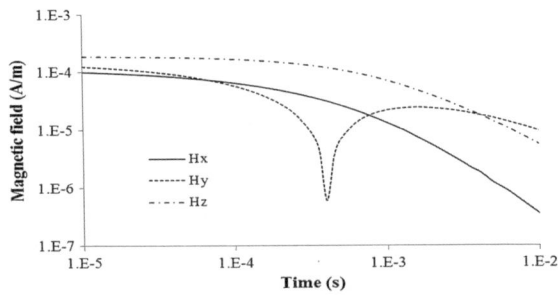

Figure 14
Response *curves* of TE, TE–TM and combined TEM magnetic field vs time at point (200, 600)

maximum values of relative difference have the same variation trend, i.e., TE–TM (E_y) and TEM-mode (E_x), so that electric fields have the same sensitivity to the resistive anomaly when the resistivity of the embedded layer is small. However, when the resistivity of the embedded layer is larger than tenfold resistivity of the first layer, the maximum value of relative difference for E_x increases faster than that for E_y with increasing resistivity of embedded layer, from which the TEM-mode electric field (E_x) is more sensitive to the resistivity of the embedded layer than that of TE–TM field (E_y).

In conclusion, the detection capability of the TE–TM electric field is higher than that of TEM electric field, whereas the detection capability of the TEM magnetic field is higher than that of TE and TE–TM magnetic fields. However, for the low-resistivity embedded layer, TEM magnetic field are less sensitive to the thickness and resistivity of the embedded layer than other magnetic fields.

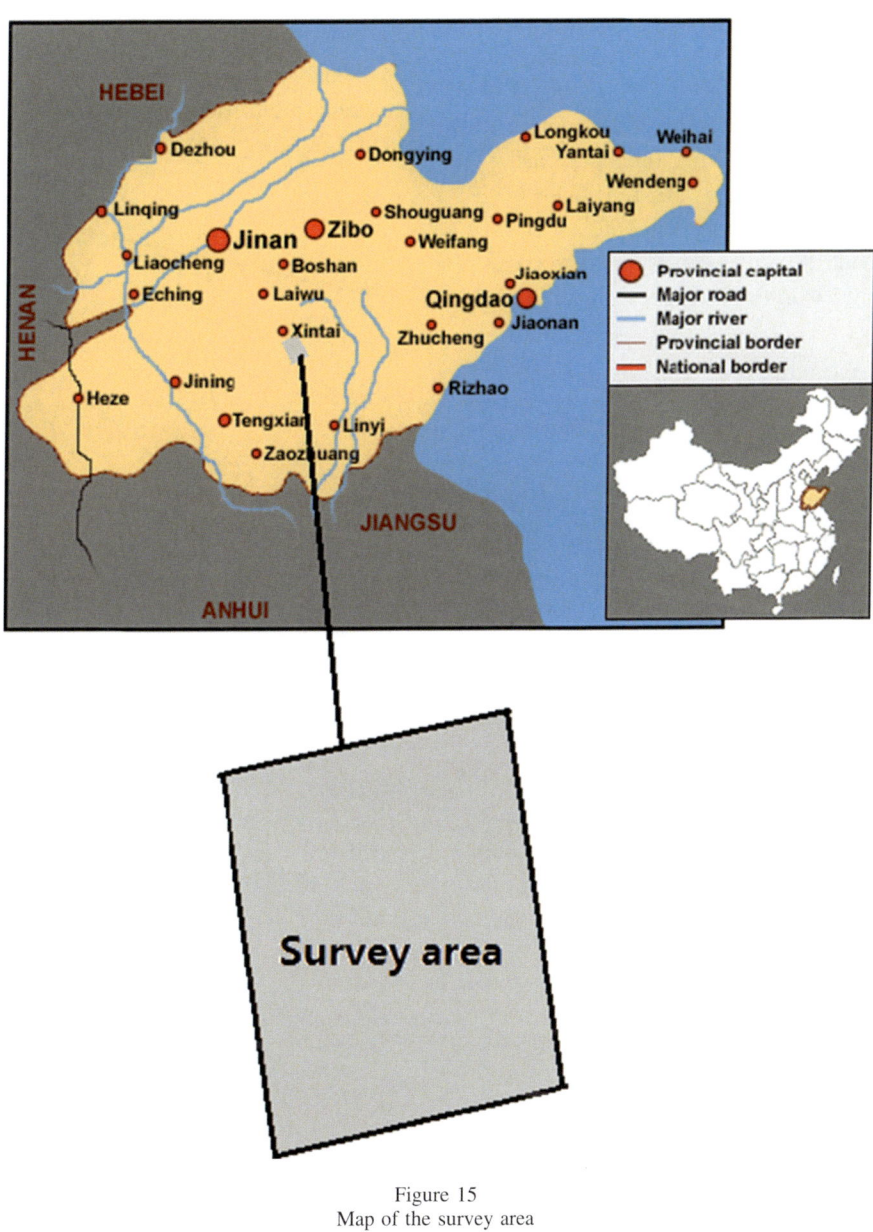

Figure 15
Map of the survey area

1.4. Detection Depth Comparison of Different-mode Fields

The investigation depth of TEM is defined as the time at which the signal decays to the noise level (SPIES 1989), but it is also related to the source moment and the resistivity of earth. Under the same noise level, the amplitude of response can also serve as an estimation factor of the investigation depth, with the large amplitude at the late time indicating a large detection depth. The investigation depth of TE, TE–TM, and combined TEM fields are compared using the forward response from the proposed 1D model.

As shown in Fig. 13, the response of TE–TM electric field is about half magnitude of the combined TEM field at early time. With time increasing, however, the response of the TE–TM field (E_y) decays more quickly than that of the combined TEM field. Therefore, the investigation depth of the TE–TM field is smaller than that of the combined TEM field.

In Fig. 14, we show the response curve of different magnetic fields and notice that the response of the TE magnetic field (H_z) is larger than that of the TE–TM (H_x) and combined TEM field (H_y) at early time. At late time, however, the TE–TM field (H_x) decays more quickly than that of combined TEM field and TE field. Therefore, the investigation depth of TE–TM field is smaller than that of combined TEM field and TE field under same noise levels.

2. Case Study

To verify the distribution characteristic and depth of different fields demonstrated in the forward model, two grounded-wire TEM surveys were carried out in a coalfield in Shandong Province and an ore deposit in Hebei Province, China.

2.1. Case study 1: Coal Mine in Shandong Province

The coalfield is located in the middle part of Shandong Province (Fig. 15). In this study, the Canadian-made V-8 electrical instrument is used, and the transmitting grounded wire is 1000 m

carrying 12A current. Copper plates with silver paper are used as grounded electrodes, and buried at the 1 m-depth underground with salt water. The transmitting power is 30 kW. The magnetic field receiving probe type is a SB-7 K with the effective area 40,000 m². Induced voltage is observed and transformed into vertical magnetic field through the method shown above. Recording time is from 0.95 to 89.83 ms. The point space is 40 m and line space is 100 m. There are in total 31 survey lines and each line consists of 26 observation points (see Fig. 16).

The measured x-, y-, and z-component impedance data along line 5 are shown in Fig. 17. There is a signal change and multiple extreme values in the measured multi-track map of the x-component impedance. The value observed at the central point

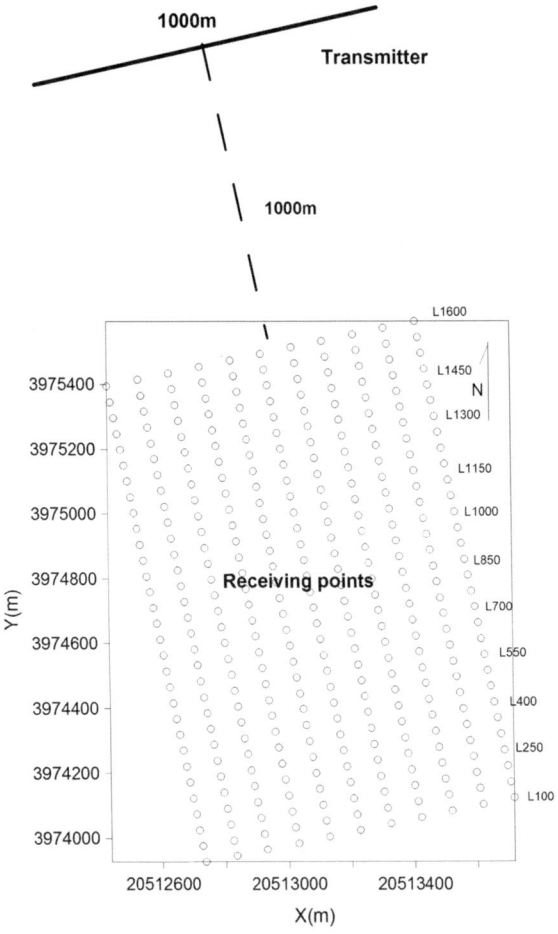

Figure 16
Design of survey lines

of the line is zero, but the observed data is symmetrical about the central point from the aspect of absolute amplitude. For the y-component impedance, a minimal value occurs at the central point, and the distribution of the field is symmetrical about the central point.

Unlike the distribution of x- and y-components impedance, the maximum value of the z-component impedance occurs at the central point. The distribution of these fields is the same with the forward results. Generally, there are more changes in the multi-track map of the x-component impedance than that in the map of the y- and z-component impedance. Therefore, the distribution of the TE–TM field is more uneven than that of TE and combined TEM fields. The TE field is more uniform than the combined TEM field.

Figure 18 shows the decay curves of x-, y-, and z-component impedance at point 1000 in line 5. It is clear that the value of the x-component is one order magnitude smaller than that of the y- and z-component at early time, so that the decay speed of the x-component impedance is higher than that of other components. Therefore, the value of the x-component impedance is far smaller than that of other components at late time. The TE and combined TEM fields have larger detection depth than the TE–TM field under same circumstances based on the definition of investigation depth with regard to time.

2.2. Case Study 2: Metal ore Deposit in Hebei Province

We applied the grounded-wire TEM method to a field example from Longhua iron ore in northern China. The Longhua region is located in the north part of Hebei Province and North China craton, and Longhua iron ore deposit is the classical Precambrian banded iron ore (ZHANG et al. 2004). There are seven 600-m-long survey lines (Fig. 19). Point 300 is the central point of the line. The length of transmitting

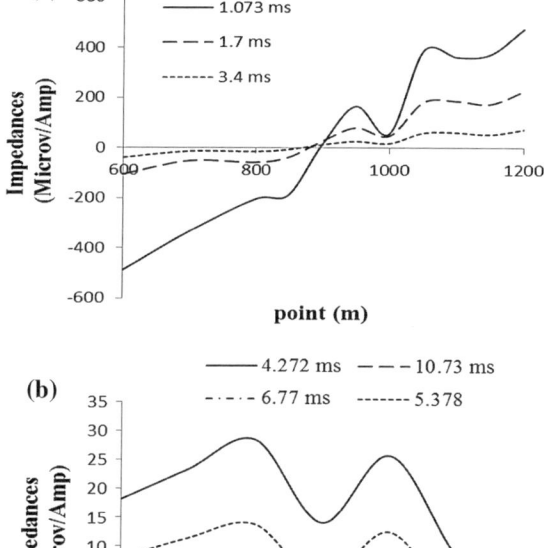

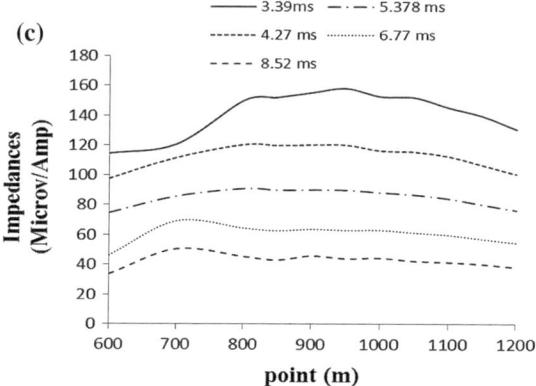

Figure 17

Measured impedances (voltage divided by current) data along line 5. **a** x-Component impedance, **b** y-component impedance, **c** z-component impedance

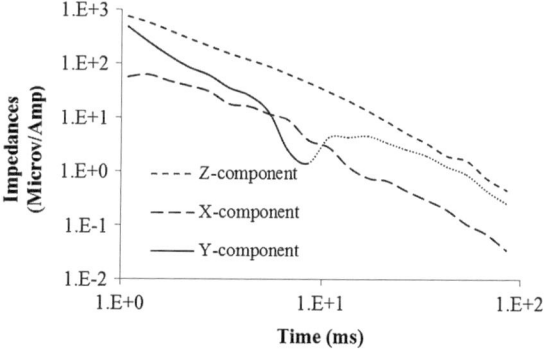

Figure 18

Decay *curves* of x-, y- and z-component impedances (voltage divided by current) at point 1000 in line 5

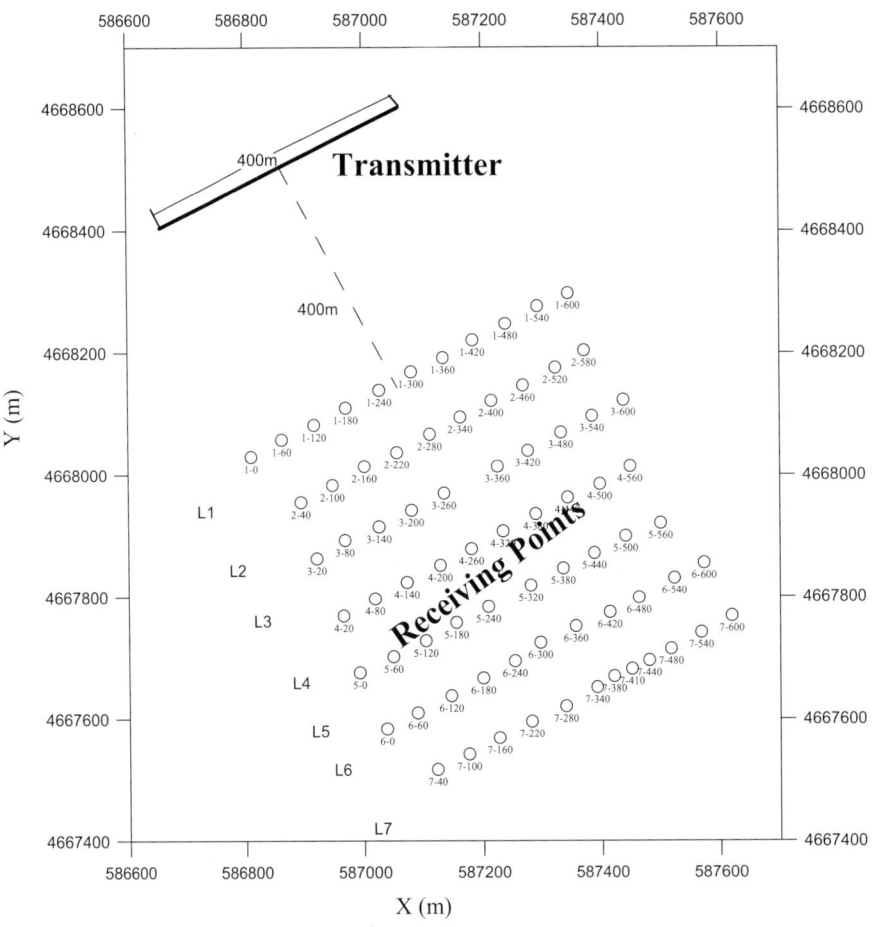

Figure 19
Sketch map of survey lines

wire is 400 m and transmitting current 16A. The power of the transmitter is 30 kW, and electric fields are observed using electrode pairs. Transmitting base frequency is 5 Hz, and recording time is from 0.4979 to 45.41 ms. The distance between the transmitter and closest receiving point is 400 m.

Figure 20 shows the decay curves of E_x and E_y (divided by current) at point 400 in line 2, in which the value of E_x is far larger than that of E_y at early time. A similar decay speed is also noticed for E_x and E_y. Therefore, the value of E_y (TE–TM field) is far smaller than that of E_x (combined TEM field) at late time. The combined TEM electric fields have a larger depth than the TE–TM field under the same backgrounds.

The measured E_x data along line 2 are shown in Fig. 21. The maximum value of E_x occurs at the

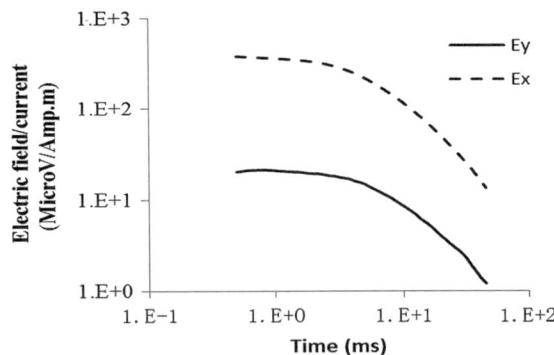

Figure 20
Decay *curves* of E_x and E_y (divided by current) at point 400 in line 6

central point of the survey line, and the distribution of the field is the same with the forward result shown above. The distribution of E_x is symmetrical about the central point.

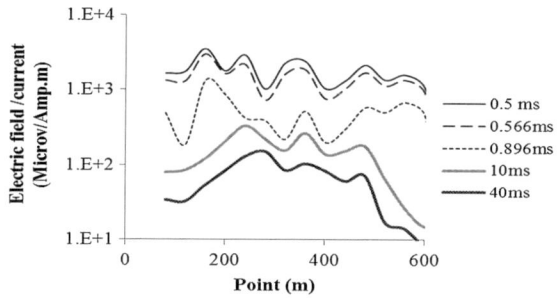

Figure 21

Measured E_x (divided by current) data along line 2 at different time gates

3. Conclusions

This study performs a comparison of distribution characteristics, detection depth, and sensitivity to anomalies between the TE, TE–TM, and combined TEM fields generating from grounded-wire source. The TE–TM electric field is more sensitive to resistive and conductive anomalies than the TE and combined TEM fields, and thereby it is most applicable for the exploration of resistive and conductive resources. However, the response amplitude of the TE–TM fields is smaller and decays faster than that of the TE and combined TEM fields, which makes the detection depth be the major limitation of using TE–TM fields. Meanwhile, the distribution of the TE–TM field is more uneven than that of the TE and combined TEM fields, so that there should be more requirements when designing the optimal field observation in practice. The combined TEM magnetic field has a high sensitivity to conductive anomalies and is distributed uniformly at a horizontal scale, indicating its great potential for detecting hydrogeological targets and conductive resources in both long-offset and short-offset zones.

Acknowledgments

This research was supported by R&D of Key Instruments and Technologies for Deep Resources Prospecting (the National R&D Projects for Key Scientific Instruments), Grant No.ZDYZ2012-1-05-04,the State Major Basic Research Program of the People's Republic of China (2012CB416605) and Natural Science Foundation of China (NSFC) (41174090 41174108, and 41130419). Finally, the authors would like to thank three reviewers, the editor for many helpful comments and suggestion, and Dr. Di H. B. for grammar and English expression.

REFERENCES

CALDWELL, T. G., and BIBBY, H. M. (1998), *The instantaneous apparent resistivity tensor: a visualization scheme for LOTEM electric field measurements*, Geophysical Journal International *135*, 817–834.

COMMER, M., and NEWMAN, G. (2004), *A parallel finite-difference approach for 3D transient electromagnetic modeling with galvanic sources*, Geophysics 69(5), 1192–1202.

COMMER, M., HELWIG, S. L., HORDT, A., and TEZKAN, B. (2005), *Interpretation of long-offset TEM data from Mount Merapi (Indonesia) using a 3D optimization approach*, Journal of Geophysical Research 110, B03207, doi:10.1029/2004JB003206.

COMMER, M., HELWIG, S. L., HORDT, A., SCHOLL, C., and TEZKAN, B. (2006), *New results on the resistivity structure of merapi volcano (indonesia), derived from three-dimensional restricted inversion of long-offset transient electromagnetic data*, Geophys. J. Int. *167*(3), 1172–1187.

COX, C.S. (1980), *Electromagnetic induction in the oceans and inferences on the constitution of the earth*, Geophysical Surveys *4*, 137–156.

EDWARDS, R. N. (1997), *On the resource evaluation of marine gas hydrate deposits using sea-floor transient electric dipole-dipole method*, Geophysics *62*, 63–74.

EDWARDS, R. N., and CHAVE, A. D. (1986), *A transient electric dipole-dipole method for mapping the conductivity of the sea floor*, Geophysics *51*, 984–987.

EVANS, R. L. (2007), *Using CSEM techniques to map the shallow section of seafloor: from coastline to the edges of the continental slope*, Geophysics 72(2), 105–116.

HORDT A., JODICKE H., STRACK K. M., VOZOFF, K., and WOLFGRAM, P. A. (1992), *Inversion of long-offset TEM soundings near the borehole Munsterland 1, Germany, and comparison with MT measurements*, Geophysical Journal International *108*, 930–940.

HORDT, A, and MULLER, A. (2000), *Understanding LOTEM data from mountainous terrain*, Geophysics 65 (4), 1113–1123.

HORDT, A., and SCHOLL, C. (2004), *The effect of local distortions on time-domain electromagnetic measurements*, Geophysics 69(1), 87–96.

KAUAHIKAUA, J. (1978), *Electromagnetic fields about a horizontal electric wire source of arbitrary length*, Geophysics 43(5), 1019–1022.

KELLER, G. V., PRITCHARD, J. I., JACOBSON J. J., and HARTHILL, N. (1984), *Mega-source time-domain electromagnetic sounding methods*, Geophysics 49(7), 993–1009.

NABIGHIAN, M.N. (1979), *Quasi-static transient response of a conducting half-space—an approximate representation*, Geophysics *44*, 1700–1705.

NESTOR, H. C., and ALUMBAUGH, D. (2011), *Near-source response of a resistive layer to a vertical or horizontal electric dipole excitation*, Geophysics 76(6), F353–F371.

PODDAR, M. (1982), A rectangular loop source of current on a two-layered earth, Geophysical Prospecting 30.101–114.

PODDAR, M. (1983), A rectangular loop source of current on multilayered earth, Geophysics 48(1), 107–109.

PODDAR, M. (1996), Grounded-source transient E-field modeling of a shallow resistive layer with 3-D inhomogeneity, Pure and Applied Geophysics 147(3), 551–566.

RAICHE, A. P., and SPIES, B. R. (1981), Coincident-loop transient electromagnetic master curves for interpretation of two-layer earths, Geophysics 46, 53–64.

RAICHE, A.P. (1987), Transient electromagnetic field computations for polygonal loops on layered earths, Geophysics 52(6), 785–793.

SCHWALENBERG, K., WILLOUGHBY, E. C., MIR, R., and EDWARDS, R. N. (2005), Marine gas hydrate electromagnetic signatures in Cascadia and their correlation with seismic blank zones, First break 23, 57–63.

SPIES, B. R. (1989), Depth of investigation in electromagnetic sounding method, Geophysics 54, 872–888.

STRACK, K. M. (1984), The deep transient electromagnetic sounding technique: first field test in Australia, Exploration Geophysics 15, 251–259.

STREICH, R., and BECKEN, M. (2011), Electromagnetic fields generated by finite-length wire sources: comparison with point dipole solutions, Geophysical Prospecting 59(2), 361–374.

VOZOFF, K., MOSS, D., LEBROCQ, K. L., and MCALLISTER, K. (1985), LOTEM electric field measurements for mapping resistive horizons in petroleum exploration, Exploration Geophysics 16, 309–312.

WARD, S. H., and HOHMANN, G. W. Electromagnetic theory for geophysical exploration. In: NABIGHIAN, N. (Ed.), Electromagnetic Methods in Applied Geophysics, Society of Exploration Geophysics (Tusla, Oklahama 1991) pp. 121–223.

WEIR, G. J. (1980), Transient electromagnetic fields about an infinitesimal long grounded horizontal electric dipole on the surface of a uniform half-space, Geophysical J. Roy. Astr. Soc. 61, 41–56.

WEITEMEYER, K. A., Constable, K., and Behrens, J. (2006), First results from a marine controlled-source electromagnetic survey to detect gas hydrates offshore Oregon, Geophysical Research Letters 33(4), 1–4.

XUE, G. Q., BAI, C. Y., and YAN, S. (2012), Deep sounding TEM investigation method based on a modified fixed central-loop system, Journal of Applied Geophysics 76, 23–32.

YANG, S. (1986), A single apparent resistivity expression for long-offset transient electromagnetics, Geophysics 51(6), 1291–1297.

ZHANG, S., ZHAO, Y., SONG, B., and WU, H. (2004), The late Paleozoic gneissic granodiorite pluton in early Precambrian high-grade metamorphic terrains near Longhua County in northern Hebei Province, North China: Result from zircon SHRIMP U-Pb dating and its tectonic implications. Acta Petrologica Sinica 20, 621–626 (in Chinese with English abstract).

ZHOU, N. N., and XUE, G. Q. (2014), The ratio apparent resistivity definition of rectangular-loop TEM, Journal of Applied Geophysics 103, 152–160.

ZHOU, N. N., XUE, G. Q., GELIUS, L. J., YAN, S., and WANG, H.Y. (2015), Analysis of the near-source error in TEM due to the dipole hypothesis, Journal of Applied Geophysics 116, 75–83.

(Received September 17, 2014, revised April 12, 2015, accepted April 15, 2015, Published online May 7, 2015)